The Anthropocene: Politik—Economics—Society—Science

Volume 29

Series Editor

Hans Günter Brauch, Peace Research and European Security Studies
(AFES-PRESS), Mosbach, Baden-Württemberg, Germany

Luis-Alberto Padilla

Sustainable Development in the Anthropocene

Towards a New Holistic and Cosmopolitan Paradigm

Luis-Alberto Padilla
International Relations & Peace Research (IRIPAZ)
Guatemala City, Guatemala

Acknowledgement The cover photograph was taken by Eduardo Sacayon, a Guatemalan photographer who also granted the permission to use it in this book.

ISSN 2367-4024   ISSN 2367-4032 (electronic)
The Anthropocene: Politik—Economics—Society—Science
ISBN 978-3-030-80398-8   ISBN 978-3-030-80399-5 (eBook)
https://doi.org/10.1007/978-3-030-80399-5

Copyediting: PD Dr. Hans Günter Brauch, AFES-PRESS e.V., Mosbach, Germany
Language Editing: Dr. Vanessa Greatorex, Chester, England, UK

This Springer imprint is published by the registered company Springer Nature Switzerland AG
The registered company address is: Gewerbestrasse 11, 6330 Cham, Switzerland

Where is the Life we have lost in living?

Where is the Wisdom we have lost in · knowledge?

Where is the Knowledge we have lost in information?

—T. S. Eliot (2014: 350)

Hegel wrote that the only thing we can learn from history is that we learn nothing from history, so I doubt the epidemic will make us any wiser. The only thing that is clear is that the virus will shatter the very foundations of our lives, causing not only an immense amount of suffering but also economic havoc conceivably worse than the Great Recession. There is no return to normal; the new "normal" will have to be constructed on the ruins of our old lives, or we will find ourselves in a new barbarism whose signs are already clearly discernible. It will not be enough to treat the epidemic as an unfortunate accident, to get rid of its consequences and return to the smooth functioning of the old way of doing things, with perhaps some adjustments to our healthcare arrangements.

We will have to raise the key question: What is wrong with our system that we were caught unprepared by the catastrophe despite scientists warning us about it for years?

—Slavoj Zizec (2020: 3–4)

My own sense in the short run – strangely enough, the most dangerous thing facing us isn't (yet) the extinction scenario, since that's centuries in the future. What scares me more is the economic effect of simple sea-level rise as a consequence of climate change. CNN did a very interesting article some years ago about how much it would cost to retrofit ports around the world for a six-foot sea-level rise. It's literally trillions of dollars. Not only docks, but every airport facing the ocean – Sydney, China, San Francisco especially. Secondly there's the effect on human food. If there's a six-foot sea-level rise, it's astonishing how much food will be wiped out – most rice is in low-lying areas, for instance. But the long term is different because these greenhouse extinctions are devastating. The most recent one was the Paleocene-Eocene Thermal Maximum, and that was caused almost entirely by methane. So the scariest thing we're seeing today is the liberation of methane from higher latitudes, and it's happening far faster than anybody ever predicted.

—Diamond (2017: 14)

The ideas that stem from Gaia theory put us in our proper place as part of the Earth system – not the owners, managers, commissars, or people in charge. The Earth has not evolved solely for our benefit, and any changes we make to it are at our own risk. This way of thinking makes it clear that we have no special human rights; we are merely one of the partner species in the great enterprise of Gaia. We are creatures of Darwinian evolution, a transient species with a limited lifespan, as were all our numerous distant ancestors. But, unlike almost everything before we emerged on the planet, we are also intelligent, social animals with the possibility of evolving to become a wiser and more intelligent animal, one that might have a greater potential as a partner for the rest of life on Earth. Our goal now is to survive and to live in a way that gives evolution beyond us the best chance.

—James Lovelock (2009:9)

Economic design in particular has cross-cutting significance because it mediates our relationship to nature and to each other. Great Transition economies would be understood as the proximate means to the ultimate ends of vibrant lives, harmonious societies, qualitative, not quantitative, development. Innovation would continue in a post-material-growth era – indeed, would likely soar. But these new economies, by whatever name, would be post-capitalist, since private profit and capital accumulation would no longer have primacy. Some places might rely on governmental controls, others on decentralized arrangements, and still others on social ownership and workers' control.

—Paul Raskin (2002: 54–55)

Sustainable development tries to make sense of the interactions of three complex systems: the world economy, the global society and the Earth's physical environment...(and) it recommends a set of goals to which the world should aspire [calling] for a world in which economic progress is widespread; extreme poverty eliminated; social trust is encouraged through policies that strengthen the community; and the environment is protected from human-induced degradation...[with] a holistic framework, in which society [aims] to achieve the economic, social and environmental objectives of the SDGs [including] good governance. Thus the normative side of sustainable development envisions four basic objectives: economic prosperity; social inclusion and cohesion; environmental sustainability and good governance by major social actors including governments and business... Achieving sustainable development on our crowded, unequal and degraded planet is the most important challenge facing our generation. The SDGs must be the compass, the lodestar, for the future development of the planet during the period 2015 to mid-century.

—Jeffrey Sachs (2015: 3)

If your plans are for next year plant wheat

If they are for the next decade plant a tree

If you think in the next century educate people

—Zygmunt Bauman (2017: 10)

Sustainable development is the central challenge of our times. Our world is under strain. Poverty continues to plague communities and families. Climate change threatens livelihoods. Conflicts are raging. Inequalities are deepening. These crises will only worsen unless we change course. We have the technologies and the know-how to succeed. With dedicated efforts of each and every one of us, we can be the first generation to end extreme poverty and the last generation to face climate change as an existential threat.

—Ban Ki-Moon (2015: xii)

Preface

Sustainable development in the Anthropocene refers to two visions that could be complementary but are paradoxically contradictory. From the perspective of preserving ecosystems and protecting the environment, this is due to the current incompatible patterns of economic growth. The author argues that humanity is leaving an ecological footprint on the planet that is unsustainable and will be impossible to rectify unless substantial changes are made to the prevailing theoretical paradigm. He therefore calls for fundamental changes to the said paradigm and advocates replacing it with a holistic paradigm of sustainable development, fostered by a vision of natural and social sciences working together in an interdisciplinary manner to identify a viable alternative which is capable of transcending the current untenable situation.

The work of researchers such as Jeffrey Sachs, Thomas Piketty, Yanis Varoufakis, Maja Göpel and Paul Mason inspired the author's reflections on sustainable development (defined here as the adequate management of the four interconnected spheres of technical and economic factors, social dynamics and terrestrial ecosystems which are a permanent component of governmental policies). The unifying argument seeks to reconcile environmental protection with social public policies which have the capacity to satisfy human needs.

The book also deals with diverse themes from contemporary reform development models and styles, including considerations about the theory of international relations. An interesting topic is the groundbreaking proposition regarding the United Nations itself – given that it is the institutional synthesis of two great theoretical paradigms of international relations, Roosevelt having improved on the Wilsonian model of 1919 – and the suggestion that case studies contrasted with the norms and principles of the UN Charter are the best way to demonstrate that hypothesis. Likewise, the author underlines the importance of Chapters VI and VII of the Charter as consecrating on the one hand the idealist approach reflected in the norms for negotiation, mediation and arbitration as peaceful means to solve controversies and on the other hand the realistic approach of collective security and peace enforcement in circumstances of aggression or threats to international peace.

Regional integration is also examined, and the book stresses the importance of European integration as a means of guaranteeing peace and sustainable development in a continent that unleashed two devastating world wars in the last century. The author also explores the existing tensions between globalization at worldwide level and the different types of economic interdependence achieved through integration processes such as the European Union and other regional experiences in Latin America.

The next chapters focus on geology and natural sciences, and the author explains why the former geological epoch of the Holocene has been replaced by the Anthropocene, including the reasons for the scientific communities' agreement with such an innovative proposal, based on the *Great Acceleration* of industrial and economic development in the second half of the twentieth century, including ever-increasing urbanization, demographic growth, the construction of physical infrastructures and the advancement of cyber technologies, all combined with environmental degradation and globalization.

Furthermore, these changes have aggravated the emission of the greenhouse gases (GHGs) that are triggering global warming and climate change – the Anthropocene's main problem. They are also increasing ocean acidification, intensifying the reduction of the stratospheric ozone layer and exacerbating other man-made predicaments such as deforestation, predatory fishing and the over-exploitation of grassland for cattle and commodity crops and so on. However, what particularly draws the author's attention is the similarity between modern scientific views and the traditional thinking or cosmovision of Latin American indigenous peoples – hence the reference in the book to concepts such as good living or *Pachamama* (a spiritual deity and not just a geophysical entity, according to the indigenous perspective) who, in the form of droughts, tropical storms, earthquakes and other natural catastrophes, could induce Planet Earth to react more frequently to human depredations on the ecosystem. Additionally, and as the Anthropocene crisis is largely considered a cultural crisis, the book emphasizes the urgent need for a transformation from the fragmented and specialized vision characteristic of particular sciences towards the holistic and multidisciplinary conception required to tackle planetary problems in the foreseeable future.

To sum up, this volume covers a variety of themes under the unifying umbrella of the holistic paradigm and calls for an energetic change of course in humankind's relations with the natural environment. The text ends with a reference to cosmopolitanism as an alternative to those attitudes that are looking to return to the anachronistic ideology of nineteenth-century nationalism, favouring instead transnational citizenship, which could open the doors to a revision of the priorities in multilateral policies, including issues such as human rights, environmental deterioration, climate change, demographic growth, gender vindication, and social and human development. Those are the main issues explored in this exceptional manuscript.

Guatemala City, Guatemala
April 2019

Gert Rosenthal

Gert Rosenthal is Former Foreign Minister of Guatemala and Ambassador to the UN in New York. He was also Executive Secretary of the UN Economic Commission for Latin America and the Caribbean (ECLAC).

Foreword by Nelson Amaro

Sustainable development is the paradigm of the future. As such, in the case of a new idea, you need as much dissemination as possible to reach all the people involved and affected by your research, practices and predictions. Professor Luis Alberto Padilla has made a thorough review of all the research that is being done in this area at planetary level and has included the debates and agreements of decision-makers in all areas. This work is immensely useful because it crosses all disciplines, overcoming the isolation resulting from specialization in specific fields. Its historical account from the peace of Westphalia in the seventeenth century to the present day is a particularly impressive narrative of how trends and decisions change, often following opposite paradigms which influence the progress of our planet.

This book will convince readers that globalization has come to stay despite ignorance and parochial interests. For this reason, research on this topic is one of the most discussed issues worldwide today. The idea of transnational citizenship and the weaknesses that geographical borders represent in an unstoppable globalization process is another of the main arguments of this book, providing a basis for formulating proposals for achieving a peaceful transition from fragmented regional scenarios to integration processes like the European Union, which until now has been a major success story in maintaining peace and putting an end to the European wars that occurred in the past (for instance, the two world wars and the recent conflict in former Yugoslavia). War is still a terrible risk that may even endanger the survival of human life on this planet, because the threat of a nuclear holocaust has not disappeared. As a corollary, we must be aware that sustainable development – especially in the environmental dimension that this paradigm includes – must replace the poorly directed human development efforts made so far, because otherwise the possibility of armed conflict threatens the life of all kinds of species in the world, not just human life. This warning about extinction prevails in the text of the entire book and alerts decision-makers to its importance.

The book can help many people involved in different activities that cross all fields of knowledge and decision-making. Its content is addressed not only to university professors in all fields but also to decision-makers in political organizations in all governments. The book's contents can also be useful for non-governmental organizations (NGOs) and all kinds of private corporations that depend on global markets. In the real world, all these communities where family, interest groups, religious associations, professionals and academics interact with this kind of knowledge, can also make a difference. Therefore, it is a call to organize networks capable of defining priorities and interventions that could raise awareness and build a better world.

At university level, this vision suggests that the knowledge condensed in this book should be taught in all disciplines. Therefore, the following question arises: How could this Sustainable Development Paradigm help science in general, different disciplines and professions? There is no doubt that engineers have been able to make use of this knowledge to reduce risks related to storms, floods, landslides, droughts and all kinds of consequences related to any type of infrastructure which they plan to build. Architects have already included so-called Green Buildings in their designs. Many of them promise, if the client agrees on the inclusion of alternative energy sources, systems different from those that contribute to greenhouse gases (GHG), which are considered the main cause of the anticipated increase in global warming by the end of this century if this is not prevented by respecting the commitments of the Paris Agreement (COP 21). Most scientists agree that in the foreseeable future climate change could be the source of one of the greatest catastrophes of our planet if those commitments are not fulfilled. However, awareness in the field of ecology is not limited to scientists or engineers. Lawyers and legislators promote laws to protect the environment, forests, water resources, norms against pollution, the trafficking of endangered species and so on. But this is still not enough to counteract the enormous number of violations of the law and violators going unpunished. This gap must also be acknowledged by law schools and the judiciaries of all countries.

Economists also need to change the way gross domestic product is calculated in order to include the Human Development Index and also environmental categories. It is known, for example, that storms such as the Stan in Guatemala (2005) erased the gross domestic product of that year that came to represent a loss of 3.5% of the total, while that year's growth was 3.3% (Loyal, Amaro, Milan and Guzmán, 2014; Banco de Guatemala, 2005). Climate scientists agree that these storms will be more frequent over time, with stronger winds as the years go by, if the effects of global warming are not prevented.

The same can be said of the main sciences related to human behaviour. This book offers information about the problems faced by political scientists, based on the derivations of the Sustainable Development Paradigm. A look at the contents of the index will be sufficient to convince readers of its relevance. The examples are multiple: international relations where all kinds of entities related to the United Nations Organizations operate; regional and geopolitical integration problems; confrontations due to the globalization process that affects the fragility of geographical boundaries; the problems created by the increase in migratory flows, etc. On the other hand, sociologists have to deal with the impact of ecological phenomena on families. For example, unexpected droughts can be a factor of expulsion for people who decide to migrate. Local groups are also affected by unexpected ecological events. In another example, small groups of citizens could join municipalities to face these threats in alliance with national interest groups, or social forces favouring or promoting sustainable development policies over time could be subject to analysis and interventions; existing social conflicts between different groups could also be a focus of analysis.

Looking at this development in institutions of higher education, it is possible to observe the emergence and foundation of centres and institutes for land studies and sustainable development in many universities and countries. Climate, biological, chemical, ecological and geological sciences come together to conduct research and learning activities on these phenomena and the way these processes affect different populations. This interest includes the need to network with other institutions in many parts of the world. Public administration organizations are also emerging with specific mandates to take care of issues that have arisen due to climate change, urban and rural pollution of the environment, the disappearance of forests and the survival of endangered species.

Finally, this trend that crosses all disciplinary fields is more multidisciplinary and interdisciplinary than transdisciplinary, but there are some interesting transdisciplinary proposals that touch on philosophical and even theological questions. The above is inferred from Professor Padilla's quotations in his book when he mentions authors such as Teilhard de Chardin, S.J., whose influence comes from Henri Bergson, both French philosophers. In addition, the word 'paradigm' is an elaboration developed by another philosopher, Thomas Kuhn, to whom the author gives deserved credit.

The book written by Professor Luis Alberto Padilla offers all disciplines background information about how the world has arrived at the main concerns of our era, and his work should cross all sciences, practices and careers. It should be a reference text for all professionals in the academic, private, governmental and civil society sectors at present. These approaches follow the Sustainable Development Paradigm, and its call touches the effort that humanity has to make to survive, considering the consequences of exploiting resources for a better quality of life. These objectives have to be combined with the finitude of the goods granted by nature and genuine human development, which is synthesized by the 17 objectives and 169 goals specified in the United Nations Declaration of Sustainable Development and its 2030 Agenda, approved by more than 193 countries in September 2015 and later. These efforts should always recognize the concern of scientists and academia to leave a better world for future generations.

Guatemala City, Guatemala   Nelson Amaro
February 2019

Nelson Amaro is Guatemalan Sociologist, born in Havana, Cuba. He is Director of the Institute for Sustainable Development at Galileo University in Guatemala City (2014–present). He holds a degree from the Catholic University of Santiago de Chile, undertook master's degree studies at Harvard and MIT in Boston, USA, and completed his doctorate in sociology at the University of Wisconsin–Madison. He has dedicated more than thirty years of his professional life to multiple international missions, such as the United Nations, USAID and the World Bank, among others. In Guatemala, he has also held the positions of Dean of Social Sciences of the Guatemala del Valle, UVG (1995–2000), and Vice Minister of Urban and Rural Development (1987–1989).

Foreword by Gabriel Aguilera Peralta

In the field of social sciences, especially in international relations, books published by American and European authors are common, which is understandable because of the historical origin of those disciplines and the presence that they have in the universities and research centres of developed countries. In Latin America, we do have an important and relevant scientific output, although its major contribution is in the analysis of case studies like international conflicts, border disputes, the foreign policy of particular countries or studies about regional integration processes or specific international policy conjunctures.

This situation also characterizes Central American countries, with the regional particularity that most of the contributions are concentrated in articles of specialized revues and less in books addressed to a wider audience. In the case of Guatemala, most articles and publications deal with affairs pertaining to the region, including the Central American integration process, the territorial dispute with Belize and foreign policy in general terms, including multilateral and bilateral affairs and the relations with the UN and regional international organisms. Few scholars address theoretical considerations or substantive themes.

An exception is Dr. Luis Alberto Padilla, a pioneering academic whose book *Peace and Conflict in the Twenty-First Century* was published in 1999 and constitutes one of the main contributions of a Guatemalan author to the debate on international relations theory. His work is nourished with his solid experience in peace research and his wide participation in academic international activities as founder and former director of the *International Relations and Peace Research Institute* (IRIPAZ), a prestigious Guatemalan institution member of the *Latin American Council of Peace Research* (CLAIP) and of the *International Peace Research Association* (IPRA).

In his new book, Dr. Padilla tackles current debates on the great themes of international relations. The text is rich in references and has succeeded in the analysis and explanation of some of the central axes of the international relations theory of our times. He uses a holistic methodology to expose concrete situations and the connections they have with philosophical views as well as perspectives emerging from particular disciplines, including economics, ecology, anthropology, history, physics and natural sciences, in a very impressive review of current bibliography in all those disciplines.

The result is an extensive book of dense reading that seeks to discuss contemporary pivotal themes of problematic international relations. His intense narrative requires the attention of readers in a provocative manner. The author starts with a discussion about the great international relations paradigms – idealism and realism – already developed in his previous book. That he has a personal bias in favour of idealism is quite evident, but this is a valid option discussed with objectivity. The first chapter presents international conflicts as seen from the paradigmatic optic of both realism and idealism, and it presents interesting observations about the way that states use paradigms as a foreign policy tool according to circumstances and national interests: Could that kind of behaviour be called another form of 'realism'?

The author then moves on to regional integration and geopolitics. The selection of those themes is a consequence of the centrality and importance that the integration process has for Central American countries and the fact that transforming the current international system in a progressive way entails deepening integration processes and transcending the Westphalian order in the direction of a new transnational order. The said order will ideally be supported by a new kind of world citizenship with a cosmopolitan conception in order to ensure global peace and security. The author's idealism appears again in these proposals.

The third chapter concerns sustainable development in the Anthropocene, and it is a novelty in the traditional perspective of international relations because it deals with the subject in a well-documented and profound discussion. Here, we enter an absolutely contemporary debate where the scientific arguments coming from different disciplines are presented in a very convincing and sound manner. The author reminds us of the risks arising from the Great Acceleration of human activity, and as a result the increase in greenhouse emissions (GHEs), the pollution of the planet's freshwater and oceans with all sorts of industrial waste and the burning of fossil fuels all enter the description, and the author reinforces his arguments in favour of sustainable and human development with the support of multiple quotations from recent scientific research, including the Encyclical Letter *Laudato Si'* of His Holiness Pope Francis.

The fourth chapter relates to sustainable development and could be the most interesting in the book because its purpose is to discuss the UN Sustainable Development Goals and the possibility of reaching the 2030 targets. The revision of each goal is interwoven with concrete situations and case analyses, some of them peripheral, like that of Mexican Zapatistas, but others quite significant, such as the experiences of constitutional law reform in Bolivia and Ecuador, including the novelty that both countries have given a legal personality to *nature* ('Pachamama') and endowed individuals who are aware of environmental problems with the right to defend *nature* through empowering them to present their claims in court.

In this chapter, the author lays down the core of his work: Is it possible to achieve the SDGs without substantial reform of the capitalist system? It is a central issue that Dr. Padilla discusses extensively, following the ideas in the German scholar Maja Göpel's book *The Great Mindshift*, in which she argues that capitalism is not going to be destroyed, as Marx believed, but that a new mentality is needed to change mainstream economics and establish a sustainable system based on social solidarity. Several theoretical proposals are reviewed, among them some alternatives originating from the microeconomic experiments of producers as well as social solidarity economy in a form reminiscent of the proposals of old utopians.

Here, the author reiterates the leitmotif of his work. He believes that democracy is a fundamental tool to reform the international system: However, what kind of democracy is he thinking about? It is the classical liberal electoral democracy, not the capitalism mainstream economics that is not compatible with the conservation of nature and the satisfaction of human needs. He recalls the Ancient Greek modalities of democracy: participative democracy where change is obtained by conviction. He refers to the German scholar Maja Göpel's description of how the mindshift of the German private sector and politicians opened the way for fossil energy to be replaced by renewables like solar energy, wind energy, geothermic energy, water power and so on.

The final chapter is dedicated to reinforcing the academic debates concerning transnational citizenship and cosmopolitism as alternatives to the capitalist hegemonic kind of globalization. It also introduces some ideas about the role of global civil society, going as far as considering the democratization of the UN system and the establishment of a world parliament. An interesting discussion of new concepts is presented, such as human mobility, including a solid defence of the right to migrate. The quotation of classic authors like Wallerstein and Castells and of the new perspectives offered by Rainer Bauböck provides the theoretical framework for the author's thinking.

A book that opens an intellectual window to the understanding of humanity's alternatives concerning sustainable development is worthy of a reflexive reading as a Guatemalan contribution to the current debate on international relations which goes beyond what has so far been done nationally. Furthermore, it could be of great utility as a textbook in universities and academic institutions worldwide.

Guatemala City, Guatemala   Gabriel Aguilera Peralta
March 2019

Gabriel Aguilera Peralta is Vice President of IRIPAZ, Former Guatemalan Vice Minister of Foreign Affairs and formerly Ambassador in Germany, Peru, The Netherlands, and the Organization of American States (OAS) in Washington DC. He has also been University Professor and Researcher at the Latin American Faculty for Social Sciences (FLACSO), the University of San Carlos and Rafael Landívar University. He is currently also Advisor to the Guatemalan Ministry of Foreign Affairs.

Acknowledgments

This book is the result of a conversation with Hans Günter Brauch after a session of the tenth congress of the *Latin American Council of Peace Research* (CLAIP) celebrated at the campus of the *National Autonomous University of Mexico* (UNAM) in January 2017. We were talking about sustainable development, the Anthropocene and climate change and its relation with geopolitics and international relations theory, and he suggested the writing of this text and finally finished after more than three years of work. Without the stimulus provided by Hans Günter throughout this time, I would not have been able to reach the last page, but also the pandemic of 2020 was an incentive to move forward and to include the issue of the sixth extinction.

I also want to express my acknowledgment to friends and colleagues who revised the text and suggested changes to the English wording or gave me ideas about the contents. Gabriel Aguilera Peralta wrote the foreword, and Gert Rosenthal helped me correct part of the initial pages; Nelson Amaro read the whole book and suggested several changes for the English version. Edmundo Vásquez read Chap. 3 and did the same, as well as my son Diego, who suggested authors and themes, and also my son Andrés, who gave me ideas and brought me useful books from Mexico and the United States. Without the technical software support of Marco Vinicio Quan and other colleagues at the Diplomatic Academy, and Maria Elisabeth Van der Linden's help with the translation, I would not have been able to advance at a good pace. Alejandra García collaborated with the elaboration of maps and tables.

To all of them my gratitude and appreciation for their effort and goodwill.

To my wife Lucrecia, who also helped me with some of the English translation, I would like to extend my thanks for her patience during the lengthy writing and preparation of the book.

Finally, thanks are due to Dr. Vanessa Greatorex for her extensive linguistic and factual corrections to the preliminary English translation.

Translator's Note

It has been an honour and a pleasure to translate this book in close cooperation with its creator, Dr. Padilla.

I have conducted the translation with the intention of both following the author's interpretation and relaying the meaning of the text to English-speaking readers. This means that, on occasion, the phraseology and paragraph structure of the Spanish and the English versions may differ. Apart from the text itself, all quotations from Spanish or French literature have also been translated into English by me and refined, where necessary, by the English language editor, unless otherwise acknowledged.

The language in this book, like the topics and issues it addresses, is complex; however, nothing less will do in order to reflect the objective with which it was written. Linking sustainable development and development studies with multilateral diplomacy is no small feat, and properly laying the groundwork to explain each paradigm requires a certain amount of manoeuvering. Consequently, the new holistic and multidisciplinary approach presented can be considered pioneering, which is why I am all the more honoured to have contributed to this project, as I believe this book deserves a wide audience, including both Spanish- and English-speaking readers.

The Hague, The Netherlands
2019

Maria Elisabeth van der Linden

Introduction

We have entered the geological epoch of the Anthropocene, which essentially means that the previous geological epoch – the Holocene – ended prematurely as a result of the great acceleration of capitalism at the middle of last century, with ruinous effects on the atmosphere and the biosphere exacerbated by the increase in *greenhouse emissions* (GHEs), global warming and the predatory behaviour of the small elite that rules an economic system that does not care about human beings, environment, nature or the planet. The irresponsibility of 'fossil capitalism' (Angus 2016) has triggered climate change, the crisis of the Earth Systems, pandemics – like the coronavirus of 2020 – and other expressions of the *revenge of Gaia* (Lovelock 2007), as Earth's response to an economic model that is destroying our *common house* (Laudato Si' 2015) has been called. Indeed, if we compare the millions of years of geological history of the planet with human history, the Anthropocene footprint is quite recent: it was only after the impact of a meteorite provoked the fifth extinction (of dinosaurs) sixty-five million years ago that the first hominids – descendants of early primates – that preceded the appearance of *Homo sapiens* began to appear around two and a half to four million years ago. Our own species is just 200,000–300,000 years old – almost nothing in geological history terms. We survived as nomadic hunters and gatherers until agriculture – generally agreed to have begun in the Fertile Crescent in the Middle East around 10,000 BCE – facilitated the permanent settlement of people. Written history of civilizations began around 3100 years before our present Christian era, and the Holocene commenced around 11,700 years ago, after the end of the last great ice age when agriculture started.

There are various theories regarding the start of the Anthropocene. Many experts believe it dates from the Industrial Revolution, which began in Britain in the eighteenth century and was reinforced by the *Great Acceleration* after the Second World War. 'Anthropocene' is therefore a very recent geological epoch. The term was coined by Eugene F. Stoermer in the early 1980s and popularized by the Nobel Prize winner Paul Crutzen, who first used it during a conference in the year 2000. The recognition of this epoch is thus a relative novelty in both natural and social sciences, and, at the time of writing, its onset has yet to be officially ratified by either the International Commission on Stratigraphy (ICS) or the International Union of Geological

Sciences (IUGS). Nevertheless, arriving at the present time, we can ask ourselves if humanity is now facing the sixth extinction because of the Anthropocene and even if the 2020 coronavirus pandemic could be an expression of it. We do not have precise answers to this terrifying question, but in the following pages some exploratory paths of research and clues will be offered in order to orientate readers in their own studies and investigations.

Along these lines, in the first chapter I argue that the classic orientation of international relations history (and theory) – being, in fact, the study of *interstate* relations – must be modified to include truly *international* relations that are much more multi-faceted, and entail a wider scope and the furtherance of what Edgar Morin called *complex thinking* using the holistic methodology of systems theory coupled with a transdisciplinary cosmopolitan paradigm. Therefore, from this perspective, IR theory can coexist with the main paradigms of realism and idealism because even if *power politics* constitutes its main focus of study there is no doubt that international law is also a fundamental part of international relations. By the same token, the relationship between the different constituents of a nation, such as ethnic groups, indigenous peoples, religions, social movements, cultural expressions within nation states and economic actors – like transnational corporations – must be part of the discipline. In other words, if the *international system* as a part of the world system is the proper object of study of IR theory, the system's components or *subsystems* – such as the political, military, economic, social and cultural subsystems – must also be included in IR theory. The importance of this holistic approach due to its overarching vision and methodology is clear. Consequently, as an interdisciplinary field of studies, IR theory automatically has to deal with political science, economics, anthropology, sociology, ecology, physics and natural sciences in order to obtain the transdisciplinary knowledge that the theory seeks. Of course, ecology and environmental sciences are the ontological support of the international system which is based on the terrestrial ecosystem. Thus, the planet itself is a fundamental part of IR theory, given the fact that all nation states, civilizations and cultures are sustained by the Earth – a circumstance that enhances the role played by environmental sciences and policies regarding ecology and sustainable development.

Consequently, *sustainable development* – as a set of policies promoted by governments and civil society – in this geological epoch of the *Anthropocene* could be considered the overarching cultural model for international relations, geopolitics, human mobility, the world economy and the environmental issues of current times. Therefore, sustainable development requires a methodology designed to produce the kind of transdisciplinary knowledge[1] which is indispensable for policy planning and

---

[1] I understand transdisciplinarity in the way this concept is understood by the French philosopher and scientist Edgar Morin and the institution created by him and Basarab Nicolescu, the *International Center for Transdisciplinary Research* (commonly known by its French acronym, CIRET), with headquarters in Paris. Transdisciplinarity is neither multiple disciplines talking together to study a problem nor the use of methodology from another discipline, but a new approach proposed by the Swiss philosopher and psychologist Jean Piaget in the 1970s that refers to a kind of knowledge that goes beyond the particular separate disciplines. Basarab Nicolescu developed the term in order to highlight the fact that beyond each particular science exists a 'unity of knowledge of the human

its implementation in the domains of foreign policy, world politics and economics. Thus, going beyond interstate relations in a way that can integrate different scientific visions within the theory of international relations is essential in order to encompass and explain the multiple and diverse issues of a social nature (such as human mobility, migrations, religious movements, cultural and ethnic relations, pandemics and so on), economic interdependence and trade (such as financial flows and markets, technology transfer, commerce, etc.), and political and military relations between nation states and world powers. As I have underlined before, academic books and articles of IR theory mostly focus on the study of *interstate* relations not *international* relations, which is a larger scope concept because nations are human conglomerates rooted in culture, language, religion, territory and diverse factors. Power politics restricts its application to the *political subsystem* of the international system; hence, other subsystems have been left outside.

Therefore, in order to reintroduce this all-encompassing approach within IR theory it is essential to update its methodology with a holistic approach based on complex[2] thinking.[3] It is not difficult to see how a holistic methodology is also required for sustainable development. For instance, the economic realm must be integrated into the social domain and natural ecosystems, but that fact implies that good data, information and knowledge about economics, sociology, psychology, law, physics, chemistry, biology, palaeontology and so on are obligatory. Therefore, it is impossible to have an accurate idea about what has to be done in matters of sustainable development without having this kind of holistic approach in each precise circumstance or case study. For instance, when a government or a corporation or a group of entrepreneurs in the private sector decides to build a hydroelectric power plant, it is absolutely necessary to be aware of national legal regulations and procedures, and additionally to have a good relationship with the local government, and in-depth knowledge of the physical and geographical place where you plan to construct the plant, the community of people living in the area, the technical and mechanical procedures that you will have to take into account to build the dam and exert control over the water current, the agency in charge of the distribution of electricity to people in the surrounding areas, and so on and so forth. If some of these factors are not properly

---

being' which essentially consists of the establishment of bridges between disciplines, through disciplines and beyond disciplines. Thus, the concept must be differentiated from *multidisciplinarity* (studying a problem using several disciplines) and *interdisciplinarity*, which consists of the transfer of methodology from one discipline to another.

[2] I use the term 'complexity' in the sense employed by Edgar Morin and the French school of complex thinking that established the *International Center for Transdisciplinary Research* (CIRET) in Paris.

[3] For instance, it is not possible to discuss why realism does not explain the current Venezuelan crisis, without an analysis of the social factors (the middle classes that oppose the regime), the political factors (the election's legitimacy) and economic issues (hyperinflation, oil production and exports, and the lack of a national industry). Of course, this complex analysis must include the use of the classic *realpolitik* approach based on geopolitics scrutinizing the role played by the US, the EU, Russia and China – and, naturally, if the role played by the UN or the OAS is added, the international law factor is involved, hence the normative or idealistic paradigm, particularly if the case is taken before the UN Security Council.

considered, the project can fail and generate problems with government agencies, community leaders or even face bankruptcy. If the community is not consulted (as prescribed by the 169 ILO Convention), people can go to court and ask the judges to stop the construction of the dam. If the existing research on the vulnerability of the zone to climate change is not sufficiently thorough, you might be surprised by a drought that will lessen or stop the water flow. If everything has gone well but you did not consider the fact that the distribution of electricity to the inhabitants of the region was going to increase the monthly bill for consumers, you will have social protest – or even illegal connections to the grid – facing constant disappointments and conflict.

Therefore, to achieve a successful outcome, it is crucial to take a holistic approach to planning, otherwise you risk failure. And – of course – what is valid at a micro-level is also valid at a macro-level. When the NAFTA negotiators agreed to exempt maize imports from the United States from trade tariffs, they did not take into account the inevitability that this decision would dramatically affect millions of Mexican peasants who would be unable to compete with the US industrialized agriculture – much more productive than theirs – and would then be obliged to cease food production and migrate to the US in large numbers.

An analogous observation can be made concerning globalization: in spite of the fact that the phenomenon is anthropogenic and not reversible, societies around the world must adapt or face social crisis and political turbulence. The acceleration of climate change is also anthropogenic, and societies must adapt to it or try to mitigate its effects, but the phenomenon cannot be rolled back. This means that both the globalization of economy and climate change must be considered in any sustainable development study or neither public policies nor development projects will meet the conditions for success.

The examples above also illustrate the need for this holistic approach to sustainable development because *transformative sciences* (Brauch 2016) are fundamental for the transdisciplinary research needed to overcome the major challenge of thinking beyond the boundaries of each academic discipline. Furthermore, we have to realize that the *world view* (the *cosmovision*) or set of ideas used to understand the world is fundamental to sustainable development because the so-called *mainstream economic paradigm* (Göpel: 2016) that is based on growth, personal gain and capital accumulation is completely inappropriate. Neither can you trust in governments' planning 'growth targets' because in a market economy the private sector – not the bureaucratic state apparatus – rules the economy, and governments are only facilitators of the economy, not rulers. In other words, as civil society must be the main agent of economic reform from the sustainable development perspective, public policy must serve essential social needs and human development and not the interests of the entrepreneurs and corporations. Hence, state planning must be at the service of society, not of the private sector or transnational corporations.

And lastly – but not less important – concerning natural ecosystem government, planners must always take into account the fact that nature functions in a cyclical manner; it is circular – not linear – and consequently the main goal of sustainability must be to preserve the natural processes and not to affect them. Pollution is unnatural (there is no waste in natural processes: everything is recycled), and therefore the main problem of industrial output is what to do with discarded goods and leftovers. One must consider how to avoid the excess of plastics going into rivers, lakes and oceans, diminish the *greenhouse emissions* (GHEs) that increase global warming, and impede the pollution of the water resources needed for industrial or mining activities.[4]

Whales and dolphins in the oceans are dying because of the ingestion of plastics. Lots of species are disappearing because of human activity. The crucial issue to be aware of is that sustainable development is a complex process that involves government action that must address social dynamics, techno-economic processes and natural ecosystems (Jeffrey Sachs 2015).

However, as governments' actions, social dynamics and techno-economic factors are influenced by ideas and the way we think, our mindsets, mentality and ideology influence decision-makers constantly. In classical philosophy, there is a difference between essence and existence, and since the times of the Greek philosophers (especially Plato), essence (the *Topus Uranus*) was considered the source of reality. The essence is the design, the idea; when making a plan you will normally have to first conceive the idea – the design – and construction follows. Constructivism in international relations theory (Wendt 1989) is based on the approach originated in philosophical idealism as well as the normative paradigm of international law with roots in the famous Grotius' principle (*pacta sunt servanda*), as will be seen in the first chapter. Although the *balance of power* realistic paradigm is also of fundamental importance, the fact remains that mindsets and ideology are the source of political behaviour and social hegemony in national states. In consequence, as governments' economic policies are influenced by ideology, the enormous weight of liberal ideology or *mainstream economics* explains why sustainable development is not being fully implemented in the world in spite of the 'wishful thinking' expressed by governments in declarations and UN resolutions since the times of the Brundtland Report (and recently consecrated by the United Nations with the 2030 SDGs Agenda approved in New York by the General Assembly in September 2015).

The question, then, becomes what kind of parameters must the paradigm have that will replace the so called mainstream economics ideology? I have already mentioned the holistic cosmopolitan paradigm whose main focus is in the use of a transdisciplinary methodology, including the Anthropocene's cultural model or ecological *world view* that regards Planet Earth as our common house – as Pope Francis has stated in his Encyclical Letter *Laudato Si'* – that can be compared with Lovelock's philosophy of *Gaia* (the Earth) as a living organism and even Teilhard de Chardin's (1955) theory about the *noosphere* as a kind of spiritual sphere that involves our planet in the same manner that the atmosphere and the biosphere do, and regards every person as a cosmopolitan world citizen entitled to universal human rights independently of governments' recognition.

---

[4] A sad example of governmental mismanagement of natural resources is the Aral Sea in Central Asia, where, during the era of the USSR, Soviet bureaucrats decided to divert the waters of the Syr Darya and Amu Darya rivers for the irrigation of cotton fields needed by the Soviet textile industries.

Thus, governments and global civil society must adopt a different vision to address sustainable development – a holistic and cosmopolitan vision that enables us to focus on global problems by overcoming nationalist ideology, which, along with neoliberalism, is another cause of the sustainable development stagnation and blockade since the last decade of the twentieth century. This *neo-nationalism* that has reappeared in Europe, the United States and elsewhere is characterized by a narrow and anachronistic world view that uses pernicious extreme right-wing ideology against globalization, interdependence, democracy (in some particular cases) and migration. Of course, cosmopolitanism does not mean that people do not have the need for a national identity that appreciates and values the culture, language, religion, landscapes, arts, customs and traditions of the homeland, because this positive cosmopolitanism is complementary – not opposite – to the need to simultaneously enhance values linked to *Mother Nature* (*Gaia*) and to the idea that you can be, at the same time, a good citizen of the world, your region, your community and your nation, which entails the fact that you can take care of your homeland as well as the planet, humanity's homeland.

Furthermore, if people's mobility is a human right – as stated in the UN *Universal Declaration of Human Rights* – it provides grounds for establishing a kind of *transnational citizenship* (which does not imply the granting of the nationality of the host country) as the legal category that could be granted to persons who must travel constantly (such as the CEOs of transnational corporations) as well as to migratory workers who move frequently between different countries. Thus, the world community as a whole must find a way to regulate human mobility without violating the human right to migrate and freely decide where you want to live, which is at the core of migratory problems worldwide. Indeed, if, as a result of globalization, both free trade and the free mobility of persons have become a necessity for global markets, then regulation of them is indispensable, as postulated by the *Global Compact for Migration* adopted at the Marrakech Conference in December 2018. The lack of workers in industrial countries is linked to demographic factors like the zero growth of native populations in industrialized countries, and the phenomenon constitutes a growing pull factor for foreign labour for qualified, skilled and unskilled workers. It is regrettable that, because of his demagogic and racist attitude, Trump invented an artificial and false problem in the US for electoral purposes by presenting immigration as a source of problems, even though the economy was in good shape, with full employment, including moderate sustainable economic growth in 2016.

And it is also true that most governments accept the free movement of merchandise (free trade) but at the same time reject the free movement of persons on the grounds of a mistaken and outdated conception of territorial sovereignty, forgetting that immigrant workers are an economic asset. Nonetheless, some countries (Germany, the rich Arab Gulf States, Canada) have legislation in accordance with this need for temporary foreign workers who are not interested in applying for citizenship of the guest country, keep their original nationality and travel frequently to their home countries. Thus, as previously stated, awarding these workers transnational citizenship – not nationality, but a visa for multiple entries in their passports, as happens with middle-class executives, professionals and tourist travel – is the best way to facilitate the mobility of these people and regulate migratory flows all over the world.

Thus, it is important to state that among the policies that can be implemented to solve the current migratory crisis, governments must make a serious commitment to implementing the norms of the *Global Compact for a Safe, Orderly and Regular Migration* adopted by the Marrakech Conference, which includes granting *transnational citizenship* as a way to make migration orderly, safe and regular. These policies could be complemented in the foreseeable future by a world summit on human mobility, similar to the UN conferences on the environment, human rights, women, populations and social development. Thus, a summit on human mobility could define the paths to be followed in multilateral negotiations in order to create a legally binding international treaty in that respect.

This book is composed of six chapters. Chapter 1 deals with the classic theory of international relations as the transformation of the interstate political subsystem since the Peace of Westphalia to the present time. A historical summary outlines the way that the great paradigms of IR were embedded in the foreign policy of great powers like the European powers that established – at the end of the Thirty Years' War in 1648 – the system of Westphalia as the first model of international order among sovereign independent states. This historical fact can be considered the source of both the idealist (*pacta sunt servanda*) and the realist (*the balance of power*) paradigms. I describe the form in which these great paradigms were applied, highlighting their usefulness in explaining the turning points in history, such as the Napoleonic Wars, the Peace of Vienna, the German Unification under Bismarck, the Great War and the Treaty of Versailles, the Second World War and the United Nations, bipolarity, the Cold War and the fall of the Berlin Wall. Maintaining the importance of a further development in IR theory accords with the fact that – as previously upheld – *relations among nations* must include social, cultural, economic and environmental phenomena, and therefore the interstate vision provided by idealism and realism is clearly too narrow. The increase in globalization, economic multipolarity – including the upsurge in transnational corporations – and the formation of a global civil society and transnational social movements and organizations after the Second World War cannot be explained by the classic nineteenth-century political science single approach, and that is why a focus on social construction (Wendt 2005), world economy (Wallerstein 1989; 2006), complex interdependence (Keohane/Nye 1988) and cosmopolitism (Held 2005; Archibugi 2011; Mignolo 2011; Raskin 2002) is needed. In other words, IR theory requires a holistic paradigm and cosmopolitan world view in order to improve the scope of its sphere of knowledge and better contribute to the transformations that the international system requires in order to adapt itself to the new geological epoch of the Anthropocene by means of sustainable development.

Chapter 2 addresses integration as a way to transcend the Westphalian order and go beyond geopolitics. Therefore, I initially refer to some theoretical aspects, such as the difference between regionalization and integration, before addressing the phenomenon of regionalization in Latin America as well as the EU regional integration process, including the roots of the EU's current financial and social crisis that have triggered *Brexit* and the increase in nationalism and extreme right-wing ideologies that are nowadays thwarting integration in Europe. Nonetheless, and in spite of the said difficulties, the European Union clearly demonstrates that geopolitics (and the importance of the Westphalian system) is receding, while geo-economics and globalization are increasing. Germany does not need the *Wehrmacht* to be hegemonic within the EU any more because its financial and industrial power is enough to exert economic hegemony within the EU. Fortunately, Berlin understands this new twenty-first-century reality, and it is precisely that correct understanding of the realities of the twenty-first century that has obliged Germany to accept its relatively diminished role in world politics since the end of the Second World War.[5]

In this chapter, I also refer to the Latin American regionalization processes, accepting Andres Malamud's theory in the sense that there is a difference between regionalization and integration and that in Latin America (as in other parts of the world) the first phenomenon occurs but not the second. At the end of the chapter, I refer to the role of the International Court of Justice in the solution of several territorial controversies in Latin America. I also make a brief reference to the present influence of Halford Mackinder in classic geopolitics regarding the US foreign policy of NATO expansion as a way to get the US military machine near the *heartland* of Eurasia, as prophetically announced by Zbigniew Brzezinski in *The Great Chessboard* (1996), a book in which he writes about the geostrategic imperatives of the United States if that superpower wants to maintain world primacy. Thus, the geopolitical conflicts of Ukraine, the Middle East (Syria and Iran), and the maritime border disputes with China (in the South and East China seas) are perfectly comprehensible in this geopolitical and geostrategic context in spite of the anachronism, a fact that does not diminish the dangers unleashed, particularly if the contenders are nuclear powers.

In Chap. 3, I deal with the fact that the world is living in the age of the Anthropocene on the grounds that this is not only a geological epoch but a cultural model and historical age, which means that it can also be regarded as a scientific paradigm. Good governance, expressed as *cosmopolocene* and *cosmopolitics* (Delanty/Mota: 2018), is necessary to address the ecological challenges and reassess the role of natural and social sciences as well as philosophy, hence the reference to Teilhard de Chardin (1955) and his ideas about the *noosphere* as a spiritual sphere of knowledge surrounding the planet that is nowadays tantamount to culture, science and technology: the substantial but somehow forgotten counterpart to what Paul Crutzen calls the techno-economic revolution that can be found at the origin of this age of the Anthropocene.

Chapter 4 is about sustainable development and the problems that the social and economic spheres face when they have to integrate sustainable natural ecosystems. Since development is linear, it is quite difficult to integrate ecosystems because they

---

[5] The two great powers defeated in the Second World War were Germany and Japan. Tokyo has followed the same economic path of Berlin that has led both countries to become the third and fourth largest economies in the world, though lacking political and military power because, as is well known, neither Germany nor Japan are permanent members of the UN Security Council and they do not have nuclear weapons.

are essentially circular and their dynamics are cyclical not linear. Thus, if policy-makers want to preserve those cyclical ecosystems, development must not interfere with them, or, if it does, sufficient action must be taken to facilitate the rapid regen-eration of ecosystems, otherwise, as stressed by James Lovelock (2007, there is a growing risk that humanity will reach a tipping point where instead of sustainable development it will be necessary to adopt a world policy of *sustainable retirement*. I extensively quote from Jared Diamond's book (2007) about collapse or 'why soci-eties choose to fail or succeed', an investigation in which the author addresses the causes of some historical and prehistoric cases of the collapse of societies – particu-larly those significantly influenced by environmental changes, the effects of climate change, hostile neighbours, trade partners, and the society's response to the chal-lenges of nature – and describes the success or failure different societies have had in coping with such threats.

The case studies chosen by Diamond are those of the Norwegians settlers in Green-land at the end of last millennium; the Maya Civilization in Yucatan and Guatemala during the classic period from the sixth to the ninth centuries; the islanders of Easter Island in the South Pacific; and the Anasazi Indians in the US. Those are cases of failure, while other examples (the island of Tikopia in the South Pacific; the Dominican Republic; the Tokugawa dynasty in Japan; and the natives of the high-land forests of Papua New Guinea) are all success stories, a fact that, according to Diamond, demonstrates the importance of cultural factors (ideology, religion, tradi-tions and customs) and governance in the decisions that societies take concerning their future.

While the bulk of Diamond's book concerns the demise of these historical civi-lizations, for him humanity collectively faces a comparable turning point in present times, but on a much larger scale, as many of the same issues are currently involved, with possibly catastrophic near-future consequences for peoples and nations if they are not able to cope with the various challenges posed by environmental threats. In this regard, the constitutional reforms of Bolivia and Ecuador are an interesting example of two Latin American countries that have approved a new constitutional standard that not only recognizes the traditional indigenous philosophy of *buen vivir* (good living) as the point of departure of a new social pact designed to put an end to the discrimination and exclusion that these peoples have suffered from oligarchic rulers since colonial times, but is also a legal reform that accepts that nature and the planet itself (the *Pachamama* or Mother Earth) have the right to be protected from human exploitation and contamination, which is a novelty in the framework of a western-type philosophy of law that usually recognizes only persons (human or legal) as entitled to have rights.

Chapter 5 concerns sustainable development as an alternative to the dominant 'mainstream economics' paradigm proposed by the Brundtland Report *Our Common Future* (1987). It has not yet been fully implemented precisely because most govern-ments of the world continue to be under the influence of neoliberalism and continue to regard growth as the quintessential parameter of development. In my view, one of the most appropriate ways to confront this predominant reductionist kind of thinking

is the sustainable development paradigm, among other reasons because the dominant economic system based on the neoliberal ideology hidden beneath 'mainstream
economics' responds to an unsustainable situation because development is not guided
by the satisfaction of human needs but by the growth and accumulation of capital
for the benefit of the world oligarchies of the super-rich. The majority of the world's
population lives in poverty, and it is important to realize that economic growth, as
shown masterfully by the French scholar Thomas Piketty (2014) in his book *Capital
in the Twenty-First Century*, does not respond to the interests of national states but to
the interests of the small world elite of super-rich people transformed into a new class
of *rentiers* in such an excessive manner that it is the main source of inequality and
concentration of wealth at world level, not just within national states. This situation
is absolutely unsustainable. It is also the root of social reactions expressed through
protests and violence (like the *gilets jaunes* of France or the social movement of
Chile in 2019). The electoral support for the extreme right nationalists in Europe,
Trump in the United States and *Brexit* can be explained by this *social malaise*. It goes
without saying that the inequality and enormous concentration of wealth provoked
by neoliberalism are the sources not just of the world's social uneasiness but also of
the governments' inaction concerning climate change and the political disarray of
the international system that personalities of the US political establishment – such
as Richard Haas (2017) and Henry Kissinger (2014) – deplore in books dealing with
the crisis in the Westphalian system.

Nevertheless, I do not believe that capitalism will explode, as leftist extremists
have always thought, or implode as Bruno Latour (2015) argues, at least not in the
mid-term. Therefore, instead of searching for alternatives to the economic system
as such, we have to look for alternatives to the narrow predominant reductionist
neoliberal conception that sees everything from the point of view of economic growth,
free markets and capital accumulation for the world oligarchies.

Thus in Chap. 5, I argue that the world is in need of a second 'great transformation'
(Polanyi 1972), comparable to that which occurred in Europe in the years of the capitalist Industrial Revolution. This great transformation is also related to the *mindshift*
required to adopt the new holistic paradigm of sustainable development proposed
by the Brundtland Report (1987) and the UN SDG commitments of 2015, which
essentially postulate that social and human development must be achieved in a form
compatible with natural ecosystems as the only way to overcome the misconception
that development is tantamount to economic growth. According to this erroneous,
biased and incomplete perspective (which is also partial, narrow and reductionist), if
there is no growth, capital accumulation or personal gain, there is no development.
Consequently, it is indispensable to have a *post-capitalist* approach, such as the ones
proposed by Paul Mason (2016), Paul Raskin (2002), John Holloway (2002), Howard
Richards (2017), James Lovelock (2007) and Ian Angus (2016), since private profit
and capital accumulation would no longer have primacy if this 'great transformation'
occurs. Some economies might rely on governmental supervision and control, as in
the Chinese system, others on decentralized arrangements or on social ownership and
workers' control, as in the classic model of socialism that Richards and Holloway
suggest in their writings.

The holistic nature of sustainable development also demands that in order to solve the problem of inequality and wealth concentration at world level it is absolutely necessary to fulfil the UN 2030 Agenda commitments. Consequently, it is crucial to combine adequately the four spheres of social dynamics, techno-economic factors, good governance and natural ecosystems (Jeffrey Sachs 2015). Hence, there is no question that the state must act decisively in implementing both fiscal reform and social policies (SDGs concerning education, health, decent work, the eradication of extreme poverty, empowering women) within the framework of a democratic system capable of creating effective, responsible and inclusive institutions at all levels, as posited by SDG 16. In addition to the appropriate combination of techno-economic factors, social dynamics and good governance, it must be kept in mind that the articulation with the sphere of natural ecosystems does not follow a linear trajectory but a circular one and must not go beyond the planetary boundaries that put a limit on economic growth, all of which essentially means getting rid of neoliberalism and – at least, and as a minimum – returning to Keynesianism.

The latter also explains the chapter's emphasis on capitalism as the world's dominant economic system. We have seen that crises like those of 1929, 1973 and 2008 call into question the survival of the system itself if the structural reforms that this mode of production requires are not made. And, of course, the same can be said about the fatal and horrible year of 2020 because, beyond being a health crisis, the pandemic was a world economic crisis that will inevitably oblige all governments of the world to proceed to complete reform of the system, not just by forsaking neoliberal ideology but by giving back to the state its role in the regulation of markets and the economy. Thus, whether the systemic crisis derives from the absence of a surplus recycling mechanism (Varoufakis 2015) or is the result of the process of concentration of wealth among a small world elite of *rentiers* (Piketty 2014) or of the coronavirus pandemic of 2020 is not the most important issue. The real issue is that both of the capitalist system reform policies proposed by Varoufakis and Piketty (the surplus recycling mechanism and appropriate taxes on big fortunes) must be implemented jointly with the energetic intervention of the state and multilateral agreed measures and policies; otherwise, the social, environmental and health crisis at world level will continue, putting at risk the survival of the species itself.

Also, Paul Mason's ideas about *post-capitalism* are mentioned as an interesting expression of the prevailing zeitgeist, especially in the microeconomic field, where new information technologies, such as the Internet and smartphones, are erasing the borders that previously separated employees from freelance workers. Additionally, information has become abundant and is no longer a scarce commodity, which has had an impact not only on the reduction in the price of electronic commerce but also on the free information provided by social networks of free collaborative work. For instance, this technological revolution has facilitated the rise of the Internet's *Wikipedia*, which not only superseded the former printed encyclopedias and dictionaries (and in some cases led to or hastened the bankruptcy of their publishers), but is based on the free, collaborative and self-programmable work of thousands of academic volunteers who do not receive any monetary remuneration, something

analogous to Howard Richards's ideas about the social solidarity economy practised in Chile and other Latin American countries at microeconomic level or John
Holloway's findings concerning the Zapatista social movement in southern Mexico.

But the need to give back more power to the state also has a worrying aspect
regarding the threats to fundamental freedoms posed by the new cyber technologies,
including computerized artificial intelligence (AI), because the fusion of info and
biotechnologies can fall under the control of authoritarian governments (or private
corporations) that do not care about individual liberties (Harari 2016). Thus, the main
reason for promoting a substantial shift towards cosmopolitan consciousness is the
need for an antidote to the new ways of manipulating human behaviour which are
hidden in these cyber technologies.

Inspired by Polanyi's *great transformation*, Paul Raskin (2002), the Stockholm
Environment Institute, the Tellus Institute and the *Global Scenario Group* (GSG)
have called this new planetary situation the *Great Transition* or the *planetary phase
of civilization* as a phenomenon that is increasing *global interdependence* as well
as the *risk of trespassing planetary boundaries* – like climate change – that have
transformed the whole planet into a unitary socio-ecological system. Hence, in spite
of the crisis of the Westphalian order, the growth of population and economies, the
collective erroneous belief in the inexhaustibility of natural resources, and disregard
for the environmental impact of the human 'Great Acceleration', the shift in human
collective consciousness has not yet become reality.

Consequently, if the planetary phase of civilization is characterized by globalization, biosphere destabilization, mass migration, the Internet revolution in world
communications and so on, unfortunately the shift in culture and consciousness –
the third significant transition in civilization after the two earlier macro-shifts from
the beginning of civilization to the modern era – has yet to occur. Thus, from the
times of the hunting and gathering communities – when spoken language was the
basic means of communication – to the shift into early civilization at the beginning
of agriculture some 12,000 years ago (which brought more structured city states and
kingdoms, settled agriculture and the invention of writing as a means of communication), we have arrived at this *planetary phase* where social organization, the
economy and communications are seeing through the lenses of the *world wide web*
but the human mind is still the same that prevailed in the past. Furthermore – and
unlike prior transitions – this actual planetary phase is linked to the new geologic
epoch of the *Anthropocene*, in which human activity has become the primary driver
of changes to the Earth ecosystems and the planet itself.

Nevertheless, the technological and social dynamics appear to be accelerating
because the longevity of each successive period is less than the previous one (100,000
years for primitive *Homo sapiens*, 10,000 years for *early civilization*), while current
modern times have had the pattern of a *Great Acceleration* in a very short period of
200 years if we take as the point of departure of this period the Industrial Revolution in the eighteenth century. Thus, we risk falling into some kind of social devolution or anarchic *barbarization* unless we overcome the hindrances to the *Great
Transition* provoked by the persistence of dominant institutions and cultural values

(including governments' attempts to address the problems with incremental small policy changes typical of a *business as usual attitude*).

What might be the way out of this world crisis? What is remarkable is that Raskin and GSG advocate a possible *Great Transition* that could develop new institutions to promote environmental sustainability, social equity and equality, and once a sufficient level of material prosperity is attained, *qualitative fulfilment* (rather than the quantitative consumption characteristic of our times) as the more important goal of human and sustainable development. Of course, these changes must be rooted in the values of solidarity, ecology and well-being that would gradually displace the *mainstream economics* triad of individualism, domination of nature and consumerism. Nonetheless, the political and cultural shift envisioned in this scenario depends on global civil society as a potential actor that will promote this *post-capitalist* model where private profit and capital accumulation would no longer have primacy or otherwise in a sudden world monumental crisis like the one provoked in 2020 by the coronavirus pandemic.

In Chap. 5, there is also a short description of the UN's Sustainable Development Goals (SDGs), including observations on the difficulties and obstacles that could arise in the implementation of them during the next decade, as well as a reference to certain paradigmatic 'niche' changes that in practice could help in the implementation of the SDGs, ranging from those that are described in Maja Göpel's book (such as the strategy of successive small changes that transformed the ideas of the ruling classes in Germany about renewable energy) to the experiments of the 'communities in transition' and the 'common goods of humanity', as well as the remarkable efforts of the small Kingdom of Bhutan in the Himalayas to achieve a 'gross inner happiness' index capable of replacing the GDP. Social experiments such as those suggested by Howard Richards and John Holloway and even the constitutional law reforms promoted by the indigenous peoples of South America are also mentioned, as well as the fact that globalization and the crisis of the Westphalian order, along with the new geological epoch – the Anthropocene – have opened the way to a cultural model that proposes more democratic and cosmopolitan world governance.

In Chap. 6, I deal with human mobility in the context of the Westphalian order crisis characterized by a considerable reduction in the capacity of national states to keep their territories under control. This happens because globalization inevitably requires 'porous' borders; consequently, the defensive/security-prone reaction of the conservative sectors against immigration is clearly condemned to failure. Since one of the most visible effects of globalization is human mobility, the resulting predicament with the increase in immigration flows is one of the most serious world problems nowadays. Thus, respecting the basic human rights of migrant workers and refugees and granting transnational citizenship to those workers are feasible alternatives for solving migratory problems. Additionally, I stress that if both globalization and interdependence presuppose the economic mobility of both merchandise and people, the free movement of workers within the global job market should be assured by regularizing their legal status by granting them transnational citizenship, visas or temporary work permits.

Thus, the regularization of migrant workers must become a priority in the foreign policy of countries that 'export' labour, bearing in mind that all measures destined to eliminate obstacles and facilitate the free movement of people are positive for both the world economy and international security – and, of course, for the individuals involved in these epic journeys. The right to emigrate is a human right, as is freedom of mobility, and the same applies to the right of individuals to live and work in any part of the planet. Hence, *human mobility* should be deemed a decisive element in the fulfilment of human rights, and it is also clear that implementing such objectives will require multilaterally negotiated policies at the United Nations designed to regularize migratory flows and put an end to policies that criminalize migrant workers and violate human rights.

In summary, although the two classical paradigms of the Order of Westphalia were capable of explaining the functioning of the international system during the nineteenth and twentieth centuries, they cannot assume the same function today because the system must be enlarged to encompass its social, environmental, cultural and economic dimensions. Henceforth, the new holistic and cosmopolitan paradigm capable of fully explaining the whole and complete nature of IR and the complex structure of the international system (both synchronic and diachronic) using this multidimensional approach corresponds to the ideas that postulate the need for a global cosmopolitan society in this *planetary phase of civilization.* However, as evolution can take different directions (depending on the ways ecological systems respond to anthropogenic stresses, such as climate change and how social institutions evolve and conflicts are resolved), the model that will finally emerge depends on which human values become ascendant and the choices people make in the critical years ahead.

The world needs a holistic approach to world governance in the age of the Anthropocene and cosmopolitan consciousness to protect human rights, cultural diversity, global public goods, biodiversity, freshwater, oxygen, the oceans, glaciers and high mountain ranges, the polar ice fields and so forth, because what is really at stake is the compatibility of economic, social and political progress with the conservation of natural ecosystems and respect for the rights of Planet Earth (to biodiversity, to an atmosphere free of pollution and greenhouse emissions and so on), otherwise humanity risks extinction.

Contents

Abbreviations

AI	Artificial Intelligence
APEC	Asia-Pacific Economic Cooperation
ASEAN	Association of Southeast Asian Nations
ASOCODE	Asociación de Organizaciones Campesinas Centroamericanas para la Cooperación y el Desarrollo [Central American Peasant Association for Cooperation and Development]
BCIE	Banco Centroamericano de Integración Económica [Central American Bank of Economic Integration]
BRICS	Brazil, Russia, India, China and South Africa
CACM	Central American Common Market
CALC	Latin American Conference on Integration and Development
CD	Conference on Disarmament
CECA	Comunidad Europea del Carbón y del Acero [European Community for Coal and Steel]
CEIPAZ	Centro de Educación e Investigación para la Paz [Centre for Education and Peace Research]
CELAC	Comunidad de Estados Latinoamericanos y del Caribe [Commonwealth of Latin American and Caribbean States]
CEO	Chief Executive Officer
CIA	Central Intelligence Agency
CICIG	Comisión Internacional Contra la Impunidad en Guatemala [UN Commission against Impunity in Guatemala]
CIRET	Centre International de Recherche et d'Etudes Transdisciplinaires [International Centre of Studies and Transdisciplinary Research]
CLAIP	Consejo Latinoamericano de Investigación para la Paz [Latin American Council of Peace Research]
COP 21	Conference of the Parties on the Paris Climate Change Agreement
CTBTO	Comprehensive Nuclear-Test-Ban Treaty Organization
DR CAFTA	Dominican Republic and Central American Free Trade Agreement [with the United States]
ECB	European Central Bank

ECLAC	Economic Commission for Latin America and the Caribbean [CEPAL in Spanish]
ECOWAS	Economic Community of West African States
EMS	European Monetary System
ERCA	Environmental Restoration and Conservation Agency
EU	European Union
FAO	Food and Agriculture Organization
FDI	Fuerzas de Defensa Israelíes [Israeli Defence Force]
FLACSO	Latin American Faculty for Social Sciences
FTAAP	Free Trade Area of the Asia Pacific
GDP	Gross Domestic Product
GEL	Grupo Editor Latinoamericano [Latin America Publisher Group]
GHE	Greenhouse Emission
GHG	Greenhouse Gas
GSA	Geological Society of America
GSG	Global Scenario Group
HCFC	Hydrochlorofluorocarbon
HYDE	History Database of the Global Environment
ICAN	International Campaign to Abolish Nuclear Weapons
ICBM	Intercontinental Ballistic Missiles
ICJ	International Commission of Jurists
IISD	International Institute for Sustainable Development
ILO	International Labour Organization (United Nations)
IMF	International Monetary Fund
INF	Intermediate Nuclear Forces (Treaty between the US and the USSR)
INGO	International Non-Governmental Organization
INGOs	International Non-Governmental Organizations
IPCC	Intergovernmental Panel on Climate Change
IPRA	International Peace Research Association
IR	International Relations
IRIPAZ	Instituto de Relaciones Internacionales e Investigación para la Paz [International Relations and Peace Research Institute]
IS	International System
IWRM	Integrated Water Resources Management
LAFTA	Latin American Free Trade Association
LASA	Congress of Latin American Studies Association
MAD	Mutual Assured Destruction
MERCOSUR	Mercado Commún Sudamericano [South American Common Market]
MINUGUA	Misión de Naciones Unidas en Guatemala [United Nations Verification Mission in Guatemala]
MIT	Massachusetts Institute of Technology
MP	Ministerio Público [Ministry of Justice]
NAFTA	North American Free Trade Agreement
NATO	North Atlantic Treaty Organization

NGO	Non-Governmental Organization
NGOs	Non-Governmental Organizations
NOAA	National Oceanic Atmospheric Administration
OAS	Organization of American States
OBOR	One Belt One Road (or the new 'Silk Route')
ODECA	Organización de Estado Centroamericanos [Organization of Central American States]
OECD	Organisation for Economic Co-operation and Development
OPANAL	Organismo para la Proscripción de las Armas Nucleares en la América Latina y el Caribe [Agency for the Prohibition of Nuclear Arms in Latin America and the Caribbean]
OPCW	Organisation for the Prohibition of Chemical Weapons
PRI	Partido Revolucionario Institucional (Mexico) [Institutional Revolutionary Party]
PT	Partido de los Trabajadores (Brazil) [Workers' Party]
RCEP	Regional Comprehensive Economic Partnership
SALT	Strategic Arms Limitation Talks
SD	Sustainable Development
SDG	Sustainable Development Goal
SDGs	Sustainable Development Goals
SICA	Sistema de la Integración Centroamerica [Central American Integration System]
SME	Système Monétaire Européen [European Monetary System]
SPD	Sozialdemokratische Partei Deutschlands [German Social Democratic Party]
START	Strategic Arms Reduction Treaty
TIAR	Tratado Interamericano de Asistencia Recíproca [Inter-American Treaty on Reciprocal Assistance]
TPP	Trans-Pacific Partnership
UN	United Nations
UNAM	Universidad Nacional Autónoma de México [National Autonomous University of Mexico]
UNASUR	Unión de Naciones Suramericanas [Union of South American Nations]
UNCLOS	United Nations Convention on the Law of the Sea
UNCTAD	United Nations Conference on Trade and Development
UNEP	United Nations Environment Programme
UNFCCC	United Nations Framework Convention on Climate Change
UNSC	United Nations Security Council
UNSD	United Nations Statistics Division
UNWTO	United Nations World Tourism Organization
US	United States
USA	United States of America
USAID	United States Agency for International Development
USSR	Union of Soviet Socialist Republics

USSR-Russia	Union of Soviet Socialist Republics – Russia
USTR	United States Trade Representative
UVG	Universidad del Valle de Guatemala [University of the Valley of Guatemala]
WBGU	Wissenschaftlicher Beirat der Bundesregierung Globale Umweltveränderung [German Advisory Council on Global Environmental Change]
WH	White House
WMO	World Meteorological Organization
WTO	World Trade Organization
WWF	World Wide Fund for Nature (formerly the World Wildlife Fund)

List of Figures, Graphs and Photographs

List of Tables

Chapter 1
International Relations Theory and the Great Paradigms

There is a major difference between the present situation and the time of the first European concerts: in the nineteenth century five predominant states excluded ten others; today seven of them exclude 185 others while the G20 excludes 173 of them. Yet the major lesson to be drawn from globalization is total inclusion, which itself has two meanings: all peoples of the world are partners in the international system, and global governance can only function if it leaves no one standing at the door...exclusion is dually blameworthy because, objectively, it diminishes any chance of regulation, and subjectively, it sows frustration, humiliation, resentment and hence violence.
Badie (2012: 33)
A country with America's idealistic tradition cannot base its policy on the balance of power as the sole criterion for a new world order. But it must learn that equilibrium is a fundamental precondition for the pursuit of its historical goals.
Kissinger (1994: 833)

1.1 Introduction

The theory of *International Relations* (IR) is essentially a transdisciplinary domain of knowledge whose main object of study is the *world system* (Wallerstein 2006), which includes the *international system* (IS). Each subsystem of the IS has its own particular field of studies. For instance, the politico-military subsystem has political science as the foremost discipline to support its research, the social subsystem has sociology and anthropology, the economic subsystem has economics, and the cultural subsystem has history and anthropology. Furthermore, as mentioned in the introduction of this book, Planet Earth itself is the ontological[1] foundation of the IS scaffolding. Clearly, therefore, knowledge derived from geology, biology, physics,

---

[1] According to philosophy, ontology is the theory of being and epistemology the theory of knowledge. Philosophical realism – as in Aristotle's and Saint Thomas's scholastic philosophy – must be differentiated from philosophical idealism – like the doctrine of Plato in ancient times or the Kantian

© Springer Nature Switzerland AG 2021
L.-A. Padilla, *Sustainable Development in the Anthropocene*,
The Anthropocene: Politik—Economics—Society—Science 29,
https://doi.org/10.1007/978-3-030-80399-5_1

chemistry and environmental sciences like ecology must also be applied in IR theory. That is what this book is all about. The transformations in the political and military subsystem of the IS are the result of historical events that occur as consequences of the behaviour of the main actors in the IS, i.e. the sovereign state members of the UN. In accordance with the normal foreign policy practices – including military action and wars – UN members are constantly introducing changes which eventually lead to the transformation of an IS that is composed of several subsystems that are studied by the different disciplines mentioned above. In this chapter I will review how these changes have come about in the history of international relations, or, in other words, in the history of the politico-military subsystem, which essentially refers to inter-state relations from the Peace of Westphalia (1648), the Peace of Vienna (1815), the Treaty of Versailles (1919) and the San Francisco Peace Conference (1945) to the end of the Cold War in 1989. It also covers the increase in globalization when the former Communist Bloc was absorbed by the capitalism that became predominant at world level and prevailed as the mode of production within the international economic subsystem. Finally, I will make a brief reference to the United Nations and the need for the foreign policies of all member states to strengthen multilateralism in order to democratize the IS.

1.2   Multilateralism and the Democratization of the International System

The transformation of the international system is one of the greatest challenges of the twenty-first century because the prevailing Westphalian order, based on the sovereignty of national states, is no longer sufficient to address global issues that require cooperation and a common effort. The environmental threats of climate change, the social and economic crisis provoked by inequality, wealth concentration and poverty, the monumental health crisis provoked by the coronavirus pandemic of 2020, and the security and military menaces of terrorism, internal armed conflicts, nuclear proliferation and the geopolitical and territorial hazards still pressing foreign policy everywhere cannot be solved by any state in isolation. Therefore, among other important tasks for international policy and governments,

---

and Hegelian rationalism of modern times – because they have epistemological implications, rationalism being the predominant theoretical framework of modern science. Nevertheless, in spite of the new paradigm which emerged from quantum physics at the beginning of last century, natural sciences tend to be realist in epistemology, thanks to the old realist dictum that considers truth to be *adaequatio rei et intellectus* (the adequation of things and intellect), as happens when – for instance – astronomers study the solar system and corroborate that Earth is a planet orbiting around the sun in 365 days. Therefore, according to astrophysics, it is quite evident that the international system (IS) can be considered – in accordance also with the realist epistemological and ontological explanation – part of the terrestrial ecosystems which in the last resort are dependent on *Gaia* (Planet Earth), and this fact explains why environmental sciences have actually become an important component of IR theory.

it is crucial to promote multilateral cooperation geared towards reinforcing global democratic governance, strengthening multilateralism, and overcoming the *democratic deficit* inside the multilateral system (the *diplomacy of connivance* of the G7 and G20 groups) that provokes humiliation and resentment, as Badie (2012) notes, particularly since the challenges of climate change, inequality, wealth concentration, the regulation of migratory flows and the implementation of the UN SDGs have not yet been adequately addressed.

Furthermore, if the SDGs – including the commitments made in 2015 at the Paris COP 21 – that are fundamental to solving world problems are not met, the human species and the planet itself are under threat, as demonstrated by the crisis of the COVID-19 pandemic in 2020. This situation requires more active multilateral negotiation dynamics within the UN system and a new proactive form of diplomacy – not the usual cooperative and associative type, but a *diplomacy of solidarity*, especially with regard to the Global South.

This is due to the fact that the SDGs depend, to a large extent, on good global governance, which in turn is articulated by the way multilateral negotiations operate – and resolutions are obtained – within the United Nations system. The UN is essentially an intergovernmental organization, which means that member states are represented by government officials, not by representatives of the peoples or nations. Thus in order to reform the UN system, promote the presence of representatives of civil society and perhaps adopt a tripartite form of representation like the International Labour Organization (ILO), it is necessary to deepen and enhance democracy at national level[2] and to create a new type of transnational citizenship on a scale of vast regional groups as a means of counteracting hegemony – the 'realist' approach – and strengthening integration and the 'idealist' approach.

At European level, for example, the Westphalian system has already been transformed considerably thanks the supranational institutions created within the integration process. The EU has gradually been moving towards overcoming the Westphalian model and even the definition of the concept of sovereignty has been undergoing novel interpretations that make it possible to think about future changes of great importance, as noted in the academic work of several researchers, among them Badie (1999a, b), a professor at *Sciences Po* in Paris.

This implies that the pre-eminence of national states within the system and the principle of sovereignty are no longer the same in the contemporary world,[3] not only because the very legitimacy of the Westphalian system is being questioned

---

[2] This subject has been considered in the books of Cortina (2010) and de Sousa Santos (2014). The democratization of the international system requires a better and more vigorous multilateralism based, among other proposals, on a new type of tripartite representation like the presence of International NGOs at the UN General Assembly to establish a World Parliament or a *United Nations Parliamentary Assembly*.

[3] The current debate – especially in the European Union – on the role of national sovereignty when supranational institutions such as the European Commission are established is quite important. It is also worth noting that in countries like Guatemala, at the request of its government, the UN established a Commission in order to support national efforts to investigate networks of organized crime. The *International Commission against Impunity in Guatemala (CICIG)* has had extraordinary success fighting corruption and strengthening Guatemalan national institutions of justice, such as

(Kissinger 2014)[4] – the great historical civilizations and ancient states such as China, India, Japan, Iran, the Mayas, Aztecs, Incas, and the Arab countries were never part of it – but because reforming the United Nations and strengthening the processes of regional integration implies the practice of a new type of multilateralism, which – forsaking traditional associative negotiations based on national interest – should be based on solidarity, at least among countries in the same region, since, as Badie points out, in addition to markets in the social sphere, transnational forms of solidarity have been developed. Thanks to the immediacy of image, information and communication, all individuals are directly involved in the internal affairs of neighbouring or distant states. It is therefore a kind of multilateralism that points to the establishment of a transnational citizenship, endowed with a cosmopolitan vision that will be fundamentally concerned not just with the interests of nation states but also with local communities, regions and humanity as a whole, which includes the novelty – as will be seen in the following pages and chapters – of considering the interests and rights of *Gaia,* the living planet (Lovelock 1979) which we inhabit and which is the sustenance ground of the international system.

In any case, the following pages provide a brief historical review of the history of international relations or – more precisely – interstate relations within the political and military subsystem of IS. The narrative begins with the system's origins in the Peace of Westphalia (1648) and ends at the present day, emphasizing factors such as the way the social and political context has influenced the way the great theoretical paradigms (realism/idealism) have been applied, although it should be remembered that events usually precede explanations, which are constructed *ex post facto.* Hence, from the methodological point of view, I maintain that the study of socio-political dynamics in the course of history is absolutely fundamental to understanding the emergence of theory, with the caveat that this sort of practical historicism does not mean disregarding the influence of the diverse streams of philosophical thought (rationalism, empiricism, positivism) in both the theory itself and the methodology that is used in the different research topics. In relation to the latter, it is essential to bear in mind that international relations constitute an interdisciplinary knowledge that aspires to have transdisciplinary findings, i.e. that aspires to go beyond the particular disciplines in order to produce a more integral and 'transformative' knowledge.

For instance, Brauch (2018) argues that environmental problems cannot be properly understood without a transformative science approach which includes both a sustainable peace and a peace ecology perspective:

---

the Public Ministry (Attorney General). Unfortunately, the good performance of the CICIG led the conservative (and corrupt) sectors of the Guatemalan political establishment to terminate the agreement with the UN in September 2019. For a theoretical debate on this subject see Gutiérrez (2016) and Villagran (2016).

[4] According to Kissinger, as the present international system is based on the Treaty of Westphalia, and only the European states of that era participated in its establishment, the states that were formed around ancient civilizations (China, India, Iran, Japan, Islamic and pre-Columbian American civilizations) do not fully identify with the current status quo and in some cases question its legitimacy.

The goal of a 'sustainable peace' may be addressed from a peace ecology perspective that integrates both peace and security studies and ecology or ecological approaches aiming at the implementation of the goal of 'sustainable development'. It is argued that this requires a shift from disciplinary and multidisciplinary research methods towards inter- and transdisciplinary approaches by moving towards a 'transformative science' aiming at a 'sustainable peace' where the needed policy changes and the actors and processes of this change towards sustainability should become a part of the research design and action research process. That transdisciplinary approach also implies the use of a comprehensive methodological approach in which we do not ignore any of the factors (psychological, geopolitical, ecological, cultural, religious, economic, etc.) involved in explaining international phenomena (foreign policy, international commerce, conflicts – interstate, interethnic, intercultural – processes of integration etc.), which implies that hermeneutic methodology, functionalism, positivism or statistics with their surveys can be used in the study of international relations, but always executed with care to make explicit the different approaches – cultural, intersubjective, sociological, legal – with which the theory works (Brauch 2018: 175).

In the following narrative I will explain how the paradigms were born and how they have been applied throughout history, including the fact that, because of strong ideological biases, in some cases they have not been applied appropriately when certain governments and their political agents have tried to justify foreign policy without social and political support due to a lack of knowledge of the context in which actions unfold. Afterwards, especially in Chaps. 3 to 6, I will address the way that environmental sciences apply to IR theory as well as economic, social and natural sciences.

1.3   A Historical Perspective of the Great Paradigms

It is well known that the concept of 'paradigm' (Kuhn 1962) is used to refer to a set of ideas and theories applied by science, so the great paradigms of international relations[5] have been generically identified as 'idealism'[6] and 'realism'. To a great

---

[5] I use the term 'paradigm' according with the definition given by Kuhn (1962), i.e. as a concept which identifies the joint set of ideas, beliefs, theories of a science or – in the field of their practical application – experiments and the methodology for solving knowledge problems which are shared by a community of scientists or decision-makers, as usually happens in IR Theory with regard to foreign policy.

[6] The use of the term 'idealism' is explained when one considers, for example, the ideals that guided the policies of former US Presidents Woodrow Wilson and Franklin D. Roosevelt, which facilitated the establishment of the League of Nations as a result of the Versailles negotiations at the end of the First World War and subsequently of the United Nations in 1945 at the end of the Second World War. However, it should be borne in mind that in the United States the policies of the Democratic Party (to which both Presidents belonged) are labelled 'liberal', which is why there are some authors who use the term 'liberal' instead of the term 'idealist' to refer to this paradigm. For my part, I prefer the concept of 'idealist' because, among other reasons, the paradigm is closely related to international law insofar as it is a normative conception of international society which supposes an ideal (deontological) view of how international relations 'should be', rather than how they are (ontological), which is one of the main postulates of the realist paradigm. On the aforementioned, see Padilla (2009).

extent these paradigms guide or inspire the decision-making of statesmen in matters regarding foreign policy, even when such decisions are made without full awareness of the theoretical (or ideological[7]) reasons that lead them to make the decisions in question. That is to say, political actors often do not fully realize what has caused them to follow certain lines of political action, or worse, they behave unconsciously under the influence of the ideology. This usually gives their actions a bias of concealment of reality, which obscures and denaturalizes knowledge and generally leads to failures in political actions, as will be shown later, following Habermas (1986).[8]

In addition, it should be borne in mind that a single paradigmatic theoretical framework is not usually the only one to give orientation to foreign policy – especially in the case of the great powers – which means that in certain circumstances and fields of action, the paradigms are usually interchanged and used alternatively. Or it may happen that different agencies of the same State follow lines of action based on different theoretical conceptions, because the circumstances and interests at stake demand this.

Hence it is possible that the US President Woodrow Wilson acted from the purest idealistic motives when he proposed the establishment of the League of Nations during the peace negotiations of Versailles (1919) at the end of World War I. Even if the League did not accomplish its goals, the creation of this first international organization of multilateral vocation and supranational intentions had positive consequences in the long term because it led to the foundation of the United Nations in 1945. Nevertheless, given that the United States was part of the Allied forces winning the war, when President Wilson participated in the decision to impose considerable war reparations on the defeated Germany, he did so within the framework of theoretical

---

[7] In general terms, ideology can be regarded as the set of ideas that people have about the society in which they live. Its purposes are to produce not knowledge (this is the main purpose of science) but images of reality and to mobilize people around objectives of a political, cultural, economic, religious, or any other nature. For example, in the field of political ideologies we can differentiate between those who seek to *preserve* the political system (conservative ideologies), those who seek to *transform* it (through reformist or revolutionary paths), and those who seek to *restore* the *status quo* when changes have already occurred but certain conservative sectors are opposed to those changes (reactionary ideologies). This fact explains why political ideologies are generally provided with a system of representations (systems of ideas) and programmes of action which support the conduct of leaders and constitute the guidelines of political parties. Science, on the other hand, is the set of knowledge obtained by observing regular patterns, reasoning and experimentation that is divided into certain academic disciplines (physics, chemistry, biology, social sciences, law, political science, international relations), and it proceeds methodically, generating questions, constructing hypotheses, and elaborating principles and general laws that give a systematic structure to knowledge. In addition, science uses different methods and techniques for the acquisition and organization of knowledge, which causes it to increase and enrich, and under certain circumstances, science even allows for predictions about future situations and scenarios (prospective science), all of which is impossible for ideology. Nevertheless, social scientists must be aware that in some particular cases and circumstances even science and technology can be used as ideology, as explained by Habermas (1986).

[8] From a different perspective, Harari (2018) argues that human freedom has always been a myth because people are under the permanent influence of unconscious algorithms that determine behaviour, so decision-making has never been a free individual choice.

realism, not of idealism. Unfortunately, Wilson's ideological doctrinal bias paradoxically not only lacked a grip on reality itself, but led Germany to the economic crisis that opened the doors to the electoral triumph of Nazism in 1933, and subsequently to World War II. This contradiction demonstrates that the paradigmatic understanding of a historical case must be nuanced by a careful reading of the decision that the actors made, since, based on the evidence, neither the realistic nor the idealistic paradigm was 'applied' by Wilson in accordance with the spirit of the League or in a manner that can be called appropriate to the context of the socio-political situation that the European nations were experiencing at that time. Hence, Wilson's idealism was somehow imposed on the governments of the European nations via the creation of the League of Nations at the same time that his realism was imposed on the defeated German and Austro-Hungarian Empires via the imposition of enormous war reparations, including the dissolution of both empires, which were completely politically transformed.[9]

On the other hand, the use of different paradigms can also be seen in the operation of different agencies of the same State. For example, when the *United States Trade Representative* (USTR) brings a contentious case to the *World Trade Organization* (WTO) it is based on international law and the treaties signed with the defendant State; but when – at the same time – the US Department of Defense and the US Air Force attack positions of terrorist groups by aerial bombardments in the territories of Iraq or Syria, such military actions are clearly situated within the framework of a vision inherent to the *realpolitik* of the realist paradigm.

Thus, the paradigms are somehow 'interchangeable' when interpreting foreign policy decisions, and realizing that it is always possible to locate them in a given context is important – as postmodern hermeneutics holds – because paradigms, when scrutinized, reveal their ideological roots, so the interpretation may change. As ideology in international relations is permanently linked to the great paradigms, every political action can be 'read' in realistic or idealistic terms, or, from the methodological point of view, more in diachronic than synchronic terms, as the decisions taken by rulers sometimes respond to idealism (when international law is applied) and sometimes to realism (when opting for military solutions). What I want to emphasize is that there will always be an ideological-paradigmatic root for decision-making, and that being conscious of that root is not necessarily the most important element of the process.

In another case, it may be recalled that during the terrible years of armed conflict experienced in Central America at the end of the twentieth century, the realist policy of the Reagan Administration – which saw the triumph of the Sandinista revolution of Nicaragua (1979) as the result of a Soviet-Cuban intervention in the context of the Cold War, and Reagan therefore trying to oppose the Sandinistas by military means – was opposed by the idealist policy promoted by Mexico and the *Contadora*

---

[9] Russia was already under the rule of Lenin and the Communist Party after the 1917 October revolution. Moscow decided to withdraw from the war with the signing of Brest-Litovsk treaty with Berlin.

Group.[10] The former considered the conflicts of countries such as Guatemala and El Salvador to have endogenous causes that should be addressed not by military means but by promoting peace negotiations in accordance with the Charter of the United Nations. Mexico believed that in these peace negotiations the claims of non-state armed groups should be taken into account, as they saw themselves as an insurgent expression of social injustice and inequalities and not as the spearhead of Communism. Therefore, at that time, insurgent groups considered it perfectly feasible to find negotiated solutions instead of insisting on the zero-sum game of armed confrontation which the Reagan Administration sought in the case of Nicaragua and the Sandinista Government.

In historical retrospect, it is clear that even if it is an 'ideal' approach (since all negotiations to end armed conflicts must conclude in peace agreements whose normative nature is clear, as rules are established on how actors *should* behave if the causes of the conflict are to be overcome), it was this *idealistic vision* of the Central American governments that prevailed over the *realistic* approach of Washington. Thus, the policy of neutrality promoted by Guatemala in 1984 and 1985[11] was boldly replaced by the 'active neutrality' of President Vinicio Cerezo (1986–1990), which led to the signing of the Esquipulas Peace Agreements in 1987, which in turn constituted the origin of the negotiations that subsequently led to the peace agreements of Nicaragua (1990), El Salvador (1992) and Guatemala (1996), all of which were achieved thanks to the mediation of the United Nations.[12]

On the other hand, the United Nations itself operates in such a way that when the Security Council decides to apply coercive measures of *peace enforcement* – which has only happened twice in its history[13] – it operates under the umbrella of realism, thanks to Roosevelt's innovation of the Wilsonian League of Nations when the Security Council of the UN was established in 1945. Furthermore, what Roosevelt did in 1945 was introduce realism into the very heart of the world organization, because the UN has the right to use military means to re-establish the status quo. The problem of the 'malfunctioning' of the Council – given the United Nations' limited ability to impose peace when this is required – is not exclusively attributable to the right of veto granted to the five permanent members, because then the elimination of this right would solve the problem, and it is not as simple as that. The right of veto is only a reflection of what happens in the international system's geopolitical reality, because as long as it continues to be anarchic (in the sense that the system is

---

[10] The Contadora Group was formed when representatives of Panama, Mexico, Colombia and Venezuela met on the island of Contadora (Panama) to seek a peaceful solution to the Central American regional conflict.

[11] However, the Guatemalan Chancellor Fernando Andrade acted with a realistic perspective when he approached the Contadora Group and proposed Guatemala's neutrality in exchange for the displacement of refugee camps located in the border state of Chiapas. In this regard, see *Peace Making and Conflict Transformation in Guatemala* (Alker et al. 2001).

[12] Negotiations under United Nations mediation constitute another novel aspect of what happened in Central America during that decade, as it should not be forgotten that Washington always preferred to hold this type of negotiation under the auspices of the OAS (Padilla 2013).

[13] In the Korean War (1950–1953) and in the Gulf War of 1991, when Iraq attacked Kuwait.

not hierarchic; there is no world government) the balance of power is what, in last instance, determines issues relating to war and peace. And the winners of World War II (the US, USSR-Russia, France, China and the UK) – all of them nuclear powers – are not going to abandon their military supremacy for the sake of idealistic UN reform projects. Furthermore, in this regard it is convenient to remember that both Japan and Germany would like to become UNSC permanent members but they have no nuclear weapons or military might precisely because they were defeated in World War II, and powers like the US and Russia distrust the defeated powers of WWII.

Thus the only theoretical formula for solving the dilemma would require the current anarchy of the international system to be transformed into a hierarchical world government, which is not only impossible at the current time but undesirable from the point of view of conservatives who would not like to change the Westphalian system. This explains why, with respect to the aforementioned strengthening of the United Nations and issues such as multilateralism, multipolarity, and international cooperation, the still prevailing bipolarity of the world military subsystem – due to nuclear weapons – would need to be dismantled to put an effective limit to the proliferation of these devices and an end to the great powers' hegemonism. In other words, despite the fact that the pursuit of hegemony remains a primordial interest of the great powers, the search for *democratic multipolarity* must not be seen as an idealistic objective whose implementation must be postponed constantly (as happens with nuclear disarmament) but as a permanent foreign policy objective of small countries that must use the multilateral system as a shield and leverage to face the great powers' permanent search for hegemony.

Furthermore, another way to strengthen multilateralism is through the processes of regional integration like the European Union or other schemes of regionalization that have occurred in Africa, South East Asia, Central America, Latin America and the Caribbean, as well as the political mechanisms for coordination and consensus-building for common political positions such as the Latin American Community of States (CELAC). Other types of multilateral negotiations have concluded in commercial accords – like the free trade agreements that seek to facilitate commerce – and some of them, like MERCOSUR in South America or ASEAN in South East Asia, seek to promote regionalization and are part of the general globalization trend towards the de-territorialization of the world economy, reducing the role of national states as the sole important actors of the international system to the benefit of regional organisms. The reaction of conservative sectors at national level has been a matter of inquietude and preoccupation that will be examined briefly at the end of the chapter.

1.4   The Peace of Westphalia

The paradigms of international relations have their origin in the Peace of Westphalia (1648), so they are closely linked to both historical events and the very emergence of *nation states* that developed through the epochs of the *Renaissance* and the

Enlightenment in Europe. Bertrand Badie has referred to this matter by pointing out that:

> The nation state, as it appears now in international law, is a unique political system invented in Western Europe, which took six centuries (from the thirteenth to the nineteenth century) to assert itself throughout the continent. When the state was born in France, Spain and England, it still coexisted with other political systems, that is, cities, the Empire and the Papacy, from which it had to emancipate itself. Later it penetrated into other areas of Western culture: the Americas, with the independence of the United States and that of Latin American societies, in which the nation state was triumphing as a mode of political organization as they reached independence. At a third moment, the model of nation state spread, partially but deeply, to empires located in the periphery near or far from Europe, victims of the emergent power of the European model. The policy of these empires was specifically to introduce selectively the recipe of the victor to settle or intend to establish one self. Thus gradually, at the end of the nineteenth century, the nationalization of the Ottoman Empire was carried out, which led to Kemal's Turkey of the 1920s. Nationalization also occurred in Persia, Afghanistan and in different systems such as the Burmese kingdom, Siam and especially the Japan of the Meiji in the nineteenth century, which would not be defeated until 1945. Lastly, the most important wave quantitatively was the decolonization of Asia and Africa in the 1950s and above all in the 1960s. It consecrated the birth of nation states that reflected the Western, and mainly French, model (Badie 2008: 3).

But once established in Europe, this system, which became dominant on a world scale, required an explanatory theory to shed light on the best way to avoid war and keep the peace. As for the paradigms that are largely the result of the modern age and the Enlightenment, with all the implications that this has from a philosophical point of view, since the illustration is linked to both the Cartesian/Kantian rationalism of the continental type and the empiricism/positivism of British roots, it must be kept in mind that they are articulated through both philosophical thought and political practice. Indeed, if one takes as a point of departure idealism, both philosophical rationalism – recall Grotius' famous maxim *pacta sunt servanda* – and concrete historical facts such as the Westphalian Peace Treaty, which in 1648 put an end to the European Thirty Years' War, are influential factors. It is also necessary to remember that on that same occasion the equally famous principle of the balance of power came to life as the main basis for sustaining any agreement that would make a *status quo* of peace and absence of war among independent states possible.

And the *balance of power* is nothing more – and nothing less – than the essential postulate of the realist doctrine: to maintain peace, no sovereign state should try to impose its hegemony on other powers of similar status. In order to avoid undesirable hegemony, the weakest states can resist the attempt of the most powerful to overthrow them and for that purpose form military coalitions, but if any such coalitions or alliances try to impose themselves on others (or to equip themselves with military resources that might eventually allow them to achieve this objective) the equilibrium is broken and war is triggered. This is the type of situation that statesmen face permanently because it does occur, regardless of good or ill will and/or the proposed political intentions. Therefore the balance of power cannot be seen as a theoretical conception disassociated from reality.

Of course, the Thirty Years' War cannot be understood without remembering what was really at stake. The main contenders, France and Austria, were two Catholic kingdoms ruled by dynasties with imperial aspirations (the French Bourbons and the Austrian Habsburgs) that paradoxically – since both were Catholic – disputed the European hegemony. The main contenders of this bloody struggle were thus France and Austria, which, in the context of the Wars of Religion, were both trying to dominate the small Protestant kingdoms and autonomous cities of a Germany that was still fragmented into multiple independent states, most of them Protestant. The Wars of Religion were initiated in Europe in 1517, when, after the writing of Martin Luther's *Ninety-Five Theses*, the Protestant movement took only two months to spread throughout Europe with the help of the printing press, overwhelming the abilities of Holy Roman Emperor Charles V and the papacy to contain it. In 1521, Luther was excommunicated, sealing the schism within Western Christendom between the Roman Catholic Church and the Lutherans and opening the door for other Protestants to resist the power of the papacy. As is well known, the German nobility and most of the European aristocracy of the time wanted to separate their kingdoms from Vatican control, stop paying tributes to the Pope in Rome, and expropriate the rich patrimony of the Catholic Church, and these facts were the real causes of more than a century of religious wars fought on the grounds of theological arguments. The roots of nation states are also in these Wars of Religion, and thus in reality the Peace of Westphalia put an end to 130 years of conflagration, not just to the Thirty Years' War between Protestants and Catholics from 1618 to 1648, and the number of casualties, according to historians like Perez de Anton (2017), was of 12.5 million people – a third of the European population of the epoch.

However, the dispute of the two Catholic kingdoms of France and Austria was in reality a dispute for the European hegemony of those powers. That is why Richelieu – the *grey eminence* of the French monarchy – coined the celebrated concept of *raison d'etat*[14] as a criterion to justify its alliances with the Protestant Dutch, German, and Nordic rulers in its quest to prevent the Habsburgs, heirs of the Holy Roman Germanic Empire, becoming the hegemonic power of a *German nation* that two centuries later Bismarck, the *Iron Chancellor*, took charge of unifying *manu militari* after beating the Danes, Austrians and French in successive wars. Or, in other words, it was a far-sighted prospective vision – two hundred years in advance – that allowed Cardinal Richelieu to perceive clearly that a unified Germany would seek hegemony in Europe to the detriment of France, and that this endeavour would lead Berlin to unleash two great World Wars in the twentieth century.

Hence it is perfectly possible to say that Richelieu was clairvoyant, among other reasons because probably his intuition enabled him to perceive – in a great prophetic vision – that France was going to follow that strategy with Louis XIV and afterwards with the Napoleonic Empire, and in consequence, it was an affaire of *raison d'etat* for France to form alliances with the Dutch, German and Scandinavian Protestants against the Austrian Catholics in order to keep Germany divided. It was a question of pure *realpolitik* (as the Germans would call *raison d'etat*) to place the interests of the State above interests of any other nature, including morality and religion.

---

[14] According to Foucault (2014: 413), the concept of '*raison d'état*' essentially refers to a technique of diplomatic and military power that consists of the consolidation and development of power through a system of alliances and the organization of armed forces in the quest for equilibrium.

This, however, does not prevent the Westphalian Peace negotiations being considered a good expression of idealism in international relations. This is because, among other reasons, the process culminated in the first historical major international law treaty that consecrated Grotius's maxim *pacta sunt servanda* (treaties must be respected). Moreover, this maxim, which seems simple, was a great achievement of the negotiators of the time,[15] from both the international law and the international relations point of view, because such negotiations had never occurred within the empires of ancient or medieval times, when emperors did not negotiate with vassal states but simply imposed their will on them. Similarly, *inter-imperial* relations had never previously existed (terrestrial and oceanic distances circumvented the need), and therefore an 'inter imperial law' never existed either, so the construction of the first treaty of international law and the establishment of an international order of sovereign states (the Westphalian system) was a remarkable achievement for the epoch.

Of course, in the seventeenth century there were no nation states as we know them today, but it was already clear that an anarchic international system par excellence[16] could only be organized on an ideal basis via mutually agreed rules, such as treaties or accords between sovereign states, and on a factual basis via armies and military balance. Both were achieved simultaneously with the Peace of Westphalia, hence its fundamental importance in the history of international relations, given that the international system as it exists today was established as a result of the Westphalian treaties, as Kissinger rightly points out:

> The Peace of Westphalia became a turning point in the history of nations because the elements it set in place were as uncomplicated as they were sweeping. The state, not the empire, dynasty, or religious confession, was affirmed as the building block of European order. The concept of state sovereignty was established. The right of each signatory to choose its own domestic structure and religious orientation free from intervention was affirmed, while novel clauses ensured that minority sects could practice their faith in peace and be free from the prospect of forced conversion. Beyond the immediate demands of the moment, the principles

---

[15] Kissinger recalls the fact that, unlike other major international agreements such as the Congress of Vienna (1814–1815) and the Treaty of Versailles (1919), the Westphalian Peace is not the result of a single conference. As a reflection of the great diversity of delegates – from Spanish to Swedish to Dutch and including the French, Austrian and several German kingdoms – peace emerged from a series of separate agreements between negotiators meeting in two cities. The 178 participants of the different states belonging to the Holy Roman Empire (Habsburg Austria) met in the Catholic city of Münster, while the Protestant powers, along with the French Catholics, met in Osnabrück, about 50 kilometres away. There were about 255 delegates involved in the negotiations. They had to stay in the houses of the local population and move from one place to another because there were no special rooms where they could all meet, nor was there any official authority for the conference. Monarchs received the title of 'majesty', and it was also then that the custom of calling ambassadors 'excellency' began. When a hall was set up for meetings, it was necessary to ensure that there were simultaneous entry doors for all the delegates so that no one could take precedence or be obliged to cede it (Kissinger 2014: 23–31).

[16] It should be borne in mind, however, that the concept of 'anarchy' is used in international relations theory with the simple meaning of 'no hierarchy', that is, that sovereign states that are part of the international system are not subject to any hierarchical order that allows any of them, however powerful, to legally impose their orders on others or exert hegemony.

of a system of "international relations" were taking shape, motivated by the common desire to avoid a recurrence of total war on the Continent…The genius of this system, and the reason it spread across the world, was that its provisions were procedural, not substantive. Hence, if a state would accept these basic requirements, it could be recognized as an international citizen able to maintain its own culture, politics, religion, and internal policies, shielded by the international system from outside interventions. The ideal of imperial religious unity – the operating premise of Europe's and most other regions' political orders – had implied that in theory only one centre of power could be fully legitimate. The Westphalian concept took multiplicity as its starting point and drew a variety of multiple societies, each accepted as a reality, into a common search for order. By the mid-twentieth century, this international system was in place in every continent; it remains the scaffolding of international order such as it now exists (Kissinger 2014: 26–27).

Hence, it is because of this historical event that the present international system could also be called – following Kissinger – a 'Westphalian order', since the fundamental principles of its functioning (respect for the sovereign free determination of each member state to equip itself with its own political, economic and cultural 'subsystems', all within the framework of international law and the balance of powers) subsist to this date, although it is also true, as Kissinger himself points out, that there were large regions of the world (with their corresponding civilizational areas) that were not part of the construction of such a European order and the states from those regions do not truly identify with it, detracting from the legitimacy to the original design.

In other words, in the establishment of the modern international system in 1648, extra-European states played no role, either because – like Russia – they were not yet fully involved in the inter-European relations of the time, or because, due to the distance, they lacked permanent political links with European powers, as was the case with China and Japan. Even considering the sporadic political contact, as in the cases of India, Iran, the Ottoman Empire and colonial empires such as those of Spain and Portugal (the British Empire was in the process of formation), an international system as such did not exist in the sixteenth century. This fact explains in part why, in current times, the legitimacy of the Westphalian order has been contested by radical Muslim terrorist groups who dream of resurrecting their ancient seventh-century caliphate. Additionally, as Badie remarks (2008), those terrorist groups that were born as a reaction to the Western bombing over the Middle East now retaliate by refusing to accept the Westphalian order that the West imposed on them.[17]

Consequently, from the point of view of legitimacy, the Chinese claims to insular territory to expand its maritime border in the Pacific Ocean, and the annexation of the Crimean peninsula by Russia are both expressions of the questioning of the legitimacy of the Westphalian order. Therefore, the explanation of those 'imperial' foreign policies must go back to the non-participation of these empires – or civilizations[18] – in the establishment of the order emanating from Westphalia, which was confined to Western Europe at the time at which it was created.

---

[17] "The bombing is counterproductive. The more they hit these 'entrepreneurs of violence' – like the Islamic State – the more they are strengthened. Terrorism is born of the bombings and Western failures" (Badie 2008: 84).

[18] For Huntington (1996) Russia and China are different *civilizations*.

The issues discussed so far mainly concern the sphere of realism. In the realm of the idealist paradigm and international law, the legitimacy of the Westphalian system is based on international law because all States which apply for membership of the United Nations – in exercise of their sovereignty – have accepted the Charter of the United Nations and the prevailing world order envisaged therein. As a consequence, nowadays the Westphalian system finds its legitimacy in the consensus that entails the acceptance of the UN Charter by all member states, even if – as stated before – that doesn't mean that, from a realist point of view, the legitimacy of the system cannot be challenged.

In any case, and returning to the historical question, it is important to emphasize that during this part of European history the Westphalian system worked quite well for the rest of the seventeenth century. It was not until the late eighteenth and early nineteenth centuries, with the French Revolution and the subsequent Napoleonic Wars, that the system was transformed, since what, in fact, was at play in Europe in the early nineteenth century was French hegemony over the rest of the continent (from the point of view of the *balance of power*).

It should also be mentioned that the French Revolution was a social movement that claimed a system of government that was directly opposed not only to the interests of the nobility and aristocratic classes of feudal origin, but also to the monarchist system of absolute power (for the King) that prevailed throughout the continent at that time and was afterwards transformed via the parliamentary political system. Hence, from the outset, the French revolutionaries had to face both the internal enemies – the French aristocracy – and a permanent coalition of opponents from outside their borders, obliging France to confront powers like Austria, Great Britain, Prussia and Russia, that later – after the end of the negotiations in Vienna – would constitute the fundamental axis of the European '*concert*', as the *balance of power* equilibrium during the nineteenth-century was called, along with the France of the restoration (under the rule of Louis XVIII), which wisely was not excluded by the victors from the Vienna peace negotiations, despite being defeated at Waterloo, because in the Austrian capital Talleyrand represented France.[19]

1.5   The Peace of Vienna and the *Balance of Power*

Napoleon's military genius enabled France not only to defeat its enemies but to expand the dominance of its empire throughout Europe in the historical period from

---

[19] At a conference in Lyon, Badie (2008) remarked that at the Battle of Jena (1806) for the first time an army of conscripts composed mainly of soldiers of French nationality defeated the Prussian army of aristocrats and multinational conscripts. That defeat was seen in the Germanic society of those times as the first occasion that 'Germany' suffered a *national* defeat. This phenomenon implies the irruption of national societies in history, and is the origin of *nationalism*. In a lecture about French-German relations at the Geneva Institute of Higher Education (2015) Joschka Fisher made similar remarks in the sense that German nationalism consciousness began to be forged at the beginning of the nineteenth century thanks to Napoleon. It should not be forgotten that Beethoven's Eroica symphony was composed in honour of Napoleon before the Jena defeat. Afterwards the great musician said he had composed it "in memory of a great man". Nevertheless, it is quite clear why European 'liberals' of all nationalities were in sympathy with the French revolution.

the last decade of the eighteenth century till the Battle of Waterloo in 1815, the year in which the French emperor was finally defeated by a coalition of European armies under the British command of the Duke of Wellington. Consequently, Napoleon's defeat led to the negotiations of the Peace of Vienna the same year. The most important outcome of those negotiations is that they led to the establishment of a new type of European equilibrium. Despite being defeated, France participated in the negotiations in the Austrian capital and was represented by the former foreign minister of Napoleon, the famous Talleyrand, although in Vienna he was acted as the representative of the Louis XVIII re-established monarchy.

The French presence in Vienna was crucial to ensure the fulfilment of the commitments undertaken. In recognizing this, the negotiators were complying with one of the essential rules of any successful peace process by allowing the vanquished to participate in the terms of the peace accords.[20] One of the results of the Vienna Congress – which had other illustrious participants, such as the Russian Tsar Alexander I in person, along with the no less famous and skilful Austrian diplomat, Prince Klemens von Metternich, in addition to the English Lord Castlereagh and the Prussian monarch Friedrich Wilhelm III – was the establishment of the so-called 'Holy Alliance' of Russia, Prussia and Austria to ensure the absence of war on the European continent. This situation was maintained long enough during the nineteenth century to provide several decades of peace, firstly until the Crimean War (1853–1856) and then until Prussia, under the command of the *Iron Chancellor*, Otto von Bismarck, set out in the 1860s to unify the sovereign states of the German nation under the umbrella of a renewed German Empire.

Additionally, it is important to recall that, thanks to the Vienna Congress, a quadruple alliance was established which brought together the three powers of the Holy Alliance and Great Britain. Because the British Isles lie beyond the coastline of Continental Europe, in political jargon Britain was regarded as the '*offshore balancer*', in charge of dealing with European equilibrium from the outside. Paradoxically, however, in certain circumstances the British role in opposing the emergence of a hegemonic power on the Continent – balancing Europe – led them to collide with

---

[20] It is interesting to note that, conversely, Germany was excluded from the Versailles peace negotiations of 1919, with the result that the period of 'absence of war' (between the two World Wars of the twentieth century) lasted only twenty years: from 1919 until 1939. During the Vienna negotiations of 1815, the rules of balance of power theory were met and the '*concert of Europe*' equilibrium was established successfully, while during the Versailles negotiations of 1919 these rules were ignored and no equilibrium was reached, then the Second World War followed in just twenty years. Of course, the explanation of the hundred years of European equilibrium (1815–1914) is the result of a complex process, because, as every system is composed of both structure and process, any balance of power is essentially dynamic, which means that in order to maintain the balance it is necessary to keep in mind that realism postulates that each State will permanently try to increase its power and capabilities. Negotiation should be preferred to war, but if violence becomes inevitable, to preserve the system it is absolutely essential to stop the fighting when one of the actors is about to be defeated. According to power politics, any *zero sum game* is not convenient for the health of the system because the pursuit of hegemony is always absolutely rejected by the rest of the players. Hence, in the event of the defeat of one of the system's essential players, the loser must be allowed to recover an acceptable role through negotiations.

Russia (when the other members of the Holy Alliance abstained from involvement). This occurred, for instance, during the Crimean War (1853–1856), during which the British and French intervened to prevent Russia from becoming the hegemonic power of the Eastern Mediterranean when Moscow was trying to control the passage of the Dardanelles and the Bosporus – with obvious strategic and geopolitical intentions – but using the protection of the Orthodox Christian population residing in the Ottoman Empire as the ideological pretext for its military intervention in the region.

Other 'small wars' fought in Europe during the nineteenth century, besides the Balkan wars at the end of the century, were occasioned by the clash of Germany against its neighbours in the process of unification. Bismarck's strategy led first to a short war against Denmark, then to a war of seven weeks' duration against Germany's Austrian 'cousins' (1866), and lastly to a war from July 1870 to May 1871 against France, which culminated in German victory over the French, France's consequent cessation of Alsace-Lorraine to Germany, and the establishment of the German Empire – i.e. a unified Germany – thanks to the strength of the Prussian armed forces as well as Bismarck's strategic military leadership and political acumen.

1.6   The First World War or the Rupture of the European Concert

As already stressed, and in spite of the aforementioned particular breaches of peace, it may be said that the Vienna Congress succeeded in maintaining the balance of power in Europe for almost a century (1815–1914), that is to say until the outbreak of the First World War, which was essentially due to the rupture of the equilibrium provoked by the attempt of the two Germanic empires (those of the Kaiser Wilhelm and the Austro-Hungarian Habsburgs) to become the hegemonic powers of the continent. This attempt was disrupted by the intervention of another powerful actor or *offshore balancer*, this time extra-continental, since Great Britain had to cede that role to Washington, given its inability to face the enormous German power without Washington's help.

Much has been discussed about the origins of the *Great War* – as it was called in Europe – but, based on an analysis of the way in which the *balance of power* theory works in practice, what is beyond doubt is that complying with its minimal rules is not an easy task. The rules are seemingly simple: avoid trying to become a *hegemon* within the system (because such attempts inevitably lead to coalitions designed to prevent it). It is clear, however, that the good or ill will of the rulers and political leaders alone was not enough, because in 1914 it was the very same structure of the system that failed. In other words, from the moment military forces began to prepare for the war, military leaders prevailed over politicians, so that the armed forces, pressing governments to order the mobilizations, determined the decisions that those political rulers took. Thus, according to Kissinger:

The astonishing aspect of the outbreak of World War I is not that a simpler crisis compared to many others already overcome was the one that finally triggered a planetary catastrophe, but that it took so long to do so…To make matters worse, military strategists had not adequately explained to politicians the implications of their task. In fact military strategy had become autonomous. The first step in this direction was during the negotiation of a Franco-Russian military alliance in 1892. Until then, the alliance negotiations had been about *casus belli*, or specific actions of the adversary that could force allies to enter at war. And almost invariably its definition depended on who had broken the hostilities. In May 1892 the Russian negotiator, assistant general Nikolai Obruchev, sent a letter to his Foreign Minister, Giers, explaining how the traditional method of defining *casus belli* had been affected by modern technology. Obruchev argued that what was important was who would be the first to mobilize, not who would make the first shot: "Taking up the mobilization can no longer be considered a peaceful act; on the contrary, it represents the most decisive action of war." The side that would be slow to mobilize would lose the benefit of their alliances and allow their enemy to defeat their adversaries one after another. The need for allies to mobilize simultaneously had become so pressing in the minds of European rulers that it became the foundation of solemn diplomatic compromises… This infernal procedure made *casus belli* out of all political control (Kissinger 1996: 197–198).

In brief, the First World War was, to a large extent, a consequence of what could be called the 'malfunction' of the realist principles of the balance of power. During the time of Bismarck the system worked quite well, thanks to the diplomatic skill of the *Iron Chancellor*, but once Bismarck was out of office, Kaiser Wilhelm II, a person who lacked Bismarck's powers of diplomacy and great statesmanship, was not up to the challenge and let the situation deteriorate to the point that in 1914 it was impossible to avoid the outbreak of war. The war detonated due to the famous Austrian ultimatum against Serbia after the attack that cost the life of Archduke Franz Ferdinand, heir to the Austrian throne, in Sarajevo, the capital of Bosnia-Herzegovina. The attack, committed by an individual terrorist, would not normally have been enough to trigger the worldwide conflagration it provoked, but, as Kissinger says, as the *casus belli* was no longer the one which initiated hostilities but the one which first mobilized troops, all the powers hastened to do just that and, obviously, if military forces are mobilized to enter into combat, it can prove very difficult to stop them and prevent hostilities or the outbreak of war.

Regarding this issue, it is interesting to note that – except in proportion – a similar situation occurred in the early 1980s in the case of the Beagle Channel between Argentina and Chile. The armed forces of both countries mobilized in order to initiate hostilities over three small, uninhabited islands whose sovereignty was claimed by the military dictators in both Santiago and Buenos Aires (Generals Pinochet of Chile and Videla of Argentina). The mediation of Pope John Paul II resolved the said conflict, but the Argentine forces, already mobilized, then invaded the Falkland Islands (Islas Malvinas).The Argentinian troops had to face the counter-attack unchained by the 'Iron Lady', Margaret Thatcher, who at the time was the United Kingdom's Prime Minister, with the end result that the Argentinians were defeated and the military junta overthrown by the subsequent crisis. Elections were held and Raul Alfonsin was elected. He convoked a referendum regarding the Pope's decision that was ratified by the people.

The consequences of the *Great War* were terrible: more than 65 million men fought in it, and more than eight million soldiers died. As a result of the First World War four empires – the German, Austro-Hungarian, Russian and Ottoman – fell and three great monarchic dynasties – the German Hohenzollerns, the Austrian Habsburgs and the Russian Romanovs – ended their days. In addition to the eight million people killed, about six million people were disabled. France, with one and a half million dead and missing (10% of its male population), was one of the countries that suffered the greatest demographic consequences of the war, a hard price to pay for the recovery of Alsace-Lorraine.

On a political level the collapse of these four empires profoundly transformed the map of Europe. The Ottoman Empire dissolved, giving way to the new republic of Turkey – territorially encompassing the Anatolian highlands and Istanbul – and several new countries appeared, including Yugoslavia in the Balkans (with Slovenia, Croatia, Bosnia-Herzegovina, Macedonia, Serbia and Montenegro as part of the new independent state), Albania, Bulgaria, and Greece, while the Austro-Hungarian Empire was dissolved. The new sovereign countries of Austria, Hungary, Czechoslovakia and Romania were also established. At the same time the German Empire ended and was replaced by the Weimar Republic, which ruled over a Germany that was territorially and economically depleted due to the payment of war reparations imposed by the Treaty of Versailles, as mentioned before. As for the vast Tsarist Empire, thanks to the Bolshevik Revolution of October 1917, the new Communist rulers decided to end the war with Berlin separately, and Lenin signed the Brest-Litovsk Treaty with enormous territorial losses for Russia. Poland, Finland and the three Baltic republics – Estonia, Latvia and Lithuania – came to life, and Russia itself was completely transformed and later became the Union of Soviet Socialist Republics (USSR), composed of fifteen new republics, some of them former colonies ruled by the Tsars in Saint Petersburg. In short, the geopolitical transformation was colossal.

1.7   The Versailles Treaty, Wilsonian Idealism and the League of Nations

When the Armistice was declared at the end of World War I, the Allies (France, the United Kingdom and the United States, as well as representatives of their allies during the war) met at the Paris Peace Conference to agree on the terms of peace with Germany. Following the principles of self determination of peoples enacted by the US President Woodrow Wilson, the now defunct Austro-Hungarian Empire was divided into the new states of Austria, Hungary and Czechoslovakia, with substantial territorial losses in favour of Romania, Italy, and Yugoslavia, while the Ottoman Empire became Turkey (now fully partitioned) and the Kingdom of Bulgaria. The Allies drafted and signed treaties by each of the defeated powers, but the Treaty of Versailles was the one that was ultimately imposed on the former German Empire. Woodrow Wilson personally attended the Versailles Conference and played a role of the highest importance in the decision taken by the Conference, especially concerning the creation of the League of Nations as a new international institution intended to guarantee collective security via a new idealist doctrine designed to replace realism.

Discussions between the victorious powers on what would constitute the conditions of peace began on 18 January 1919, and the result was presented to Germany in the month of May as a unique alternative: rejection of it would have involved the resumption of hostilities. The day after the acceptance of the Treaty was a kind of day of mourning in Germany because its signature became the 'original sin' that was going to be paid in very expensive terms by the newly formed Weimar Republic. The Treaty of Versailles was thus signed on 28 June 1919 in the Gallery of Mirrors of the Palace of Versailles (reinforcing the symbolism of the French rematch) five years after the Sarajevo attack. The negotiation of the Treaty excluded German representatives, and one of its most unjust decisions established that Germany and its allies accepted all moral and material responsibility for having caused a war which – as explained above – was the result of the mobilizations ordered by the military commanders of all the powers involved, not just the central powers, as the last resort of the malfunctioning of the realist balance of power doctrine.

Thus the stipulations for German disarmament, along with the important territorial concessions to the victors and the payment of exorbitant pecuniary compensation to the victorious states, were imposed by the victors on the vanquished, who did not receive any opportunity to negotiate the conditions of the new international order, in contrast to what happened in the case of France at the end of the Napoleonic Wars during the Congress of Vienna negotiations. Therefore, it is not surprising that the Treaty of Versailles was undermined early and repeatedly violated, given that one of the basic rules of balance of power theory was not respected, since the defeated was not permitted to participate in the peace negotiations.[21]

For these reasons both the German delegation and the German government considered the Treaty of Versailles an imposition (*diktat*) by the winning powers. Although a German delegation was present at Versailles, from the outset its members realized that their presence was only 'decorative'. The German delegation protested that it was unfair to attribute all responsibility for the war to Germany. As one might expect, unilateral blame for the war became an element of permanent tension in domestic politics, because the right-wing parties and nationalist groups disagreed with the Social Democrats and the Liberal centre parties. In the end, the political right-wing tendencies prevailed in the 1933 elections, bringing Hitler and the Nazi party to power.

We could ask: what is the meaning of the so called "irruption of idealism" at the Versailles Conference? In light of what has been described, it would seem altogether inappropriate to speak of idealism at the Versailles negotiations. However, 'idealism' becomes evident if we recall that, due to the initiative of the US President Woodrow

---

[21] As a historical curiosity it is interesting to note that, owing to financial collapse, Hitler's refusal to pay, World War II, and newly negotiated deals over deadlines and amounts, it took Germany ninety-two years to pay off its war reparations. It was eventually agreed that the back interest would not be payable until Reunification (3 October 1990). The final interest repayment was made on the twentieth anniversary of the Reunification in 2010.

Wilson, the League of Nations was the first international organization in history which sought to arbitrate in international disputes and prevent future wars through the use of the new doctrine of *collective security*, according to which, in the case of aggression by one member State against another, all the members (in this case of the League of Nations) must act to assist the attacked State by rejecting and sanctioning the aggressor.

On the basis of this principle, Wilson believed that international relations could overcome and render ineffective the principle of *balance of power* used until then by governments under the influence of the realistic doctrine. Unfortunately, neither the United States (by the negative vote of the House of Representatives under the control of the Republicans), nor – at the beginning of the decade – Germany or the Soviet Union were part of the League, and, in any case, it lacked a coercive mechanism for implementing its resolutions, such as the one which was subsequently introduced by the United Nations with the establishment of the Security Council. Consequently, the ineffectiveness of the League led to its disappearance without pain or glory on the outbreak of the Second World War, very shortly after its creation.

The novelty of Woodrow Wilson's idealist doctrine lay in the firm belief that the solution of international conflicts had to go through an inter-state institution based on international law that would be in charge of avoiding wars and maintaining peace. Contrary to the realist dictum – if you want peace, prepare for war, or *si vis pacem para bellum* – the idealist principle consists of this firm belief in international law and negotiation as the means of resolving conflicts. Thus, bringing this idealist principle (besides the 14 points[22]) to the negotiation of the Treaty, and finally obtaining an agreement to establish the *League of Nations* paved the way for the idealist paradigm

---

[22] Wilson's 14 points referred first to the need to avoid *diplomacy and secret treaties* between powers and to ensure that all negotiations were public and culminated in open agreements. It also demanded the absolute *freedom of navigation* in peace and war outside jurisdictional waters, except when the seas were closed by some international agreement; the disappearance, as far as possible, of the economic barriers to free trade; the establishment of adequate safeguards for the reduction of national armaments (disarmament); the 'readjustment of colonial claims', so that the interests of peoples receive equal consideration with the legitimate aspirations – if any – of the colonizing governments; the evacuation of Russian territory, Russia being given "full opportunity for its own development with the help of the powers" (this point has to do with Western intervention in the Russian civil war after the Bolshevik revolution of 1917); the full restoration of Belgium's complete and free sovereignty; the liberation of the entire French territory and reparation for the damages caused by Prussia in 1871 (mainly concerned with the return of Alsace-Lorraine to French sovereignty); the readjustment of Italian borders in accordance with the principle of nationality (because a large part of the northern territories – the Tyrol and others – were under Austrian sovereignty); the provision of "*an opportunity for an autonomous development of the peoples of the Austro-Hungarian Empire*" (with the disappearance of the Habsburg Empire, Czechoslovakia and Hungary emerged as independent states; the collapse of the Habsburg Empire is related to the evacuation of troops from Romania, Serbia and Montenegro as well as the issue of granting access to the Adriatic sea to Serbia); the "*settlement of relations between the Balkan States in accordance with their sentiments and the principle of nationality*", which led to the creation of Yugoslavia; the security of autonomous development for the non-Turkish nationalities (such as the Bulgarians) of the former Ottoman Empire (which also disappeared) and the pass through the Straits of Dardanelles; the declaration of Poland as an independent state with access to the Baltic Sea; and, finally, the question of the League of Nations, which, as the 14 points pointed out, consisted of the creation of a

(based essentially on international law) in international relations – if you want peace, prepare for peace (*si vis pacem para pacem*) – although the treaty did not yield the expected results for world peace and security because, paradoxically, the United States didn't became a member of the League and other important powers were excluded from it.

The other point of great importance within the 'idealist' policy of the American President was the famous question of the self-determination of peoples and nationalities. As the empires of the time (such as the Habsburg and Ottoman dynasties) were multinational in character, Wilson believed that the best way to guarantee peace was to allow these peoples to have their own independent state – like the new sovereign states of Czechoslovakia, Hungary, Poland, Romania, Bulgaria and Yugoslavia – where they would speak their own language and promote their own culture, allowing them to govern themselves with their own national institutions.

This entire population of great ethnic, cultural, religious and linguistic heterogeneity was relocated within new territorial circumscriptions, some of them acquiring the status of new sovereign nations. Nonetheless, it is a paradox that at the end of the twentieth century history would repeat itself and demonstrate – especially in the region of the Balkans, where again terrible wars exploded, forcing the dissolution of Yugoslavia – that deciding 'from above' the regrouping of populations based on the principle of self-determination for distinct nationalities was much less successful in maintaining peace than the old imperial order established by the Austrians and Turks under the iron control of the Habsburg and Ottoman dynasties.

Nevertheless, what Kissinger says about the European reordering of Versailles is, indeed, quite accurate: "Rarely has a diplomatic document so missed its objective as the Treaty of Versailles. Too punitive for conciliation, too lenient to keep Germany from recovering, the Treaty of Versailles condemned the exhausted democracies to constant vigilance against an irreconcilable and revanchist Germany as well as a revolutionary Soviet Union" (Kissinger 2014: 84).

Moreover, since neither the United States nor France was in a position to guarantee or enforce the application of the clauses of the treaty, the situation could not have been worse, leaving Germany resentful due to the loss of territory and the imposition of the payment of war reparations in the context of a serious economic crisis. This simultaneously took place in a European context which the newly born Soviet Union, with its own vision of what the new world order should be, imagined to be based on some kind of 'proletarian internationalism' (which explains the establishment of the international organization of workers with the Communist Party and Moscow as the vanguard) that would be in charge of promoting, through Communist parties, a world revolution to replace the Westphalian order based on state sovereignty. Thus, lacking legitimacy and equilibrium, the Treaty of Versailles could not even serve as reference point for the Locarno Pact (1925). Although, via this Pact, Germany accepted the demilitarization of the Rhine area and the new frontiers with France, giving assurances that it would not seek to recover Alsace-Lorraine, at the same time

---

"general association of nations" that was to be set up by "specific pacts" with the aim of "mutually guaranteeing political independence and territorial integrity for both large and small States".

it refused to recognize the new borders with Czechoslovakia and Poland, making explicit its ambitions regarding territorial expansion, as was proved with the arrival of Hitler to power in 1933. Indeed, Hitler and the Nazis, in addition to putting an end to the Weimar Republic and with the inauguration of the Third Reich, the *Führer* reassembled Germany in less than six years, re-occupying the Rhine area. Then he began to dismantle the Central European states one by one. Austria was first 'victim' of the *Anschluss*, followed by Czechoslovakia and finally Poland, thus triggering the Second World War. So between one war and another, there was actually only a truce of twenty years, largely because of the shortcomings of the negotiation of the Treaty of Versailles that failed to establish a balance of power and sought to transfer responsibility for peacekeeping to an international entity devoid of any power.

1.8   The United Nations as an Expression of the Idealist Paradigm

If the League of Nations is the irruption of idealism in international relations, the United Nations could be seen as idealism in progress and the point of departure of multilateralism as a new kind of diplomacy to be practised at world level, with the US President Franklin D. Roosevelt as its symbolic father. World War II was to a large extent the rupture of a truce and the continuation of the *Great War*, as Germany, feeling unjustly treated by the victorious powers, both in territorial and economic losses (due to war reparations) in addition to the disarmament and demilitarization obligations in its own territory (the border area with France on the Rhine), did not rest until it had the means to take revenge. Although a heinous character, Hitler was, in that sense, a clear manifestation of the feelings of humiliation and resentment against France, Great Britain and the United States that prevailed in the German population of the epoch.

One could speculate about what might have happened if someone with less 'demonic' characteristics – to use Kissinger's expression – had been elected in Germany in 1933, and moreover, ask why there were no statesmen or German citizens capable of preventing the Nazis coming to power or "*maintaining an international order capable of deterring them if they do*". The whole debate about the Western powers' policy of *appeasement* and their inability to confront Hitler at the Munich summit is also connected with this problem. However, here I am making a speculative analysis similar to my conjectures (Padilla 2013) about what would have happened in Guatemala in 1954 if President Arbenz had had the diplomatic means in New York (at the United Nations) to oppose the US-sponsored military intervention effectively, instead of following the mistaken course of action of purchasing armaments in Czechoslovakia. But, in retrospect it is not possible to know whether the course of history would have changed, or whether, taking into consideration the hypothetical alternatives, Hitler would have no longer attacked Poland, or whether, concerning Guatemala, the country could have avoided the guerrilla warfare (from

1960 till 1996 with more than 200,000 deaths) if Washington had supported an opposition candidate in the 1956 elections instead of launching a military operation to overthrow Arbenz's government. The well-known writer and Nobel Prize winner Vargas Llosa (2019) argues that the *coup d'état* against Arbenz was fundamental to the loss of confidence in democracy and the beginning of insurgent armed movements across Latin America.

So, in order to draw conclusions about how the great paradigms of international relations 'functioned' (or not), which is the main purpose of this investigation, what follows is an analysis of the consequences of that second world conflagration. Alternatively, what I want to emphasize is the fact that effective international relations theory is the result of what happens in the real world and not vice versa. This, of course, is closely related to the *theory of knowledge* – or epistemology – and although, for obvious reasons, I will not enter into that philosophical academic debate,[23] my position in these pages is that it is essential to discern when a statesman's political action is explicitly or implicitly guided by knowledge. In some circumstances, when a person is guided by ideological reasons, the risk of obtaining negative or even catastrophic results is much greater because when the political action is a consequence of a doctrinal-ideological vision and not the result of knowledge of the socio-political, economic, or cultural context in which the foreign policy of any State is manifested, the risk of errors increases.

For example, as Morgenthau (1986) relates in his classic book *Politics among Nations*, on which I have previously commented (Padilla 2009: 22–34), the misuse of the idealist paradigm (*pacta sunt servanda*) could have impeded the Allied victory in the Second World War, allowing the Axis powers to triumph. Indeed, Morgenthau recalls that, in compliance with their military commitments with Finland, France and England could have entered into war against the USSR in the winter of 1939 in order to defend Finland when it was attacked by Stalin. The sending of Franco-British troops to help the Finnish was prevented by the refusal of Sweden – a neutral country – to allow the passage of the military through its territory.

In the same work Morgenthau also refers to the unsatisfactory US foreign policy decision not to grant recognition to the government of the People's Republic of China for more than two decades after the triumph of Mao's Communist revolution in 1949, instead maintaining diplomatic relations with the nationalist Republic of China government installed in Taipei (Taiwan). That policy was finally modified in the 1970s thanks to the advice of Henry Kissinger, which persuaded the administration

---

[23] The whole debate of philosophical realism/idealism lies in the question of the importance of the material world – ontology, or the theory of being – because the 'ontological realism' of Aristotle, Thomas Aquinas and both English empiricism and Marxist materialism are the opposite of the platonic 'world of ideas' and the importance of reason as a precondition of knowledge. So the idealism of philosophers like Plato, Descartes and Hegel, and even Kantian rationalism, Husserl's phenomenology and Heidegger's postmodern view, have a role to play in the formation of human knowledge. Contemporary approaches are much more complex insofar as the theory of systems and physics has introduced their views into the epistemological debate of philosophy. My own position is that it is possible to accept the existence of both material reality (in spite of the fact that the material nature of subatomic particles is challenged by quantum physics) and spiritual reality (as postulated by Buddhist philosophy), and that both are equally important and valid in epistemology.

of the Republican President Richard Nixon to adopt a more appropriate 'realistic' vision.

Naturally, the opposite – the incorrect application of realism by the White House – has occurred on many occasions as well.[24] Two of the clearest examples have already been mentioned here: the overthrow of the democratic President Jacobo Arbenz in Guatemala in 1954, and the Reagan administration's *roll back* policy against Nicaragua in the 1980s. Many other examples of this failed US policy of *regime change* could be added, from the overthrow of Mossadegh in Iran in the 1950s to coups in the Dominican Republic in the 1960s, the *contra* war against the Sandinista regime of Nicaragua in the 1980s and the invasion of Iraq sanctioned by President George W. Bush in 2003. In all of those cases, the remedy (intervention by violent means) was worse than the disease (the supposed interference of the USSR in the American zone of influence). Both the twenty-six years of armed conflict in Guatemala and the insurgency of the 'contras' in Nicaragua led to destruction and the innumerable deaths of civilians (especially innocent people who were victims of war crimes and violations of human rights and humanitarian law) which destroyed social structures, traumatizing the entire society. An interesting analysis of this erroneous US policy of regime change and its disgraceful legacy is offered by Kinzer (2019) in his review of Lindsey O'Rourke's (2018) book on this topic.

On the contrary, the negotiated outcome of the Central American conflicts (to which must be added the Salvadoran conflict), which, thanks to the invaluable contribution of United Nations mediators, was driven by the Esquipulas peace process and by Presidents Vinicio Cerezo (Guatemala) and Oscar Arias (Costa Rica), demonstrates the usefulness and correct application of the idealist paradigm established by Chapter VI of the United Nations Charter on the negotiated settlement of disputes, as I have previously explained (Padilla 2009: 75–84) in an analysis of the Central American peace processes under the mediation of the United Nations, with the fulfilment of agreements supervised by UN missions such as ONUCA, ONUSAL and MINUGUA.

Later, in treating the whole geopolitical issue, I will provide other examples of the ideological justification of foreign policy proposed by Mearsheimer (2014a, b), a well-known professor at the University of Chicago, who has strongly criticized the US intervention in Ukraine against Russia within the geopolitical US policy of NATO expansion, as well as in the Middle East in the wars of Syria and Iraq (2016), but for the time being the focus remains on exposing the vicissitudes of the theory in recent history by addressing the issue of the role played by President Franklin Roosevelt in the creation of the United Nations.

In line with Woodrow Wilson's idealistic thinking, President Roosevelt set out to give continuity to the League of Nations – though under another name and organizational structure – but improved it by introducing within the future United Nations a

---

[24] It should be borne in mind that "incorrect application" does not mean that diplomatic agents, or those who are responsible for making foreign policy decisions, are fully aware of the paradigmatic elements that can be found in their actions: these are implicit and it is very rare that they are made explicit in analytical documents produced by the government agencies themselves. Usually these analyses are made *ex post* by professors in the academic community and not by government officials.

realistic component that sought to overcome Wilson's excessive hopes through international law: the Security Council, as the entity in charge of enforcing its resolutions, should this become necessary.

Since coercive application may involve the use of military force within the framework of the *collective security* doctrine, the Charter establishes rules to regulate the obligations of United Nations member states, which must provide military forces in order to ensure that the Council's resolutions are fulfilled, or to oblige an aggressor State to restore the *status quo*. Indeed, Chapter VII of the Charter provides the legal framework for the Security Council to take coercive measures, although to this end the Council must first determine "the existence of any threat to peace, breach of peace or act of aggression."

Recommendations or decisions are made on which measures involving military action or the use of armed force are to be employed "in order to maintain or restore international peace and security". For a closer look at this, see the United Nations 'Repertory', which includes references to Chapter VII, in particular Articles 39 to 51 of the Charter, as well as the various documents of the Security Council. It also includes case studies in which the Council applied the respective articles of Chapter VII when examining specific situations included in its agenda, although a detailed and specialized examination of this problem is beyond the scope of this section.[25]

Throughout its 75-plus years of history, the United Nations has had a multitude of peace operations[26] in many different countries, concerning conflicts that are also of

---

[25] To obtain an idea of the normal work of UNSC, I recommend the article of the former Guatemalan ambassador Rosenthal (2016) on the participation of Guatemala in the UN Security Council in 2012 and 2013. Concerning the complexity of the issue of collective security and the procedures of the Security Council, Article 39 states that before taking coercive measures the Security Council must verify whether there are threats to peace, a breach of the peace, or an act of aggression between States. Article 40 refers to the provisional measures that can be taken in order to prevent a situation from worsening, and provides for measures including the withdrawal of armed forces, the cessation of hostilities, concerted action or compliance with the ceasefire, and the creation of conditions for the provision of humanitarian assistance. With regard to measures that do not involve the use of armed force (Article 41), there are sanctions ranging from general economic and commercial sanctions to more selective measures, such as arms embargoes, travel bans, and both financial and diplomatic sanctions. Those measures could also include the establishment of international tribunals (such as those established for the former Yugoslavia and Rwanda in 1993 and 1994) or funds for the payment of compensation for the damage caused by war. Article 42 empowers the Council to use "other measures to maintain or restore international peace and security" if it considers that *measures that are not of a military nature may be inadequate or have proved to be inadequate*. In view of the fact that the United Nations has no armed forces at its disposal, the Council uses Article 42 to authorize the use of force through peacekeeping operations, multinational forces or interventions by regional organizations. Member states are obliged to accept Council decisions because they are binding. Another important article of the Charter is Article 51, which establishes *the right to individual or collective self-defence* not only as an exception to the prohibition of the use of force (Article 2, paragraph 4, of the Charter) but also so that, in the case of an armed attack, member states can defend individually or collectively by communicating immediately to the Council. Measures will then be taken as prescribed by the United Nations Charter.

[26] Peacekeeping operations are of a very different nature with respect to coercive measures and peace enforcement, since their usual aim is to separate contending parties and maintain ceasefire situations, as in Lebanon/Israel, Syria/Israel, India/Pakistan, and Cyprus, or in internal armed conflicts of the

a very diverse nature. Although, on balance, its performance in the field of peace and international security is favourable, the same cannot be said of the decisions of the Security Council aimed at ending international conflicts via coercive implementation. For in many world conflicts (Vietnam, Afghanistan, the Middle East, the Balkans and many others), it has been impossible for the five permanent members of the veto-led Council (US, UK, France, Russia and China) to agree, the most dramatic recent cases being those of the wars of Iraq (2003) and Syria in current times.

In fact, there are only two cases in which it was possible to mount coercive peace enforcement operations: the Korean War in 1950 and the first Persian Gulf War – on the occasion of Saddam Hussein's invasion of Kuwait in 1991 – which are worth mentioning briefly.

In the Korean War, the requirements established by the Charter were clearly met because when one member state is attacked by another (South Korea by North Korea), the Security Council meets to hear the case and, in the absence of the USSR – the Soviet delegate was not present in the room when delegates proceeded to vote – [27] it was decided to mobilize military forces to intervene in the peninsula and reject the aggressor. The United States provided the largest military contingent, which was placed under the command of General Douglas MacArthur, a great strategist and hero of the Pacific War against Japan, who quickly controlled the military situation, pushed back the invaders, and even penetrated into the territory of the north by taking the capital (Pyongyang), but was forced to withdraw south of the 38th parallel by a North Korean counterattack with the assistance of Chinese 'volunteers'.

The decision of the Democrat administration (Truman, who replaced the late President Roosevelt in 1945) regarding the terms of the Security Council resolution that essentially sought to restore the status quo – not gain military victory over North Korea – opened the way for the establishment of an armistice (1953) that maintained the border between the divided Koreas in the aforementioned 38th parallel and also established a demilitarized zone of four kilometres between the two contending countries. One may have ideological sympathies for one side or the other, but it must be admitted that in this case the concept of a *collective security* doctrine worked well and an aggressor state was rejected by the United Nations military forces, restoring the situation (status quo) that prevailed prior to the invasion.

The other exemplary case of peace enforcement by the UN is that of the first Persian Gulf War in 1991, when Saddam Hussein, at that time Iraq's president, attacked Kuwait, a sovereign and independent state which was also a member of the

---

type that have been suffered by countries such as the former Yugoslavia as well as several African countries (Liberia, Mozambique, Rwanda, Sudan, the Democratic Republic of the Congo) and Haiti. Therefore its implementation is less controversial and much more likely to be approved by the Council.

[27] The Soviet delegate was absent from the session of the UNSC at which the decision was taken to reject the North Korean aggression in protest at the Council's decision to maintain in the seat of China the representative of the nationalist government of Chiang Kai-shek (in exile on the island of Taiwan and protected by the US) instead of giving it to the new delegate appointed by the Communist government installed in Beijing after the triumph of the Maoist revolution the previous year (1949).

United Nations. The Security Council ordered the re-establishment of the status quo, as in the case of the war of the two Koreas (Padilla 2009, 308–316).

The origin of the war lies in the Iraqi claim to the territory of Kuwait, which historically belonged to the Ottoman province of Basra in Iraq and was a British protectorate from 1899 to 1961. Since its independence, British troops and the Arab League prevented its annexation by Iraq, although it should be remembered that the latter never recognized Kuwait's independence or the borders that separated the two states. After the bloody war which Iraq faced against Iran, the Iraqi dictator Saddam Hussein (who was supported by the United States in its war against Iran) claimed as a *"price for the Arab blood spilled"* in its fight against the Iranians the remission of its foreign debt, the increase of its quota of petroleum production, and the facilities to create a deep water port in Kuwaiti territory. The US ambassador in Baghdad told the dictator in a meeting that the conflict between Iraq and Kuwait was considered by Washington to be a 'bilateral problem'. As a result, Saddam Hussein misjudged the likely response of America if he took unilateral action to achieve his objectives, and on 2 August 1990 Iraqi troops invaded Kuwait.

Due to the chaotic situation in the Soviet Union that year (the previous year the Berlin Wall had been demolished and the USSR would collapse the following year), the international context was decisive for Washington's positions being accepted within the Security Council by a majority of the members, with Resolution 678 adopted on 29 November 1990 (with the affirmative vote of twelve members, abstention from China and votes against by only Yemen and Cuba) containing an *ultimatum* that set a deadline of 15 January 1991 for Iraqi forces to leave Kuwaiti territory and restore the *ex ante* situation.

Furthermore, on 3 August 1990 the Security Council decided to adopt *unanimously* resolutions 660, 661 and 662, requiring Iraq to withdraw "immediately and unconditionally" from the invaded territory. The Arab League also condemned Saddam's aggression, and on the 6th of the same month the Security Council imposed economic sanctions by enacting an embargo against Iraq. The use of force to make Iraq evacuate its troops was authorized on 25 August, and on 29 November Resolution 678 set 15 January 1991 as the deadline for the evacuation of Iraqi forces from Kuwait.

The deployment of American and British troops began on 8 August, and on 12 August the Arab League decided to send troops as well. Saddam Hussein attempted various tactics to curb the military attack: taking hostages among foreigners living in Kuwait, linking Iraq's withdrawal to Israel leaving the occupied Palestinian territories, and even calling for an opportunist 'jihad' against the infidels despite the fact that Saddam's government and policies were never religious. Finally, on 16 January 1991, as decided by the UN Security Council, Operation Desert Storm was unleashed with 800,000 troops, more than half a million of which were Americans and the rest from different countries, including several Arab nations. After brutal and continuous bombing, on 24 February the coalition troops advanced to reach their targets in just four days and forced Iraqis to withdraw from Kuwait.

It is noteworthy that, as with the Korean War, the United States respected the mandate of the United Nations Security Council, so Saddam Hussein's regime remained in Baghdad, with the Republican administration of George H.W. Bush (father) regretting that internal rebellions against the dictatorship in Iraq (Kurds in the north and Shiites in the south) could not be supported. From then on, Iraq was subjected to a UN sanctions regime geared towards the dismantling of its nuclear and chemical weapons-building potential, and part of its territory to the north and south remained a 'no-fly zone' patrolled by British and American air forces.

The lack of a definitive solution to the problem posed by Saddam Hussein's Iraq does not undermine the historical importance of the Gulf War as the second example of a Security Council operation to enforce coercive measures which was successful and complied with the regulations established by the Charter. With regard to matters such as the failure to use the Military Staff Committee, although this was a well-founded and correct criticism, the fact remains that because the Soviet Union did not send military contingents and China abstained from voting on the resolution, it was perfectly understood that the United States required Operation *Desert Storm* to be under its command. This would explain why the Secretary-General of the United Nations at the time, the Peruvian Javier Pérez de Cuéllar, argued that, despite not being a war led by the United Nations, the Gulf War was a "legal war, in the sense that has been authorized by the Security Council" (Merle 1991). This was not in any sense the case with the second war against Iraq, launched by George W. Bush (son), who acted unilaterally without the authorization of the Council in 2003. The situation created by Saddam Hussein's *fait accompli* in 1990–91 was such that rapid and energetic action by the UN was needed, as the experience of applying sanctions in other cases (South Africa during the *apartheid* era, for example) indicated that they took too long to start delivering results. With regard to the statute of limitations of the Charter which obliges a negotiated settlement of disputes to be sought, such a negotiation between an occupying (Iraq) and occupied (Kuwait) State was simply impossible.

So it seems that realism was not 'surreptitiously' introduced in the Charter of the United Nations, because it is clear from certain articles in Chapter VII concerning threats, acts of aggression and other breaches of the peace that, within the framework of collective security, "peace enforcement" can be carried out via military force provided it is directed by the Security Council itself (via the Staff Committee). Thus we have to accept that this was always the intention of Roosevelt and other founders of the world organization. This intention to provide the UN and the Security Council with 'realistic' means, which is to say making the military forces of member states available to the Council, placed the Council in a position to enforce its resolutions in a coercive manner. The necessity for the use of force would arise in the event that any of the member states refused to abide by a Council resolution on a voluntary basis, as occurred in the case of Iraq in 1991, when the latter refused to vacate Kuwait despite the ultimatum issued by the Council in its above-mentioned Resolution 678. This indicates that realism was not introduced surreptitiously but was clearly and intentionally sought by the founders of the world organization, since the theory of the balance of power was expressed in the very structure of the Council, judging by the presence of the five permanent members with the right to veto, all of which have nuclear weapons.

On the other hand, the foregoing demonstrates the need for an integral and innate reform of the Council so that countries like Germany and Japan, the powers that were defeated in the Second World War, can occupy a permanent place in the Security Council, like other emerging powers, such as India, Brazil and South Africa, who are also demanding a permanent seat nowadays. So the idea that the United Nations constitutes "idealism in progress" – as affirmed at the beginning of this section – is only half true because the global organization is a synthesis of the two great paradigms of international relations, especially with respect to the maintenance of international peace and security, since all the problems of coercive measures – which include the use of military forces – are to do with maintaining the balance of power among the permanent members. This means that the use of the right of veto is, in fact, an expression of the balance of power (the "correlation of forces") that exists on a global scale.

Based on what has been stated, it is clear that the absence of Japan and Germany, the powers that were defeated in World War II, and of some emerging powers of the Global South (India, Brazil and South Africa) causes an imbalance that has an impact on the outbreak of armed conflicts or prevents the conflicts being stopped in accordance with the principles of collective security provided for in Chapter VII. In both circumstances, the importance of reforming the current structure of the Council is evident.

In any case, what must be borne in mind is that, despite the fact that international law (and therefore the idealist paradigm) has been the backbone of the functioning of the United Nations throughout its history, realism too (the balance of powers) has always been present in the rules of the Security Council. This has, of course, made its performance particularly difficult, given the impossibility of avoiding the question of the correlation of forces in each circumstance or difficult conjuncture, unless chance and inexperience play their role in an unexpected way, as happened in the case of the Korean War and then the first Gulf War, when an irresponsible dictator committed acts of aggression against another member of the world organization, giving rise to a *casus belli* of such remarkable clarity that it was able to obtain the affirmative vote of twelve member states of the Council (including four of the permanent members with veto rights, the abstention of China and only two votes against) for his evident violation of the principles of the Charter.

1.9 Bipolarity and the Cold War

After the end of the Second World War in 1945 the international system changed from the multipolar power structure of the nineteenth century and the first decades of the twentieth century to a bipolar structure, as the United States and the Soviet Union were the only two hegemonic powers of their respective spheres of influence. Even if the newly independent decolonized states intended, through a *Non Aligned Movement,* to play a significant role in this bipolar structure of the Cold War, they were not really able to express a non-aligned position in the world scenario.

Apart from the fact that the great powers were reduced to two (the US and the USSR), the main characteristic of this bipolar world was that the possession of nuclear weapons was at the base of the 'balance of terror' which was maintained during this entire period. The classic geopolitical issue, according to which the contiguity of the territory is a determinant of the foreign policy actions of the states – as will be seen in the following chapter – ceased to be of maximum importance due to the geostrategic planning which could now be done by disregarding considerations of a classic geopolitical-territorial proximity nature. This is possible because both the terrestrial *intercontinental ballistic missiles* (ICBMs) and those launched by submarines or aircraft equipped with nuclear bombs allow long-range attacks to be carried out from one continent to another or from aerial and marine mobile vectors and do not depend on spatial (territorial) contiguity, a condition that gives great flexibility and manoeuverability to military strategists, even allowing them to formulate hypotheses for the possibility of winning a nuclear war.[28]

That is why the emphasis of equilibrium was transferred to the field of negotiations on the basis of the *Strategic Arms Limitation Treaty* (SALT), which refers to ICBM intercontinental missiles in the 1970s and subsequently to their reduction (START) in the 1990s. However, in these years of the Cold War the most significant achievement was the elimination of short and intermediate range projectiles, which was accomplished in Europe thanks to the Intermediate Nuclear Forces (INF) agreement signed between Presidents Reagan and Gorbachev in 1987 in order to eliminate all short and medium range missiles on European soil.[29] These achievements also include other disarmament negotiations carried out at the *Conference on Disarmament* (CD) in Geneva.[30]

---

[28] Nuclear armament is different from conventional weaponry not so much because of the shockwave or fire of the explosion as because of the very nature of nuclear energy (radioactivity), which prevents precise military objectives being set. Some nuclear explosions on scattered targets in a certain territory trigger uncontainable radioactivity that would affect human life on a larger geographic scale than would suffice, which makes it difficult, if not impossible, to aim nuclear weapons at precise targets (e.g. silos that hide missiles). This makes any hypothetical nuclear war incompatible with limited political objectives. Therefore, a nuclear attack can only be a zero sum game, a total war. That is why it has been said that a nuclear war would be an 'all for all' scenario, making men subordinate to nuclear weapons, a situation aggravated by the fact that any human failure would be absolutely irreparable, which holds true regardless of any possible security mechanisms. The uncertainties and difficulties involved in using nuclear weapons, as well as the extent and duration of the damage that their use would cause, explain why it would be madness to use them, since doing so would trigger *mutual assured destruction* (MAD) with no winners (Rousset 1987).

[29] Trump created a very dangerous situation at world level with his decision to abandon the INF.

[30] The Conference on Disarmament (CD) is a negotiation forum on multilateral disarmament founded in 1979 as a result of the first Special Disarmament Session of the General Assembly of the United Nations, which took place in 1978 (it started life as the Committee on Disarmament, but was renamed in 1982). The CD, as a forum for the negotiation of multilateral agreements on arms control and disarmament, depends on a representative appointed by the Secretary-General of the United Nations. It operates by consensus, and has successfully negotiated the Biological Weapons Convention, the Chemical Weapons Convention and the Comprehensive Nuclear-Test-Ban Treaty. The headquarters of the Comprehensive Nuclear Test-Ban-Treaty Organization (CBTBO) are located in Vienna, and the Organisation for the Prohibition of Chemical Weapons (OPCW) is based in The

By its very nature, nuclear disarmament has both a realistic dimension – maintaining the balance between nuclear powers, for which negotiations like START that seek to maintain a balanced number of nuclear warheads and their vectors in the hands of the two superpowers are crucial – and an idealistic dimension that refers to the ideal of complete nuclear disarmament in the future, as an essential factor for world peace and security and a long-term objective. Nuclear disarmament, in principle, can be negotiated at the CD in Geneva, but can also be promoted by multilateral action, such as the International Campaign to Abolish Nuclear Weapons (ICAN).[31]

Given that the USA and Russia still possess most of the nuclear arsenal of the nine countries with atomic weapons,[32] it could be said that, as far as nuclear weapons are concerned, the structure of the world's *military subsystem* continues to be bipolar, which also means that the Cold War between the two major nuclear powers is not necessarily at an end but has only reached a truce. Because the danger of nuclear confrontation between the United States and Russia is a permanent risk that has not disappeared with the collapse of the USSR, I have argued (Padilla 2015) that in order to end the tension between the two great nuclear powers and guarantee peace in the long term, it is essential to continue the nuclear disarmament negotiations at the Geneva *Conference on Disarmament* (CD) in which the two major nuclear powers participate (as well as other minor nuclear powers like France, China and the UK).

Furthermore, it would be best if such negotiations – as requested by the Community of Latin American and Caribbean States (CELAC) at the January 2015 summit in San José (Costa Rica) – were continued within the framework of the CD because the bipolar military international subsystem – as well as the multipolar one – has functioned on the basis of the two great paradigms of international relations that, when well understood, are not contradictory but complementary, as can be seen from the peace enforcement operations of 1950 and 1991. However, it should be born in mind that this is due to the difficulties inherent in the balance of powers *within* the Security Council, since the five permanent members occupy a place equivalent to the five-pole structure of the period between the peace of Vienna (1815) and the First World War, with the aggravation of the exclusion of the powers vanquished during the Second World War (Japan and Germany) and the absence of other potential permanent members who could reflect the new structure and dynamics of the system in the contemporary world.

It is evident then, that in order to properly face the challenge of the transformation of the international system, nuclear weapons should be forbidden, hence the importance of disarmament negotiations. Furthermore, at least while these negotiations are unfolding, there must be a guarantee of full enforcement and compliance

---

Hague. The CD is currently also negotiating the signing of a treaty to ban fissile materials and nuclear weapons in space in order to achieve its objective of complete nuclear disarmament.

[31] ICAN proposes the abolition of nuclear weapons through the signing of a treaty similar to the one implemented for the prohibition of military antipersonnel mines a few years ago. It has held several international meetings (in Norway, Mexico, and Austria) with growing strength and drive. ICAN won the Nobel Peace Prize in 2017 and the new Treaty has already been signed by a number of states.

[32] United States, Russia, France, United Kingdom, China, India, Pakistan, Israel and North Korea.

with treaties such as the non-proliferation treaty (NPT), which is key to avoid the dangerous increase in the number of countries in possession of such weapons, as has been the recent case of the last of those admitted to the Nuclear Club (North Korea).

On the other hand, it should be borne in mind that the processes of regional political concentation constitute a highly positive phenomenon for the cause of peace in general and nuclear disarmament in particular. In the case of Latin America, this is not only because of the signing of the Treaty of Tlatelolco, which declares the subcontinent a nuclear-weapon-free zone, and the subsequent establishment of an organization to ensure compliance with this objective (Organismo para la Proscripción de las Armas Nucleares en la América Latina y el Caribe/OPANAL), but also because, with proper orientation, these regional integration processes can contribute to the consolidation of regional actors that could have greater influence by seeking in principle to enhance sustainable development and democracy. Hence promoting the formation of these new regional actors on the international stage should be an objective of global civil society actors (social movements) as well as governments guided by the holistic and cosmopolitan paradigm that is considered in the final chapter and epilogue of this book, which means that regional actors constitute a fundamental component of any process aimed at the transformation of the international system.

Furthermore, this transformation of the international system (the dismantling of the bipolar nuclear military subsystem through nuclear disarmament) could also be supported by the parallel processes in which countries such as Mexico, Austria and Norway have been involved in order to get nuclear weapons banned because of the danger that ths kind of weapon represents for the survival of the human species on the planet. The facts that the 2017 Nobel Peace Prize was awarded to the *International Campaign to Abolish Nuclear Weapons* (ICAN) and that the proscription treaty has started to be signed and ratified encourages the continuation of work in that direction and gives a reason to be optimistic for the foreseeable future.

Democratizing the international system thus means transforming it in order to consolidate the role of multilateral negotiations in a new multipolar political structure. To achieve this, in addition to strengthening regional integration processes which would reduce the tendency towards hegemonism, multilateralism must be reinforced in order to give legitimacy to a global reordering based on democratic consensus. The first essentially means working within the United Nations system to strengthen it, which would entail reforming the current situation, especially the Security Council. As for the second, judging from the incipient Latin American integration processes of UNASUR, CELAC and SICA, and other regions' integration processes – such as those of the EU – it is crucial to become aware of what integration means for the democratization of the international system, as will be seen in the next chapter. We need a new kind of multilateralism in a renewed multipolar structure as well as consolidation of the regional integration processes.

1.10 Summary

In this chapter I have presented an outline of the two classic paradigms of international relations theory that have predominated since the establishment of the prevailing system of international relations, that is to say the Westphalian system. My main conclusion is that both paradigms are important in the history of international relations and have been applied according to the different circumstances and historical context, sometimes adequately and successfully, like realism in the Treaty of Vienna (1815), and some times erroneously, like idealism in the Treaty of Versailles (1919). Another important conclusion is that international relations theory is essentially an interdisciplinary inquiry that until present has been reduced mainly to an investigation of inter-state relations, thus it is necessary to enlarge its scope in order to explain how nations and culture, the economy and ecology – for instance – are also part of the discipline and must be used to better understand its problems and give direction to its research. Therefore, the idea that I will enlarge upon in the following chapters is that the international system is a complex social construction, composed of different subsystems – political, social, cultural, economic, military – and that each one has to be addressed in a proper methodological manner using the tools of the different disciplines involved, such as political science, sociology, economy or ecology. Consequently, a holistic vision of the international system presupposes an overarching approach for all those fields of knowledge. For instance, in order to explain why the territorial issue was so important in international relations during the nineteenth and twentieth centuries but in current times its importance has diminished, or, in other words, why geo-economy is more important than geopolitics nowadays, it is necessary to use the tools of economy, sociology and ecology. Otherwise the task will not be addressed properly. And, of course, the resulting knowledge will become an element of international relations theory.

In this chapter I have also dealt with the transformation of the Westphalian system. Transforming the international system entails, in the political subsystem, making it advance from the current situation – in which there is a single great power in search of global hegemony – towards a multipolar scenario similar to the one that already prevails in the economic field, and promoting the democratization of international politics through a new type of multilateral negotiation within the UN political subsystem. Furthermore, concerning the military international subsystem – which is by far the most dangerous to peace and security, given the existence of nuclear weapons – I have also argued in favour of the continuity of multilateral disarmament negotiations with the aim of eventually ending nuclear bipolarity and reaching nuclear disarmament, as requested by ICAN and the proscription treaty. Hence transformation of the international military subsystem will not be possible without going through the enormous task of total nuclear disarmament ('option zero' for the nine nuclear powers), which must, inevitably, be considered a long-term project.

What has been seen throughout this chapter is the way in which, through wars, diplomatic agreements and the conclusion of treaties, the great paradigms of international relations have influenced and changed the Westphalian system, moving from multipolar situations in the seventeenth and eighteenth centuries to the penta-polar situation (Austria-Hungary, Germany, Russia, France and Great Britain) that prevailed in Europe during the nineteenth and early twentieth centuries and led to military bipolarity during the Cold War. Given the overwhelming nuclear status of the two great powers (the US and Russia), this situation still prevails in the military subsystem. Nevertheless, even though, on a political level, the US continues the search for unattainable world hegemony, at the same time the economic subsystem clearly shows that the world has evolved towards a condition of multipolarity, with several economic powers besides the United States. Of course, what I am calling the 'economic subsystem' is not the equivalent of the political subsystem, and there-fore it is clear that multipolarity in the latter requires political and military power in addition to economic power. Thus far, apart from the US, only China seems to be in that situation, hence multipolarity does not yet exist at world level. But the tendency towards that end is clear, and traditional powers like Russia, Germany, Japan and India could evolve in that direction.

As will be seen in the next chapter, the other pillar of the democratization of the international system is the regionalization and integration process. In addition to the common market, the customs union, monetary union and other economic para-phernalia, the European integration process has built supranational political scaf-folding (the European Commission, the European Parliament, the European Court etc.) and large-scale social integration (like the Schengen agreement) which is abso-lutely unprecedented in other similar processes. However, its greatest achievements have been attaining sustainable peace with no more wars between the great Euro-pean powers; the democratization of countries like Greece, Portugal, Spain and the former East European Communist Bloc countries; and, in general terms, the human, social and sustainable development of poor and backward regions and countries in the continent.

Seen from the perspective of regional integration, another interesting feature of the democratization of the international system is that the transformation of political power within the system makes it more rigid and authoritarian in schemes of few actors and much more open and flexible in schemes of several or multiple actors. For democratization to occur, it must be recognized that small and weak states are more likely to exercise their sovereignty and independence in a framework of breadth and flexibility than rigidity and hegemonic authoritarianism. A multiplicity of actors fosters freedom of action for small powers and the possibility of implementing autonomous policies without being subject to foreign directives, as happens among the twenty-seven member states of the EU. Conversely, a reduction in the number of international actors limits and reduces the autonomy of small states. Consequently, multipolarity favours the democratization of the system and benefits both middle-sized and small states like the Central American and island nations. Hence promoting multipolarity is a desirable foreign policy goal for most countries of the world.

Concerning the need to overcome the classic approach of IR based on realism/idealism, it is also clear that weak and small states do not fit within the Westphalian system and that the balance of powers theory is not sufficient to explain their interrelationship, especially in light of certain cases. For instance, the small

Baltic States should not feel 'safe' because they have become members of the North Atlantic Alliance, since that does not provide any guarantee of survival in the framework of a nuclear confrontation between the great powers. The Inter-American Treaty of Reciprocal Assistance (*Tratado Interamericano de Asistencia Recíproca* [TIAR]) was of no use to Argentina in the Falklands War against the UK, and, thanks to the 1962 nuclear missile crisis, Cuba needed a bilateral agreement between Washington and Moscow to avoid being invaded by American troops. Hence, and even from a *realpolitik* perspective, the security of small states depends on military alliances like NATO, or on the umbrella provided by great powers in different regions of the world. Take the Latin American countries as another example: it is clear that the international security of the Latin American countries is much more related to the superpower military balance at world level than to a hypothetical balance of small or middle-sized regional powers, as can be seen in the case of Venezuela, where neither the neighbouring countries nor 'almighty America' has yet been able to unleash a military intervention for regime change, such as the one the Bush Administration implemented in Iraq in 2003 or his father initiated in Panama in 1989 to overthrow the Noriega regime, not forgetting the intervention of Eisenhower and the Dulles brothers' in Guatemala in 1954 to topple the democratic regime of President Jacobo Arbenz.

There are many more examples, but what I want to highlight is the fact that realism is not useful for those states that do not qualify for the 'great power'[33] category, which is why the option of 'idealism' to sustain foreign policy in international law is absolutely necessary for the majority of the world's sovereign states. This is why this chapter emphasizes the importance of multilateralism and working with the United Nations as an essential factor in the democratization of the international system. But, as previously pointed out, this democratization of the system also requires people to realize that it is composed of several subsystems. In this chapter I have just referred to the dynamics of the political subsystem – that is to say the inter-state and intergovernmental relationship – not to the social and economic dimensions, which provides an idea of the complexity of the task. Thus, the relationships between nations, peoples, cultures, social movements, ethnic groups and so on must be also considered and incorporated into IR theory. Hence *international relations theory* (IR) must include the study of relations between social movements, cultures, religions (remembering that all religions transcend national boundaries, which essentially makes churches, religious leaders, priests and believers *a transnational structure*), and migratory flows, as well as the study of trade relations and

---

[33] In his lecture in Lyon (France) Bertrand Badie compared the international system to football divisions: some states are part of the major leagues while others are in the inferior ones. Obviously, the great powers belong to the major leagues, but the minor ones are not in any condition to access that category unless they increase their military spending, which is problematic for global security and erroneous from a sustainable development perspective. Saudi Arabia and Iran are (bad) examples of middle powers that have spent large amounts of money on conventional armaments in order to maintain a 'regional balance of power', but the Middle Eastern context, where other powers like Israel, Turkey, Syria and Egypt are also important actors, is fortunately a very particular geopolitical scenario that applies only to that region of the world.

exchange between nations under the umbrella of governments (including transnational corporations), financial flows, and – in general terms – all the matters of the complex interdependence and international regimes that have been studied by Robert Keohane and Joseph Nye in their classic books, not forgetting constructivism and the contributions of Alexander Wendt, among others, that explain how power politics is a social construction, not an 'ontological' (existential) reality, as most realist are convinced.[34]

On the other hand, as will be seen in Chap. 3 about the Anthropocene, the environmental dimension of international relations is fundamental because climate change – the most important issue of our times, since this natural phenomenon is placing both humanity and the planet as a whole at risk – does not recognize borders or the abstract and artificial creation of States (or politicians, hence also 'social constructions') designed to exert control over territories and populations. As will be seen later, this fact also requires a new kind of multilateralism within the international system – a multilateralism that should replace the idea prevailing among diplomats that negotiation is a means to obtain advantages based on national interests. This new "solidarity" multilateralism (Badie 2008) must be open to protecting the planet's ecosystems and human development, which means fulfilling the commitments of the UN Agenda 2030 (the Sustainable Development Goals [SDGs]) and conserving common public goods. In other words, the negotiations must be fundamentally aimed at the protection of the planet we inhabit: the atmosphere, the oceans, the ice sheets in the polar regions, the forests, the species diversity, the fresh water (including rivers and lakes), the cultural patrimony, and so on.

As Brauch (2016) argues, the world needs a peace ecology perspective based on transformative science with an underlying vision of *sustainable peace*, a concept that should be understood as "peace with nature in the Anthropocene". This means that governments must address the economic causes of greenhouse gas concentration in the atmosphere, as has been done by the Paris COP 21. The actions taken by concerned individuals, families and local communities must be combined with the public policies of national states, international organizations, international NGOs and social movements. According to Brauch, sustainable peace could be addressed from a peace ecology perspective that integrates peace and security studies in a form that encompasses ecology and sustainable development. The methodology suggested by Brauch also requires a shift from disciplinary and multidisciplinary research methods towards a transdisciplinary approach of transformative science that looks for the establishment of a sustainable peace that demands policy changes in the mindsets

---

[34] As is well known, complex interdependence is characterized by the study of the multiple channels of exchange, relationships and action that exist between societies in interstate, trans-governmental, and transnational relations, the absence of a hierarchy of issues with changing agendas, and a decline in the use of military force and coercive power in international relations. Nye/Keohane (1988) argue that the decline of military force as a policy tool and the increase in economic and *other forms of interdependence* should also enhance cooperation between states. Therefore they criticize realism for ignoring the social nature of relations between states and the social fabric of international society, which makes interdependence a sociological perspective in international relations theory. Regarding these issues see also Padilla (2009) and Wendt (2005).

of actors and decision-makers. This goal may be considered utopian (as it requires a fundamental change in the dominant mindsets of decision-makers, practitioners and citizens), but it is necessary if society is going to move towards an alternative sustainable approach (Brauch 2016).

The next chapters will extensively refer to the need for this change of the neoliberal mindset (Göpel 2016) that, from my perspective, could be replaced by cosmopolitanism as a political philosophy and guiding principle. One of the positive results of globalization is precisely this kind of philosophy together with the surge of a new type of social strata formed by people who have a planetary consciousness and are permanently preoccupied with what happens to our planet at a global level, and who, through global civil society, have been able to organize a transnational social movement that seeks the promotion of planetary concerns and makes policy proposals to the United Nations in order to develop, in addition to national policies, a Planet Earth policy or *Gaia politics* (Latour 2015).

The foregoing means seeing globalization from a new perspective, a true *alter globalization* (as demanded by the World Social Forum), and realizing that it essentially consists of economic *interdependence* between national states and that this interdependence or interconnection demands not only the free flow of information, free trade and the free movement of people, but also multilateral agreements to solve global problems concerning the environment (the COP 21) and social, economic and political problems (the UN SDGs of the 2030 Agenda), and to guarantee the freedom for humans to move about the planet – migrate, transmigrate – via the regularization of human mobility, as demanded by the Global Compact for Migration approved by the Marrakech Conference in 2018, and so on. This is why globalization transcends the national sphere, becomes *transnational* and needs to be addressed in a different way, according to which the central concern of a *world citizen* who looms on the horizon must be *geo-ecological* and *cosmopolitan* with permanent environmental and planetary concern for the interests of humanity as a whole, not just the sectoral interests of the groups that control governments, which are ultimately what predominate in each nation state.[35]

This is the reason why globalization is making national borders less and less important and increasingly 'porous', provoking the reaction of conservative political sectors against migrants and leading to calls for protectionism and a return to

---

[35] This cosmopolitan citizenship could be seen as part of the idealist utopia. However, research in the field of psychology explains how the individual ontogenetic evolution of human beings transits from stages of pre-conventional mentality (egoistic) to conventional (ethnocentric or socio-centric) and finally reaches a post-conventional geocentric or *worldcentric* cosmopolitan stage of mindsets and mentalities. As Kohlberg (1983, 1984) and other researchers have demonstrated, these individual stages of evolution are tantamount to phylogenetic processes in society that run along similar paths of cultural evolution, leading to the formation of a significant segment of society with a *cosmopolitan world-view and mentality*, which proves the existence of the phenomenon at the collective level. The pioneer in these investigations is Piaget (1932), but in addition to Lawrence Kohlberg other exponents include Gilligan (1977), Gebser (1991) and Wilber (1991). Concerning cosmopolitanism as a political philosophy, there are a number of authors, such as Held (1995, 2005, 2006, 2010), Archibugi (2011), Mignolo (2011), Smith (2017), Delanty (2012, 2017), and Beck (2009), among others, who have carried out investigations and published important contributions in this field.

the nationalist ideology of the defence of territory and national sovereignty. The intervention of international agencies or NGOs in defence of human rights, indige- nous peoples, women's rights and the environment (*the rights of Gaia*), and against corruption, is considered by these nationalist groups to be an "intervention in internal affairs" and hence their reaction in supposedly in "defence of national sovereignty", including the rejection of globalization, which, in spite of being a flag of leftist polit- ical groups at the end of the twentieth century has now become the new opposition flag of the political right, while paradoxically the Chinese president, Xi Jinping, attends the annual meeting of the elite of world capitalism – the World Economic Forum at Davos – as a great supporter of globalization and free trade.

To conclude, it is clear that the transformation of the international system requires a new perspective to address the theory of international relations. This must be complemented with the new perspectives that were introduced by political and social sciences, since globalization has also had positive consequences, such as the emer- gence of a cosmopolitan world-view, as well as very important regional integration processes, such as the European Union. Thus globalization can have benefits in spite of the fact that the concentration of wealth has increased poverty at world level. Ultimately, these negative features can be attributed to an anarchic form of wild capi- talism (neoliberalism) that must be placed under the control of a renovated system of global governance, preferably by multilateral agreements negotiated by sovereign states under the UN umbrella. And, of course, as the international system is based on the terrestrial ecosystem, it is also indispensable to be guided by the limits posed by planetary boundaries (Sachs 2015).

References

Archibugi, Daniele, 2011: *Cosmopolitan Democracy: A Restatement* (Rome: Italian National Research Council).
Alker, Hayward; Gurr, Ted Robert; Rupesinghe, Kumar 2001: *Journeys Through Conflict, A Study of the Conflict Early Warning Systems Research Project of the International Social Science Council* (Boulder, CO: Rowman & Littlefield Publishers).
Badie, Bertrand, 1999a: *Un Monde sans Souveraineté* (Paris: Fayard).
Badie, Bertrand, 1999b: *Les Mutations de l'Etat-Nation dans Europe à l'Aube du XXIe Siècle* (Strasbourg: Council of Europe).
Badie, Bertrand, 2011: *La Diplomatie de Connivence* (Paris: La Découverte).
Beck, Ulrich, 2009: *A Critical Theory of World Risk Society: A Cosmopolitan View* (London: Blackwell Publishing Ltd.).
Brzezinski, Zbigniew, 1998: *El Gran Tablero Mundial: La Supremacía Estadounidense y sus Imperativos Geoestratégicos* (Barcelona: Paidós).
Brauch, Hans Günter; Oswald Spring, Úrsula; Bennett, Juliet, Serrano Oswald, Serena Eréndira (Eds.), 2016: *Addressing Global Environmental Challenges from a Peace Ecology Perspective* (Cham: Springer International Publishing).
Cortina, Adela, 2010: *Etica Aplicada y Democracia Radical* (Madrid: Tecnos).
Delanty, Gerard, 2012: *Routledge Handbook of Cosmopolitanism Studies* (New York/ Oxford: Routledge).

Delanty, Gerard; Mota, Aurea, 2017: "Governing the Anthropocene: Agency, Governance, Knowledge", in: *European Journal of Social Theory*, 20,1: 9-38.

Derek Heater, 2002: *World Citizenship: Cosmopolitan Thinking and its Opponents*, (London: Continuum).

Fisher, Joschka, 2013: *Peligra la Unidad Europea*, in: *El País* (Madrid: 3 May).

Foucault, Michel, 2014: *Seguridad, Territorio, Población* (México: Fondo de Cultura Económica [FCE]).

Gebser, Jean, 1985: *The Ever-Present Origin* (Athens: Ohio University Press).

Gilligan, Carol, 1977: "In a Different Voice: Women's Conceptions of Self and Morality", in: *Harvard Educational Review*, 47,4: 417–481

Göpel, Maja (2016): *The Great Mindshift: How a New Economic Paradigm and Sustainability Transformations Go Hand in Hand* (Cham: Springer International Publishing).

Gutiérrez, Edgar, 2016: "La CICIG: Un Diseño Nacional y una Aplicación Internacional", in: *Política Internacional* (Guatemala City: Diplomatic Academy).

Harari, Youval Noah, 2018: *21 Lecciones para el Siglo XXI* (Barcelona: Penguin Random House).

Held, David, 1995: *Democracy and the Global Order* (Cambridge: Polity Press).

Held, David, 2005: "Principles of Cosmopolitan Order", in: *Anales de la Cátedra Francisco Suárez*, 39: 153–169.

Held, David, 2006: *Models of Democracy* (Stanford University Press

Held, David, 2010: "Global Democracy: A Symposium on a New Political Hope", in: *New Political Science*, 32,1: 83–121.

Huntington, Samuel, 1996: *The Clash of Civilizations and the Remaking of World Order* (New York: Simon & Schuster).

Huntington, Samuel, 2004: *¿Quiénes somos? Los Desafíos de la Identidad Nacional Estadounidense* (Barcelona: Paidós).

International Organization of Migrations, 2016: *Encuesta sobre Migración Internacional de Personas Guatemaltecas* [IOM Report] (Guatemala City: OIM).

Kaldor, Mary (Ed.), 2003: *Global Civil Society 2003* (New York: Oxford University Press).

Kissinger, Henry, 2014: *World Order* (New York: Penguin Press).

Kohlberg, Lawrence, 1984: *The Psychology of Moral Development: The Nature and Validity of Moral Stages* (New York: Harper & Row).

Kohlberg, Lawrence et al., 1983: *A Longitudinal Study of Moral Judgment: Monographs of the Society for Research in Child Development* (Chicago: University of Chicago Press).

Kuhn, Thomas, 1962: *The Structure of Scientific Revolutions* (Chicago: University of Chicago Press).

Lacoste, Yves, 2009: *Geopolítica: La Larga Historia del Presente* (Madrid: Editorial Síntesis).

Latour, Bruno, 2015: *Face a Gaia: Huit Conférences sur le Nouveau Régime Climatique* (Paris: La Découverte).

Matul, Daniel; Cabrera, Edgar, 2007: *La Cosmovisión Maya*, 2 vols. (Guatemala City: Liga Maya de Guatemala; Real Embajada Noruega en Guatemala; Amanuense Editorial).

Mearsheimer, John, 2014a: *The Tragedy of Great Power Politics* (New York: Norton & Co.).

Mearsheimer, John, 2014b: "Why the Ukraine Crisis is the West's Fault: The Liberal Delusions that Provoked Putin", in: *Foreign Affairs* (September/October): 61–73.

Mearsheimer, John; Walt, Stephen: "The Case for Offshore Balancing: A Superior US Grand Strategy", in: *Foreign Affairs*, 95,4 (July/August).

Merle, Marcel, 1991: *La Crise du Golfe et le Nouvel Ordre International* (Paris: Économique).

Mignolo, Walter, 2011: "Cosmopolitan Localism: A De-colonial Shifting of the Kantian's Legacies", in: *Localities*, 1: *Cosmopolitan Localism*: 11–46.

Morgenthau, Hans: 1986: *Política entre las Naciones: La Lucha por el Poder y la Paz* (Buenos Aires: Grupo Editor Latinoamericano [GEL]).

Nye, Joseph; Keohane, Robert, 1988: *Poder e Interdependencia, la Política Mundial en Transición* (Buenos Aires: Grupo Editor Latinoamericano [GEL]).

O'Rourke, Lindsey, 2018: *Covert Regime Change: America's Secret Cold War* (New York: Cornell University Press).

Padilla, Luis Alberto, 2009: *Paz y Conflicto en el Siglo XXI: Teoría de las Relaciones Internacionales* (Guatemala IRIPAZ).

Padilla, Luis Alberto, 2012: *La Cosmovisión Maya y el Desarrollo Sostenible*, Paper presented at the FLACSO anniversary conference, Guatemala City.

Padilla, Luis Alberto, 2013: "Guatemala City: Relaciones Internacionales y Contexto Geopolítico Mundial, 1954–1996", in: *Guatemala: Historia Reciente* (1954–1996): pp.97–144 (Guatemala City: FLACSO).

Padilla, Luis Alberto, 2015: "El Conflicto de Ucrania a la luz de los Paradigmas Clásicos de la Teoría Internacional", in: *Espacios Políticos, Review of the Faculty of Political Sciences* (Guatemala City: Rafael Landívar University).

Piaget, Jean, 1932: *The Moral Judgment of the Child* (London: Trench).

Popkin, Eric, 2005: "The Emergence of Pan Mayan Ethnicity in the Guatemalan Transnational Community Linking Santa Eulalia and Los Angeles", in: *Current Sociology*, 53,4 (Monograph 2): 675–706.

Portes, Alejandro, 1996: "Transnational Communities: Their Emergency and Significance in Contemporary World System", in: Korseniewics, Roberto Patricio; Smith, William C. (Eds.): *Latin America and the World Economy* (Westport, CT: Greenwood Press).

Pries, Ludger, 2017: *La Transnacionalización del Mundo Social: Espacios Sociales más allá de las Sociedades Nacionales* (Mexico City: Deutscher Akademischer Austauschdienst [DAAD]; El Colegio de México [Centro de Estudios Internacionales]).

Raskin, Paul, 2002: *Great Transition: The Promise and Lure of the Times Ahead* (Cambridge, MA: Tellus Institute).

Rosenthal, Gert, 2016: "Participación de Guatemala en el Consejo de Seguridad de Naciones Unidas 2012–2013", in: *Política Internacional* (Guatemala: Academia Diplomática).

Rousset, David, 1987: *Sur la Guerre: Sommes Nous en Danger de Guerre Nucleaire?* (Paris: Éditions Ramsay).

Sachs, Jeffrey, 2015: *The Age of Sustainable Development* (New York: Columbia University Press).

Smith, William, 2017: "Cosmopolitanism", in: *International Studies* (London: International Studies Association/Oxford University Press); https://doi.org/10.1093/acrefore/9780190846626.013.133

Sousa Santos, Boaventura, 2014: *Democracia al Borde del Caos* (México: Siglo XXI editores).

Vargas Llosa, Mario, 2019: *Tiempos Recios* (México: Alfaguara).

Villagran, Carlos Arturo, 2016: "Soberanía y Legitimidad de los Actores Internacionales en la Reforma Constitucional de Guatemala El Rol de CICIG", in: *Politica Internacional* (Guatemala City: Diplomatic Academy).

Wendt, Alexander, 1992: "Anarchy is What States Make of it: The Social Construction of Power Politics", in: *International Organization*, 46,2: 391–425.

Wilber, Ken: 2006: *Sex, Ecology and Spirituality: The Spirit of Evolution* (Boston, MA: Shanbhala Publications).

Wilber, Ken, 2018: *The Religion of Tomorrow: A Vision for the Future of the Great Traditions: More Inclusive, More Comprehensive, More Complete* (Boulder, CO Shambala Publications Inc.).

Internet Links

Badie, Bertrand: "Conference", at: https://www.youtube.com/watch?v=uY5LaUrgaP4&t=113s.

Global Citizens Movement; at: https://en.wikipedia.org/wiki/Global_citizens_movement.

Fisher, Joschka, 2018: "interview", at: https://www.euractiv.com/section/euro-finance/interview/jos chka-fischer-stabilise-the-eurozone-to-defuse-hurricane-brexit/.

Habermas, Jürgen, 2003: "Theory of Cosmopolitanism", in: *Wiley Editing Services;* at: https://onl inelibrary.wiley.com/doi/abs/10.1046/j.1351-0487.2003.00348.x.

Kinzer, Stephen, 2019: "America's Legacy of Regime Change", at: http://stephenkinzer.com/cat egory/1470%20bookreviews

Chapter 2
Integration: A Means to Transcend the Westphalian Order and Go Beyond Geopolitics

> *World peace can not be achieved without creative efforts tantamount to the dangers it confronts. The contribution that an integrated Europe can give to civilization and peace is fundamental. But European integration will not be the result of a single undertaking but of concrete small steps addressed to promote a type of solidarity that goes beyond the secular opposition between France and Germany.*
> Schuman (1984: 268)

> *The main purpose of a politician is not to be right but for people to acknowledge that he is on the right path. A successful person is the one who lights the candle in the direction of the wind.*
> Adenauer (1984: 273)

> *Geopolitically, America is an island off the shore of the large landmass of Eurasia, whose resources and population far exceed those of the United States. The domination by a single power of either of Eurasia's two principle spheres – Europe or Asia – remains a good definition of strategic danger for America, Cold War or no Cold War. For such a grouping would have the capacity to outstrip America economically and, in the end, militarily. That danger would have to be resisted even were the dominant power apparently benevolent, for if the intentions ever changed, America would find itself with a grossly diminished capacity for effective resistance and a growing inability to shape events.*
> Kissinger (2014: 813)

> *Latin America and the Caribbean can make important progress in the goals that each country has proposed, and as a region as a whole within the framework of the 2030 Agenda on Sustainable Development. This will make it possible to meet the goals established therein, based on a new paradigm of development, with a strong focus on rights of a participatory nature, promoting democratic governance as an essential instrument for peace.*
> Aravena (2017: 32)

© Springer Nature Switzerland AG 2021
L.-A. Padilla, *Sustainable Development in the Anthropocene*,
The Anthropocene: Politik—Economics—Society—Science 29,
https://doi.org/10.1007/978-3-030-80399-5_2

2.1   Introduction

The first chapter showed that IR theory, as an area of interdisciplinary research, is not complete unless the social, economic and environmental dimensions of this field of study are taken into account. The two great paradigms were appropriated to explain historical events of the political and military subsystems, but unappropriated for the explanation of phenomena and situations that are part of the socio-economic and ecological international subsystems. Integration theory is essentially a component of the latter, and, in my view, the integration and regionalization processes are of the utmost importance for *international relations* (IR) in the twenty-first century. Furthermore, they are fundamental to the transformation of the system towards multi-polarity and its democratization, including the rejection of extreme right xenophobic nationalism and the adoption of a cosmopolitan world-view and transnational European citizenship. For instance, the EU is in the process of transforming *the practice* of the Westphalian order because the *balance of powers* is no longer the key concept (as it was in the nineteenth century) in the way the new integrated European system is functioning. Consequently, it is necessary to resort to Integration Theory and cosmopolitanism to explain the transformations that have already occurred. Thus, this chapter analyses how the EU system is functioning (or malfunctioning), and special attention is drawn to the fact that the way the EU is currently functioning is not the result of a military balance of powers within its twenty-seven member states – since the whole of Europe is under the umbrella of NATO as a last resort for its "security and defence" from powers outside the states' territory – but of the way the monetary union, the Schengen agreement and the Lisbon Treaty of 2007 are put into practice by Frankfurt, Strasbourg and Brussels. I have also pointed out that the dictatorships of Spain, Greece, Portugal and the former East European Communist Bloc have disappeared not just because of institutions like the European Parliament and the transnational citizenship that will be examined in Chap. 6, but also because of internal democratization. The presence of multiple actors – both governmental and from civil society – also makes the integrated EU system much more democratic than Latin American, South-East Asian and new regional systems in the process of formation, such as the African and Eurasian Unions. Thus, my principal aim in this chapter is to stress the way that regionalization and integration are helping to: (1) transform the Westphalian system based on a narrow nationalist concept of sovereignty; (2) transform the Westphalian system based on realism and the principle of the *balance of powers* and guarantee a sustainable peace, as is happening in the current EU situation and even in regions where regionalism is fragile and in crisis;[1] (3) promote democratization both inside each country and among them as members

---

[1] But even in the case of Latin American regionalism, as can be seen in the social and political crisis of Venezuela, the neighbouring countries have not applied a realist *balance of power* conception. This explains why they have abstained from military intervention, despite the flow of three million refugees to Brazil, Colombia, Ecuador, Peru and Chile and the rejection of the political legitimacy of Maduro's regime by Western countries and the Lima Group. And the hemispheric hegemon (the US) has not yet decided on military intervention, possibly because the Pentagon understands that the Russian and Chinese support for the Caracas regime is not a real geopolitical threat.

of the regional system, despite economic and social crisis, as is happening in the EU at present; and (4) make the geopolitical struggle for the control of territories recede, as is happening in Latin America, thanks to the fact that most Latin American countries have decided to resolve territorial disputes via mediation or the International Court of Justice. And despite the fact that Mackinder's (1904, 2004,2010) concept ofgeopolitics still seems to be embedded in the US's foreign policy with regard to the *heartland* of Central Asia (as can be seen from Kissinger's quotation above), the intelligent decision of China to opt for the *one belt one road* (OBOR) – or the new '*silk route*' – initiative could be interpreted as a geo-economic response to the US exerting geopolitical power in Central Asia via Washington's policy of *regime change* that, when applied in Ukraine, led to the disastrous result of Kiev losing the Crimean peninsula and having to face a separatist movement in the Donbass region.

2.2 Regionalization and Integration

A distinction has to be made between the concepts of regionalization, regionalism and integration. Usually regionalism is promoted by states that have geographic contiguity, therefore spatial proximity is a factor of paramount importance in the regional government's decisions that initiate and maintain these processes which offer the possibility of speaking generically of a European, Central American, Southern Cone, South-East Asian or West African regionalization, to give some precise examples. Malamud (2018) argues that *regional integration* must be differentiated from *regionalism,* which consists of a formal process led by States, and from the concept of *regionalization*, which is broader, with economic and social roots, and as such refers to the increase in intraregional *interdependence*, being an informal process by which exchange flows are increased between a set of territorially contiguous countries whose "main engine does not reside in the State, but in the market and, secondarily, in civil society" (Malamud 2018: 77).

In any case, according to this theoretical perspective, regional integration is a concept that can be applied only to the European process, while in Latin America (and other world regions) integration could have been the declared purpose or objective, but the concrete fact in both Central America and South America is that what exists so far is regionalism – not integration – among other reasons because no supranational institutions have been established as in the case of the EU. Thus, as Dabène (2012: 18–19) states, this part of the world has seen waves of integration intents, each with its own peculiarities; however, after the failed attempt of integration by the substitution

---

Evidently, a peaceful way out of that crisis requires new elections (which could be organized – for instance – under the UN Security Council umbrella) and the installation of a genuine process of sustainable development by a democratic regime, not the military 'solution' implicit in those who are demanding the *Interamerican Treaty of Reciprocal Assistance* (TIAR by its Spanish acronym) to be applied, to give another example. Thus, it is clear why a new approach to IR theory must include an environmental, social, economic and cosmopolitan perspective. Geopolitics and the *realpolitik* of recent history must not be the exclusive parameters of international security.

of imports promoted by Raúl Prebisch and the *United Nations Economic Commission for Latin America and the Caribbean* (ECLAC) during the 1950s and '60s, most were not achieved. And when the second and third waves came (from the 1970s to the 1990s), what they achieved, with difficulty, was the establishment of free trade zones, and neither the Andean subregion nor the Southern Cone could move towards the establishment of a customs union or a common market that, according to the old postulates of experts like Balassa, are indispensable requirements of any economic integration (Malamud 2011).[2]

Concerning political and social integration, although the fourth wave of integration was supposedly going in that direction, lacking a basis of support in the economy and caught in Dabène's "trap of sovereignty", it has not been able to reach very far, although there are some hopes that it can move forward in the future with experiments like SICA in Central America. This pessimistic perception is a consequence of the fact that although, according to the theory of integration, sovereign states need to voluntarily relate to their neighbouring countries and agree to diminish the scope of their own sovereignty (losing certain attributes while acquiring new techniques to solve their conflicts together) while simultaneously creating permanent common institutions which are capable of making binding decisions for all members, none of this has happened in Latin America. Therefore, the economic elements that are present in SICA, MERCOSUR, UNASUR and the like (e.g. the increase in commercial flow, the promotion of contact between elites, the facilitation of encounters or communication between people across national borders, the invention of symbols representing a common identity etc.) can be useful for genuine integration later on, but they are not its equivalent. As Malamud (2011: 221) points out, since none of the processes that took place in the subcontinent have been characterized by this "loss of factual attributes of sovereignty" or by advances in economic or social integration, there are many pending tasks to be carried out in the foreseeable future regarding the integration of the Latin American countries, because what exists at present is regionalism, not integration, as Malamud affirms.

It is worth being aware of the academic debate regarding the incomparability of the processes that occur in different regions of the world.[3] Hence, if integration implies sharing sovereignty and establishing supranational institutions that deal with

---

[2] According to Bela Balassa, economic integration has four stages: (1) The zone of free trade, a territorial scope in which the products of any member country can enter others without paying tariffs, as if they were sold anywhere in the country of origin. (2) The customs union that establishes a tariff that they will pay for products from third countries; this implies that the Member States form a single entity in the field of international trade. (3) The common market, customs union to which is added the free mobility of capital and labour, that is, of workers to the already existing mobility of goods and (eventually) services. These advances require the adoption of common trade policies, the coordination of macroeconomic policies and the harmonization of national legislations. (4) The economic union that consists of the adoption of a single currency and monetary policy, such as the Euro in the EU (cited by Malamud 2011: 221).

[3] For example, Hettne (2002: 956) argues that "It is understandable that the most advanced case of economic integration, the European one, is used as a paradigm to compare what happens in other regions, but in light of current regional experiments it is more important to look without prejudice to the formation of a region anywhere in the world and justify the peculiarities of the context. In

economic, social and political issues, as in the EU, and none of these features are present in practice, what exists in Latin America is regionalism. Revisions can be made to notions about the stages through which these diverse processes have taken place in Latin America. Hettne (2002), Serbin (2012), Sanahuja (2012), Aguilera (2015) and Dabène (2012) agree that the first stage (or first wave) of the attempts to carry out regionalism since the post-war decade in the twentieth century is characterized by the influence of Raúl Prebisch and ECLAC. These efforts were at the beginning protectionist and sought to create internal markets by substituting imports and establishing a national industry for the domestic markets of each country in order to promote without prejudice a sort of free trade within the countries that were part of the different schemes that were launched: basically, the *Latin American Free Trade Association* (ALALC), the Andean Pact and the *Central American Common Market* (CACM).

After the ECLAC stage, a period followed in the 1980s and 1990s during which a newly proposed concept called '*open regionalism*' was used, paradoxically by the same ECLAC. From the beginning this was predominantly influenced by neoliberal ideology and distinguished by the abandonment of protectionism and the policy of imports substitution, all in the context of an increase in the process of globalization, especially at the end of the Cold War in 1989.[4] As the ideology that guided this period was a neoliberal one – those were the years of Reagan and Thatcher – the *Washington Consensus* was the economic ideology to be applied, and its sacrosanct principles (liberalization, privatization, stabilization) were considered to be dogmatic principles which were not to be discussed but applied by governments under the attentive surveillance of the IMF and the World Bank.

While Washington did not regard the integration process with sympathy, the White House observed the possibilities offered by regionalism with interest, as a way to insert the Latin American states into the world market managed by large transnational corporations, or, in other words, to insert their national economies into the *hegemonic globalization* under the US umbrella. Given the 'open' nature of this type of regionalism, the priority was assigned to external markets and, again, to the export of *commodities* and the promotion of the so-called 'comparative advantages' of each country. Unprotected and non-competitive domestic industries gave way to imports from abroad, and the domestic market ceased to have importance it had enjoyed during the ECLAC epoch, with the social consequences that such policies entail: inequality, unemployment and poverty increased while the *leitmotiv* of economic activity again became foreign investment, competitiveness, the spillover of economic production, and *trickle down* as the best way to create employment and

---

addition, the moment has arrived to give entrance to a theorization of the new regionalism based on comparative studies and post-structuralist theories."

[4] As previously seen, globalization does not respond to any malevolent conspiracy of the world capitalist elites; it is an economic phenomenon. However, the fall of the Berlin Wall provoked the absorption of the immense geographic area of the former Communist Bloc – including China, which had already embarked on Deng Xiao Ping's economic reforms – into capitalism, and the increase in the globalization process was accompanied by the rise of new information technologies (internet, cellular telephony) and air and maritime transport facilities.

diminish poverty. The transnational corporations asked for deregulation and flexibility of the labour markets and gave free rein to the 'invisible hand', all of which led to the abandonment of the industrialization policies promoted by the original ECLAC model. Obviously, true integration policies were absent during this entire period.[5]

The neoliberal wave lasted until the end of the century, when the electoral rise to power of a series of left-leaning governments (the PT of Lula in Brazil, the *Frente Amplio* in Uruguay, the Kirchners in Argentina, Evo Morales in Bolivia, Rafael Correa in Ecuador, Hugo Chávez in Venezuela and a coalition of centre-left parties in Chile) began to give rise to significant changes in foreign policy regarding the issue of integration. Consequently, there was a proliferation of initiatives of a political nature, from the ALBA of Venezuela to UNASUR of Brazil through the transformation of the former Rio Group into the Community of Latin American and Caribbean States (CELAC), which laid the foundations for a general reorientation of national policies towards social issues and, in general, exhibited a trend to recover an inward kind of national development instead of the outward orientation characteristic of open regionalism.

This new wave of integration attempts has been called 'post-liberal' by researchers like Sanahuja (2012, 2018), but it has not yielded the expected results in the social and political terrain in part due to the divergence between Brazil and Venezuela and because none of the Latin American countries wanted to share sovereignty or move towards the construction of supranational integration institutions. According to Sanahuja, the region started to face a 'trilemma', opposing the old nationalist ideology of territorial sovereignty with the aspirations to effective regional integration, which would also clash with the search for greater autonomy in the face of US hegemonic power. Therefore, the initiatives of ALBA and UNASUR, which are part of the search for autonomy from Washington, have the following characteristics:

> Both express a 'return of politics' in foreign relations and development policies, with less attention to the trade agenda and economic liberalization that has dominated both policies in recent years, which is no stranger to the coming to power of different governments of the left, their nationalist discourses and the attempts of Brazil and Venezuela to exercise their leadership in the region. They also express the return of the 'development agenda', inscribed in the broader framework of the 'post-consensus of Washington'. This implies policies that try to distance themselves from neoliberalism, open regionalism and their focus on trade liberalization. Both express the 'return of the State' to politics, particularly in external relations and economic and social development. This means a greater role for state actors in the markets in the face of the predominance of private actors characteristic of the model of open regionalism. … The search for greater autonomy vis-à-vis the market, in the field of development policy and against the foreign policy of the United States, is also an explicit goal of this stage of 'post-liberal regionalism'… [It involves] g) Increasing attention to social issues and the reduction of asymmetries in development, linking regional integration with reduction of poverty and inequality, in a political context in which social justice has acquired

---

[5] It was during this period that the North American Free Trade Agreement (NAFTA) was signed by Mexico, the United States and Canada in 1994. It was also in these years that Washington promoted the signing of a free trade treaty with the entire region, but the attempt did not prosper. Instead, Washington opted for the signing of bilateral treaties with different countries, such as Colombia, Peru, Chile, Central America and the Dominican Republic – the so-called DR-CAFTA.

a new and greater weight in the political agenda of the region. h) The search for formulas to promote greater participation of non-state actors and the social legitimization of integration processes (Sanahuja 2012: 32–33).

But this is history. In 2019 the migratory crisis in Central America and Mexico, the socio-political crisis of Venezuela and the election of Bolsonaro in Brazil put a brake on these 'intents of integration' described by Sanahuja. From another perspective, the French expert on integration, Olivier Dabène, argues that in the happier times of this 'fourth wave' there was also an interesting change towards a neo-structuralist paradigm, although the most significant element seems to be that the entire process, in spite of the constitutional integrationist rhetoric of some leftist south American governments, could not prevent itself falling – as usual – into a 'sovereignty trap':

> What this twenty-year history shows is that the Left clearly introduced a paradigm shift regarding regional integration. Its components are defensive, claiming that integration must shield Latin America from imperialist threats, and proactive, resurrecting some elements of ECLAC doctrine and envisioning a deeper institutionalized integration. Three countries inserted this renewed conception of regional integration into their new constitution: Venezuela in 1999; Ecuador in 2008; and Bolivia in 2009. Just to mention some examples, the 1999 Venezuelan constitution includes an article (153) setting the objectives of creating a "community of nations" and "granting supranational organizations, by mean of treaties with the exercise of necessary competences to achieve regional integration". The Ecuadorean one is probably the most "integrationist" of all Latin American constitutions. It also aims to promote regional integration in a wide array of issues (economy, environment, law, cultural identity) and also mentions the possible creation of supranational bodies, and incursions into the idea of deep integration, which the Bolivian constitution mentions as well. These constitutions, as well as the São Paulo Forum reflections, seem to be caught in a double bind. On the one side, there is the undeniable attraction of supranational integration. But on the other, regional integration is defensive, driven by a preoccupation with imperialist threats. The latter spurs an insistence on sovereignty while the former requires ceding or pooling sovereignty. Admittedly, this leftist conception of regional integration entails a collective defence of sovereignty, while at the same time it compels individual countries to cede sovereignty. Yet, I claim that leftist governments in the last fifteen years have not addressed these contradictions properly; they have been caught in a sovereignty trap (Dabène 2012: 18–19).

In other words, this fourth wave of post-neoliberal integration, with its deep social and political content, confronted the opposition of conservative sectors that have already achieved the defenestration of Dilma Rouseff and the election of the far-right candidate Bolsonaro in Brazil, the electoral defeat of Peronism in Argentina, and the election of the neoliberal Macri. To that appalling situation must be added the failed governments of Maduro and Ortega in Venezuela and Nicaragua, so for the time being Latin America is again installed in the 'trap of sovereignty' and the intents of 'integration' – regionalism, in fact – (Malamud 2018) are already over.

However, and regardless of the value judgements that can be made about the results obtained by the intent of the fourth wave, what seems to be clear is the impossibility of solving the three aspects of the 'trilemma' simultaneously, or in other words, a government cannot pursue the defence of sovereignty, face the US and other governments of the region that do not share its political ideology, and at the same time try to increase regionalism with policies in the social and political arenas, even if it is perfectly aware of the debate concerning these issues within the EU in Brussels and accepts that common policies are needed in order to escape from the sovereignty trap.[6]

On the other hand, both the Pacific Alliance of Peru, Colombia, Mexico and Chile – which seems to emphasize free trade arrangements – and the Central American free trade agreement with the United States (the DR CAFTA) are mainly commercial arrangements and not integration processes. Such agreements do not include the free movement of people, in spite of the fact that the main 'export item' of Central America (except Costa Rica and Panama) is currently workers, not merchandise, and that their money remittances are by far the main source of foreign exchange for these countries.[7]

Despite the fact that the orientation of PRI governments was nationalistic at the beginning of the 1990s and in line with the neoliberal period, Mexico took a radical turn towards the Washington *Consensus* and signed the free trade agreement with Canada and the United States (NAFTA), which distanced it from Latin American integration processes. On the instructions of the Trump administration, the agreement suffered a revision that was accepted by the former Mexican government of Enrique Peña Nieto.

It now seems pertinent to offer a brief commentary on the Latin American political mechanisms of coordination that are mainly addressed through multilateral policies under the umbrella of CELAC. Because, in the collective social imaginary, the ideal of integration is always a factor that must be taken into account, it was not appropriate to exclude Mexico, the Caribbean and the other Central American countries from UNASUR. Therefore, during Lula's PT government, Brazil summoned all the Caribbean and Latin American countries, including Mexico, to the First Summit of *Latin America and the Caribbean on Integration and Development* (CALC), which met in Salvador, Bahia, Brazil, on 16 and 17 December 2008. The official purpose of this summit was to promote discussions about regional integration processes and development, especially in the face of challenges such as the financial crisis that was unleashed the same year by Wall Street, energy problems, security, climate change, and other issues. But the main merit of CALC (and of Brazil) was to sow the seed that later led to the establishment of the *Community of Latin American and Caribbean States* (CELAC), which includes the thirty-three sovereign states of the region and is established on the foundation left by the extinct Rio Group.

---

[6] In the academic debate regarding these problems, some scholars maintain that there is no cession of sovereignty even in the supranational institutions of the European Union because what exist in reality are *common policies* "that mean management arrangements of some sovereignty but they are not cession ... an objective analysis of the EU shows a scheme not very different from the flexible and multidimensional regionalism of Southeast Asia" (Nájera 2017). For some analysts, this means that integration has ceased to be a way to advance towards political union or some kind of federalism (like the Swiss-style confederation suggested by Joschka Fisher) and is only a means to strengthen multilateral alliances on strategic matters.

[7] In Guatemala alone, according to data from the Banco de Guatemala in 2016, US $7,159,967 million entered the country with an annual growth of 13.9% compared to 2015. Maquila textiles, which constitute the second category for foreign currency income, generate a little more than a billion dollars.

The *Rio Group* was established in the 1980s as a result of the mediation of the *Contadora Group* (Mexico, Colombia, Venezuela and Panama) in the Central American armed conflicts of the epoch. It was later supported by Brazil, Argentina, Uruguay and Peru, and subsequently became the Rio Group, composed of all Latin American and Caribbean countries. It established a mechanism for dialogues and the coordination of the Latin American countries' foreign policy in the multilateral sphere.[8]

As for CELAC, being essentially an intergovernmental mechanism for dialogue and the political coordination of the thirty-three Caribbean and Latin American states that are "committed to advancing the gradual process of integration of the region" (as officially stated on its webpage), it is expected that it will eventually enable the region to achieve a balance between the necessary integration and the huge diversity of its 600 million inhabitants. CELAC's declaration of intentions also states the following:

> Since its implementation, in December 2011, CELAC has helped to deepen respectful dialogue among all the countries in the region, in areas such as social development, education, nuclear disarmament, family farming, culture, finance, energy and the environment. [Likewise,] CELAC has encouraged Latin America and the Caribbean to envisage itself as a community of nations, capable of dialogue and building consensus on issues of common interest. By mandate of the Heads of State and Government, CELAC is the unified voice of the region on issues of consensus (CELAC website).

CELAC is claimed to be the only interlocutor that can promote and project a concerted voice from Latin America and the Caribbean in the discussion of major global issues, with the aim of seeking a better insertion and projection of the region in the international arena.

Judging by this rhetoric, CELAC could be situated in Dabène's fourth wave or in Sanahuja's post-liberal period.[9] Whether it will be able to overcome the simple

---

[8] The Rio Group arose from this search for reaffirmation of the autonomy of Latin American states against the hegemonic power, which in those years sought to overthrow the Sandinista government of Nicaragua. As for the origins of CELAC, the Unity Summit (2nd CALC summit) held in Cancun in February 2010 approved the Cancun Declaration, whose operative clauses point to the creation of an entity such as CELAC. Its fourth clause states the need to: "Promote an integrated agenda, based on the assets of the Rio Group and the agreements of the CALC, as well as existing integration, cooperation and coordination mechanisms and groups, which together constitute a valuable regional asset that is based on shared principles and values, with the purpose of giving continuity to our mandates through a work programme that promotes effective links, cooperation, economic growth with equity, social justice, and in harmony with nature for sustainable development and the integration of Latin America and the Caribbean as a whole" (CELAC Internet Portal); at: http://www.sela.org/celac/quienes-somos/que-es-la-celac/la-calc-simiente-de-la-celac/.

[9] An attempt was made in the 1970s to establish a regional consultation and coordination body called the *Latin American Economic System* (SELA). It was supposed to be permanent and it was based in Caracas. Its purpose was to promote intraregional cooperation via a permanent system of consultation and coordination for the adoption of common positions and strategies on economic and social issues in international bodies and forums as well as before third countries and regional groupings. It was greatly influenced by the ideas of that time about the possibility of a "new international economic order" (NOEI), but SELA never managed to have a prominent presence in the Latin American scene, and in practice the coordination of policies has been absorbed by CELAC.

condition of 'unified voice of the region' that deals with the coordination of the positions of the thirty-three countries in the major multilateral forums of the United Nations (which is no small achievement and requires enormous tasks of dialogue and communication) and go further in order to work effectively towards the objective of "advancing the gradual process of integration of the region" is a different matter. Nevertheless the declaration itself is a wager in which all the members of CELAC are involved, so the citizens of the region can retain the hope that, in the long term, it could become a reality.

2.3   Central American Regionalization

The Central American regionalism (usually called the "integration process") is one of the oldest in the world because it began in the 1950s with the establishment of the *Organization of Central American States* (ODECA)[10] and constituted the first steps towards the establishment of a common market, as the signing of the treaty of Central American Integration in the 1960s led to the establishment of its technical secretariat in Guatemala (the Secretariat of the Central American Economic Integration Treaty [SIECA by its Spanish acronym]). Regarding the historical origins of this process, after gaining independence from Spain (September 1821) the five provinces of what in colonial times was the General Captaincy of Guatemala included the now independent states of El Salvador, Honduras, Nicaragua and Costa Rica that constituted the 'Federal Republic of Central America', which dissolved in the mid-nineteenth century as a result of the instability generated by the various and innumerable conflicts. However, it always maintained what came to be called the *'unionist'* ideal that even gave rise to some attempts to re-establish the Central American Union by military means, such as the attempt of the Guatemalan *caudillo* Justo Rufino Barrios, who lost his life in combat against Salvadoran troops during the battle of Chalchuapa (1885).

The ups and downs of this particular 'integration process' in the years that preceded the outbreak of armed conflicts in Guatemala, El Salvador and Nicaragua in the 1970s and 1980s (which coincided with the 'first wave' of Latin American regionalism) were highly influenced by the ECLAC vision that promoted the substitution of imports and the strengthening of domestic markets but was violently ended by the guerrilla warfare. The nature of those conflicts has already been widely studied and explained in a quite extensive bibliography available to those who wish to better inform themselves about this period of regional history; see, for instance, books and articles by Aguilera Peralta (1998), Aguilera Peralta/Gabriel, 2016), Torres (1969,

---

[10] However, according to a personal communication from the expert on Central American integration, Rubén Nájera, the programme of economic integration requested by the governments of the ECLAC region in those years preceded the foundation of ODECA by five months, and that explains why ODECA was never defined as an integration project. In Najera's view, ODECA was more an associative model ("a small-scale UN") that not only lacked the objective of integration but was originally conceived as "antidote to federalism".

1998, 2011) and Sanahuja et al. (2007). Concerning the 'integration', the process was halted, and then reinitiated after the signing of the Esquipulas Peace Agreements in 1987. This opened the way for the mediated negotiations that took place under the UN umbrella, and their importance lies in the fact that the peace process led to the signing of the Tegucigalpa Protocol in 1991, which replaced the ODECA Charter. Despite the fact that Central America (like the rest of Latin America) was already under the influence of the neoliberal paradigm, the *Central American Integration System* (SICA) was under the influence of the 'fourth wave' of integration due to the significant fact that, in order to resolve their internal conflicts, the governments of Nicaragua, El Salvador and Guatemala had to resort to the United Nations as a mediator. Therefore, those years had characteristics such that – at least from the theoretical point of view – they perfectly fit within the post-liberal period of integration (the fourth wave).

ECLAC and the first wave of integration influence is also clear in the objectives of SICA that, as stated by the official SICA internet portal, are to achieve the integration of Central America to constitute "a region of peace, freedom, democracy and development" whose purposes are "to consolidate democracy" and "strengthen its institutions on the basis of the existence of governments elected by universal, free and universal suffrage, and of strict respect for Human Rights".[11] Nonetheless, it is curious that, in matters of regional security, SICA says that it seeks to "concretize a new regional security model based on a reasonable balance of forces"[12] at the same time that constant references are being made to the need to strengthen civil society, overcome extreme poverty, promote sustainable development, protect the environment, and establish a regional system of economic and social welfare and justice.

SICA's mention of the need to achieve an economic union in order to strengthen the financial system and promote the region as an economic bloc demonstrates the influence of the UN paradigm of human development as well as the visions of the UN and the EU that have supported the Central American peace process. Of course, these are mainly goodwill declarations that reflect the intentions and ideals of the text drafters, not the social and political regional realities, because idealism and theory itself have shown themselves to be very far from crystallization, or – in other words – theory and practice differ enormously in the Latin American region of the world. To give an example of this rhetoric, SICA's website declares that SICA seeks to:

> reaffirm and consolidate Central America's self-determination in its external relations, through a unique strategy that strengthens and expands the participation of the region, as a whole, in the international arena... [and] to establish concerted actions aimed at preserving

---

[11] SICA's official internet portal; at: https://www.sica.int/sica/sica_breve.aspx.

[12] This reference to a *"reasonable balance of forces"* is curious wording that could be interpreted as an exception to the ideological predominance of the idealist paradigm in the SICA system (because it can only be understood as an expression of the classic realistic doctrine of the balance of power) and thus as a extrapolation of nineteenth-century Westphalian Europe to twentieth-century Central America (in a regional context where Costa Rica lacks armed forces). However, probably it just expresses a lack of knowledge of international relations theory.

the environment through respect and harmony with nature, ensuring a balanced development and rational exploitation of natural resources, with a view to the establishment of a new ecological order in the region.[13]

It is also interesting to underline the fact that the above-mentioned declaration of principles allowed SICA to subscribe to what was called the *Alliance for Sustainable Development* based on the Brundtland report,[14] even though the region was not in a position to facilitate the adoption of either effective environmental practice or coordinated political positions on environmental issues across the whole region.[15] In brief, the SICA is an example of classic Latin American rhetoric, full of magnificent intentions, where the influence of the idealist paradigm of the United Nations is clear and locates the SICA in the *post-liberal period*. Unfortunately, SICA practice has had enormous deficiencies, ranging from its inability to face crises, such as the one provoked by the Honduran *coup d'état* in 2009 against President Zelaya and the brutal repression of Nicaragua's President Ortega against his own citizens in 2018, to its difficulties in articulating both the common market and the customs union in the face of policies from Washington that seek regional insertion – rather than integration – in hegemonic globalization, something reflected in the migratory crisis provoked by thousands of people fleeing to the US to escape from insecurity and poverty in recent years. Thus, despite the rhetoric and the 'goodwill' contained in a number of official statements, the actuality of neoliberalism and the so-called Washington Consensus remain the guiding principles of the process, because, as Gabriel Aguilera remarks:

> [F]ar from seeking the creation of protected economic spaces to promote domestic industrialization and internal markets with the help of the State, the Washington Consensus proposed the opposite: opening national economic spaces to the forces of the international market with the free flow of goods, services and capital, minimizing the role of the State in the economy, privatizing the State's means of production, maintaining maximum macroeconomic stability without considering its social effects with the (erroneous) belief that investments and the creation of jobs would boost development. This meant the transition to the new regionalism (as the SICA) sought to adapt (the region) to the economic order of globalization, simultaneously trying to preserve its own integration. It did so by adopting ECLAC's new paradigm, known as open regionalism, a hybrid formula that sought to preserve the interdependence derived from the preferential trade agreements contained in the integration agreements with the interdependence born of the dynamics of international markets, accepting that in order to achieve development it was required to increase competitiveness. Contrary to past formulas,

---

[13] SICA's official internet portal; at. https://www.sica.int/sica/sica_breve.aspx.

[14] The *Brundtland Report* was prepared by the former Prime Minister of Norway, Gro Harlem Brundtland, (and a team of international experts) at the request of the United Nations. The report criticizes the concept of economic development by proposing the alternative concept of environmental sustainability, thus rethinking economic development policies which tend to insert countries into globalization without care for the high environmental costs of industrialization and consumerism. It was in this report that the term "sustainable development" was used for the first time, defined as the kind of development which meets the needs of the present without compromising the needs of future generations (Brundtland 1987).

[15] At the COP 21 negotiations in Paris the government of Nicaragua refused not only to be part of a common regional association but even to sign the multilateral agreement on climate change.

this time it required the reduction of tariffs and the opening of national markets to international trade (Aguilera Peralta 2017: 99–100).

Therefore, the common external tariff of the Central American common market was abandoned *de facto*, as noted by the Spanish expert Pedro Caldentey quoted by Aguilera in his article, thanks to the negotiation of the free trade agreements with the United States (the so-called DR CAFTA) which impeded the progress towards a customs union.

It is also interesting to note that in the middle of this *de facto* wave of neoliberalism, the sequence of the consolidation of common tariffs thwarted that progression towards the customs union, given the fact that the various free trade agreements signed bilaterally with the SICA countries established different terms and conditions for the tariff reductions. Thus the customs union had to wait for the deadlines of the CAFTA. However, the *Association Agreement* (ADA) signed with the European Union is a curious exception to this wave of neoliberalism introduced by the DR CAFTA. This is because the Europeans have been more permeable to the social aspects of integration – the so-called 'pillars' of cooperation and political dialogue – which they introduced in the ADA treaty. These pillars include, among other things, the commitment made by the countries of the region to advance the integration process, including the obligation to realize the customs union within a period of five years and incorporate Panama into the SIECA. However, at the time of writing, in terms of the customs union between Guatemala, Honduras and El Salvador, only partial advances have been made.

The absence of Nicaragua and Costa Rica from these custom union endeavours could be explained, in the case of Nicaragua, by the short-sighted neo-authoritarianism of Ortega, but in the case of Costa Rica the reasons are linked to the fact that because that country is much more developed and democratic, it has its own views about what is really happening in the Central American integration process. Thus, at an academic conference delivered at the University of San Carlos in Guatemala City, the Costa Rican President Solís (2016) was emphatic in pointing out several of SICA's deficiencies, among which he mentioned the institutional weakness, lack of adequate coordination and deficient operative capacity of both SICA and its specialized entities, the poor effectiveness in driving, building consensus and monitoring agreements, and what he called the structural asymmetries of the Tegucigalpa protocol. In his view the institutional weakness of SICA fundamentally stems from the fact that Central American states have been reluctant to endow SICA with supranational powers, a refusal which prevents it from issuing binding resolutions. This situation in turn transforms the SICA decision-making processes into merely symbolic acts, most of which are only applied voluntarily and therefore asymmetrically (Solís 2016: 185), and the same situation is hindering both the *Central American Parliament* (PARLACEN) and the Central American Court of Justice. The institutional weakness is also the result of the lack of resources that has even led SICA to search for funds from international cooperation. The Secretary-General of SICA has never been able to fulfil its coordinating functions adequately, due to the reluctance of the member states to confer more autonomy on him and to the fact that some former

Secretary-Generals have consciously chosen to maintain a low profile, which, in the opinion of the former Costa Rican President (Solís 2016: 185), ultimately conspires against the required versatility of the position.

As for the poor effectiveness of consensus-building and compliance with presidential agreements, according to Solís this is due to the absence of a monitoring mechanism, a fact that highlights the irrelevance of the resolutions and questions the nature of decision-making itself. The former President of Costa Rica argues that this situation would require the reform of a consensus rule that in practice is used as a veto, as well as the procedure of *pro tempore* presidencies, because often the country temporally in charge subordinates the regional agenda to national priorities through lack of time or leadership.

As far as the structural asymmetries between the countries of the isthmus are concerned, Solís recommends solving the problem by drawing inspiration from the European example, recalling that, through compensation funds, the German-French axis allowed backward economies to be incorporated into the European integration process, facilitating the entry of countries such as Spain, Portugal and Greece to the best practices and economic standards of the relatively more developed partners. Finally, he refers to what he calls the need to reform the Tegucigalpa Protocol through a renegotiation to update it with the incorporation of the changes derived from the new paradigm supposedly adopted in the region since the inception of the sustainable development alliance at the end of last century, rethinking issues such as democracy, development and even the kind of integration that is sought and that has been frustrated by the adoption of the neoliberal *Washington Consensus* agenda. In addition, the emergence of new regional groupings – such as the Pacific Alliance or CELAC – would require Central America to reassess where it is located in the geopolitics of the second half of the twenty-first century, and to redefine the legal framework of integration and the opportunities for civil society to participate in the process.

The lecture of Luis Guillermo Solís is important because he tackled issues such as supra-nationality, the *de facto* presidential right to veto, the lack of resources and European-style compensation funds, the need to reform and update the Tegucigalpa Protocol, and a long list that demonstrates the urgent need for an energetic approach to the numerous pending tasks of the process that are quite similar to the reports on the *State of the Region* (ERCA) that, financed by the *Central American Bank of Integration* (BCIE), are periodically made in Costa Rica.[16]

---

[16] The various reports on the state of the Central American region (ERCA) have been financed by the BCIE and numerous experts from all countries participated in the creation of them, but the result is not "a photograph of reality, but a selective documentation of processes which specify and detail what a number of social, economic, political and institutional actors did in the recent past, and the footprints they left in the development of the Isthmus when seeking to re-establish [the appropriate way to deal with] the challenges and opportunities that transcend territorial borders between countries and to lay the foundations for processes of social and political dialogue to promote sustainable human development in Central America," says the BCIE website; at https://www.bcie.org/novedades/eventos/evento/quinto-infome-estado-de-la-region-2016/.

2.4   The European Integration Process

Both the transformation and the democratization of the international system need the contribution of regional integration processes to be taken into account when heading in the direction of an emerging multipolar world whose governance is assured thanks to the multilateral negotiations that are permanently held in the multiple international organizations of the United Nations system or in the framework of regional organizations like the OAS and regional coordination mechanisms like CELAC – in other words, due to the strengthening of multilateralism. Hence the importance of expressing views on a regional integration process like the European one. On the following pages I will therefore comment on what, from my point of view, led to phenomena like the financial crisis and the rise in unemployment and social discontent, including the surge of a neo-nationalist movement that opposes integration and the regrettable British decision to leave the EU.

The EU is the result of the most successful regional integration process of the world with regard to the economic, social, environmental and political dimensions. All these achievements are due to the establishment of supranational political institutions like the European Commission, the judiciary bodies, the European Parliament, the monetary union (the Central Bank and the Euro), and the free movement of goods (the Common Market and the Customs Union) and people (the Schengen agreement) within the Union, to briefly mention some of its remarkable accomplishments. However, the fundamental and most important achievement – which must never be taken for granted – is peace and the replacement of the Westphalian Order amongst its members with a system whereby international security is based on sharing sovereignty based on the legislation enacted by the European Parliament, international law and cooperation, including respect for the principle of collective security as understood in the UN Charter and doctrine.

Hence, rather than a description of the principles, institutions and specific achievements of the EU, what follows is an essay which comments – from a Latin American perspective – on the reasons why this process has now entered a critical situation that must be overcome in order to maintain not just sustainable development – which is crucial to preserve the good health of the planet in this epoch of the Anthropocene – but the general global cosmopolitan trends and processes in the field international relations.

Therefore the establishment and success of the new holistic and cosmopolitan paradigm as a great ideal looming on the horizon, as well as the attainment of *ecological peace* in the framework of the UN SDGs and the commitments of both the 2030 Agenda and the Paris Climate Conference (COP 21), are really the minimum that governments must do to save the planet from the sixth extinction, but they are dependent to a large degree on the consolidation of the European integration process and on the EU as a global actor in the field of IR. In brief, if the EU expects to continue playing the important role it has so far performed in global affairs, it is absolutely indispensable to overcome its current crisis provoked by nationalism, the coronavirus and *Brexit*.

Furthermore, as pointed out when discussing the European integration process, it is important to keep in mind that, beyond sustainable development and socio-political integration, its main success has been the attainment of *sustainable peace* in a region dogged by centuries of constant war and in which two world wars originated. This really is a tremendous achievement. The fact that integration has truly appeased powers such as France, Germany and Great Britain for more than half a century with such solid foundations that European citizens have internalized a peaceful mentality – in other words, the fact that the vast majority of people in contemporary Europe are convinced that the possibility of renewed Franco-German, Franco-British or German-British conflagrations is simply unimaginable, unthinkable, unconceivable – is a great achievement, something really formidable and of intrinsic value. But neither the world nor European citizens must take this good result as open-ended or for granted.

Thus concerning the issue of war and peace, the establishment of the European Union signifies that the best guarantee for peace – understood in both the negative form of peace as simply the absence of war and the positive sense of diminishing the structural violence caused by the absence of development (Galtung 1987) – among its member states is a reality and this *mindshift* (Göpel 2016) is an outstanding achievement and clearly shows that something of fundamental importance –sustainable peace – is perfectly possible. This change of mind or mentality among the majority of citizens is not only due to education in the framework of what UNESCO calls a 'culture of peace', but is also the result of a true *social construction* (Wendt 2005) concerning a new cultural type of *hegemony* (Gramsci 1932) deep-rooted in people's mentality, which allows European nations to be convinced that cooperation and respect for international law is crucial to guarantee that the catastrophic wars that devastated the continent in the past will never be repeated. Hence it is possible to affirm with pertinence that this new mentality (Göpel 2016) and cultural hegemony are the best assurance of lasting peace between the ancient arch-enemies on both sides of the English Channel, or – more important for the EU in this *Brexit* era – on both sides of the Rhine, Germany and France must act in partnership towards the preservation of a sustainable and ecological peace as one of their main responsibilities as founding fathers of the Union.[17]

---

[17] The German-French axis is the pivot of the EU because the whole integration process started in the 1950s due to the clairvoyance of statesmen like the French statesman Robert Schumann and the German Chancellor Konrad Adenauer. The process started on the basis of solid economic foundations of the Franco-German (and Benelux) community of coal and steel (CECA) on the old battlefields of the Rhine region, and that is why peace is the best achievement of the European Union. However, the current European crisis merits quoting *in extenso* an analyst who does not share my optimistic view about this fundamental issue: "The old flashpoints of Europe, the Rhine Valley, the English Channel and the rest, remain generally quiet. Franco-German tension is growing but it is far from reaching a boiling point. But underneath the surface, the engine of conflict – a romantic nationalism that challenges the legitimacy of transferring authority to multinational institutions and resurrects old national conflicts – is stirring. The right-wing parties are just the tip of the iceberg, although they must not be dismissed in themselves, but beneath the surface, the generalized unease with the consequences of transfers of sovereignty in economic matters is intensifying. For the moment the flashpoints are on the frontier of the European Union, but that

Peace is also based on social integration because sustainable development (the social market economy, as the Germans call it) has been facilitating the improvement of living standards in such a way that the *welfare state* gives sufficient credit to the idea that sustainable development will continue and violent conflict among the members of the Union is an implausible hypothesis. Thus, by extension, this supports the consolidation of a peace that, in this case, is not based on the – always fragile – *balance of power*. It finds itself based on this mindshift concerning war and peace that prevails among the majority of the European citizenship but also in the new institutions established as a result of the integration process as well as in the new regional integrationist international law institutionally embedded in Strasbourg, Brussels, Luxembourg and Frankfurt. This is why I am optimistic and profoundly convinced of the importance of the fact that the idealist paradigm is at the very core of the existence of the Union, which in turn has brought into crisis the old Westphalian order, whose main point of support is both state sovereignty and the balance of power.

Union itself is crumbling. There are four European Unions. There are the German states (Germany and Austria), the rest of northern Europe, the Mediterranean states, and the states in the borderlands by Russia. The Mediterranean Europeans face massive unemployment, in some cases greater than the unemployment experienced by Americans in the Great Depression. The northern European states are doing better but none are doing as well as the Germans. The dramatic differences in the conditions and concerns of the different parts of the European Union represent the lines along which it is fragmenting. Each region experiences reality in a different way, and the differences are irreconcilable. Indeed, it is difficult to imagine how they may be reconciled. There are four Europes, and these four are fragmenting further, back to the nation-states that composed them, and back into the history they wanted to transcend. In the end, the problem of Europe is the same problem that haunted its greatest moments, the Enlightenment. It is the Faustian spirit, the desire to possess everything even at the cost of their souls. Today their desire is to possess everything at no cost. They want permanent peace and prosperity. They want to retain their national sovereignty, but they do not want these sovereign states to fully exercise their sovereignty. They want to be one people, but they do not want to share each other's faith. They want to speak their own language, but they don't believe that this will lead to complete mutual understanding. They want to triumph, but they don't want to risk. They want to be completely secure, but they don't wish to defend themselves. But there is another Europe, as there has always been – the landlocked mainland that is never quite defeated and never quite secure. The story of modern Europe began in 1991, when the Soviet Union died and the European Union was born. In 2014 Russia reemerged as the flashpoint between it and the European Union came alive, and history began again. It is striking how short-lived were Europe's fantasies about what was possible. It is also striking that the return of Europe's more dangerous flashpoints occurred in 2014, one hundred years after the First World War began, one hundred years since Europe began its descent into hell. It has emerged from that hell. But where Faust was willing to sell his soul for perfect knowledge, modern Europe wants perfection without paying a price. There is always a price, and nothing is more dangerous that not knowing what the price is, except perhaps not wanting to know...Europe is no longer the centre of the world, but a subordinate part of the international system...But the idea that Europe has moved beyond using armed conflict to settle its issues is a fantasy. It was not true in previous generations and it will remain untrue in the future. We already see the Russian Bear rising to reclaim at least some of its place in the world. And we see Germany struggling between its own national interests and those of the EU in a world where the two are no longer one. The Europeans are still human....They will have to choose between war and peace (Friedman 2015: 257–258).

The American author George Friedman (2015) is right when he realizes that these changes could have not happened without provoking the reaction of the neo-nationalist conservatives of the extreme right – including those certain undemocratic and relatively underdeveloped regions (the 'flashpoints') situated on the borders of the old Communist Bloc – because these conservatives regard the EU institutions as an attack on national sovereignty, claiming that sovereignty and the principle of territoriality must be 'defended'. In my view, this is a *badly understood* idea of sovereignty because it is based on the anachronistic definition of the concept according to the interpretation of the strictly territorial nineteenth-century Westphalian parameters that are not appropriate in the twenty-first-century globalization, interdependence and regional integration realities that are diminishing the importance of geopolitics and enhancing geoeconomy. Thus, the electoral increase of the extreme right in almost all countries of the EU is partly explained by this kind of undemocratic ideology typical of neo-nationalist reactions, although socio-economic factors are also important, as will be seen below.

Indeed, like any process, integration is essentially dynamic – not static – diachronic and non-synchronic. It is also clear that legal-political structures are not sufficient to legitimize the functioning of an integration process such as that of the European Union, since this requires citizen consensus, which in turn depends on the way people do or do not perceive the economic and social benefits of belonging to it. And these benefits have been reduced as a result of two issues that have affected the smooth functioning of the EU in recent years: firstly, the reform of social policies in order to increase the competitiveness of European industry, a policy that originated in Germany at the beginning of the first decade of this Century under the Social Democratic government of the SPD party; and secondly, the repercussions of Wall Street's 2008 financial crisis in Europe.

The reform of social policies – in order to decrease the benefits for the working classes – initiated by the German government in the Gerhard Schröder era under the rule of the Social Democratic party (SPD) continued afterwards under the conservative rule of the CDU and Angela Merkel. With regard to social issues, this 'austerity' policy remains to this day, with predictably negative consequences for the EU. Furthermore, according to the Spanish scholar Navarro (2015), to these reforms should be attributed not only the electoral defeat of Schröder and his party against the CDU but also the collapse that the rest of European social democracy has suffered since. The latter is because, due to Germany's great influence over the EU, these policies have been applied everywhere, with results that explain the decrease in electoral support for socialist parties in Spain, France, Italy and other countries. For Navarro, the policies of austerity and the restriction of social benefits have been applied by most European governments due to the influence of Berlin, owing to which Germany stands out as responsible for the European social crisis, since the socialist claim to represent the interests of the working classes is not seen as genuine. In particular, the legitimation of Social Democratic parties in countries like Germany no longer seems valid.

According to Navarro, the evolution of the electoral behaviour of the lower and middle classes as well as the working classes in EU countries demonstrates that the socialist and centre-left parties lost their electoral support and were expelled from governments because they applied neoliberal public policies that affected the welfare of these popular classes on the erroneous basis that these measures were indispensable for the economic recovery of the EU from the point of view of exports and avoiding recession, as the German example showed. During the period 2002–2005 (under the rule of the SPD in coalition with the Greens) Germany reduced social protection, lowered wages (or prevented them rising despite the increase in productivity), decreased unemployment and health benefits, lowered pensions, and introduced neoliberal 'market flexibility', easing dismissal procedures and so on, with good results for the German private sector because, as a result of these policies, unemployment, which had reached 11.3% of the active population in 2005, fell, and exports rose in such a way that the trade balance went from being in a deficit of 1.7% of GDP to a surplus of 7.4% of GDP. These are the figures that support the 'triumph' of those policies. Unfortunately for the socialist and centre-left parties, the European working and middle classes did not agree with that point of view because, as Navarro indicates:

> what is not said when this supposed success is presented is 1) that such success was created based on the great sacrifice of the labour force: salaries remained stagnant, well below what they deserved for productivity growth; 2) that a third of the labour market was affected by low wages and conditions of great precariousness; 3) that optimized exports, which today represent no less than 52% of GDP, occurred at the expense of domestic demand, with a reduction in that demand caused by a decline in the purchasing power of the working population; 4) a reduction in the imports of the countries of the Eurozone and a stop to the stimulus and economic growth in other countries of the Eurozone, and very particularly in the peripheral countries such as Greece, Portugal, Spain and Ireland; 5) that they were forced to reduce their own salaries in order to compete with Germany. This led to a downward trend in wages in Europe in order to increase competitiveness, causing economic stagnation in these peripheral countries and throughout Europe. The evidence that supports each of these points is enormous (Navarro 2015).[18]

In other words, although on a domestic economic level the structural reform of the SPD (the *Agenda 2010*) at the beginning of the century in Germany enabled Germany's manufacturing industry to regain competitiveness, build up surpluses and reduce unemployment, on a social level it caused the stagnation of wages which reduced domestic demand and in turn resulted in the discontent of the workers and the loss of political power in favour of the CDU, which – paradoxically – has governed for more than a decade maintaining the same policies initiated by the *Social Democratic Party* (SPD). The negative consequences have also been at European level, which reduced exports from the Eurozone, hurting the rest of the member states, especially the peripheral countries, with enormous social costs and the reduction or stagnation of wages.

An even worse consequence was the reproduction of these austerity policies by other countries, like France, without their application being translated into the good results that the industrial exporters obtained in Germany. This provoked the social

---

[18] Vincenç Navarro is Professor of Public Policies at Pompeu Fabra University (Barcelona) and a professor at John Hopkins University in the USA.

crisis – the rejection of austerity policies – as well as the political crisis, with rise of the extreme right anti-European and nationalist movements, like the National Front in France, the party of Geert Wilders in the Netherlands and even the AfD in Germany, to mention just a few continental examples in major countries, although the reasons that account for the outcome of Britain's *Brexit* referendum in Britain are not very different.[19]

Unfortunately, social benefits have not only been restricted as a result of austerity policies but also because of the financial crisis. It is well known that this crisis almost forced Greece to leave the Eurozone, but although I will examine this problem further in the chapter on sustainable development, it is sufficient to point out for the moment that the crisis has been transferred to the field of politics in such a serious way that it is endangering the very existence of the Union due to the rise of neo-nationalist movements and parties which have already caused *Brexit* and the appearance of separatist movements in countries like in Spain. This is why, in the debate about possible outcomes, there are proposals ranging from the *Movement of Democracy in Europe* (DiEM) by Ianis Varoufakis to the divergent positions of Joschka Fisher and Oskar Lafontaine in Germany and of Jean Luc Melenchon and the current French president, Emmanuel Macron, who, fortunately, is in favour of the EU and of strengthening the integration process.

Although both are reformist, in the German case there is a difference in the positions of Fischer and Lafontaine.[20] For Fischer, the reform of the political structures of the EU must reinforce the legitimacy of the Union by establishing a federal or confederal government (given the social rejection of the technocratic style of governance practised by Brussels). In his view, this is more important than monetary reform, while Lafontaine argues that the Euro is one of the causes of the crisis and that it will be convenient to return to the old European monetary system as well as to a more decentralized style of government. Nevertheless, both socialist leaders agree that austerity policies should be abandoned and that Germany should adopt a policy of solidarity with countries that are suffering from the crisis, even if this means Germany sharing the costs of debt reduction.

---

[19] Speaking about *Brexit*, the former German vice-chancellor Joshka Fisher said in an interview that for the EU it would be a storm but for the UK leaving the EU was going to be a hurricane "with a grim outlook for Britain". For Fisher the vote at the referendum in 2016 was "all about emotions and kicking the ruling government in the ass". He added that, as emotions are always unpredictable, it is important to be aware that referendums are always a risk for democracy: "What we see now is that one of the most dynamic economies, a flourishing economy, one of the biggest economies in the world was crashed against the wall for no real reason. And the guys who are responsible of this misery have disappeared in a miraculous way. It is a political drama. To accept that Britain is not any longer a member of the EU is very hard"; at: https://www.euractiv.com/section/euro-finance/interview/joschka-fischer-stabilise-the-eurozone-to-defuse-hurricane-brexit/.

[20] Both were ministers of the Social Democratic cabinet of Gerhard Schröder (1998–2005). Oskar Lafontaine was Finance Minister and resigned to show his disagreement with the measures of neoliberal structural reform mentioned in Navarro's analysis, founding his own party *Die Linke* (The Left). Fischer was the federal Vice-Chancellor and in that capacity held the post of Foreign Affairs Minister in the coalition government.

Although this issue will be addressed again in the chapter on sustainable development, where I also deal with the problems of developing countries without taking a position on divergent approaches (such as those of Fischer and Lafontaine or those that exist in the French left and in other EU countries), everything seems to indicate that the exit from the social and political crisis requires, as a *sine qua non* condition, the abandonment of the neoliberal economic model – including the so-called 'austerity' social policies – and involves the deepening of democracy as a necessity (especially in these awful years of coronavirus) in order to recuperate the democratic social policies that are on standby due to the impossibility of solving Rodrik's 'trilemma'[21] (developing the integration process required by globalization, retaining the sovereignty of each member of the Union and the need to reintroduce the welfare democratic social policies), as Sanahuja corroborates.[22] Furthermore, it is likely that a transnational social movement at European level – like the one that Varoufakis founded – will arise, with militants, independent of their nationality or their partisan ascriptions or ideologies, which is good for the development of transnational cosmopolitan citizenship, as will be explained in the final chapter.

In line with that, I am convinced that what is truly required now to give renewed impetus to the European process of integration (much more than the deepening of democracy at national level) is a social transnational movement of cosmopolitan

---

[21] According to the economist and Harvard professor Rodrik (2014), the "inescapable trilemma" of globalization resides in the fact that democracy, national sovereignty and regional integration are incompatible. You can combine any two of those elements but not have all three simultaneously. Nation states in the age of globalization face a stark choice. They could pool their sovereignty in an integration process like the one of the EU, tailoring their politics to the needs of the regional market and the supranational institutions of Brussels, Strasbourg and Frankfurt. However, precisely that "pooling of sovereignty" of the EU members has been the source of the political reaction of the nationalist and conservative extreme right, for whom electoral support is increasing. However, in my view, deepening democracy through citizens' participation in the European Parliament, for instance, in favour of a return to the welfare state could give the leverage needed to national government and that could be the way out of that "inescapable trilemma". Of course, if neighbouring countries with a shared cultural and historical inheritance like those of the EU struggle to integrate, what hope is there for integration processes on a global scale? Another option, according to Rodrik, is that nation states could retain their sovereignty but make the pursuit of regional integration their over-riding policy objective, to the exclusion of other domestic goals. But that seems incompatible with the practice of democratic politics, which is all about making trade-offs between competing aims or goods. In my view, the remaining course of action would not be to sacrifice some measure of regional integration in the interests of sovereignty and democracy (as Rodrik suggest), but on the contrary, to forsake the neoliberal model and deepen democracy through social policies that could provide support to governments genuinely committed to regional integration.

[22] Sanahuja explains the Rodrik 'trilemma' as follows: "In a well known approach Rodrik (2014) says that in the context of globalization State actors face a 'trilemma' between deep economic globalization, nation states and democratic policies which it is not possible to solve because these three goals cannot be satisfied simultaneously and therefore the only way out is to combine just two of them. The financial crisis has shown that in conditions of deep globalization and with nation states as centres of political power its agency is weakened and governance only seems possible if they accept the demands of global markets, postponing electoral mandates related to social rights that are part of advanced democracies" (Sanahuja 2018: 42).

citizens within the EU[23] to push for the indispensable economic reforms needed, because the EU crisis is essentially a social crisis with roots in the dismantling of the health system due to privatization policies, as the worsening of the coronavirus infection in Italy and Spain clearly shows. Hence, given the widespread perception of citizens that the reduction in the *welfare state* is a consequence of the austerity policies decided in Brussels by non-elected technocrats without a corresponding mandate and who do not really care about human needs or social demands, it is crucial for European leaders to realize the necessity of forsaking 'austerity' policies and reinstating policies that prioritize human needs.

Indeed, although integration has facilitated and promoted social development as well as the reduction in poverty and inequality both between and within countries (thanks to social cohesion programmes and other similar mechanisms), a majority of European citizens have expressed dissatisfaction with Brussel's unelected technocracy that takes decisions that have had negative social repercussions. This disaffection has been manifested in various ways, such as the negative vote of past referendums in different countries[24] and the *gilets jaunes* protest in France, to mention two obvious examples.

EU member states of the Mediterranean region (Greece, Italy, Spain) and even Portugal and Ireland, as well as countries from the former Eastern Europe Communist Bloc which have suffered from high rates of unemployment and other social problems resent the policies of austerity imposed to pay a debt mainly originated in the purchase of industrial exports from a Germany whose economy is in good shape, thanks to the same exports within the EU's common market. Therefore, austerity policies are now perceived as part of the problem and not the solution.

European citizens consider such policies illegitimate because they do not come from national parliaments and are somehow seen as Berlin's responsibility because they were drawn up under the influence of the European Central Bank, whose headquarters, coincidentally, are situated in Frankfurt. Obviously, the Bank is not enthusiastic about the idea of debt reduction for debtor countries. Additionally, the European Union's promise of peace has faded due to terrorist attacks, and the situation has been

---

[23] For instance, the already mentioned movement called DiEM (Democracy in Europe Movement) organized by Yanis Varoufakis, the former Greek Minister of Finance.

[24] The complete series of referenda with their respective percentages (on the number of participants) is as follows: in 1992, 50.7% of the Danish people voted against the Maastricht Treaty (although the consultation was reopened to obtain their approval). In 2001, 53.9% of the Irish voted against the Treaty of Nice. In 2005, 55% of the French and 61% of the Dutch rejected the European Constitutional Treaty. They were not invited to vote again (it was too risky), but in a new consultation – which eliminated the most controversial aspects of the Constitutional Treaty – the Lisbon Treaty was approved. In 2015, 61.3% of Greek citizens voted against the austerity policies promoted by Germany, although without effect because Prime Minister Alexis Tsipras and his party had to bow to the demands of Brussels. In 2016, 61.1% of the Dutch rejected the EU Association Agreement with Ukraine, which effectively prevented the opportunity for the Ukrainian government to continue insisting on what was, *de facto,* a *casus belli* with Moscow. In 2016, in the Brexit referendum, 51.9% of Britons voted to leave the EU, precipitating the resignation of Prime Minister David Cameron, and that same year 59.4% of Italians rejected a constitutional reform that touched on issues related to the EU, causing the resignation of Prime Minister Matteo Renzi.

aggravated by fear of the wave of refugees fleeing from civil wars, climate change, terrorism and poverty in the Middle East, sub-Saharan Africa and other countries. The whole continent (and, indeed, the world) is suffering from the effects of the COVID-19 crisis, which has reminded humanity of the fragility of our species when facing the possibility of extinction due to exceeding planetary limits, as will be seen in the next chapter.

Therefore, it is unfortunate that a large number of European citizens have been following the banners of the far right, retreating to the nationalism of the past or maintaining an ostrich policy against the resurgence of xenophobia and racism, instead of supporting social democratic parties and appropriate policies to mitigate the growing social risks demonstrated by the current health crisis. In this regard, it is interesting to recall a premonitory article published by the Spanish newspaper *El País* seven years ago. It was written by the former German Minister for Foreign Affairs, Joschka Fischer, who warned:

> In the past, Europe has had a political order based on competition, distrust, conflict of powers and, ultimately, war between sovereign states. That order fell on 8 May 1945, and in its place emerged another system based on mutual trust, solidarity, the rule of law and the search for negotiated solutions. But now that the crisis is undermining the foundations of this order, trust turns into distrust, solidarity succumbs to old prejudices (and even to new hatreds between the poor south and the rich north) and negotiated solutions give way to external imposition. And once again, Germany plays a fundamental role in this process of disintegration. This is because, in order to solve the Eurozone crisis, Germany (which is by far the strongest economy in the EU) imposed the same strategy that worked for it at the beginning of the millennium, but under completely different internal and external economic conditions. For the countries of southern Europe hit by the crisis, the formula that Germany defends, with its mixture of austerity and structural reforms, is turning out to be deadly because they lack two other fundamental components: debt reduction and growth. Sooner or later, some of the major European countries in crisis will choose political leaders who do not accept the imposition of austerity measures from outside any longer. Even now, at election time, national governments promise more or less openly to protect their citizens from Europe because Germany has taken care that the main ingredients of the recipe to resolve the crisis are austerity and structural reform. The effects of the thesis that southern Europe had to be treated with 'severity' for its own sake… are in sight. Very severe was the treatment that has caused a quick economic contraction, massive unemployment (over 50% among young people) and continuous deterioration of the fiscal situation due to the increase in the cost of interest on the debt (Fischer 2013).

To the disastrous socio-economic panorama described above can be added the almost-collapse of the Franco-German axis in the conduct of the EU's general policies, the economic and social upheavals that Brexit will cause, the political crisis in Spain as result of Catalonia's separatism,[25] the difficulties in the relationship between the EU and the USA (though hopefully this will be mitigated under the Biden administration), and, of course, the economic and social crisis that will result from the above-mentioned pandemic in 2020. To that sombre landscape it is important to add that in the world's geopolitical scene the Russian presence in the Middle

---

[25] Indeed, this separatist movement could also result in a sort of *'Catalexit'* from the European Union, because Madrid could hardly accept an independent Catalonia as an independent member of the Union.

East has been strengthened and the Syrian regime seems consolidated, thanks to the support of Russia and Iran. At the same time, the number of refugees fleeing from civil war is increasing and will not cease in the foreseeable future.[26]

Furthermore, recent analyses of the European problems suggest that there is a direct relationship between the general incompatibility of market logic and the logic of social levelling or the welfare state (Poch 2017), as the essential contradiction of the integration process at present lies in the fact that both democracy and sovereignty are internal issues, but due to the treaties signed within the framework of the EU, important political decisions are made in Brussels, Strasbourg or Frankfurt. The example of the recent Greek economic crisis is dramatic because the Euro prevents adjustments and devaluations that could have been effected by the Hellenic government if they had kept their own currency, and it is precisely this situation that prevents Athens from lowering the costs of its tourist destinations, which is one of its main sources of income. Therefore, what is needed is a new set of social and solidarity policies supported by enlightened transnational and cosmopolitan citizens aware that the benefits of the process greatly exceed the costs.

This long list of the vicissitudes of the European Union is associated with the general problem concerning the transformation of the international system to make it more democratic and less subject to the influence of the traditional hegemonic powers. From my point of view, regional actors are crucial to promote such democratization, and it is regrettable that the EU – the only regional actor able to become one of those centres of power and global governance, given its important and exemplary integration process – is now confronting such difficulties and obstacles.

In order to get out of the impasse, the EU will have to find the means to recover its loss of legitimacy in its social bases, i.e. within the European citizenship. To achieve this it is necessary for Brussels' institutions – the Commission but also the Council and the Parliament – to abandon neoliberal policies. The European institutions must work together towards the establishment of a system like the confederal Swiss model – to mention a successful political example on a small scale – which would imply a

---

[26] This is the balance of the wars that hit the entire region: in Afghanistan after fifteen years of war and 230,000 dead the Taliban have not been defeated, insecurity prevails and democracy is absent. It was n this country that Al Qaeda was born, which gave rise to American intervention (via NATO) after 9/11/2001. Despite the capture (in Pakistan) and extrajudicial execution of Bin Laden, the organization still exists. In Iraq after thirteen years of the illegal war started by the Administration of George W. Bush and more than a million dead, the country is divided between Kurds in the north, Sunnis in the centre – many of whom became terrorists – and the Shia who rule in Baghdad and in the south of the country. where they are the majority. The conditions of life and the situation of violence and insecurity are, of course, much worse than during Saddam Hussein's dictatorship. As for the countries that experienced the Arab Spring, Libya has been submerged in chaos for over nine years, with more than 40,000 dead and living conditions much worse than those under Gaddafi, with the additional aggravation of the chaos resulting from the destabilization of sub-Saharan Africa by the Libyan crisis. Egypt returned to dictatorship and Yemen continues to be submerged in civil war, while in Syria after eight years of war and more than a quarter of a million dead, needless to say the general situation of the country is a thousand times worse than before the rebellion that pretended to overthrow Bashar El Assad, a rebellion that could have not prospered without the support of the US, Turkey, Saudi Arabia and some other Western powers. The only case in which there has been a relatively successful democratization result of the Arab Spring seems to be that of Tunisia.

cession of national sovereignty by each member state to a much greater degree than has occurred so far, because what is really at stake is the overcoming of the old and outdated Westphalian system that is the result of the clash of geo-economics with geopolitics or, in other words, of globalization with the principle of territoriality and the outdated nineteenth-century conception of sovereignty. In my view, that would be the best way to solve Rodrik's infamous 'trilemma'.

On the other hand, it is necessary to point out that Fischer's globalization approaches coincide with those of Braudel, Wallerstein and Sousa Santos, which will be discussed later. Of course, globalization is a phenomenon whose origins can be traced back to the sixteenth century and therefore cannot be regarded as the result of policies designed by the elites of world capitalism because it consists of complex interdependent economic relations resulting from factors like the acceleration and facilitation of world trade, human mobility, transportation (including maritime, air, and land) and is nowadays achieved in a much more efficient manner, thanks to the impressive development of new information and communication technologies like the internet and smartphones that seek the inclusion of all countries in a globalized market and a global civil society.

This means that the distances and the territory (the space geographically delimited by national borders) are less and less important for large transnational companies whose economic operations take place in the world market. Nevertheless, as nation states operate under the rules of the Westphalian system, one of the central concerns of governments continues to be maintaining security within their territories and borders as well as preserving their independence and sovereignty. Therefore geopolitical questions retain their importance, as will be seen in later sections, but the struggle to be independent of space (territory) is a result of globalization, commerce, human mobility, but also – and this is a very important factor – the economic performance of transnational corporations at world level. These corporations have struggled since the last quarter of the twentieth century, as their decision centre and the financial calculations that consequently and inexorably underlie such decisions are released from the territorial limitations imposed by the locality and the national territories a phenomenon called the *war for space* by some analysts of globalization (Bauman 2017).[27]

---

[27] Bauman cites a promoter of globalized capitalism in whose opinion a corporation belongs to "people who invest in it" and not to their employees, suppliers or the town where it is located. It is important to realize that the term 'belonging' is used not in a *legal* sense but in the sense of *space*. The underlying meaning is that "...employees come from the local population that are held back by family duties, home ownership and other related factors and therefore can hardly follow the company when it moves elsewhere. Suppliers must deliver their merchandise, and the low cost of transportation gives locals an advantage that disappears as soon as the company moves. As for the locality, it is evident that it will stay where it is, it will hardly follow the company to its new address. Among all the people called to have a voice in business management, only the 'people who invest' – the shareholders – are not at all subject to space; they can buy shares at any stock exchange and from any broker, and the proximity or geographic distance of the company will probably be the least of their considerations when making the decision to buy or sell. In principle, there is no spatial determination in the dispersion of the shareholders; they are the only factor authentically free of it. The company 'belongs' to them and only to them. Therefore, they are responsible for moving it

Therefore, governments must cooperate in order to establish principles and norms to ensure the functioning of such corporations according to transnational legislation that takes into account the interests of those who have been excluded by the constant spatial relocation of corporations, especially in these days of climate change and pandemics where the interests of terrestrial ecosystems and world population must prevail. The market must be regulated according to the interests of humanity as a whole, global public goods (such as tropical rainforests, boreal forests, glaciers and arctic ice fields, oceans, fresh water, the oxygen we breathe, a pollution-free atmosphere, etc.), and the environment on a planetary scale, which alludes to Earth or the *Pachamama*, as Andean indigenous peoples call the planet.

The foregoing, however, implies a capacity for action by the governments of each nation state that, given the economic power of transnational corporations, is only possible if there is leverage coming from global civil society, or, in other words, if a national government wants to establish regulations on transnational corporations this standard must be put in place as a matter of global public policy. Hence the fundamental importance of the great multilateral agreements such as those of the COP 21 or the SDGs, as well as the importance of mobilizing transnational civil society organizations regarded as an expression of cosmopolitan citizenry in the framework of counter-hegemonic globalization (Santos 2010) but also as an obligation imposed by the interests of *Gaia*, the living planet that we inhabit and that otherwise can take revenge on humankind, as Lovelock (2007) said in one of his books.

2.5 The Geopolitical Issue

Precisely because regionalization, regionalism, integration processes and multilateralism are crucial processes at the current time, the importance of globalization can ultimately be understood as a phenomenon of 'de-territorialisation' of the economic sphere that has led to a reformulation of the concepts of sovereignty and territorial integrity as they were conceived during the nineteenth century (an epoch when the international order was understood exclusively according to Westphalian principles). Despite the fact that geopolitics continue to determine foreign policy since sovereigns must respect the interests of humanity and the interests of the planet, regional integration (like the EU) is one of the more important ways to subdue national governments when they oppose superior principles and objectives.[28] This similarly applies to the sphere of human rights; sovereign states must respect them because they are

---

where they discover or anticipate the possibility of improving dividends, leaving the others – who are tied to the locality – the tasks of licking wounds, repairing damage and dealing with waste. The company is free to move; the consequences cannot – they remain in place. Whoever has the freedom to escape from the locality has it in order to escape the consequences. This is the most important booty of the victorious *war for space*" (Bauman 2017: 13–15).

[28] Just imagine what the policy of the former Soviet Bloc Communist states might have been without the moderate 'overpower' of the EU regarding democracy and human rights. Victor Orban, the authoritarian leader of Hungary, can give us an idea.

obliged to by international – not merely national – norms and principles, and the UN or regional organizations or institutions have the right to intervene when national governments do not respect human rights or international humanitarian law.

As a consequence, it is important to examine the territorial issue from the geopolitical stance, as it is clear that national governments are conservative regarding this issue and maintain permanent concern and vigilance (which is the *raison d'être* of the armed forces in each country) designed to protect territorial integrity. However, territorial disputes and issues are decreasing in importance in these times of globalization, human rights, humanitarian law and states' Responsibility to Protect (R2P). Notwithstanding this, in this section I deal with some of the main geopolitical problems in Eurasia and Latin America. Hence, geopolitics is important because unless these problems (separatism, security threats concerning the potential use of military force, and territorial claims and controversies in general terms) are solved, regional integration processes face greater difficulties, which is an obstacle to both integration and regionalization, and troublesome from the perspective of peace and sustainable development.

I will begin with Eurasia, which has been called by that name by the geopolitical school because, from a geographical perspective, the European continent adjoins the great Asian continent, making a gigantic joint 'world island' – as Mackinder called it – surrounded by oceans, thus constituting the largest territorial expanse of the planet in terms of physical space. In addition, the concept of Eurasia is justified in view of the absence of natural borders between Europe and the enormous Russian territorial space, boundaries that do exist between Africa and Europe (the Mediterranean Sea) or America and Australia (the oceans). Indeed, the Ural mountain range is of low altitude and not comparable to the real natural border between China and the Indian subcontinent constituted by the mountain ranges of the Himalayas and the Karakoram (with highest peaks in the world). Hence the differences between Europe and Asia are cultural, linguistic, ethnic and religious, but not geographical in the strict sense of the word.

The problems that globalization can cause due to its disconnection with the spaces under state control are quite considerable, but the problems of Eurasian geopolitics are still strictly territorial, thus the foreign policy of European countries should always heed the fact that Russia has a vital geopolitical interest not in trade but in the defence of Russian territory that, for as long as can be remembered in historical terms, has been invaded by Asian or European powers. For that very reason, the geopolitical role this global actor[29] plays is fundamental for the whole Eurasian continent. Additionally,

---

[29] Some argue that since the collapse of the USSR, Russia has ceased to be a global power and that its role has been reduced to that of a regional power. I do not agree with these opinions. Russia is one of the five permanent members of the UN Security Council and, as such, has global responsibilities. Furthermore, its nuclear arsenal is the only one in the world comparable to that of the US, hence it is the only power capable of confronting the US, to the extent that, from the perspectives of the world military subsystem and nuclear power, it could be said that we continue to live in a bipolar world. Moreover, due to the war in Syria, Russia continues to be a major player in the Middle Eastern conflict, so it is easy to imagine that the analysts of the US 'intelligence community' worry permanently about the role played by Moscow at world level. This is also true at the American level,

at the end of the Cold War the two great military alliances (NATO and the Warsaw Pact) should have been dissolved, but only the Soviets complied with this obligation. Instead of dissolving, NATO was consolidated and expanded by incorporating all the countries that had been part of the Communist Bloc, while the same policy was followed in relation to the enlargement of the European Union. Consequently, it is easy to see why the foreign policies of Western powers – even if conveniently involved in democratic ideals, and it does not matter whether they are sincere or hypocritical – do not exactly enjoy great credibility and confidence in the eyes of Moscow.

Hence, when examining recent paradigmatic cases, we should start by remembering the way in which the support of the Western powers for the democratization movement of Ukraine was received by the Kremlin, and the later even less favourable blow to Moscow when pro-western groups came to power in 2014. What followed is history, but what I want to highlight here is the geopolitical edge of this problem, for the reason that if geopolitics deals with the way geography determines foreign policy, as two classic authors of this school of thought – Mackinder (2010: 301–319) and Haushofer (2012: 329–336) – have always maintained, it is crucial to explain the conflict using this analytical framework. In other words, as the principle of *territorial integrity* and thus of respect for borders – including their delimitation and 'inviolability' – is a central and permanent concern of every sovereign State, especially since the end of the Second World War and the foundation of the UN, there is no doubt that regional integration[30] is closely related to the transformation of the *Westphalian system* – the main features of which were described in the previous chapter – and the political struggles provoked by such changes (due to globalization and regionalization/integration processes) within the States. However, in the Eurasian case, the said territorial principle has meant an often violent type of relationship, but with deep cross-cultural characteristics among the peoples, civilizations and former empires that have historically made up this supercontinent.

In addition, the geography of the Eurasian continent as a whole extends from the Atlantic Ocean to the Pacific and includes both the relatively small European continental 'peninsula' and the vast areas of Siberia (whose northern limit is the permanent ice of the Arctic Ocean), as well as the giant deserts of the Takla Makan and the Gobi in central Asia, and the Tibetan plateau, including the world's highest mountain ranges, the Himalayas and the Karakoram, where all fourteen summits with an altitude of more than 8,000 m are located. To the latter description must be added the Indian subcontinent, the peninsulas of South-East Asia, the enormous continental space of the People's Republic of China, and the Korean and Kamchatka peninsulas. It is therefore relatively easy to understand why Mackinder called this

---

as can be seen in the turmoil provoked by the investigation into the alleged Russian intervention in the 2016 US election. Finally, in another example, the role of Moscow and the Russian presence in Latin America continue to be a matter of permanent concern in US foreign policy, especially when the Russians provide armaments to countries like Venezuela, which has become an ally in Putin's geostrategic designs.

[30] The mobilizations in Maidan Square in Kyiv began with a protest against the refusal of the Yanukovych government to sign the Association Agreement with the European Union.

huge mass of land the *geographic pivot of history*, even when Asian island nations like Japan, Taiwan, Indonesia and the Philippines are omitted from the equation.

From Mackinder's perspective and despite the fact that Britain was essentially a maritime power, it is also easy to understand why controlling the geographical pivot or Eurasian heartland became so appealing to the British Empire at the beginning of the twentieth century. Britain maintained a permanent rivalry with Tsarist Russia. If the vast continental Saint Petersburg empire (which had expanded its territory to the Pacific Ocean, where the port of Vladivostok was built, which is reached via the portentous Trans-Siberian railway that crosses eleven time zones) had continued to grow, it could have constituted a risk for the British, who already controlled Hong Kong and Shanghai in China and the whole Indian subcontinent. So if the Russians descended in the direction of the 'warm waters' of the Indian Ocean, acquiring full control of the pivot zone, in the eyes of the Foreign Office only London – with its maritime strength – could limit the terrestrial expansion of the Kremlin, thanks to the British presence in the Middle East, Iran, Afghanistan and the Indian subcontinent. Thus, since the nineteenth century, for Russia the essence of the geopolitical issue in the Eurasian continent has been to stop the expansion of Western powers, while for the West it has been to acquire as much territory (or influence) as possible.

In other words, as a terrestrial power, Russia had all of Eurasia in which to expand during the nineteenth century, although China and Japan were a barrier its imperial dreams towards the east (the Russians were defeated by Japan in the 1906 war) and India – at that time under the rule of the British Empire – was a barrier in the south, which explains the concerns of Mackinder and the British worries about the 'heartland' of Eurasia.

The reason for mentioning these Mackinderian ideas in a chapter which is mainly about the importance of regional integration processes is the role played by the European Union in the conflict in Ukraine as well as the confrontation between Brussels and Moscow in connection with this conflict. This is where geopolitics comes into play – not only because, in the long term, integration processes tend to lead to the cession of sovereignty at the behest of supranational integration institutions, but also because the territorial question always assumes the proper vision of the realist paradigm, and implicit in this, as stated in the opening chapter, is the question of the balance of power and the use of armed forces for the defence of territory. My view is that if we take into account that Russia has no natural borders and that throughout history it has suffered invasions from both the East (the Mongols, the Tatars and many people with Turkish ethnic roots) and the West (the Vikings, the Teutonic Knights, the French troops during the Napoleonic wars and the Germans during the two major World Wars) it is easy to understand the strictly geopolitical reaction that the Kremlin has had in relation to the Ukrainian crisis, which explains both the very nature of the conflict and the distrust and suspicion that the expansion of NATO and EU enlargement arouses in the eyes of Moscow, although the latter concerns the Kremlin on a much smaller scale (Padilla 2015).

While for Moscow the expansion of both NATO and the EU is essentially a geopolitical issue,[31] for the EU it is an integration process whose objectives are the consolidation of democracy through free trade and development. Thus, it is quite clear that there are divergent perceptions of the same issue in Russia and in the Western countries, but the Europeans were thoroughly aware of what they were doing before proposing an *Association Agreement* with Ukraine. In other countries where the EU has made this kind of agreement (like Chile, Mexico, and the Central American countries) this would not awaken geopolitical suspicion; however this was certainly not the case in Russia.

Regarding this issue, it is pertinent to quote an article by the well-known conservative professor of Chicago University, Mearsheimer (2014), who argued that the Ukraine crisis was "the West's fault" while presenting a very interesting analysis of the "liberal delusions that provoked Putin". According to Mearsheimer, the Russians interpreted the West's policy as disguised aggression because, for the Russians, Western powers' expansion into the East was 100% geopolitical. Great powers have

---

[31] Moscow was extremely disappointed with the expansion of NATO. For some analysts of Russian foreign policy, the concept of 'civilizational realism' is a way to oppose the US's Mackinderian drive towards the East and search for multipolarity: "An even more decisive shift in Russian foreign policy thinking, however, took place after Putin was re-elected president in 2012. At his annual meeting with leading Russia specialists in Valdai in 2013, he went beyond his usual criticism of unipolarity and, for the first time, explained what values Russia stood for. Echoing the remarks of 19th century Russian foreign minister Alexander Gorchakov, Putin said that Russia was returning to its core values, while remaining open and receptive to the best ideas of both East and West. These core values were rooted in the values of Christianity and other world religions. More importantly, the unipolar and increasingly secular world that some in the West seek to impose on the rest of the world, Putin argued, is 'a rejection … of the natural diversity of the world granted by God'. Russia, he intoned, will defend these Christian moral principles, both at home and abroad. Since then, the war in Georgia and the ongoing tragedy in Ukraine have reinforced negative stereotypes both in the United States, where many blame Russian intervention; and in Russia, where many see the same events as the culmination of Western policies targeting regime change in the former Soviet Union. Since the chances of resolving the Ukraine issue are virtually non-existent, influential observers like presidential advisor Vladislav Surkov have recently suggested that Russia's attempts to become part of the West may now be over. The values-based contours of the present East-West conflict are thus clearly defined. Russia opposes the very idea that Western cultural values are the standard for international behaviour. It regards such rhetoric as nothing more than self-serving unilateralism, and believes that a multipolar world order based on pluriculturalism (diversity among nations) is preferable. Focusing on values sheds a rather different light on some widely held Western assumptions about Russian foreign policy. The first is that Russia rejects the post-Cold War international order. This is not quite correct. Russia fears global chaos and believes that America's efforts to preserve its global hegemony are leading to such chaos. Far from rejecting the international system, Russia believes it is stabilizing it by working against hegemony within the current framework, while also seeking to expand it to include new actors. A second erroneous assumption is that Russia is determined to undermine the 'liberal US-led order.' In fact, Russia expects the US to remain the leader of the liberal, western model of global development, but argues that it must also learn to co-exist with other models. As Russia sees it, in the future, there will be multiple centres of power, each believing itself no less moral than any other. Notions like 'American exceptionalism' only serve to isolate the United States, says Russia, and while US leadership will no doubt fade as the world becomes more pluricultural, liberalism need not be dragged down with it. How, then, does Russia define itself? Russia defines itself as that part of the West that has understood

always been sensitive – not logical – and this kind of action always provokes a reaction – that is essentially a matter of 'feelings' and emotion, not rational calculations – towards any potential threat to its territory: "Can you imagine the outrage for the US if China established a military alliance with Canada and Mexico against the US?" (Mearsheimer 2014).

Incidentally, scholars from Macalester College in Saint Paul in a different US State (Minnesota) arrived at similar conclusions:

> This paper concludes that there is indicative evidence to suggest that the theory is conducive – in the context of competitiveness over resources and geostrategy – to explaining the attitudinal and behavioral conducts, as well as the geopolitical and strategic factors that together characterize the geopolitics of Central Asia. Consequently, the Central Asian region, replete with oil and natural gas resources, is indeed a target for foreign policy that follows Halford J. Mackinder's model of the Heartland Theory. Whether the world-powers are indeed cognizant of the geopolitical significance of their policies is difficult to verify. However, regardless of the intentions, the geopolitical framework of these ambitious policy strategies remains entrenched in Mackinderian philosophy. Using Mackinder's 'Geographical Pivot' thesis as analogous to contemporary policy regarding Central Asia, this paper has shown that the literature around the US, Russia, the EU, and China deals greatly with Mackinderian geostrategy in their foreign policy discourse and as such reveals that the Heartland Theory is still influential in their foreign policy outlook... The US and the EU, on the other hand, are building up alliances with regional countries in order to maximize their economic power and political influence. Thus, the degree to which Central Asian energy resources are made accessible or are closed to the US and the European states is of increasing importance in the foreign policies of those powers. In the end, it is clear that Mackinder's Heartland Theory, whether acknowledged directly or in principle, is quintessential to the understanding of foreign policy relations in contemporary Central Asia (Scott/Alcenat 2008).

As we can see in the map below based on Brzezinski's 1997 book *The World Chessboard*, the 'conquest' of Ukraine has long been a very important strategic goal of the US foreign policy (Fig. 2.1).

As a consequence, it was clear to Moscow that the true stakeholder in the expansion of the 'Western Alliance' over territories that were formerly part of the Soviet Union – and of Russia itself, in the case of Ukraine – is Washington.[32] Furthermore, this is easy to conclude by simply reading Brzezinski's book,[33] where the US intention to maintain American supremacy at world level – according to the US geostrategic interests – in the twenty-first century is quite explicit. Hence, for the former US

---

the futility of liberal fundamentalism, and that seeks to establish a framework for global leadership around the values that the West shares with non-Western states. Russian political theorist Boris Mezhuev calls this 'civilizational realism'. Civilizational realism differs from classical realism, in that it embraces the importance of values in international affairs. It also differs in that it sees value in the diversity of cultural communities, as well as individuals. Russia's approach should therefore be described not as opposition to liberalism, but as a different form of liberalism, one that is divorced from Western hegemony, and open to non-western traditions and influences" (Nicolai 2018; at: https://jia.sipa.columbia.edu/online-articles/russias-mission).

[32] Without the mediation of France and Germany in the Ukrainian conflict, the Minsk Agreements establishing the ceasefire in the conflict zone could not have been reached.

[33] In his classic book on geopolitics this scholar of Polish origin argues that Ukraine is a pivotal country of European security policy, and predicts the actions of the Americans in relation to Kyiv and against Moscow (Brzezinski 1997).

Fig. 2.1 Beyond 2010: the fundamental core of European security. *Source* Own elaboration based on Brzezinski (1998: 92)

National Security Advisor, the prospective planning of US foreign policy in the first decade of the twenty-first century should have included a series of foreign policy actions addressed at maintaining American 'primacy', which includes Ukraine as a "fundamental core of European security beyond 2010" (Brzezinski 1998: 92). Thus, it is not strange to see that the policies implemented by Washington – regardless of whether the administrations are Democrat or Republican – followed the foreign policy guidelines that were so clearly outlined by Brzezinski since those years of the post-Cold War, and that were being implemented by the Trump administration in such a spectacular manner that it almost led to the impeachment of the White House incumbent because of its failures, not its success.

Therefore, in President Putin's view, American intentions concerning Ukraine have nothing to do with the promotion of democracy or European integration, but, simply and plainly, with the expansion of the territorial hegemony of one power at the expense of another, and – in the case of Ukraine – with the aggravating circumstance that it is not about the territories of countries that were satellites of Moscow during the Cold War but of historically Russian territories (especially in the case of Crimea). Putin's response to that was, essentially, not that of a politician but that of a military strategist. However, not that of any strategist, but that of the great strategist that Putin – no doubt – is, because, with the Russian military base near Latakia as leverage, he also decided to break NATO's military siege of Russian territory by providing air power to the government of Syria. Furthermore, the 2014 crisis provoked by Assad's use of chemical weapons was solved thanks to the Russian initiative to force Assad to deliver those weapons to the OPCW in The Hague. Since then outside actors against Assad have not been able to continue their support for rebel forces without engaging in dialogue with Russia in order to manage the conflict, and the way Moscow had succeeded in improving its relationship with Ankara in that regard is really exceptional, especially compared with the clear deterioration of the United States' relationship with the Turkish president, due to the Pentagon's and the CIA's support for the Kurdish Peshmerga fighters in the region.

As a consequence, some would argue that Washington has been militarily stopped by Russia in both Ukraine and Syria for geopolitical reasons. Therefore the question that arises is what US options for the future should be considered and whether the approaches of some academics – like Mearsheimer – coincide with the proposal that the US should focus its attention not on Russia but on the centres of greatest geostrategic interest for the US – the Middle East (including the Persian Gulf) and North-East Asia (China, Korea, Japan). This is why, according to Mearsheimer, both the Ukrainian question and everything concerning the problems with the European Union would pass to a second order of priorities. For Mearsheimer, Russia was a strategic mistake of the Obama administration, given the fact that China – according to him, the real US adversary in terms of geo-economics – is much more important than Moscow in terms of geopolitics. Therefore, the rapprochement with Russia that Trump was looking for during his electoral campaign could be explained by this perspective (and also because he was planning personal economic investments), but the political commotion raised by his internal adversaries over the issue of Russian interference in the US electoral process prevented Trump's initial intent of rapprochement with the Kremlin gaining any continuity.

Although China could also be seen as a potential rival of Russia, both Beijing – with the creation of the Shanghai Cooperation Association – and Moscow – with the Eurasian Union – are currently engaged in improving their relations with countries of the region. Both Moscow and Beijing are also involved in the bloc of countries made up of *Brazil, Russia, India, China and South Africa* (BRICS) in an unprecedented process of global extra-regional policy aimed at strengthening multipolarity.

I am pointing out the latter because, in addition to the influence of the world geopolitical context on what happens in Europe, it should be said that after more than seventy years of global dominance, Washington is not only badly prepared for the change of policy that new world realities are requiring (including the growing tendency towards multipolarity), but its situation worsened with the arrival in the White House of a personality as unpredictable and temperamental as Trump. However, he was not in a position to make any substantial change to a still bipolar military foreign policy that, in the last resort, is decided by the Pentagon and the US intelligence community, not by the White House or even the State Department.

Therefore, although it is possible that in the geopolitical terrain Washington will eventually modify the policy of confrontation with Moscow over control of the Eurasian 'heartland', it remains to be seen if the US will choose to follow the guidelines of those who maintain that the emphasis of foreign policy should be changed from Russia to China by opening negotiations with Beijing on the respective spheres of influence in the regional order, as suggested by Kissinger (2014), or seeking to balance it from the outside (offshore balancing) as Mearsheimer proposed (2016), or receding in the trade war against China initiated by Trump in 2018. If this policy change occurs it will be the result of the more intelligent policies of both Moscow and Beijing that have probably already had favourable results (in terms of the balance of power theory) in their rejection of the military siege of their most immediate borders (Ukraine and the East and South China Seas).

Consequently, although the geographical importance of Central Asia should not be minimized – as will be seen in later references to the new 'silk route' of China (also called the *One Belt One Road* [OBOR] by the Chinese) – it is important to note that the centres of world capitalism have gradually been relocated to the Asia-Pacific basin, away from Europe and North America. This has nothing to do with the *Asia Pacific Economic Cooperation* (APEC) or the *Association of South East Asian Nations* (ASEAN), because Asia Pacific is a basin not a region. That fact explains why APEC is just a multilateral forum – not a regionalization process and still less an organization – even if it was inspired by ASEAN in the search for effective economic cooperation across the Pacific Rim. Its members include countries that belong to the Forum because of its location in the basin, e.g. Mexico, Peru and Chile in Latin America; Canada and the United States in North America; Russia, Japan, South Korea and China in East Asia; Singapore, Thailand, Malaysia, Vietnam, Brunei, Papua New Guinea, Philippines and Indonesia in South East Asia; and New Zealand and Australia in the South Pacific. Economic cooperation is not a process of regionalization, consequently APEC also has to be differentiated from geopolitical blocs (like the Eurasian Economic Union, the Shanghai Cooperation Organization and BRICS) inspired by multipolarity and rejection of the US goal of world hegemony.

We must remember – because this is not a new phenomenon – that the increasing importance of the Eurasian Pacific *rimland* (as Nicolas Spykman, the North American geopolitician of Dutch origin, would have called it) reflects the geo-economic movement of world capitalism. Its first locations were to be found in the Mediterranean (Venice, Genoa) in the fifteenth and sixteenth centuries, then there was a geographical shift of influence and predominance to the Baltic and the North Sea (Hamburg and Amsterdam) in the seventeenth and eighteenth centuries, to the North Atlantic (London and New York) in the nineteenth and twentieth centuries, and to its current location in the Pacific basin (California, Tokyo, Shanghai, Hongkong, Singapore).[34] Finally, as Joschka Fischer stated in a clairvoyant speech at a conference held at the Geneva Institute of Higher Studies of Development:

> How did we get here? Looking back 26 years, we should admit that the disintegration of the Soviet Union – and with it, the end of the Cold War – was not the end of history, but rather the beginning of the Western liberal order's denouement. In losing its existential enemy, the West lost the foil against which it declared its own moral superiority. The years 1989–1991 were the start of a historic transition away from the bipolar world of the post-World War II era toward today's globalized world, a familiar place, but one that we still do not fully understand. One thing is clear: political and economic power is shifting from the Atlantic to the Pacific, and away from Europe. This leaves many open questions: which power (or powers) will shape this future world order? Will the transition be peaceful, and will the West

---

[34] The first time I heard this idea was in the 1980s during a conference presentation by Attali (2006), then counsellor to the French President François Mitterrand, in Mexico City.

survive it intact? What kind of new global-governance institutions will emerge? And what will become of the old Europe – and of transatlanticism – in a "Pacific era"? This might be Europe's last chance to finish the project of unification. The historic window of opportunity that was opened during the period of Western liberal internationalism is quickly closing. If Europe misses its chance, it is no exaggeration to say that disaster awaits it. European politicians today present voters with a choice between modest pragmatism and blustery nationalism. But what Europe needs now is a third way: political leadership that can think creatively and act boldly for the long term. Otherwise, Europe is in for a rude awakening (Fischer 2016: 3).

The globalized world of today "still does not understand" and "one thing is clear, political and economic power is moving from the Atlantic to the Pacific, away from Europe". These wise observations of Fischer were made during his lecture on Franco-German relations, when he remarked on the erroneous perspective of France (and also of many other countries and academics) for seeing globalization as kind of 'conspiracy' when it is becoming a fact, a phenomenon of the world's geographical – hence geopolitical – reality, similar to the resurgence in the world arena of powers such as China. The same occurs with the Chinese attempt to recover the predominance that it has always had in its adjacent seas, which is related to another fact that should be remembered but is often forgotten: besides being a *nation state*, China is a millenarian civilization.[35] This does not mean, of course, that Beijing has not also considered a geopolitical strategy of development towards the Eurasian *heartland*, as evidenced by all its undertakings towards the new silk route (the OBOR initiative or "One Belt One Road"), as can be seen in the cartography below of the railway infrastructure, highways, and gas and oil pipelines (Fig. 2.2):

The Chinese strategy for the development of railway infrastructure, highways and pipelines in Central Asia is related to Beijing's dependence on oil and gas supplies, given that most of these supplies come from the Middle East and must be transported by sea. Another strategy to get rid of the risks of Chinese tankers having to travel through the Strait of Malacca – between Indonesia and Malaysia, which are potential enemies in disputes over maritime borders and island territory, which I will discuss later – can be seen in the Chinese proposal for the construction of a new inter maritime channel across Myanmar and Thailand to go directly from the South China Sea to the Indian Ocean (and the Andaman Sea), as can be seen at a glance in the map of the plan for a Sino-Myanmar pipeline below (Fig. 2.3).

With regard to China's pending disputes with the countries bordering its eastern and southern seas, including Japan, from a geopolitical stance it is worth bearing in mind that although China maintains these maritime border disputes with island territories in the South China and East China Seas, the contention in fact has more to do with the US than with the coastal countries, as the US wants to prevent the hegemony of Beijing in the region. Therefore, according to Mearsheimer and Walt,

---

[35] In a curious anecdote of the US presidential debates of 2016, when the Republican candidate accused the Democrat administration of Obama of implementing incorrect foreign policies that allowed the surge of China as a world power, the Democrat candidate (Hillary Clinton) reminded her opponent (Donald Trump) that China was already a great civilization a thousand years ago, when America as a continent was not yet discovered, and that the US as an independent state was only born in 1776.

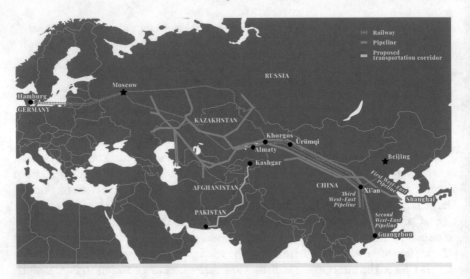

Fig. 2.2 China-Central Asia infrastructure development: the new silk route. *Source* The author's elaboration based on Stratfor data

if Beijing does not act with sufficient diplomatic ability and prudence it is possible that the friction over maritime boundaries with Vietnam, the Philippines, Malaysia, Brunei and Indonesia could transform into a confrontation with Washington. John Mearsheimer and Stephen Walt state this clearly:

> If China continues its impressive rise, it is likely to seek hegemony in Asia. The United States should undertake a major effort to prevent it from succeeding. Ideally, Washington should rely on local powers to contain China, but that strategy might not work. Not only is China likely to be much more powerful than its neighbours, but these states are also located far from one another, making it harder to form an effective balancing coalition. The United States will have to coordinate their efforts and may have to throw its considerable weight behind them. In Asia, the United States may indeed be the indispensable nation (Walt/Mearsheimer 2016: 81).

Thus, there are not only controversial points regarding the trade deficit or the supposed overvaluation of the yuan in the US's relationship with China, but there also exists a complex geopolitical situation that could lead to the confrontation of the two great powers unless the differences are solved through diplomatic channels (see Padilla 2016) or overcome thanks to the geo-economic strategies that both Putin and Xi Xinping[36] have been developing with regional bodies like the Economic Eurasian

---

[36] Thomas Wright, an expert at the Brookings Institution, explains the interests of Russia and China in Eurasia in the following manner: "In Donald Trump America has a rogue President who has a 30 year track record of opposing key elements of the (liberal) order, including free trade and alliances. Vladimir Putin wants to overthrow the order because he believes it poses a direct threat to his regime. Xi Xinping's China benefits from the open global economy but he would dearly like to replace the United States as the pre-eminent power in East Asia" (cited by Serbin 2018: 15).

Fig. 2.3 The Sino-Myanmar pipeline. *Source* The author's elaboration based on Stratfor data

Union and the Shangai Cooperation Association in order to counter the US presence in the Asia Pacific[37] and Eurasian regions.

---

[37] As already stressed, the Asia Pacific basin is not precisely a 'region' in terms of integration theory, and that is why, after quoting Thomas Wright in the article cited above, the Argentine scholar Andres Serbin mentions that in the November 2017 APEC summit in Da Nang (Vietnam) the three different narratives of globalized regionalization combined with different geopolitical and geo-economical approaches and priorities were presented respectively by Trump, Putin and Xi Xinping. According to Serbin, the US narrative promoted an 'Indo-Pacific' group of states (the US, Japan, India, Australia) as a counterweight to China and its OBOR, RCEP and FTAAP strategies, while Russia referred to the Eurasian Economic Union (EEU). Serbin adds: "In March 2017 during the meeting in Beijing of the first international forum of the ambitious OBOR Project – a Project that reactivates the commercial corridor of the silk route after 300 years of its dissolution and reconnects China with Europe – in spite of the emergence and current primacy of Asia Pacific, a new geostrategic world parameter has developed with the reactivation of Eurasia as a potential factor of economic dynamism and a key geopolitical pivot of the international system. While the

As the best way to understand any geopolitical problem is through cartography, the following maps illustrate these disputes over insular territory between China and neighbours like Vietnam, Philippines, Malaysia, Brunei and Indonesia as well as Japan, which logically has to do with Beijing's claim of sovereignty over the exclusive economic zone of 200 nautical miles under the terms of the *United Nations Convention on the Law of the Sea* (UNCLOS).

The controversy with Japan concerns the Senkaku/Diaoyu Islands, and, as can be seen in Figure 2.4, it also involves exclusive economic zone problems. Furthermore, the proximity of the US military base on the Japanese island of Okinawa is noteworthy (Fig. 2.5).

The problem with the geopolitical vision of the world is that everything is viewed from a territorial perspective. Since this applies to every sovereign State, the aspects which concern population and governance are relegated to the background. Whereas integration is a process that essentially concerns people, the emphasis of the world economy is on the exchange of goods and services free from territorial (customs) obstacles. It is therefore clear that integration and geopolitics are frequently on a collision course. Even though in this book I argue in favour of integration, while the Westphalian system still prevails at world level it is necessary to take into account all the realities of geopolitics – and the national states' foreign policies based on geopolitics – around the world.

This also seems true from the social point of view, because what is involved in geo-economy is the free transit and constant spatial relocation of people, thus it is obvious that both elements relate to people and to the legal procedures for regulating migratory flows that are, of course, a sovereign faculty of governments, which administer such exchanges and such mobility of goods and people. Accordingly, it could be said that geopolitics and integration may be opposed because borders (which delimit the said territories) hinder and do not facilitate either trade or migratory flows. At the political level, most of the problems of sovereignty derive from this issue. That is why both free trade and human mobility are currently the *bête noire* of conservatives in Europe, highlighted and emphasized by right-wing nationalists and populists and generally stressed by those who oppose free trade policies (protectionists) and are hostile to immigration (xenophobes and racists).

---

strategic maritime presence and the influence of China grows in the Pacific and particularly in the South China Sea – deepening the security concerns of the US and countries of the zone – the OBOR initiative is more complex and ambitious in its continental projection, overarching several regions and including Europe" (Serbin 2018: 13). It is interesting to remark in passing that, due to these differences in approaches and narratives within the Forum, the November 2018 APEC summit in Port Moresby (Papua New Guinea) was not even able to produce a common statement (and Trump did not attend), a fact that, undoubtedly, was positive for Xi Xinping as the 'rising star' of the meetings of APEC.

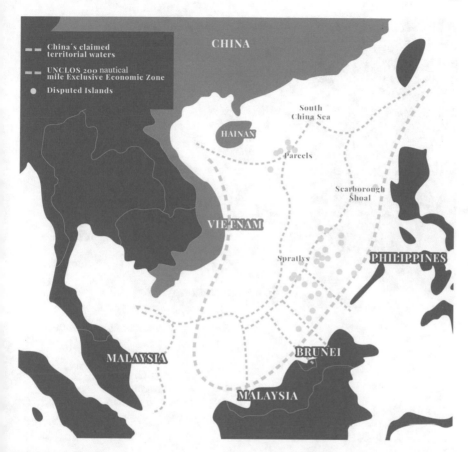

Fig. 2.4 The controversy over insular territory in the South China Sea. *Source* The author's elaboration based on the *United Nations Convention on the Law of the Seas* (UNCLOS) and on the CIA

2.6 Latin American Geopolitics

To a large extent, then, it can be said that these are geopolitical considerations which have also affected the integration processes in Latin America that have been launched in the region since the second half of the last century. Territorial disputes have always been present in nearly every country of the region, although, fortunately, and to the credit of some Latin American statesmen and legal experts, a good number of them have been resolved by peaceful means – mediation, arbitration or jurisdiction – as established in the Charter of the United Nations. In the following pages and as an example that may be useful in other regions of the world devastated by the same kind of geopolitical – territorial – disputes, I will review some of the most interesting cases that have been settled by peaceful means, among other reasons because they illustrate the form in which the idealist paradigm of IR works and functions.

Fig. 2.5 Chinese maritime claims to the Senkaku Diaoyu Islands. *Source* The author's elaboration based on data in the public domain

2.6.1 The Islands of the Beagle Channel

Three relatively small, uninhabited islets bearing the names of Picton, Lennox and Nueva, located in a small channel at the southern tip of Patagonia (below the island of Tierra del Fuego) were the subject of a dispute between Chile and Argentina towards the end of the 1970s, which nearly led the two countries into a major war. Bear in mind the fact that both governments were ruled by military dictators that faced leftist insurgent groups (the MIR in Chile, the Montoneros in Argentina), but nevertheless Pinochet and Videla were ready to fight for the sacred 'territorial integrity' of their respective nations. It was only the decision of the Chilean Chancellor to resort to papal mediation that prevented the imminent conflagration taking place, and it was thanks to the intervention of Pope John Paul II that the controversy was resolved. As the ruling favoured Chile, the Argentinian military rulers were not willing to comply. However, in a desperate manoeuvre, given that their troops were ready to attack Chile, they changed their military objective and instead chose to invade the Falkland Islands, a British Overseas Territory which Argentina calls the Malvinas and wanted to claim from the United Kingdom. The response of the British Prime Minister, Margaret Thatcher, was vigorous, and the Argentinians were defeated and forced to surrender. This defeat resulted in the fall of the military government and the holding of elections. The newly elected president, Raúl Alfonsín, submitted the papal decision to a referendum and in this manner achieved approval for the ruling, thereby resolving the conflict.

The following map shows the delimitation obtained by means of the papal mediation (Figs. 2.6 and 2.7):

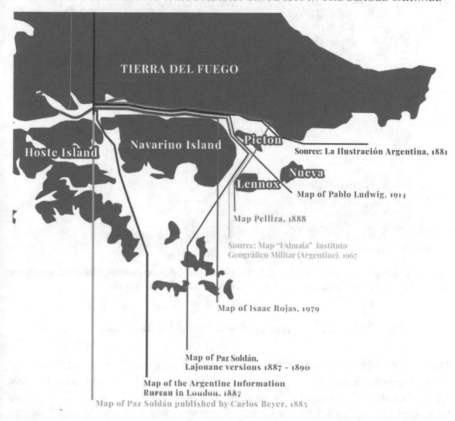

SUMMARY OF ARGENTINE CARTOGRAPHY SINCE 1881 IN THE BEAGLE CAHNNEL

TIERRA DEL FUEGO

Hoste Island   Navarino Island   Picton   Source: La Ilustración Argentina, 1881

Nueva

Lennox   Map of Pablo Ludwig, 1914

Map Pelliza, 1888

Source: Map "Ushuaia" Instituto
Geográfico Militar (Argentine), 1967

Map of Isaac Rojas, 1979

Map of Paz Soldán,
Lajouane versions 1887 - 1890

Map of the Argentine Information
Bureau in London, 1887

Map of Paz Soldán published by Carlos Beyer, 1885

Fig. 2.6 Latin American Geopolitics. Summary of Argentine Cartography since 1881 in the Beagle Channel with the delimitation as a result of Papal mediation. *Source* The author's elaboration based on public domain information

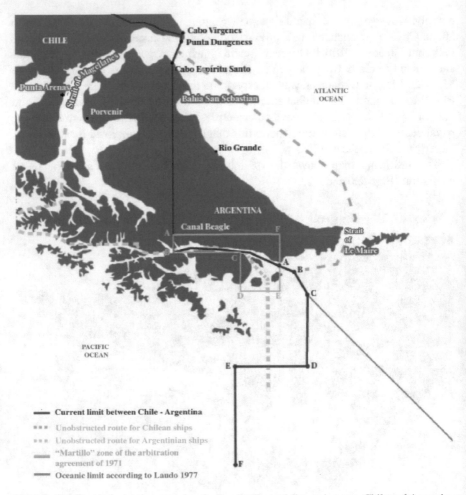

Fig. 2.7 The Pope's decision concerning the Beagle Channel dispute between Chile and Argentina. *Source* The author's elaborations based on information in the public domain

2.6.2 The Peru-Chile Case of Maritime Delimitation

The War of the Pacific (1879–1883), in which Chile defeated Bolivia and Peru in the late-nineteenth century, led to a series of territorial disputes. The solution to these various bones of contention is still pending. One sticking point is the Bolivian claim for a sovereign exit to the Pacific Ocean. Another, in which Lima opposed Santiago at the beginning of the first decade of the present century, was finally solved by the *International Court of Justice* (ICJ) in a ruling on the maritime delimitation that can be observed in Fig. 2.8, which shows the maritime delimitations decided by the judges of the UN Court.

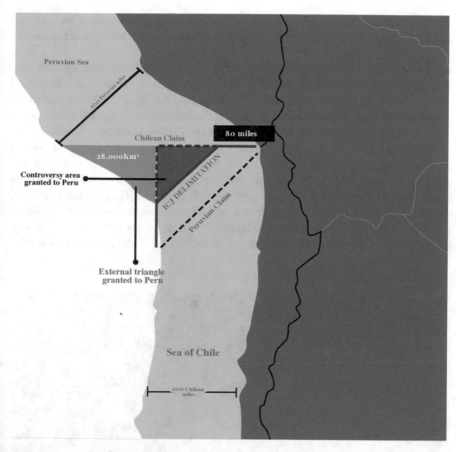

Fig. 2.8 Map showing the delimination of the International Court of Justice. *Source* The author's elaboration based on a map of the International Court of Justice.

Peru claimed not to have a maritime boundary delimitation, and announced that if there was to be one, it should be drawn in accordance with the requirements of the *United Nations Convention on the Law of the Sea* (UNCLOS), applying an equidistant demarcation. Chile, on the other hand, responded by arguing that the demarcation came from a treaty signed in 1928, according to which the horizontal line of the corresponding parallel should be taken as the demarcation line. The judgment issued by the International Court of Justice on 27 January 2014 finally resolved the controversy concerning the maritime delimitation initiated by Peru on 16 January 2008. It involved a maritime area – and its corresponding airspace – of approximately 67,139.4 km², of which some 38,000 km² were considered to be Chilean sea and 28,471.86 km² high seas. The ruling declared that the starting point of the maritime boundary between Peru and Chile is the intersection of the geographic parallel that crosses 'Hito No. 1' at the low-water line, and that the maritime boundary follows the parallel that passes Milestone No. 1 to a point located 80 nautical miles away. At the

Court's discretion, that decision is based on legally binding agreements and bilateral practice between both countries and proves the existence of a maritime delimitation made by both parties.

This decision of the Court satisfied Chile, but it immediately established a new limit applicable after 80 miles, a line that continues in an equidistant direction from the coasts of both countries until its intersection with the limit of 200 nautical miles measured from the baseline of Chile and that subsequently continues south to the point of intersection with the limit of 200 nautical miles measured from the baselines of both countries. The Court issued its ruling without determining the precise geographic coordinates, provided that the parties themselves proceeded to determine such coordinates in accordance with the ruling, and this was carried out by both parties on 25 March 2014. The ruling is fairly equitable, but it is clear that Peru, obtaining about 50,000 km^2 of maritime area (equivalent to the territory of Costa Rica) as an exclusive economic zone, was the winner of the litigation. As already stated, Figure 2.8 presents the delimitation according to the decision of the Court of The Hague (Fig. 2.9).

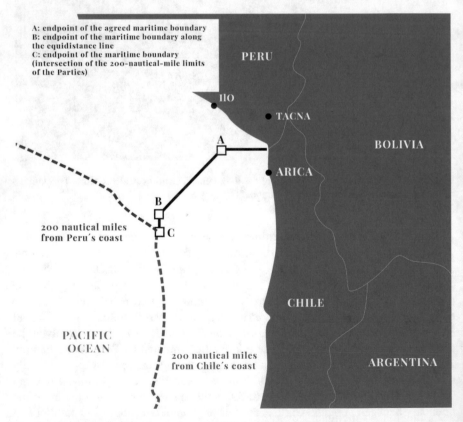

Fig. 2.9 Course of maritime border between Chile and Peru as decided by the International Court of Justice. *Source* The author's elaboration based on information in the public domain

2.6.3   The Case of Nicaragua Versus Colombia

This case is interesting because its origins go back to the intervention of the United States in Panama at the beginning of the twentieth century, which led to the secession of that country from Colombia. Later, during the government of Anastasio Somoza in Nicaragua (whose dictatorship always enjoyed Washington's support), it motivated a kind of North American compensation to Colombia for the loss of Panama. Indeed, the Sandinista government, which initiated the process against Colombia, argued that the Nicaraguan government was forced to consent to the Treaty (1928) that granted Colombia sovereignty over the islands of Providencia, Santa Catalina and San Andrés in the Caribbean Sea, due to the fact that the President of that time, the dictator Anastasio Somoza (against whom the successors of the insurgent commander Augusto César Sandino [1895–1934] were fighting), was under the rule of the Americans when the US invaded Nicaragua. According to Managua, sovereignty over these island territories should be returned to Nicaragua.

In the 2012 verdict, the judges ruled that Colombia could continue to exercise sovereignty over the island territories and the waters immediately adjacent to them. This included the islands of San Andrés, Providencia, and Santa Catalina, and seven smaller islands which Nicaragua claimed were included within its own continental shelf. In this way, the ruling acknowledged the validity of the Esguerra-Bárcenas Treaty of 1928 (a boundary treaty regarding the islands), which in 1980 was declared null and void by Daniel Ortega, the *de facto* ruler of Nicaragua who came to power when the Sandinista National Liberation Front overthrew Somoza in 1979. Managua argued that the Treaty was null and void because it was signed during the US military occupation of Nicaragua. Regarding this point, Colombia's right to sovereignty over the islands was upheld.

However, the Court's equitable judgement also ruled that Nicaragua had the right to a larger area of the exclusive economic zone. Thus, Colombia lost control of an important segment of maritime area east of the 82nd meridian, because the judges decided to grant sovereignty over the marine territory to Nicaragua. Despite the fact that a considerable part of the ocean in the latitude of the islands of San Andrés and Providencia remained within the jurisdiction of Colombia, two marine enclaves with a radius of twelve nautical miles around the smaller islands of Quitasueño and Serrana were enclosed by Nicaraguan waters, because, as the ICJ understands, according to UNCLOS, each island, however small, must have a territorial sea attached. It is worth mentioning that these marine territories are rich in oil, gas, and edible sealife, with an abundance of fish, snails, and especially lobsters.

As a result, the 2012 Court ruling, while recognizing the validity of the Esguerra-Bárcenas Treaty and ratifying Colombia's sovereignty over the islands, also granted Nicaragua exclusive economic zone rights to the east of the 82nd meridian, which displeased Colombia, as the maritime waters of Nicaragua are now enlarged. This provoked a rejection by Colombia of the ruling (it 'obeys' but does not 'comply', according to the Colombian government's declarations) and the denunciation of the

Pact of Bogotá by which the member countries accept the compulsory jurisdiction of the Court.

This situation has given rise to two new cases by Nicaragua against Colombia, one for failure to comply with the 2012 Court ruling and the other for the expansion of its continental shelf. Faced with these two lawsuits, Colombia asked the Court to declare its incompetence, but in March 2016 the International Court of Justice declared itself competent.

Faced with this last judgment of the Court, Colombia's position has been ambiguous: on the one hand, its government declares its intention not to appear before the court in the lawsuits filed by Nicaragua on which the ICJ declared itself competent, but on the other hand it is willing to defend Colombia's rights. It gives as an example cases in which the five permanent members on the UNSC have not appeared before the Court either, although, obviously, comparing themselves with the United States or China is not a very edifying decision. It remains to be seen what the new course of events will be, but what I wish to highlight is the way in which the geopolitical conflicts of the subcontinent have been approached by the ICJ in a very appropriate manner, which has favoured the normative paradigm of international law, that is, idealism, demonstrating the effectiveness and value of the UN in judiciary affairs.

The maritime delimitation of the year 2007 can be seen in figure 2.10.

As the 2007 ruling had not specified the delimitation of the maritime boundary, the International Court of Justice continued to deal with this matter and finally failed in 2012. The result (see Fig. 2.10) was an extension of the border of the Nicaraguan maritime system, as can be seen on the map. All the dark green areas are under Nicaraguan sovereignty as they are west of the 82nd meridian (except for an area in the upper part), while the light green area corresponds to the new areas of maritime territory granted by the ICJ to Nicaragua to the detriment of Colombia, as they are located to the east of the said meridian and leave the Colombian islands 'enclosed'. The dark blue area corresponds to the Colombian territorial waters as well as the projection of its continental shelf over the Caribbean Sea, while a small area to the north (light blue) corresponds to the 'common regime' decided by the Court.

2.6.4   Other Territorial Disputes

There are many other cases that could be cited as examples of these 'civilized' methods of pacific dispute resolution, in accordance with international law, to resolve the innumerable territorial-geopolitical disputes of Latin American countries. These include two recent cases of Costa Rica against Nicaragua regarding free navigation in the San Juan River and occupation of a small portion of Costa Rican territory at the mouth of the river; and a case of Nicaragua against Costa Rica for ecological damage caused to the river basin due to the construction of a highway by the Costa Rican government, which was ultimately settled in favour of Costa Rica.

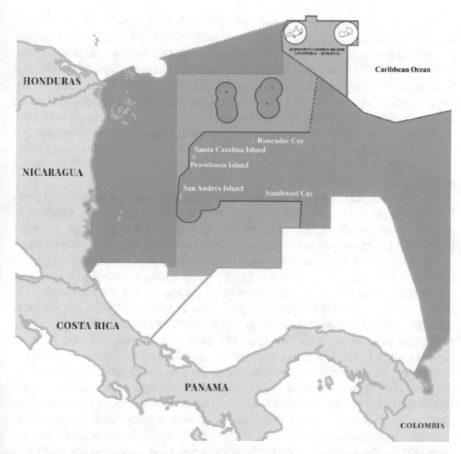

Fig. 2.10 Maritime border of the Nicaragua-Columbia dispute as decided by the International Court of Justice (2012). The exclusive economic zone of Nicaragua is in green and the Columbian zone is in blue. The main islands of Providencia, San Andrés and Santa Catalina remain under Columbian sovereignty. *Source* The author's elaboration based on information in the public domain

There are also several cases from other Central American countries, including El Salvador vs Honduras and Honduras vs Nicaragua, details of which I am not in a position to describe in the framework of this book. Other processes, such as that of Bolivia against Chile, demanding that the Court declare the obligation of Santiago to negotiate a sovereign exit to the Pacific Ocean with La Paz, have already been solved (in the example cited, the ICJ ruled that Chile was not obligated). And thanks to the good offices of the OAS in 2008, a Special Agreement relating to an unresolved dispute between Guatemala and Belize was signed by both countries, committing them to submit the dispute before the International Court of Justice, which was initiated at the end of 2019.[38]

---

[38] The delay was due to difficulties regarding the realization of a referendum in both countries requesting the acquiescence of the citizens to take the case to Court. Guatemala held its referendum

There are also other cases resolved through mediation (guarantor countries), such as Peru and Ecuador, and unresolved cases, such as that of Venezuela and Guyana concerning the Essequibo territory, and one concerning the southern ice field between Argentina and Chile. However, as previously stated, I am not in a position to discuss them in detail. In any case, I have fulfilled my objective of highlighting how geopolitical conflicts can be resolved through the United Nations jurisdictional bodies, mediation, arbitration, good offices or bilateral negotiation, all within the normative framework of the Charter of the United Nations. All of the above demonstrates that the idealist paradigm of international relations can be very effective in resolving such conflicts, which means that the realist claim that international law is frequently unworkable or ineffective is not true.

Nevertheless, it is also clear that, due to sovereignty concerns, geopolitical problems create obstacles to integration processes and make it difficult for them to progress smoothly. Thus, opting for the jurisdictional way to press for the negotiation of a sovereign exit to the sea, as Bolivia has done with Chile, is the correct way to resolve such a dispute, and, more importantly, it is in accordance with international law.[39] However, it is interesting to compare the Bolivia-Chile case with that of Austria, which is not only a member of the European Union but also an old empire that dominated the entire northern region of Italy and the former Yugoslav countries. Yet it has never filed a claim before the ICJ demanding Rome or Zagreb to conduct negotiations with Vienna on the establishment of a sovereign Austrian haven in the Adriatic Sea. This clearly shows that integration (which is essentially a matter of geo-economics, not geopolitics) is the way out of geopolitical territorial disputes. For Austria, the above claim simply does not arise thanks to its membership of the EU. Thus, as pointed out at the beginning of this chapter, one of the most important achievements of the European integration is sustainable peace. Consequently, when geopolitical problems of this kind simply disappear, it is clear that integration is the best way to solve these territorial nightmares that are absolutely anachronistic in twenty-first century international relations, even taking into consideration the fact that the historical roots of such disputes cannot be underestimated and must be addressed properly, that is to say in accordance with the UN Charter and the procedures of international law.

References

Aguilera Peralta, Gabriel, 2016: "El Regionalismo Latinoamericano entre la Unión y la Integración", in: *Revista Oasis* (Universidad Externado de Colombia): 89–105.

---

in April 2018 and Belize followed suit afterwards. At the time of writing, the process is taking place at the ICJ.

[39] In 2018 the ICJ decided not to accept the Bolivian claim on the grounds that the government of Chile is not obliged to negotiate the establishment of a sovereign port for Bolivia in the Chilean coastal territory bordering the Pacific Ocean.

Aguilera Peralta, Gabriel; Torres, Edelberto, 1998: *Del Autoritarismo a la Paz* (Guatemala City: FLACSO).

Attali, Jacques, 2006: *Une Brève Histoire de l'Avenir* (Paris: Fayard).

Badie, Bertrand, 2004: *L'Impuissance de la Puissance* (Paris: Fayard).

Bauman, Zygmunt, 2017: *La Globalización: Consecuencias Humanas* (Mexico City: Fondo de Cultura Económica [FCE]).

Dabène, Olivier, 2012: *Explaining Latin America's Fourth Waves of Regionalism: Regional Integration of a Third Kind,* Paper for the Congress of the Latin American Studies Association (LASA), Panel on "Waves of Change in Latin America History and Politics", San Francisco, CA, 25 May.

Fischer, Joschka, 2013: "European Unity is in Danger. This crisis threatens to destroy the EU and the only way to save it is to apply solidarity on the debt and, in general, to yield more sovereignty. It is not known whether France or Germany are willing to do so", In 2020 remittances from Guatemalan working in the US were 12% of the GNP, more than 11,000 millions of US.$ in: *El País*, 3 May.

Friedman, George, 2015: *Flashpoints: The Emerging Crisis in Europe* (New York/London: Doubleday).

Haushofer, Karl, 2012: "Los Fundamentos Geográficos de la Política Exterior", in: *Geopolitica* (Universidad Complutense de Madrid), 3,2: 329–336.

Hettne, Bjorn, 2002: "El Nuevo Regionalismo y el Retorno a lo Político", in: *Revista de Comercio Exterior* (Mexico), 52,11: 956.

Lovelock, James, 2007: *The Revenge of Gaia: Why Earth is Fighting Back and How We Can Still Save Humanity* (London: Penguin Books).

Mackinder, Halford, 2004 [1904]: "The Geographical Pivot of History" or the way Eurocentric thinking takes form in former colonial countries, *The Geographical Journal*, 170,4 (December): 298–321.

Mackinder, Halford, 2010: "El Pivote Geográfico de la Historia", in: *Geopolitica* (Universidad Complutense de Madrid), 1,2: 301–331.

Malamud, Andres, 2011: "Conceptos, Teorías y Debates sobre la Integración Regional", in: *Revista Norte América*, 6,2 (July–December): 220.

Malamud, Andres, 2018: "Overlapping Regionalisms, No Integration: Conceptual Issues and the Latin American Experiences", in: *Academia Diplomática* (Guatemala). 46–59.

Mearsheimer, John, 2014: "Why the Ukraine Crisis is the West's Fault: The Liberal Delusions that Provoked Putin", in: *Foreign Affairs*, 93,5 (September/October): 77–84.

Nicolai, Petro N., 2018: "Russia's Mission", in: *Journal of International Affairs* (Columbia University, NY [November])/*American Committee for East-West Accord* [ACEWA]; at: https://jia.sipa.columbia.edu/online-articles/russias-mission.

Padilla, Luis Alberto, 2015: "Neutralidad y Equilibrio de Poder en el Conflicto de Ucrania", in: *Espacios Políticos: Revista de la Facultad de de Ciencias Políticas y Sociales* (Guatemala City: Universidad Rafael Landívar).

Padilla, Luis Alberto, 2016: "Asia Pacífico, Eurasia y la Nueva Rivalidad Geopolítica de China con Estados Unidos", in: *Política Internacional: Revista de la Academia Diplomática* (Guatemala City: MINEX), (June): 91–107.

Rodrik, Dani, 2014: *Una Economía, Muchas Recetas: La Globalización, las Instituciones y el Crecimiento Económico* (Mexico City: Fondo de Cultura Económica).

Sanahuja, José Antonio, 2012: "Regionalismo Postliberal y Multilateralismo en Sudamérica: El Caso de UNASUR", in: *Anuario Regional* (Buenos Aires: Coordinadora Regional de Investigaciones Económicas y Sociales [CRIES]), 9.

Sanahuja, José Antonio, 2018: "Crisis de Globalización, Crisis de Hegemonía: Un Escenario de Cambio Estructural para América Latina y el Caribe", in: Serbin, Andres (Ed.): *Poder, Globalización y Respuestas Regionales* (Barcelona: Icaria Editorial [CRIES]).

Sanahuja, José Antonio; Sotillo, José Angel, 2007: *Integración y Desarrollo en Centroamérica: Más allá del Libre Comercio* (Madrid: La Catarata).

Serbin, Andrés, 2018: "La Configuración de la Gran Eurasia y su Impacto en la Gobernanza Global", in: *Revista Política Internacional* (Guatemala City: Academia Diplomática), 6: 7–21.

Sloterdijk, Peter, 2006: *Le Palais de Cristal: À l'Intérieur du Capitalisme Planétaire* (Paris: Maren Sell).

Solís, Luis Guillermo, 2016: "Reflexiones sobre las Relaciones de Guatemala y Costa Rica en el Marco del Proceso de Integración Centroamericano", in: *Política Internacional: Revista de la Academia Diplomática* (Guatemala City: MINEX), 1: 181–190.

Torres, Edelberto, 1969: *Interpretación del Desarrollo Social Centroamericano* (Santiago de Chile: Editorial Pla).

Torres, Edelberto, 1998: *Historia General de Centroamérica* (San José: FLACSO; Brussels: European Community).

Torres, Edelberto, 2007: *La Piel de Centroamérica: Una Visión Epidérmica de Setenta y Cinco Años de su Historia* (San José, Costa Rica: FLACSO).

Torres: Edelberto, 2011: *Revoluciones sin Cambios Revolucionarios: Ensayos sobre la Crisis en Centroamérica* (Guatemala City: F&G Editores).

Walt, Stephen; Mearsheimer, John, 2016: "The Case for Offshore Balancing: A Superior US Grand Strategy", in: *Foreign Affairs*, 85,4 (July/August): 70–83.

Internet Links

BCIE: https://www.bcie.org/novedades/eventos/evento/quinto-infome-estado-de-la-region-2016/.

CELAC: http://www.sela.org/celac/quienes-somos/que-es-la-celac/.

Fischer, Joschka: https://www.euractiv.com/section/euro-finance/interview/joschka-fischer-stabil ise-the-eurozone-to-defuse-hurricane-brexit/.

Kupchan, Charles, 2018: "Trump's Nineteenth Century Grand Strategy", in: *Foreign Affairs* (26 September); at: www.foreignaffairs.com.

Lafontaine, Oskar, 2015: "Let's Develop a Plan B for Europe", in: *International Journal of Socialist Renewal*, at: http://links.org.au/node/4573.

Navarro, Vincenç, 2015: "¿Por qué la Socialdemocracia no se Recupera en Europa?".

Nicolai, Petro: https://eastwestaccord.com/nicolai-n-petro-russias-mission/.

SICA: https://www.sica.int/sica/sica_breve.aspx

Scott, Margaret; Alcenat, Westenley, 2008: *Revisiting the Pivot: The Influence of Heartland Theory in Great Power Politics* (Macalester College USA); at: file://E:/GEOPOLITICA/Mackinder'sPivot%20in%20US%20Foreign%20Policy_Alcenat_and_Sc

Chapter 3
The Anthropocene: Are We in the Midst of the Sixth Mass Extinction?

Geophysiology, the discipline of Gaia theory, had its origins in the 1960s Gaia hypothesis. Geophysiology sees the organisms of the Earth evolving by Darwinian natural selection in an environment that is the product of their ancestors and not simply a consequence of the Earth's geological history. Thus the oxygen of the atmosphere is almost wholly the product of photosynthetic organisms, and without it there would be no animals or invertebrates, nor would we burn fuels and so add carbon dioxide to the air. I find it amazing that it took so long for biologists even grudgingly to acknowledge that organisms adapted not to the static world conveniently but wrongly described by their geologist colleagues, but to a dynamic world built by the organisms themselves.
Lovelock (2009: 48)

Ecology studies the relationship between living organisms and the environment in which they develop. This necessarily entails reflection and debate about the conditions required for the life and survival of society, and the honesty needed to question certain models of development, production and consumption. It cannot be emphasized enough how everything is interconnected. Time and space are not independent of one another, and not even atoms or subatomic particles can be considered in isolation. Just as the different aspects of the planet – physical, chemical and biological – are interrelated, so too living species are part of a network which we will never fully explore and understand. A good part of our genetic code is shared by many living beings. It follows that the fragmentation of knowledge and the isolation of bits of information can actually become a form of ignorance, unless they are integrated into a broader vision of reality.
Pope Francis, Laudato Si' (2015: 166)

Earth is a very particular planet because at the same time that she has expressed her marvellous disposition to give birth to the living world she is a global ensemble and a complex system The knowledge of Earth requires the use of resources of all its diverse constituent parts. In other words, to understand our planet it is necessary to go from the parts to the whole and from the whole to the parts. And in the field of knowledge, at the moment, this fact is absolutely illustrative and exemplary.
Morin (1999: 454)

© Springer Nature Switzerland AG 2021
L.-A. Padilla, *Sustainable Development in the Anthropocene*,
The Anthropocene: Politik—Economics—Society—Science 29,
https://doi.org/10.1007/978-3-030-80399-5_3

3.1   Introduction

As we have seen in Chaps. 1 and 2 of this book, despite being essentially an interdisciplinary field of studies, international relations theory has traditionally been treated as if its main purpose was the study of *inter-state* relations, not *inter-national* relations, because the focus of both the realist and the idealist paradigms is power politics and therefore the relationships between peoples, ethnic groups, religions, cultures and the trade and commerce characteristic of the world economy are not usually considered within its classic perspectives. Nonetheless, if we accept that the international system is the appropriate field of studies of the discipline, then it is logical to acknowledge that the different components of the said system are the political, military, economic, social, and cultural *subsystems*, hence any theory of IR must include not only the political science approach but also the economic, anthropological, cultural and social science standpoints. In other words, a holistic paradigm is necessary in order to generate a satisfactory IR theory. This understanding of the discipline also embraces environmental sciences and ecology as disciplines that contribute to this field of knowledge because the base of sustenance of the international system is the planetary system, the Earth itself. Consequently, all the states, nations, peoples, cultures, religions, transnational corporations and so on exist and operate thanks to the fact that humanity has the planet (*Gaia* in Lovelock's perspective) as its *"common house"*, as Pope Francis calls it in his Encyclical Letter *Laudato Si'*. This explains why the Anthropocene represents much more than a new geological epoch and why its importance is emphasized in these times of great menaces like climate change and health pandemics. Is our species facing the sixth mass extinction of life on the surface of the planet? Do religion and science now coincide in their way of comprehending this marvellous planet that we all inhabit?

3.2   The Anthropocene: Planetary Boundaries, Climate Change and Loss of Biodiversity

What is the Anthropocene? In technical terms, within geology, the term has been formalized according to recommendations made by the Anthropocene Working Group (AWG) in 2019, when a proposal was adopted on the following basis:

1. It is being considered at the series/epoch level (and so its base/beginning would terminate the Holocene Series/Epoch as well as Meghalayan Stage/Age).
2. It would be defined by the standard means for a unit of the Geological Time Scale, via a *Global Boundary Stratotype Section and Point* (GSSP), colloquially known as a 'golden spike'.
3. Its beginning would be optimally placed in the mid-twentieth century, coinciding with the array of geological proxy signals preserved within recently accumulated strata and resulting from the 'Great Acceleration' of population growth, industrialization and globalization.

4. The sharpest and most globally synchronous of these signals that may form a primary marker are the artificial radionuclides spread worldwide by the thermonuclear bomb tests from the early 1950s (Brauch 2019: 5).

On the other hand, and concerning the sixth mass extinction that could be provoked by the Anthropocene, in an article published in the *Proceedings of the American National Academy of Sciences* (PNAC) the palaeontologists Wake/Vredemburg (2008) opened a debate about the mass extinctions that have occurred in the geological history of the planet, which are related to the loss of biodiversity as a consequence of human predatory economic insatiability and greed and also to the risk of the disappearance of the human species if neoliberal ideology continues to permeate the mentality and behaviour of the 'private sector' and prolongs its transgression of planetary boundaries. According to these scientists, it is important to bear in mind that the planet existed without humans for millions of years, that the first *Homo sapiens* date from about 200,000 to 300,000 years ago, and that civilization began with the introduction of agriculture, which took place approximately 10,000 years BC (roughly 12,000 years ago) at the beginning of the Holocene geological epoch and initiated the substantial changes to the environment caused by man. Before these interferences of mankind in nature, the changes in matters of biodiversity (extinctions included) can be considered to be non-anthropogenic.

Therefore, it is a possibility that humanity is nothing more than a mere "Anthropocene chapter" in the geological history of the planet, because even though cultural evolution may have replaced biological evolution in humans (as noted in the studies of Piaget and Kohlberg etc. mentioned in previous chapters), natural selection is still shaping our biology in response to environmental change, so we may be superseded by a superior species more intelligent and friendly towards the environment and *Gaia*. Another possibility is that humankind will intervene in biology with genetic engineering or *Artificial Intelligence* (AI) and produce a new type of non-organic life (cyborgs) that could surpass organic life as it has been known since the beginning of life on Earth (Harari 2016).

Hence, *Homo sapiens* is not necessarily the last word of *Gaia* in evolutionary terms, and there is no guarantee that we humans will still be here in the long term. This possibility is also connected with the palaeontological evidence of the geological history of the planet concerning the crucial fact that during the millions of years of the evolution of life on Earth before our species appeared there were five mass extinctions of biodiversity, the most recent being the disappearance of dinosaurs some 65 million years ago, when a meteorite crashed into the planet, causing cataclysmic climatic change so violent and enormous that the giant reptiles that had dominated the planet were unable to survive. When the previously overshadowed mammalians no longer had to compete with dinosaurs, new processes enabled them to diversify into numerous species, amongst them the anthropoids that are our ancient ancestors.

Alternatively, according to Sachs (2015), the mass extinction of biodiversity is related to the transgression of planetary boundaries in the age of the Anthropocene. These boundaries are the 'anthropogenic pressures' on the Earth System that have reached a point where abrupt global environmental change can no longer be excluded.

Sachs says that a new concept of global sustainability is needed, in which planetary boundaries are defined as the borders or limits within which it is expected that humanity can operate safely. Thus, transgressing one or more planetary boundaries may be catastrophic for the planet (and human life) "due to the risk of crossing thresholds that will trigger non-linear, abrupt environmental change within continental- to planetary-scale systems". Sachs identifies seven planetary boundaries which are not to be transgressed:

> *Climate change* (CO_2 concentration in the atmosphere <350 ppm and/or a maximum change of +1 W m^{-2} in radiative forcing).
> *Ocean acidification* (mean surface seawater saturation state with respect to aragonite ≥80% of pre-industrial levels).
> *Stratospheric ozone* (<5% reduction in O_3 concentration from the pre-industrial level of 290 Dobson Units).
> *Biogeochemical nitrogen* (N) *cycle* (limit industrial and agricultural fixation of N_2 to 35 Tg N yr^{-1}) and *phosphorus* (P) *cycle* (annual P inflow to oceans not to exceed 10 times the natural background weathering of P).
> *Global freshwater use* (<4,000 km^3 yr^{-1} of consumptive use of run-off resources).
> *Land system change* (<15% of the ice-free land surface under cropland).
> *Rate at which biological diversity is lost* (annual rate of <10 extinctions per million species).

Two additional planetary boundaries for which a boundary level has not yet been determined are *chemical pollution* and *atmospheric aerosol loading*. Humanity is estimated to have already transgressed three planetary boundaries: climate change, rate of biodiversity loss, and changes to the global nitrogen cycle. Planetary boundaries are interdependent, because transgressing one may either shift the position of other boundaries or cause them to be transgressed. The social impacts of transgressing boundaries will be a function of the social-ecological resilience of the affected societies (Sachs 2015: 314).

Regarding the sixth extinction, Wake/Vredemburg (2008) sustain that:

> It is generally thought that there have been five great mass extinctions during the history of life on this planet….In each of the five events, there was a profound loss of biodiversity during a relatively short period. The oldest mass extinction occurred at the *Ordovician–Silurian boundary (≈439 Mya)*. Approximately 25% of the families and nearly 60% of the genera of marine organisms were lost. Contributing factors were great fluctuations in sea level, which resulted from extensive glaciations, followed by a period of great global warming. Terrestrial vertebrates had not yet evolved. *The next mass extinction was in the Late Devonian (≈364 Mya)*, when 22% of marine families and 57% of marine genera, including nearly all jawless fishes, disappeared. Global cooling after bolide impacts may have been responsible because warm water taxa were most strongly affected. Amphibians, the first terrestrial vertebrates, evolved in the Late Devonian, and they survived this extinction event…*The Permian Triassic extinction (≈251 Mya)* was by far the worst of the five mass extinctions; 95% of all species (marine as well as terrestrial) were lost, including 53% of marine families, 84% of marine genera, and 70% of land plants, insects, and vertebrates. Causes are debated, but the leading candidate is flood volcanism emanating from the Siberian Traps, which led to profound climate change. Volcanism may have been initiated by a bolide impact, which led to loss of oxygen in the sea. The atmosphere at that time was severely

hypoxic, which likely acted synergistically with other factors…Most terrestrial vertebrates perished, but among the few that survived were early representatives of the three orders of amphibians that survive to this day…*The End Triassic extinction (≈199–214 Mya)* was associated with the opening of the Atlantic Ocean by sea floor spreading related to massive lava floods that caused significant global warming. Marine organisms were most strongly affected but terrestrial organisms also experienced much extinction…*The most recent mass extinction was at the Cretaceous-Tertiary boundary (≈65 Mya)*; 16% of families, 47% of genera of marine organisms, and 18% of vertebrate families were lost…Most notable was the disappearance of non-avian dinosaurs. Causes continue to be debated. Leading candidates include diverse climatic changes (e.g., temperature increases in deep seas) resulting from volcanic floods in India (Deccan Traps) and consequences of a giant asteroid impact in the Gulf of Mexico (Wake/Vredemburg 2008).

But independently of the causes (which will continue to be debated, as both palaeontologists recognize), it is important to be aware that extinctions have occurred in the geological past of our planet and consequently they remain a source of preoccupation and danger, and – what is worse – in current times the human species is inducing the kind of changes that will accelerate extinction.

3.3   Are We in the Midst of the Sixth Mass Extinction?

According to Wake and Vredemburg, the principal cause of the extinction danger that humanity is currently experiencing is closely linked to human intervention in terrestrial ecosystems and biodiversity. The growth of the human population has had profound implications because of the demands placed on natural resources, which have increased dramatically since the Industrial Revolution and are "connected to nearly every aspect of the current extinction event". The two authors mention the growing disappearance of amphibians that have been severely impacted by habitat modification and destruction "which has frequently been accompanied by the use of fertilizers and pesticides" besides many other pollutants that are by-products of human activities and have been direct or indirect agents for the introduction of exotic organisms. This is dangerous because of the clear relation with the 2020/21 world health crisis provoked by coronavirus because, with the expansion of human populations into new habitats, "*new infectious diseases* have emerged" that have real or potential consequences for humans.

The two authors refer likewise to the fact that the most profound human impact on the environment is related to the human role in climate change, for the reason that, even if the effects of that role may have been relatively small so far, they will increase dramatically in a short period of time. As shown by the research on amphibians, many factors are contributing to their global extinction and decline, including disease, global warming and the increase in climatic variability that is intensifying the vulnerability of high-risk species. As they say:

[M]ultiple factors acting synergistically are contributing to the loss of amphibians but we can be sure that behind all of these activities is one weedy species, *Homo sapiens*, which has unwittingly achieved the ability to directly affect its own fate and that of most of the other species on this planet. It is an intelligent species that potentially has the capability of exercising necessary controls on the direction, speed, and intensity of factors related to the extinction crisis. Education and changes of political direction take time that we do not have, and political leadership to date has been ineffective largely because of so many competing, short-term demands. A primary message from the amphibians, other organisms, and environments, such as the oceans, is that little time remains to stave off mass extinctions, if it is possible at all (Wake/Vredemburg 2008).

On the other hand, Hans Günter Brauch has pointed out the following:

From 1950 to 2000 the percentage of the world's population living in urban areas grew from 30% to 50% and continues to grow strongly. … The pressure on the global environment from this burgeoning human enterprise is intensifying sharply. Over the past 50 years, humans have changed the world's ecosystems more rapidly and extensively than in any other comparable period in human history … The Earth is in its sixth mass extinction event, with rates of species loss growing rapidly for both terrestrial and marine ecosystems … The atmospheric concentrations of several important greenhouse gases have increased substantially, and the Earth is warming rapidly … More nitrogen is now converted from the atmosphere into reactive forms by fertilizer production and fossil fuel combustion than by all of the natural processes in terrestrial ecosystems put together (Brauch 2019: 10).

Thus, I am not in any doubt that the whole world has now entered the sixth period of mass extinction of species, that this phenomenon could include *Homo sapiens*, and, even worse, that the main source of the extinction is human agency. Nevertheless, it is also clear that this unconscious, irresponsible and predatory behaviour is not perpetrated by humankind as a whole, but by its ruling capitalist classes. Within that segment of society rentiers oligarchies are the main culprits because they have been managing the economy in an unsustainable way for the sake of the accumulation of capital and not of human needs, and still less with adequate regard for the conservation of nature. Neoliberalism (the ideology that preaches the worshipping of 'markets') produces a mindset which rationalizes this kind of behaviour and justifies an economy based on "fossil capitalism" (Angus 2016) through its dependence on natural gas, petrol and carbon as the main combustibles for transport, the plastics industry, heating, energy and so on.

Thus 'rampant capitalism' is the real culprit of the climate change hecatomb to come in the foreseeable future, a catastrophe of gigantic proportions that could include the disappearance of *Homo sapiens* itself as the dominant species on the planet. Thus avoiding this hecatomb is a matter of life and death for humanity. That is why I have given this book such an arresting and interrogative title that is related to the fact that, as we are already living in the Anthropocene, either we must put an end to neoliberal capitalism or be obliged to face our own extinction. I will discuss some philosophical aspects of this situation on the following pages.

3.4   Is the Anthropocene the End of Nature?

It is a long time since Fukuyama (1992) provoked the turmoil of social scientists and historians by proclaiming the end of history. Regardless of the fact that his assertion

– based on Hegelian philosophy – may or may not have been accurate, or could have been designed solely to attract attention and provoke academic debate (which it did), we may now face a similar situation. The Anthropocene could provoke the real end of history (and not just the end of the opposition between capitalism and communism as in Fukuyama's book) for the reason that if our species itself is at risk of disappearing, history will share the same fate as the *end of nature* (Arias Maldonado 2017), which could become a terrible reality with the extinction of all living beings.

While it is just a metaphorical expression that seeks to illustrate the idea that the anthropogenic intervention on Earth is so extreme and serious that nature is also at risk of disappearing through urbanization and industrialization, it cannot conceal the fact that the ecological footprint left on nature by human beings since the beginning of the industrial era in the eighteenth century is of such magnitude that there could already be a geological mark on the rock layers of the planet. Even though the *International Stratigraphy Commission* has yet to verify the existence of such layers, it is important to be aware that the scientific community – represented by the *Anthropocene Working Group* (AWG) – has finally recommended accepting the term on the basis that the *Great Acceleration* of industrialization, economic growth, demographic explosion, greenhouse emissions, and global warming etc. has left significant hoofmarks on the planet's surface since the Second World War. Hence the term has been officially proposed but not yet accepted by the hierarchy of geological sciences.[1]

Returning to the geological mark, it is assumed that it would indicate a change of the epoch within the Holocene, a period of climate warming initiated after the last ice age some 12,000 years ago. The Holocene, as can be seen below in the graph of the American Geological Society, belongs in turn to the Quaternary period that is located within the Cenozoic era in the geological history of the Earth (which is measured in millions of years), and it is thanks to those less cold temperatures that the human species could extend its habitat throughout the planet as well as obtain "possession of the domain of nature", as modern positivist and rationalist philosophies have erroneously called it.

However, the landmarks in the geological strata of the Earth are of relatively minor importance in relation to the social, political and cultural implications of this change of epoch because this would also signal the end of the concept of nature as our species has understood it until now. That is to say, until now nature has been regarded as something separate and distinct from the human being, and also at the disposal of humanity to be exploited for collective benefit without worrying about what may happen to future generations or to the planet (Fig. 3.1).

---

[1] It is worth pointing out that acceptance of the term is "defined by the standard means for a unit of the Geological Time Scale, via a *Global Boundary Stratotype Section and Point* (GSSP), colloquially known as a 'golden spike'"; that "its beginning would be optimally placed in the mid-twentieth century, coinciding with the array of geological proxy signals preserved within recently accumulated strata and resulting from the Great Acceleration of population growth, industrialization and globalization"; and also that "the sharpest and most globally synchronous of these signals, that may form a primary marker, is made by the artificial radionuclides spread worldwide by the thermonuclear bomb tests from the early 1950s", as quoted by Brauch (2019: 5).

Ultimately, what is truly at stake is the question of the sustainability of development, because if we continue to think that nature is at our disposal and its resources are inexhaustible and unlimited, economic development is not only unsustainable, but would lead us to an ecological catastrophe. Therefore, as with the end of history, the scientific and philosophical debate is not truly about the 'end of nature' but the end of an incorrect conception (and perception) of nature. Furthermore, apart from being the end of that erroneous instrumental vision, it is also the end of the idea that the human being is above nature and not part of it. This has also had implications for the relationship between social and natural sciences, because at present the knowledge produced by a transdisciplinary paradigm has arrived at the conclusion that natural and social sciences are in an integral, holistic, and indissoluble relationship, which highlights the fact that the paradigmatic change in the natural sciences produced by

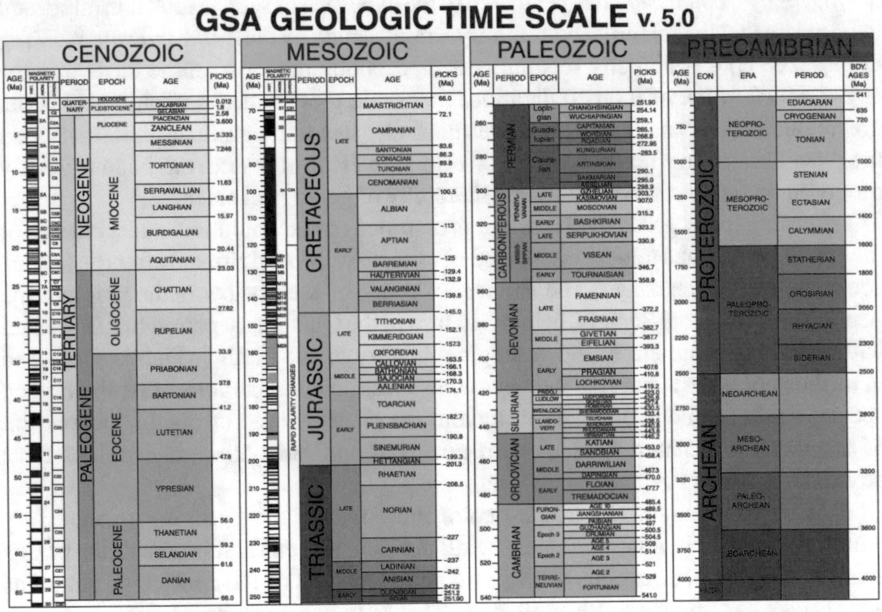

Fig. 3.1 Geological time scale (in millions of years). The epoch of the Holocene, initiated roughly 12,000 years ago, appears in the upper part of the Quaternary period, during the Cenozoic scale. *Source* https://www.geosociety.org/documents/gsa/timescale/timescl.pdf[2]

---

[2] "The Pleistocene is divided into four ages, but only two are shown here. What is shown as Calabrian is actually three ages – Calabrian from 1.80 to 0.781 Ma, Middle from 0.781 to 0.126 Ma, and Late from 0.126 to 0.0117 Ma. The Cenozoic, Mesozoic, and Paleozoic are the Eras of the Phanerozoic Eon. Names of units and age boundaries usually follow the Gradstein et al. (2013, updated) compilation. The numbered epochs and ages of the Cambrian are provisional. A " ~ " before a numerical age estimate typically indicates an associated error of ±0.4 to over 1.6 Ma." Walker, J. D., Geissman, J. W., Bowring, S. A., and Babcock, L. E. (compilers), 2018: "Geologic Time Scale v.5.0" (Geological Society of America 2018).

quantum physics a century ago has had important repercussions not only for physics, chemistry and biology but also social sciences.

Indeed, the links between natural and social sciences have been regarded very differently by the academic world since the deep overlap between the two spheres of knowledge was corroborated (Padilla 2018). For instance, in his well-known essay, "The Discourse of Science", Santos (2009) proposes a thesis – a novel approach from an epistemological point of view – which argues that the traditional positivist-rationalist paradigm of scientific knowledge has been profoundly modified by the scientific revolution initiated by Einstein at the beginning of the twentieth century, and subsequently pursued by quantum physics, since this scientific revolution has made the traditional distinction between natural sciences and social sciences disappear. Santos wrote:

> The dichotomous distinction between natural science and social science ceased to have meaning and utility. This distinction rests on a conception of matter and nature, which is contrasted with presupposed evidence, the conceptions of the human being, culture and society. The recent advances in physics and biology question the distinction between the organic and the inorganic, between living beings and inert matter and even between the human and the non-human. The characteristics of the self-organization of metabolism and self-reproduction, previously considered specific to living beings, are attributed today to the pre-cellular molecular systems. And whether we want it or not, in other places we recognize properties and behaviours previously considered as specific to human beings and social relationships. The theory of the dissipative structures of Prigogine, and the synergistic theory of Haken already cited, but also the theory of the 'implicate order' of David Bohm, the theory of the matrix-S of Geoffrey Chew and the philosophy of the 'bootstrap' that underlies it, and even the relationship between contemporary physics and the oriental mysticism studied by Fritjof Capra; all of them are of holistic vocation and some are specifically aimed at overcoming the inconsistencies between quantum mechanics and Einstein's theory of relativity. All these theories introduce into the subject the concepts of historicity and process, freedom, self-determination and even consciousness that society had previously reserved inwardly. It is as if men and women had thrown themselves into the adventure of knowing the most distant and different objects of themselves (in outer space), and once having arrived there they discovered them reflected as in a mirror (Santos 2009: 41).

Consequently, since – based on the new paradigm of quantum physics – exponents of the natural sciences are disposed to accept that notions such as those of historicity, indeterminacy and even consciousness can be applicable to what has always been considered the *material world* by philosophers and scientist such as Wallace (2003), Wilber (2007), Laszlo (2006), Capra (1975, 1982, 1996), Ricard/Thuan (2000), Morin (1999) and even the Dalai Lama (2005), certain categories of the natural sciences are perfectly applicable to issues that were previously considered exclusively within the domain of the social sciences, as posited by Lovelock's 'Gaia Theory'.[3] One example in current times is the issue of climate change, because one

---

[3] James Lovelock's 'Gaia theory' refers to the confluence of the best available scientific understanding of Earth as a living system with the social science and cultural philosophy that conceives society as a "seamless continuum of that system". Lovelock's theory posits that the organic and inorganic components of the planet have evolved together as a single living, self-regulating system that has always automatically controlled global temperature, atmospheric content, ocean salinity, and other factors in a way that maintains its own habitability in a comparable manner to other living

of the main causes of the warming of the Earth's climate is the increase in greenhouse gas emissions produced by man, ergo, anthropogenic. In other words, natural phenomena that were previously considered unrelated to the actions of mankind now prove to be connected with human history. Lovelock's theory about the planet as a living being, and the philosophy (or 'cosmovision') of the Andean indigenous people who believe that *Pachamama*, or 'Mother Nature', is entitled to rights even if it is neither a person nor a collective group of persons on whom people's collective rights are based, demonstrate that this new – or, more accurately, revived – approach to the philosophy of law[4] that arises from both the social sciences (sociology of law, legal sciences) and the natural sciences (geology, physics, biology, ecology) could be fruitful.[5]

In other cases, the close relationship between the natural sciences and the social sciences is manifested in a true interdisciplinary and methodological 'coupling'. As Pope Francis says in his Encyclical Letter *Laudato Si'* quoted at the start of this chapter:

> It cannot be emphasized enough how everything is interconnected. Time and space are not independent of one another, and not even atoms or subatomic particles can be considered in isolation. Just as the different aspects of the planet – physical, chemical and biological – are interrelated, so too living species are part of a network which we will never fully explore and understand. A good part of our genetic code is shared by many living beings. It follows that the fragmentation of knowledge and the isolation of bits of information can actually become a form of ignorance, unless they are integrated into a broader vision of reality (Pope Francis 2015).

In another example of this interrelation, the journal *Science* (September 2007) highlighted six cases in which this interrelation was put into practice in research projects in which a multidisciplinary team of more than twenty scholars used methodology and techniques from both natural and social sciences. In addition, work was carried out in both developed and developing countries – two in the USA (Wisconsin and Washington), one in Sweden (Kattenriket), one in China (Wolong Shenshuping Giant Panda Base), one in Kenya and another in Altamira, Brazil. The conclusions reached by the authors are of great importance:

> These studies share four major features. First, they explicitly address complex interactions and feedback between human and natural systems. Unlike traditional ecological research that

---

organisms like plants, animals and human beings. This means that life permanently maintains the conditions suitable for its own survival. As a consequence, Lovelock's main hypothesis is that the living system of Earth can be considered analogous to the workings of any individual organism that regulates body temperature, blood salinity, etc. So, for instance, even though the luminosity of the sun – the Earth's heat source – has increased by about 30% since life began almost four billion years ago, the planet's living system has reacted as a whole to maintain temperatures at levels suitable for life. The Gaia theory has been supported since the outset by Lynn Margulis, an influential microbiologist at the University of Massachusetts (Lovelock 1979).

[4] It is pertinent to remember that in the domain of the philosophy of law, both the Greek philosophers – like Plato and Aristotle – and Catholic theologians of the Middle Ages (e.g. Thomas Aquinas) spoke about "natural law" as the source of the norms enacted by kings and rulers.

[5] Readers interested in these issues can find out more from an extensive bibliography by various authors, including Barié (2017), Gudynas (2017), Boff (2018) and Matul/Cabrera (2007).

often excluded human impacts or social research that generally ignored ecological effects, these studies consider both ecological and human components as well as their connections. Thus, they measure not only ecological variables (e.g., landscape patterns, wildlife habitat, and biodiversity) and human variables (e.g., socio-economic processes, social networks, agents, and structures of multi-level governance) but also variables that link natural and human components (e.g., fuel-wood collection and use of ecosystem services). Second, each study team is interdisciplinary, engaging both ecological and social scientists around common questions. Third, these studies integrate various tools and techniques from ecological and social sciences as well as other disciplines such as remote sensing and geographic information sciences for data collection, management, analysis, modelling, and integration … Fourth, they are simultaneously context-specific and longitudinal over periods of time long enough to elucidate temporal dynamics. As such, these studies have offered unique inter-disciplinary insights into complexities that cannot be gained from ecological or social research alone. In coupled human and natural systems, people and nature interact reciprocally and form complex feedback loops (Liu et al. 2007).

Returning to the concept of Anthropocene, this term was proposed by the outstanding Dutch scientist – with a Nobel Prize in Chemistry – Paul Crutzen, of the Max Plank Institute of Chemistry, together with the American scientist Eugene Stoermer of the University of Michigan, who, in a joint article (Crutzen/Stoermer 2000) explain the reasons that led them to make the proposal of the Anthropocene as a new geological epoch:

The expansion of mankind, both in numbers and per capita exploitation of Earth's resources has been astounding … during the past 3 centuries human population increased tenfold to 6000 million, accompanied e.g. by a growth in cattle population to 1400 million … Urbanization has even increased tenfold in the past century. In a few generations mankind is exhausting the fossil fuels that were generated over several hundred million years. The release of SO_2, globally about 160 Tg/year to the atmosphere by coal and oil burning, is at least two times larger than the sum of all natural emissions … 30–50% of the land surface has been transformed by human action; more nitrogen is now fixed synthetically and applied as fertilizers in agriculture than fixed naturally in all terrestrial ecosystems; the escape into the atmosphere of NO (nitrogen oxide) from fossil fuel and biomass combustion likewise is larger than the natural inputs, giving rise to photochemical ozone ('smog') formation in extensive regions of the world; more than half of all accessible fresh water is used by mankind; human activity has increased the species extinction rate by thousand to ten thousand fold in the tropical rain forests … and several climatically important 'greenhouse' gases have substantially increased in the atmosphere: CO_2 by more than 30% and CH_4 by even more than 100%. Furthermore, mankind releases many toxic substances in the environment and even some, the chlorofluorocarbon gases, which are not toxic at all, but which nevertheless have led to the Antarctic 'ozone hole' and which would have destroyed much of the ozone layer if no international regulatory measures to end their production had been taken. Coastal wetlands are also affected by humans, having resulted in the loss of 50% of the world's mangroves. Finally, mechanized human predation ('fisheries') removes more than 25% of the primary production of the oceans in the upwelling regions and 35% in the temperate continental shelf regions. Anthropogenic effects are also well illustrated by the history of biotic communities that leave remains in lake sediments. The effects documented include modification of the geochemical cycle in large freshwater systems and occur in systems remote from primary sources. Considering these and many other major and still growing impacts of human activities on Earth and atmosphere, and at all, including global, scales, it seems to us more than appropriate to emphasize the central role of mankind in geology and ecology by proposing to use the term 'Anthropocene' for the current geological epoch (Crutzen/Stoermer 2000: 17).

In that same article Crutzen and Stoermer remind us that it was the famous philosopher and Jesuit priest Pierre Teilhard de Chardin (1965) who used the term "noosphere" to name the sphere of human knowledge that manifests itself together with the biosphere – the sphere of life – that, like the atmosphere, surrounds the Earth. It is therefore a term intimately linked to the notion of the Anthropocene, since this *sphere of knowledge* or culture has a spiritual dimension from which man becomes a geological force transforming the planet (and nature) due to the use of science and technology. For the French palaeontologist and philosopher, the *noosphere*, then, is a spiritual phenomenon comparable to a *universal consciousness*, one of whose manifestations is the human mind, but whose origin, like that of plant and animal life on the planet, can be explained by the theory of evolution formulated by Charles Darwin. Evolution, according to Teilhard, transits from the *geosphere* – where the geological evolution of matter takes place – to the *biosphere* – where the biological evolution that gives origin to *Homo sapiens* during the course of millions of years takes place – culminating in the *noosphere* or higher spiritual stage of evolution, which is the one that leads to the universal consciousness or *omega point* as well as to the energy released in the act of thought that interconnects the spiritual energy generating this universal consciousness. Therefore it is possible to affirm that human evolution, as well as consciousness and science and technology (culture), is a source of the anthropogenic change that is modifying nature – a phenomenon that in geological terms is tantamount of the Anthropocene. Thus if the psychic or spiritual side of nature is determinant in explaining not only the origin of life but also the culmination of an evolutionary process where the Earth's noosphere is replaced by the *universal consciousness* or *Omega Point* that consecrates realization of the spirit on the planet, it is clear that material things and spirituality are closely linked. Hence, for Teilhard de Chardin, the *Omega Point* can be understood as the "harmonized collectivity of consciences, which amounts to a kind of super-consciousness. The Earth being covered not only with grains of thought, counting by myriads, but enveloping itself from a single thinking envelope until it forms no more than a single and wide grain of thought, on a sidereal scale. The plurality of individual reflections grouped and reinforced in the act of a single unanimous reflection" (Chardin 1965: 383).

The novelty of Teilhardian thought is its philosophical approach, which aims to reconcile the theory of evolution with Catholic theology, although, as might have been expected during the conservative papacy of Pius XII, these ideas generated difficulties for the Vatican's orthodox doctrine. Nevertheless, both Paul VI and John Paul II valued the ideas of the Jesuit later on. In a work on Catholic theology in 1987 Cardinal Ratzinger (later Benedict XVI) acknowledged that the pastoral document of the Second Vatican Council *Gaudium et Spes* had significant influence on the theological debates of that conclave. Certainly, there are those who remember that the ideas of the Jesuit philosopher were inspired – besides by Darwin – by the theory of Russian scientist Vernadsky (1997), who was the first to use the concept of *noosphere* to refer to the third stage of a succession of phases in the evolution of the Earth, ranging from the geosphere or physiosphere – which consists of inanimate matter – to the biosphere or field of life. Thus, just as the emergence of life transformed the geosphere, the emergence of knowledge and science transformed the biosphere.[6]

Consequently, and making an extrapolation that seems applicable from its philosophical conception, it could be argued that thought modifies the biological and geological processes, as a consequence of the expansion of culture (which includes science and technology). The result is that for thousands of years humanity has been modifying ecosystems through its ideas, altering flora and fauna, extinguishing species and creating new domestic varieties, but – above all – it is altering natural ecosystems due to the 'Great Acceleration' of human action on the planet. This began with the Industrial Revolution in eighteenth-century England, but increased from the 1950s onwards, as can be seen in Figs. 3.2, 3.3 and 3.4 that give a scientific basis to ideas that – like those of Teilhard de Chardin or Russell (2007) – have tried to make that synthesis between philosophy and science. Now, thanks to the concept of Anthropocene, these ideas have posed a new challenge for geologists and palaeontologists accustomed to looking for stratigraphic marks on the rocky crust of the planet in order to determine how those linkages between natural and social sciences operate in the field of geology.

But referring (again) to the temporal dating from which the geological epoch of the Anthropocene began, since when did this influence of the noosphere begin to leave a physical imprint on the planet? Crutzen and Stoermer propose the beginnings of the industrial age (early nineteenth century) that, as we know, runs parallel to the development of science and technology in the modern age, while other researchers have suggested the marker of the start of the Anthropocene to be the presence of radioactive isotopes in the atmosphere as a consequence of the nuclear bombs that were blown up by the US on Hiroshima and Nagasaki in 1945 and the subsequent cycle of atomic tests of the great powers that lasted until such tests were banned by the *Convention To Ban Nuclear Testing* (CTBNT). That is the most widely accepted hypothesis, despite the suggestion that the date could be traced back to the beginning of agriculture on Earth (Ruddiman 2005: 46–53).[7]

---

[6] However, in contrast to the biological vision of James Lovelock (for whom the Earth itself – *Gaia* – is a living organism), or to the spiritual conception of Teilhard, for Vernadsky (1997) the determining factor of the origin of life is physical matter itself. According to him, humanity, thanks to the scientific thought that permeates the noosphere, can modify and take control of nature from a perspective in tune with the *dialectical materialism* that predominated in Soviet Russia at that time.

[7] Ruddiman (2005) argues that the Anthropocene was initiated about 8,000 years ago with the onset of agriculture because anthropogenic methane and carbon dioxide emissions started then and created the conditions to prevent another global cooling. Agriculture spread throughout the world thanks to migrations across all continents at the beginning of the Neolithic revolution, during which time humans developed agriculture and the domestication of animals that led to livestock of all kinds. This also allowed hunting and gathering to be replaced as the main source of food supply. Such innovations were not purely positive, because they were followed by a wave of extinctions, beginning with large mammals and land-birds. This wave was driven by the direct activity of human beings (such as hunting) as well as by indirect consequences arising from the change in land use for agriculture, but there are those who claim that such extinctions did not have an anthropogenic origin.

Be that as it may, in 2008 a large group of British geologists and researchers stated that – even taking into account the fact that scientific method demands evidence that must be provided by studying the geological strata – to a large extent the formal adoption of the term depends on its usefulness. In an article signed by a score of scientists from the Stratigraphy Commission of the *Geological Society of London* published in 2008 by the journal of the *Geological Society of America* (GSA) they affirmed the following:

> The term Anthropocene, proposed and increasingly employed to denote the current interval of anthropogenic global environmental change, may be discussed on stratigraphic grounds. A case can be made for its consideration as a formal epoch in that, since the start of the Industrial Revolution, Earth has endured changes sufficient to leave a global stratigraphic signature distinct from that of the Holocene or of previous Pleistocene interglacial phases, encompassing novel biotic, sedimentary, and geochemical change. These changes, although likely only in their initial phases, are sufficiently distinct and robustly established for suggestions of a Holocene-Anthropocene boundary in the recent historical past to be geologically

Socio-economic trends

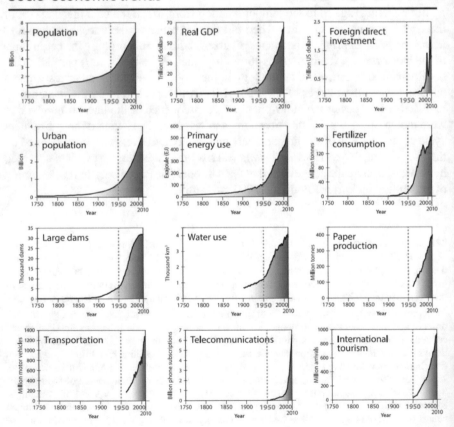

◄**Fig. 3.2**   Socio-economic trends: population, GDP, FDI (foreign investment), urbanization, energy, fertilizer consumption, dams, water use, paper production, transportation, telecommunications, and international tourism. *Source* Steffen et al. (2015): "The Trajectory of the Anthropocene: The Great Acceleration", in: *The Anthropocene Review*: 1–18. Explanation: Trends from 1750 to 2010 in globally aggregated indicators for socio-economic development. (1) Global population data according to the HYDE (*History Database of the Global Environment*) 2013 data base. Data before 1950 are modelled. Data are plotted as decadal points. (2) Global real GDP (*Gross Domestic Product*) in year 2010 US dollars. Data are a combination of Maddison for the years 1750–2003 and Shane for 1969–2010. Overlapping years from Shane data are used to adjust Maddison data to 2010 US dollars. (3) Global foreign direct investment in current (accessed 2013) US dollars based on two data sets: IMF (*International Monetary Fund*) from 1948 to 1969 and UNCTAD (*United Nations Conference on Trade and Development*) from 1970 to 2010. (4) Global urban population data according to the HYDE database. Data before 1950 are modelled. Data are plotted as decadal points. (5) World primary energy use. 1850 to present based on Grubler et al. (2012); 1750–1849 data are based on global population using 1850 data as a reference point. (6) Global fertilizer (nitrogen, phosphate and potassium) consumption based on *International Fertilizer Industry Association* (IFA) data. (7) Global total number of existing large dams (minimum 15 m height above foundation) based on the ICOLD (International Committee on Large Dams) database. (8) Global water use is sum of irrigation, domestic, manufacturing and electricity water withdrawals from 1900 to 2010 and livestock water consumption from 1961 to 2010. The data are estimated using the WaterGAP model. (9) Global paper production from 1961 to 2010. (10) Global number of new motor vehicles per year. From 1963 to 1999 data include passenger cars, buses and coaches, goods vehicles, tractors, vans, lorries, motorcycles and mopeds. Data 2000–2009 include cars, buses, lorries, vans and motorcycles. (11) Global sum of fixed landlines (1950–2010) and mobile phone subscriptions (1980–2010). Landline data are based on Canning for 1950–1989 and UN data from 1990 to 2010, while mobile phone subscription data are based solely on UN data. (12) Number of international arrivals per year for the period 1950–2010. *Source* (1) HYDE database; Klein Goldewijk et al. (2010). (2) Maddison (1995, 2001); M Shane, Research Service, *United States Department of Agriculture* (USDA); Shane (2014). (3) IMF (2013); UNCTAD (2013). (4) HYDE database (2013); Klein Goldewijk et al. (2010). (5) A Grubler, *International Institute for Applied Systems Analysis* (IIASA); Grubler et al. (2012). (6) Olivier Rousseau, IFA; IFA database. (7) ICOLD database register search. Purchased 2011. (8) M Flörke, Centre for Environmental Systems Research, University of Kassel; Flörke et al. (2013); aus der Beek et al. (2010); Alcamo et al. (2003). (9) Based on FAO (Fisheries and Aquaculture Department online) online statistical database FAOSTAT. (10) International Road Federation (2011). (11) Canning (1998); *United Nations Statistics Division* (UNSD) (2014). (12) Data for 1950–1994 are from UNWTO (*United Nations World Tourism Organization*) (2006), data for 1995–2004 are from UNWTO (2011), and data for 2005–2010 are from: http://www.commonhomeofhumanity. org/

reasonable. The boundary may be defined either via Global Stratigraphic Section and Point ("*golden spike*") locations or by adopting a numerical date. Formal adoption of this term in the near future will largely depend on its utility, particularly to Earth scientists working on late Holocene successions. This datum, from the perspective of the far future, will most probably approximate a distinctive stratigraphic boundary … (and to conclude): Sufficient evidence has emerged of stratigraphically significant change (both elapsed and imminent) for recognition of the Anthropocene – currently a vivid yet informal metaphor of global environmental change – as a new geological epoch to be considered for formalization by international discussion. The base of the Anthropocene may be defined by a GSSP in sediments or ice cores or simply by a numerical date (Zalasiewicz et al. 2008: 45).

Consequently, if the Earth has undergone changes of such magnitude, we would be facing a new epoch at least from the beginning of the Industrial Revolution (around 1750) because such changes are sufficient to leave a "global stratigraphic signature" that is different from the one of the Holocene and the previous interglacial phases of the Pleistocene, which entail new changes of a biotic, sedimentary and geochemical nature. Although the stratigraphic frontier (from the beginning of the Anthropocene) could be established by means of nail – or "golden spike" – sediments in the polar ice, it is also possible to use a numerical date to mark the beginning of the said geological time. However, what really matters in the adoption of the term is its usefulness for scientists working on the latest changes occurring within the Holocene.

Earth system trends

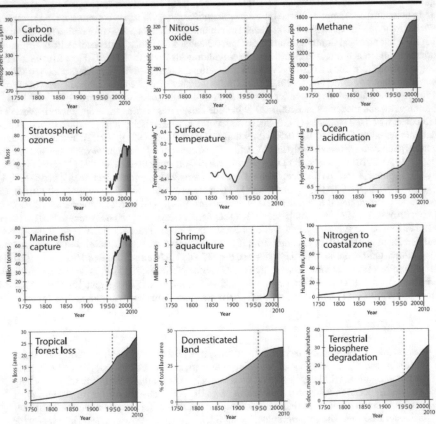

◄**Fig. 3.3** Trends in the terrestrial ecosystem (earth system trends): carbon dioxide, nitrous oxide, methane, stratospheric ozone, surface temperature, ocean acidification, depredation of fish and crustaceans, nitrogen in coastal areas, deforestation, land increase for agriculture, degradation of the terrestrial biosphere. Explanation: trends from 1750 to 2010 in indicators for the structure and functioning of the Earth System. (1) Carbon dioxide from firn and ice core records (Law Dome, Antarctica) and Cape Grim, Australia (deseasonalised flask and instrumental records); spline fit. (2) Nitrous oxide from firn and ice core records (Law Dome, Antarctica) and Cape Grim, Australia (deseasonalised flask and instrumental records); spline fit. (3) Methane from firn and ice core records (Law Dome, Antarctica) and Cape Grim, Australia (deseasonalised flask and instrumental records); spline fit. (4) Maximum percentage total column ozone decline (2-year moving average) over Halley, Antarctica during October, using 305 DU, the average October total column ozone for the first decade of measurements, as a baseline. (5) Global surface temperature anomaly (HadCRUT4: combined land and ocean observations, relative to 1961–1990, 20 yr Gaussian smoothed). (6) Ocean acidification expressed as global mean surface ocean hydrogen ion concentration from a suite of models (CMIP5) based on observations of atmospheric CO_2 until 2005 and thereafter RCP8.5. (7) Global marine fishes capture production (the sum of coastal, demersal and pelagic marine fish species only), i.e. it does not include mammals, molluscs, crustaceans, plants, etc. There are no FAO data available prior to 1950. (8) Global aquaculture shrimp production (the sum of 25 cultured shrimp species) as a proxy for coastal zone modification. (9) Model-calculated human-induced perturbation flux of nitrogen into the coastal margin (riverine flux, sewage and atmospheric deposition). (10) Loss of tropical forests (tropical evergreen forest and tropical deciduous forest, which also includes the area under woody parts of savannas and woodlands) compared with 1700. (11) Increase in agricultural land area, including cropland and pasture as a percentage of total land area. (12) Percentage decrease in terrestrial mean species abundance relative to abundance in undisturbed ecosystems as an approximation for degradation of the terrestrial biosphere (FAO-FIGIS 2013). (8) Data are from the FAO Fisheries and Aquaculture Department online database FishstatJ (FAO 2013). (9) Mackenzie et al. (2002). (10) J. Pongratz, Carnegie Institution of Washington, Stanford, USA; Pongratz et al. (2008). 1700–1992 is based on reconstructions of land use and land cover (Pongratz et al. 2008). Beyond 1992 is based on the IMAGE land use model. (11) J. Pongratz, Carnegie Institution of Washington, Stanford, USA; Pongratz et al. (2008). 1700–1992 is based on reconstructions of land use and land cover (Pongratz et al. 2008). Beyond 1992 is based on the IMAGE land-use model. (12) R. Alkemade, PBL Netherlands Environmental Assessment Agency: modelled mean species abundance using GLOBIO3 based on HYDE reconstructed historical land-use change estimates (until 1990) then IMAGE model estimates (Alkemade et al. 2009; available at: www.globio.info; ten Brink et al. 2010). Steffen et al.: "The Trajectory of the Anthropocene: The Great Acceleration" (2015), in: *The Anthropocene Review*: 1–18; at: http://www.commonhom eofhumanity.org/pdf

3.5 The Great Acceleration

Despite the fact that the changes suffered by the planet "have had the ability to leave that global stratigraphic signature" behind, as British researchers claim, what makes the Anthropocene qualify as a formal geological epoch from (according to Crutzen) the beginning of the Industrial Revolution is that it is characterized by the so-called 'Great Acceleration', which includes the following changes: (a) the extinction of plants and animals at rates which have increased above the historical average of the Earth (the most encouraging predictions state that if this continues), 75% of species – including *Home sapiens* – could become extinct in the foreseeable future; (b) the excessive increase in CO_2 emissions in the atmosphere, which makes

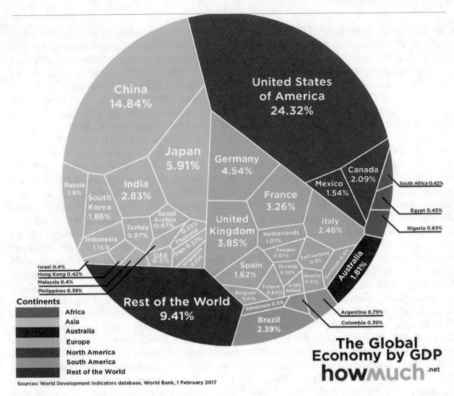

Fig. 3.4 Gross domestic product of the world economy according to the World Bank. *Source* Image: World Bank; at: https://www.weforum.org/agenda/2017/03/worlds-biggest-economies-in-2017/

it possible to foresee a global warming of great magnitude; (c) the presence of carbon in the atmosphere has reached 400 parts per million because of industrial emissions, whereas 'normal' is between 200 and 300; (d) the plastic poured into rivers, lakes and seas is not only seriously contaminating the oceans and affecting marine life, but also leaves micro-plastic particles that will inevitably cause fossil traces in subsequent generations; (e) the use of fertilizers has caused the level of nitrogen and phosphorus to rise to twice its usual level, leading to a record and unprecedented modification to the nitrogen cycle in more than 2 billion years; (f) the burning of fossil fuels is producing aerial particles like black carbon that literally stain glaciers and other parts of the Earth.

Thus, even if the dilemma of the Anthropocene could have been interpreted in the past as a clash between the different visions of social and natural sciences, contemporary epistemology is designed to transcend that kind of dichotomy through the transdisciplinary approach and methodology. Hence the use of this holistic paradigm and transdisciplinary approach to explain the 'Great Acceleration' in a great synthesis that includes the relation and interconnections between the socio-political and economic sphere with the natural ecosystems is fundamental. It is, of course, important to be mindful that the noosphere hypothesis is not presented as science *strictu sensu* but as philosophy, in a similar way that Teilhard de Chardin's scientific role as a palaeontologist was combined with his vocation as a Catholic priest.

This makes it necessary to go beyond concrete phenomena (such as climate change, acid rain, and the hole in the ozone layer) in order to correlate the numerical golden spike with the historical moment when the Anthropocene began. Therefore, it is possible to decide – as suggested by the collective research in which Zalasiewicz and Williams participated – that the "cultural spike" for the beginning of the Anthropocene was the Industrial Revolution around the mid-eighteenth century (the invention of the steam engine was in 1784), but finally an agreement was reached to take the Great Acceleration after the Second World War as its point of departure (as illustrated by the graphs of Steffen et al.). This implies that the structure and functioning of the planetary ecosystem could be also related to the viewpoints of Teilhard de Chardin, Lovelock and Vernadsky about the *biosphere* and to Lovelock's *Gaia* Theory – largely already accepted by the scientific community worldwide – but also to the philosophical or "metaphysical" (in the Ancient Greeksense of the concept) response to ideas like Teilhard de Chardin's *noosphere* and the *cosmopolocene* and *cosmopolitics* (Delanty/Motta 2018) that will discussed in Chap. 6.

In other words, what can be verified with those famous twelve iconic graphics of the Great Acceleration is that humanity's intervention in nature (Teilhard's *noosphere*, or human agency from Lovelock's perspective) has influenced terrestrial ecosystems in such a way that – and this is perfectly quantifiable by empirical science – the growth of the population, economy, urbanization, consumption of fertilizers and construction of hydroelectric plants, and the increase in water consumption, the production of paper, telecommunications, transport, international tourism and so on, have had an impact on the increase in carbon dioxide, nitrogen, methane and ozone in the atmosphere with the consequent increase in the temperature of the Earth's surface, the acidification of the oceans, the depredation of fish forests and so on, to such an extent that, independently of a stratigraphic mark, we can say that we are now undoubtedly living in the Anthropocene and that the Anthropocene as a cultural and normative model also has a reality that must be taken into consideration.[8]

---

[8] The scientists who worked on the elaboration of these graphs explain it in the following way: "What have now become known as the 'Great Acceleration' graphs were originally designed and constructed as part of the synthesis project of the *International Geosphere-Biosphere Programme* (IGBP), during the 1999–2003 period. The synthesis aimed to pull together a decade of research in IGBP's core projects, and, importantly, generate a better understanding of the structure and functioning of the Earth System as a whole, more than just a description of the various parts of the Earth System around which IGBP's core projects were structured. The increasing human pressure on the Earth System was a key component of the synthesis. The project was inspired by the proposal in 2000 by Paul Crutzen, a Vice-Chair of IGBP, that the Earth had left the Holocene and entered a new geological epoch, the Anthropocene, driven by the impact of human activities on the Earth System (Crutzen 2002; Crutzen/Stoermer 2000). Crutzen suggested that the start date of the Anthropocene would be placed near the end of the eighteenth century, about the time that the Industrial Revolution began, and that that a start date would coincide with the invention of the steam engine by James Watt in 1784. As part of the project, the synthesis team wanted to build a more systematic picture of

Hence – and this is the core point of my thesis – regardless of whether or not geologists can find traces or stratigraphic marks in the geological structure of the Earth, the truth is that in metaphysical terms the beginning of the Anthropocene can be located at the moment when human beings, thanks to science, technology, culture, thought, universal consciousness, or – as Teilhard de Chardin calls it – the *noosphere* (as a sphere of knowledge that covers the Earth in addition to the biosphere, geosphere and atmosphere) started to directly affect the natural terrestrial and oceanic ecosystems. This can be verified by making a correlation between the data of the Great Acceleration derived from the socio-economic field and the data that strictly refers to 'natural ecosystems', meaning the levels of carbon dioxide, methane, ozone and nitrogen present in the atmosphere.

Consequently, along with the method of elaboration and sources used for comparison, the aforementioned twelve iconic graphs of the Great Acceleration are reproduced on these pages. The first group of them concerns socio-economic trends; the second contains a comparison between the member countries of the OECD, BRICS (Brazil, Russia, India, China, and South Africa) and the rest of the countries of the world with the same twelve trends (except primary use of energy and international tourism); and the third series is related to trends in the terrestrial and oceanic ecosystems, that is to say it refers to the figures that demonstrate the impact that the great industrial, agricultural and population acceleration since the 1950s has had on the planet's ecosystems (IGBP 2015).

To conclude, it is important to bear in mind that what truly matters is not whether the evidence of a stratigraphic change is sufficiently clear and widespread for geologists, but the fact that the Anthropocene – this "vivid but informal metaphor" of the change in the global environment, as Zalasiewicz thinks; or this new "cultural model" as Delanty and Motta argue – has already been accepted by the scientific community as a new geological epoch. In the following pages I will briefly refer to some of the more severe issues, such as the acceleration in population growth; the prevalence of the idea of unlimited growth as a criterion for defining development; consumerism; and the "culture of discarding", which leads to the accumulation of waste and toxic waste, as well as to problems concerning energy, water, food and climate change.

---

the human-driven changes to the Earth System, drawing primarily, but not exclusively, on the work of the IGBP core projects. The idea was to record the trajectory of the 'human enterprise' through a number of indicators and, over the same time frame, track the trajectory of key indicators of the structure and functioning of the Earth System. Inspired by Crutzen's proposal for the Anthropocene, we chose 1750 as the starting date for our trajectories to ensure that we captured the beginning of the Industrial Revolution and the changes that it wrought. We took the graphs up to 2000, the most recent year that we had data for many of the indicators" (Steffen et al. 2004).

3.6 Demographic Explosion

The disease that afflicts the Earth is not just climate change – manifested by drought, heat, and an ever-rising sea. Added to this there is the changing chemistry of the air and the oceans, and the way the sea grows acidic. Then there is the shortage of food for all consumers of the animal kingdom. As important is the loss of that vital biodiversity that enables the working of an ecosystem. All these affect the working of the Earth's operating system and are the consequences of too many people. Individuals occasionally suffer a disease called polycythaemia, an overpopulation of red blood cells. By analogy, Gaia's illness could be called polyanthroponemia, where humans overpopulate until they do more harm than good.

Lovelock (2009: 233)

For thousands of years the world population maintained a relatively low and stable growth rate, thus it was not until the first years of the nineteenth century that the world population grew beyond one billion human beings, and it was not until 1959 that this amount surpassed 3 billion. Since then, the Great Acceleration in demographic matter has been taking place. In the very short period of 53 years, it took a gigantic leap. The amount of people comprising the world population grew from 3 to 7.2 billion by 2012, of which 1,350 million were inhabitants of China, 1,260 million of India, 241 million of Indonesia and 194 million of Brazil. These are the four countries with the largest populations in the developing world, which means that more than 3 billion people live in just those four countries, tripling the number of inhabitants that had been around the planet at the beginning of the nineteenth century. At the beginning of 2021 the world population had risen to more than 7.8 billion. The projections for the year 2050 (as can be seen in Table 3.1) are truly worrying because, if the population growth is not reduced, the planet will, by then, be inhabited by almost 10 billion human beings.

Table 3.1 More populated countries at the global level

Country (year 2012)	Millions of people	Year 2050	Year 2021
China	1,350	1311	1,394,016,000
India	1260	1691	1,326,093,184
United States	314	423	332,639,104
Indonesia	241	309	267,026,368
Nigeria	180	402	214,028,304
Pakistan	180	314	233,500,640
Brazil	194	213	211,715,968
Bangladesh	153	226	162,650,848
Total (world) population	7,200	9.730	4,141,670,416

Source The author's elaboration based on data from Emmanuel Todd's book: *Apres l'Empire* and Index Mundi

As can be seen, overpopulated countries like India, whose fertility rate was 5.8% in 1960 but by 2012 had fallen to 2.5%, will continue growing but not at the high rate that is predicted for its more conservative neighbours with Muslim populations, such as Bangladesh and Pakistan. China, thanks to anti-natalist policies promoted by the State but also because of the rise of the middle class – which rose because of social policies and economic development promoted by the State – has managed to reduce its population growth rate to a minimum of 0.5%. This rate accounts for the annual birth of some 6 million additional babies despite the mortality rate.

However, according to its projection, the population would have decreased by 2050, which means that the development and growth of the middle class constitutes the best and most effective policy for reducing the world's population. Nonetheless, it should also be borne in mind that the countries with the highest population growth rates are usually those in which, in addition to the fertility rate, there is a high influx of migrants due to the high demand for labour. This is the case in the small oil-rich countries of the Persian Gulf (Kuwait, the Emirates, Qatar, Bahrain), as well as Saudi Arabia and other countries with a relatively higher degree of development, such as Panama, Costa Rica, the Dominican Republic, Chile, and Argentina in Latin America; and of course the United States, Canada, Australia and the European Union are among the countries with job markets because of globalization and the lack of workers due to low fertility rates.

What is also remarkable is that the demographic transition in countries such as China and India has taken place in peace, without major upheavals and without the violence that, according to demographers, usually occurs when conservative sectors that oppose the change in cultural and religious traditions react in a violent manner. This violence is due to opposition to the change of mentality that education and the new medical procedures in terms of reproductive health tend to provoke in conservative sectors, especially in countries where religion is part of the state institutions, as is usually the case in Islamic countries, and can be seen in Table 3.1 concerning the demographic projections for 2050 in very conservative Muslim countries like Indonesia, Nigeria, Pakistan and Bangladesh.[9]

---

[9] According to the French demographer and sociologist Emmanuel Todd, an explanatory hypothesis of terrorism in Islamic countries could be formulated on the basis the resistance to modernization that exists in many Muslim countries with high fertility rates (Libya 3.9, Qatar 3.9, Syria 4.1, Kuwait 4.2, Sudan 4.9, Iraq 5.3, Pakistan 5.6, Saudi Arabia 5.7, Nigeria 5.8, Afghanistan 6.0, Oman 6.1, Mali 7.0, Yemen 7.2, Somalia 7.3, Niger 7.5, Bahrain 2.8, Indonesia 2.7). The influence (or presence) of Islamist terrorism in those countries would then be explained by the rejection of modernization and the cultural change of mentality regarding the reproductive health of women, in a manner analogous to the violence that this demographic transition unleashed in Western countries during the seventeenth century (the English Civil War), the eighteenth century (the French Revolution) and the beginning of the twentieth century (the Bolshevik revolution in Russia), which assumes that these phenomena of violence and terrorism should be seen as part of the general process of the development of societies, and thus the solution is to promote education and social development, not repression or military means. Another of Todd's interesting observations is that in the countries of central Asia which were part of the USSR, the education promoted by the Soviet state produced a demographic transition that has caused the Muslim religion to lose influence, and therefore the radical Islamist movements also lack strength there. This can be demonstrated by the relatively low

The causes of population growth are multiple, but in general terms it can be said that when the fertility rate of women (the number of children they have) is high, which is considered to be more than three children, it has causes related to poverty, illiteracy and, to some degree, with the local customs and religious beliefs. In addition, for many people living in poverty, having large families is a 'survival strategy' because, before the widespread application of vaccines and antibiotics by health services, infant mortality rates were very high (Table 3.2).

Table 3.2 Some global fertility rates (number of children per woman)

	1981	2001	2021		1981	2001	2021
United States	1.8	2.1	2.0	India	5.3	3.2	2.28
Canada	1.8	1.4	1.7	Sri Lanka	3.4	2.1	2.00
UK	1.9	1.7					
France	1.9	1.9	2.04	Argentina	2.9	2.6	2.20
Germany	1.3	1.3	1.48	Mexico	4.8	2.8	2.17
Italy	1.7	1.3	1.47	Bolivia	6.8	4.2	2.45
Spain	2.5	1.2	1.51	Peru	5.3	2.9	2.02
				Brazil	4.4	2.4	1.73
Guatemala	6.27	4.49	2.67	Colombia	3.9	2.6	2.14
Romania	2.5	1.3	1.38	Venezuela	4.9	2.9	2.24
Poland	2.3	1.4	1.39				
Russia	2.0	1.2	1.60	South Africa	5.1	2.9	2.20
Ukraine	1.9	1.1		Rwanda	6.9	5.8	3.42
				Zambia	6.9	6.1	4.63
Japan	1.8	1.3	1.38	Zimbabwe	6.6	4.0	3.91
China	2.3	1.8	1.60	Kenya	8.1	4.4	3.36
Taiwan	2.7	1.7	1.07	Tanzania	6.5	5.6	4.45
South Korea	3.2	1.5	1.09	Ethiopia	6.7	5.9	4.07
North Korea	4.5	2.3		DR Congo	6.1	7.0	5.70
Vietnam	5.8	2.3	2.06	Ivory Coast	6.7	5.2	4.65
Thailand	3.7	1.8	1.54	Sierra Leone	6.4	6.3	4.04
Philippines	5.0	3.5	2.89	Liberia	6.7	6.6	4.32

Source The author's elaboration based on data from Todd (2004: 40); *Population and Societies* and the World Web (World Population Review)

---

birth rates in the former Soviet republics of Azerbaijan (2.02), Turkmenistan (2.68), Kyrgyzstan (2.9), Tajikistan (2.4) or Uzbekistan (2.38) in 2021. The same applies to the Maghreb countries that were colonized by France, such as Tunisia (2.13), Algeria (2.94) and Morocco (2.37) or by the English, such as Malaysia (1.80), United Arab Emirates (1.39), Jordan (2.69) or even Saudi Arabia (2.27) which, although still relatively high, are lower than those countries such as Iraq (3.58), Pakistan (3.42) or Syria (2.7). And, a propos, Iran (a country not exposed to radical internal terrorism) had a birth rate of 2.14 in 2020.

The other reason is economic, since poor families, in both the countryside and the city, require their children to work at the earliest age allowed in order to secure income, this being one of the reasons why child labour is so difficult to eradicate in countries and regions where there is a high percentage of extreme poverty.

Low fertility rates, on the other hand, are related to a high level of education (or at least literacy) and to the degree of autonomy that women are able to acquire in relation to their own reproductive health, which includes the use of contraceptives. The demographic transition consists, then, of the passage of high fertility rates (more than 3 children) to low fertility rates (1–2 children on average), as well as a change of mentality in people, thanks to education, which allows them to understand the importance of children to be free from work during childhood, so that they may dedicate themselves to receiving a good education in the school system of each country as well as play activities appropriate for each stage of childhood. It is also clear that, in addition to the differences that exist between developed and developing countries in relation to fertility rates – with low rates for the former and high for the latter[10] – there is a considerable difference within the countries in terms of fertility rates according to social classes, as large families tend to be in the poor classes, while the middle classes have a different type of reproductive behaviour (Table 3.3).[11]

The close relationship between fertility and education is a fact that is verified by correlating literacy and fertility, as demonstrated by countries such as Afghanistan, which has been at war for almost four decades and is considered one of the 'sanctuaries' of world terrorism, and which has a literacy rate of just 13% while its fertility rate was as high as 6.5 in 1981, 6.00 in 2001 (the year of the attack of September 11) and 4.27 in 2020. In another country at war, the Democratic Republic of the Congo with a literacy rate of 56% the fertility rate was 6.1 in 1981, 7.0 in 2001 and 5.70

---

[10] Even though this data is from 2001, it allows comparisons between the fertility rates of, for example, Canada (1.4), the United Kingdom (1.7), France (1.9), Germany (1.3), Italy (1.3), Spain (1.2), the USA (2.1) and Japan (1.3) with the Philippines (3.5), India (3.2), Mexico (2.8), Peru (2.9), Brazil (2.4), Colombia (2.6), Venezuela (2.9), and Argentina (2.6); or, more seriously, those of Africa, because if there is no sustainable development, Africa is the continent that will have the largest number of inhabitants in 2100: 42% of humanity – about 4 billion people – will live in it, given the fertility rates – to cite the example of some countries, though all have high indicators – of 7.0 in the Democratic Republic of the Congo, 6.1 in Zambia, 6.3 in Sierra Leone, 6.7 in Liberia, 5.8 in Rwanda, and 5.9 in Ethiopia, with the lowest fertility rate being in South Africa – one of the most developed countries on that continent – at 2.20 (2020), almost the same as Mexico in Latin America. Figures from: *Populations et Societés*, 151 (July/August), quoted by Todd (2002: 40) and of the World Population Review in the Web.

[11] Let's take as an example the case of a country like Guatemala, which is one of the most backward in Latin America in terms of indicators of inequality and social development (its social development indexes are barely better than those of Haiti). Although its fertility rate decreased from 3.8 in 2005 to 2.9 in 2015, there are notable differences when examining the figures by area. The areas with a majority indigenous population still have very high fertility rates (Alta Verapaz 4.8, Huehuetenango 3.6, Quiché 3.7, San Marcos 3.2, Totonicapán 3.0, Baja Verapaz 4.3) while the area of the department of Guatemala, where the capital city is located, has lower poverty rates because there is a greater number of people who belong to the middle class, and its fertility rate is 2.3 (hence below the national average of 2.9). *Source* Statistics from 2015 according to data from the *National Institute of Statistics* (INE) of Guatemala at: https://www.ine.gob.gt/index.php/estadisticas/tema-indicadores.

Table 3.3 Fertility rates in Muslim countries (children per woman)

	1981	2001	2020		1981	2001	2020
Azerbaijan	3.1	2.0	2.02	Libya	7.4	3.9	2.19
Turkmenistan	4.8	2.2	2.68	Qatar	7.2	3.9	1.84
Tunisia	5.0	2.3	2.13	Syria	7.2	4.1	2.7
Kyrgyzstan	4.1	2.4	2.9	Kuwait	7.0	4.2	2.07
Tajikistan	5.6	2.4	3.52	Sudan	6.6	4.9	4.41
Lebanon	4.7	2.5	2.09	Iraq	7.0	5.3	3.58
Turkey	4.3	2.5	2.07	Pakistan	6.3	5.6	3.42
Iran	5.3	2.6	2.14	Saudi Arabia	7.2	5.7	2.27
Indonesia	4.1	2.7	2.28	Senegal	6.5	5.7	4.52
Uzbekistan	4.8	2.7	2.38	Nigeria	6.9	5.8	5.28
Bahrain	7.4	2.8	1.95	Palestine	6.9	5.9	3.5
Algeria	7.3	3.1	2.94	Afghanistan	6.9	6.0	4.27
Malaysia	4.4	3.2	1.80	Mauritania	6.9	6.0	4.47
Bangladesh	6.3	3.3	2.02	Oman	7.2	6.1	2.89
Morocco	6.9	3.4	2.37	Mali	6.7	7.0	5.88
Egypt	5.3	3.5	3.33	Yemen	7.0	7.2	3.67
United Arab Emirates	7.2	3.5	1.39	Somalia	6.1	7.3	6.07
Jordan	4.3	3.6	2.69	Niger	7.1	7.5	6.91

Source The author's elaboration based on data from Todd (2004: 40); *Population and Societies* (Paris: INED), 151 (September 1981); 370 (July/August 2001) and the World Web (World Population Review)

in 2020; in Sudan (divided into two states in 2011 as a result of an internal armed conflict) the literacy rate is 60% and the fertility rate 6.6 in 1981, 4.9 in 2001 and 4.45 in 2020. In another country devastated by war like Yemen the fertility rate was 7.0 in 1981 and it even increased to 7.2 in 2001, before its fall to 3.67 in 2020, and so on. In contrast, Australia, with 96% of the literate population, the fertility rate is 1.9; in Germany with 99% it is 1.3 and in the United States, with 99% of the population literate the fertility rate is 2.0. In China itself, with 91% of its population having attended school – a huge achievement given its more than 1.3 billion inhabitants – the fertility rate has dropped to 1.8 children per woman according to Juniper (2016: 22–23) and it is expected that with high rates of growth and economic development currently under way, it will become approximately 0.5 for 2050, as previously stated.

Reviewing the evidence, the solution to the population explosion is clearly social and human development, and as part of this, specifically education as a primary factor as well as the fulfilment of the United Nations Sustainable Development Goals (SDG), as will be seen in the next chapter. There are no 'demographic' objectives in the 2030 Agenda, which is probably due to opposition from conservative and religious countries during the multilateral negotiations at the UN, but despite their absence it is perfectly possible to obtain a reduction in population growth simply as an expected result of the fulfilment of the SDGs. As is well known, the SDGs are interrelated and have an important 'social pillar' that extends from putting an end to poverty in all its forms around the world to the goals in the fields of health (guaranteeing a healthy life and promoting well-being for all in all ages) and education (ensuring inclusive and equitable education and promoting lifelong learning opportunities for all) and the gender objectives of achieving gender equality and the empowerment of women and girls. Fulfilling these goals would be enough to attain a significant reduction in the fertility rate of women, thus achieving the purpose of guiding the public policies of all countries towards transition and demographic stabilization.

In other words, given that the 193 members of the United Nations formally committed themselves to complying with the 17 SDGs at a summit in New York in September 2015 and are therefore prepared to direct their public policies accordingly, they will be working towards the demographic transition indirectly, i.e. through reducing poverty, expanding and improving the quality of health and education services, and promoting gender equality, all of which will accomplish a reduction in fertility rates.[12] Hence there is no need to establish explicit demographic objectives, as social development (vertical mobility in the social strata as well as education, the empowerment of women and knowledge of reproductive health) leads couples to limit the number of children, and therefore also leads to a decrease in the female fertility rate.

3.7   The Technocratic Paradigm

The main problem today in the relationship of human beings with nature is the false idea of nature as a different and separate entity that allows science and technology to be used to manipulate the former without consideration for the planet's carrying capacity or the needs and rights of nature, understood as *Mother Earth* (the *Pachamama* of indigenous peoples or *Gaia*, Lovelock's living planet). This is precisely the kind of misconception rooted in the 'technocratic paradigm', as Pope Francis calls it in his encyclical letter *Laudato Si'*, in which he states the following:

> The basic problem goes even deeper: it is the way that humanity has taken up technology and its development according to an undifferentiated and one-dimensional paradigm. This paradigm exalts the concept of a subject who, using logical and rational procedures, progressively approaches and gains control over an external object. This subject makes every effort to establish the scientific and experimental method, which in itself is already a technique of possession, mastery and transformation. It is as if the subject were to find itself in the presence of something formless, completely open to manipulation. Men and women have constantly intervened in nature, but for a long time this meant being in tune with and respecting the possibilities offered by the things themselves. It was a matter of receiving what nature itself

---

[12] Again by way of example, SDG 3.7 clearly states that signatories will commit themselves to ensuring by 2030 "universal access to sexual and reproductive health-care services, including for family planning, information and education, and the integration of reproductive health into national strategies and programmes", as stated by the United Nations, General Assembly resolution A/69/L.85.

allowed, as if from its own hand. Now, by contrast, we are the ones to lay our hands over things, attempting to extract everything possible from them while frequently ignoring or forgetting the reality in front of us. Human beings and material objects no longer extend a friendly hand to one another; the relationship has become confrontational. This has made it easy to accept the idea of infinite or unlimited growth, which proves so attractive to economists, financiers and experts in technology. It is based on the lie that there is an infinite supply of the Earth's goods, and this leads to the planet being squeezed dry beyond every limit. It is the false notion that an infinite quantity of energy and resources are available, that it is possible to renew them quickly, and that the negative effects of the exploitation of the natural order can be easily absorbed (Pope Francis 2015: 78–80).

As Pope Francis says, human beings and material objects no longer extend a friendly hand to one another and the relationship has become confrontational. This process is done in spite of the fact that in the past, even if humanity has constantly intervened in nature, this meant being in tune with and respecting the possibilities offered by nature, because "it was a matter of receiving what nature itself allowed, as if from its own hand" (Pope Francis 2015: 78). We have to go back to this spontaneous and non-destructive relationship by putting an end to the erroneous idea of human beings as separate from nature, as such an attitude currently allows manhood to exploit and abuse natural resources at will, which could lead to the sixth mass extinction and the end of humanity and multiple forms of biodiversity.

The Pope's Encyclical Letter also underscores the importance of overcoming the ideology of infinite or unlimited growth, which is not only a lie, as Pope Francis says, but part of the false assumption that energy is unlimited and that natural resources are capable of immediate regeneration. This is precisely the problem posed by economic expansion as it is understood within the framework of the capitalist neoliberal economy, and this situation must be changed or otherwise we could face our own extinction.

Moreover, the current false ideological belief in infinite growth partly originates in the modern philosophical tendency to regard technology as a neutral phenomenon that need not question the values applied or the end it serves. The trauma of Einstein when he realized the manner in which the Pentagon was going to use the atomic bomb – largely made possible due to his discoveries – by launching it on Hiroshima and Nagasaki illustrates this ethical issue very well. Hence, the Pope's 'technocratic paradigm' takes on precisely that supposedly neutral position of the technocrats of the economy in the face of social and environmental problems, in which the purpose of economic activity – serving human beings for the satisfaction of their basic needs – is not promoted and still less achieved. Competitiveness guides the performance of entrepreneurs seeking to maximize capital gain on pain of bankruptcy and market disappearance. As a result, capitalist economy functions for the sake of growth without values or purposes, as if it were a cyclist who does not know where he is going but must keep pedalling because if he stops the bike loses its balance and falls. But obviously, if the cyclist does not know where he is going, he risks falling into an abyss.

The philosophical problem of technology is quite complex and difficult to solve because if – as was conceived by positivism and modernity – it is neutral and value-free, then this vision is instrumentalist, separates means and ends, and is orientated by the liberal faith in a linear progress that in the end does not know where is going, as in the cyclist analogy. In contrast, if technology is perceived to be embedded in nature – as was the case in Ancient Greek philosophy – it would be value-laden, and means and ends would be an integral part of the system because all existing things have an end according to its ideal essence, as in Plato's philosophy of *Topus Uranus*.

The rationalism and positivism of modern philosophy, as expounded by philosophers like Descartes, Leibniz, Hobbes, Locke, Hume, Bacon, Kant, Hegel and even Marx, was criticized by postmodern philosophies such as phenomenology (Husserl), hermeneutics (Gadamer, Ricoeur) deconstructivism (Derrida), existentialism (Sartre) and by one of the most influential German philosophers of the twentieth century, Heidegger (1977), author of *Sein und Zeit*. According to him, technology is a way to understand the world, and the instrumentalist liberal conception must be rejected because when you choose to use technology you do not just make your existing way of life more efficient, but you really are choosing a different way of thinking and living. Technology is thus not simply instrumental to whatever values you have. It carries with it certain values and mindsets similar to religious belief. But technology is even more persuasive than religion since no belief is required to recognize its existence or follow its commands.

Therefore once a society follows the path of the technocratic paradigm – as Pope Francis says – it will be inexorably transformed into a technological society dedicated to values such as efficiency, expansion of power and control over nature, as had happened with industrialization phenomenon. Actually, this conception of technology functions in the same way as neoliberal *mainstream economics* (the so called *invisible hand of the market*, the individual permanent search for enrichment) and growth as the most obvious way to measure development. Hence, technology ceases to be an instrument and becomes the centre of our lives, which is analogous to viewing money as the main purpose of our lives. Normally one may believe that money is a neutral instrument of our objectives in life, but on closer examination it is clear that for a lot of people making money has become the main purpose of life. Even though there are clearly things money cannot buy (such as the recognition of others, or love and happiness), people do try to buy them constantly, with inevitably disappointing results. Those who base their whole lives on the acquisition of money have poor lives that usually corrupt and diminish people with this kind of belief. Like technology, money has implicit values, and basing a way of life on it is absolutely wrong.[13]

As I have mentioned before, Heidegger (1977) argued that modernity is characterized by the predominance of technology over every other value. In Greek philosophy the concept of *techne* ('doing/being') was related to the practical application of knowledge to get things done, and the German philosopher argued that this starting point culminates in modern technology. Whereas Greeks such as Plato regarded *techne* as the model of being in the world of ideas, we have ontologically transformed the notion of being from the platonic concept of *topus uranus* as a world of ideas

---

[13] Money is part of the social imaginary that facilitates the exchange of goods and cooperation but it can also be addictive and cause greed and corruption. For an interesting analysis of the function of money in collective imaginaries see Harari (2014, 2016).

to existence in the material world (*Dasein*), which in practice consists of the technical conquest of Nature. This conquest transforms everything into raw materials for technical processes, including human beings themselves (and that situation explains – for instance – the establishment of departments of 'human resources' in every corporation or bureaucratic institution in order to manage 'human capital'). From Heidegger's perspective, although we may manage the world through technology, we do not command our own obsession with control, that becomes an end in itself alienating human beings. Fortunately, postmodern critical theory (Habermas 1986; Feenberg 1999) does not share Heidegger's pessimism, and the critique of instrumentalism does help us to understand that technologies are not neutral, meaning that tools and goals, means and ends are connected in a form that can be understood if we use a holistic approach.

Thus even if human control of technology is possible, according to critical theory, technologies must be considered not as simple tools but rather as useful frameworks for interpreting (or understanding) the way people behave.[14] Therefore the choices open to us are situated at a higher level than the instrumental level. That is why – to give an example of US politics quoted by Feenberg (2003) – the instrumentalist slogan of the US National Rifle Association regarding the terrible frequency of killings at High Schools in the US (*"guns don't kill people; people kill people"*) is absolutely false. The possibility to purchase even army assault rifles in the US creates a social misconception, a mindset that is quite different from the mentality of citizens in a peaceful society (like Bhutan, Sweden, Norway and others) because in those countries people do not have that kind of choice. Hence, through legislation, society can choose which world we wish to live in by making the possession of guns either legal or illegal. But this is not the sort of decision that can be made by a government that only 'pretends' to have technology under control. This is what you might think of as a 'meta-choice' (as Feenberg says) – a choice at a higher level determining which values are to be embodied in the 'technical framework' of our lives. For Feenberg, a critical theory of technology opens up the possibility of thinking about such choices and submitting them to more democratic controls, thereby enabling people affected by technological change to protest (or even make innovations) in a form that promises greater participation by citizens and opens

---

[14] The views of Andrew Feenberg about the political importance of technology are comparable to the views of Jeffrey Sachs on the political sphere of sustainable development – and to my own thinking concerning democracy and cosmopolitism (discussed in the final chapter) – when Feenberg states that the idea of technology as an autonomous force separated from society is erroneous because: "[T]his conception of technology is incompatible with the extension of democracy to the technical sphere. Technology is the medium of daily life in modern societies. Every major technical change reverberates at many levels, economic, political, religious, cultural. Insofar as we continue to see the technical and the social as separate domains, important aspects of these dimensions of our existence will remain beyond our reach as a democratic society. The fate of democracy is therefore bound up with our understanding of technology." As Harari and others (like Henry Kissinger) have said concerning internet technology for the purpose of civil society surveillance applied by authoritarian states (like China), because of the coronavirus pandemic in 2020/21 it is absolutely essential to be aware of these issues, otherwise "these dimensions of our existence will remain beyond our reach as a democratic society" (Feenberg 1999: 7).

the door to a more democratic control of technology in the future. It is easy to understand the pertinence of this philosophical approach in our contemporary world where internet and the technologies of information, particularly smartphones, are threating our freedom with their all-encompassing capacity to control our lives, as Harari (2014, 2016) and postmodern philosophers (Heidegger, Habermas, Feenberg) have highlighted in recent times.

Where it used to be possible to silence all opposition to technical projects by appealing to progress, today communities mobilize to make their wishes known, as happens with their opposition to the construction of hydroelectric power plants or the mining industry in developing countries or with civil society opposition to nuclear power plants in developed ones, like Germany and others. On the other hand, technological innovations can also be the result of a "civil society choice". For instance, as pointed out by Feenberg, *email* technology was introduced to the Web by skilled users but it did not figure in the plans of the original designers of the internet.

Computer technology has become so entwined with our daily lives that our activities have begun to shape its development. However, while this is a positive and democratic 'choice' of global civil society (because nobody could deny that *email* is one of computer technology's most important contributions to our daily lives), critical observers of these new technologies of information and communication have also detected a disturbing trend towards greater control of global citizens by governments, powerful transnational corporations and even computerized *artificial intelligence* (AI). This is a problem which must be addressed. Participation of cosmopolitan citizens in shaping this global civil society appears to be expanding slowly to encompass technical issues that were formerly viewed as the exclusive domain of experts, making it possible for the pernicious tendency to abuse and misuse information technology to be regulated by a world citizenship able to exercise democratic and conscious control over the technical framework of our daily lives (such as smartphones) instead of allowing technology to control us.

Thus, if the ways in which a capitalist economy functions can be explained by a competitive desire to maintain an economic activity whose purposes (apart from the accumulation of capital and the distribution of profits to the shareholders of large companies) are not known, there is a clear necessity to reform the current world economy, which will be discussed further on in this book. For instance, if we selected some figures from world economic data and tried using them to assess global economic growth (guided by competitiveness and capital accumulation) using the classic category of the increase in per capita income at national level and on a global scale,[15] according to experts we would find that personal income has increased from US \$3,305 in 1960 to US \$9,472 per capita in 2010 to reach US \$14,000 in 2017

---

[15] This exercise involves adding together the GDP of each country to obtain a 'world GDP' and then dividing that figure by the number of inhabitants of the planet – a methodology which does not yield an accurate picture of the reality as it fails to take into account the enormous gulf which currently exists between the richest and the poorest people in the world.

(Sachs 2017). From an axiological point of view (implicitly because economy technocrats do not base their decisions on ethical judgements), this would be considered something 'good' or 'positive'. However, that would be an erroneous and incomplete (reductionist) perception because from the perspective of natural ecosystems or social welfare, in a world where nature is affected by waste and pollution as well as concentration of wealth with the consequent inequality, the increase in growth (GDP) and per capita income does not appear to have those positive connotations. This is because poverty and inequality result in a permanent social crisis at world level, as demonstrated by the massive increase in migratory flows, the proliferation of armed conflict, the intensification in the number of refugees and social turmoil, the surge of terrorism, organized crime and the like.[16]

It is clear from the pie chart of the World Bank regarding the global economy by GDP in page, which shows how the GDP of the whole world is distributed, that the United States has the biggest share with 24.32%, followed by China with 14.84% and Japan with 5.91%, who thus occupy the second and third place. Germany (4.54%) occupies the fourth place while the United Kingdom (3.85%) and France (3.26%) are in the fifth and sixth positions. India (2.83%), Italy (2.46%) and Brazil (2.39%) are in the seventh, eighth and ninth places, and then there are a series of countries with smaller percentages as well as the 'Rest of the world' with 9.41%.[17]

If this reductionist approach that does not take into account the specific weight of demographic factors is complemented with the criterion of *per capita income* to position the countries, as occurs in the IMF's ranking of countries of per capita income in 2017, the United States falls to the tenth position while Norway (US $70,665), Luxembourg (US $107,736), Switzerland (US $61,014) and Ireland (US $72,529) are in higher places. Using this criterion they appear to be richer than the US because they are less populated,[18] but in reality the US obviously continues to be in the first place. This distortion of reality produced by economic figures also occurs with countries like Singapore, the United Arab Emirates, Kuwait, Qatar and Brunei, which, as oil producers with very small populations, all have a per

---

[16] To obtain an idea of this monstrous concentration of wealth we can refer to Jeffrey Sachs (advisor to the UN Secretary-General for sustainable development and director of the Earth Institute at Columbia University in New York), who, during the high-level segment of the UN's Economic and Social Council (ECOSOC) in New York (October 2017), stated in a very energetic and critical manner that if the $127 trillion World GDP (with per capita income of $14,000 for each inhabitant of the planet) were appropriately distributed, it would be enough to finance the attainment of the UN 17 SDGs, but the problem lies in the iniquitous concentration of wealth. Just to give an idea of this heinous distribution, some $20 trillion are deposited in tax havens and just 2,043 of the world's richest individuals own $7 trillion between them, while $13 trillion is wasted on wars and armaments. With regard to this, it is possible to watch a recording of the *43rd meeting* of the high-level segment of the 2017 session of the UN's *Economic and Social Council* (ECOSOC), posted by the United Nations Information Service; see at: webtv.un.org.

[17] Hence, from an ethical perspective, Trump's campaign slogan (*Make America great again*) was part of his own 'fake news', since the US economy has been in the first place at world level since the late nineteenth century.

[18] But the explanation also has to do with the fact that Luxembourg is an EU financial centre, Switzerland is a tax haven and Ireland has a policy of fiscal incentives for transnational corporations.

capita income higher than the US and other developed countries such as Germany, Sweden, Denmark, Austria, Canada and Australia that are evidently richer and more important states when benchmarks like territory, population, armed forces, industry, trade, culture, and so forth are introduced.

The relativity of these economistic quantifications is of such magnitude that China, which has the world's second-largest economy, does not even appear in the IMF's table (Fig. 3.5), although Hong Kong and Taiwan are there.[19] Its absence can be explained by population figures. To calculate per capita income the GDP has been divided by the number of inhabitants of each country – i.e. by more than 1,300 million in the Asian giant, but merely 320 million in the US. According to this calculation, one would receive the impression that China is not doing well economically speaking; however the opposite is true. Nevertheless, the aforementioned evidence also shows that such figures are merely theoretical elaborations, the use of which depends on the ideological purposes of the economists who work constantly with these kinds of figures. However, given that technocracy is supposedly neutral, the goals (and moral values) of economists are never explicitly articulated. Hence the relativity of economic statistics and figures in general terms is quite clear. This means that – as has already been done via the Human Development Index and other similar efforts – when evaluating the performance of countries, it is important to use a different methodology based on the sustainable development framework, which is the category that truly matters from the perspectives of both human development (social dynamics sphere) and environmental sciences (natural ecosystems sphere), as will be seen below Jeffrey Sachs's theoretical approach is examined (Table 3.4).

Therefore, in order to analyse why wealth and poverty exist, it is necessary to introduce value judgements to evaluate the functioning of the economic system and see how it favours or prevents social development and the redistribution of wealth. At the end of the day, most of the benefit from the fruit of labour goes to entrepreneurs, not workers. In order to change that, governments must implement social and fiscal redistribution policies. Consequently, if we want to know how wealth is redistributed within a country, we can use the Gini[20] coefficient that at least has the virtue of pointing out which countries are more egalitarian (more democratic from the social point of view) and which countries are less egalitarian (less democratic), which in turn could help correct the vision that presents supposedly neutral strictly technocratic figures, and in this way provide balance to the usual economistic approach of technocrats thanks to the use of value judgments related to equality, as required by the UN SDGs.

---

[19] Both Hong Kong and Taiwan appear in the list as states separated from China but strangely without an official number of positioning, because – I suppose – they are, in China's view, both 'officially' provinces of the People's Republic of China and not independent states. And curiously, Hong Kong has a higher per capita income than the US (US $60,553 for Hong Kong while the US has 'just' $59,609).

[20] This coefficient measures the inequality in the distribution of wealth in such a way that a perfect distribution would be equal to zero. Lower figures indicate less inequality, while larger numbers correspond to the countries or situations of greater inequality.

Table 3.4 Ranking of countries by per capita income according to the *International Monetary Fund* (IMF 2017)

Position	Country	Per capita income (dollars)
1	Qatar	129,112
2	Luxemburg	107,736
3	Singapore	90,724
4	Brunei	76,567
5	Ireland	72,529
6	Kuwait	71,306
7	Norway	70,665
8	United Arab Emirates	68,424
9	Switzerland	61,014
–	Hong Kong	60,553
10	United States	59,609
11	Saudi Arabia	55,477
12	The Netherlands	53,139
13	Iceland	52,496
14	Bahrain	51,956
15	Sweden	51,377
16	Australia	50,817
–	Taiwan	49,901
17	Germany	49,814
18	Austria	49,370
19	Denmark	49,364
20	Canada	47,771

Source International Monetary Fund (IMF)

In other words, if we apply this value approach in order to compare, for instance, the United States with China, it is evident that the US is more democratic from the political point of view (it favours freedom) but less democratic from the social point of view (it is very much unequal). By contrast, China is more democratic in social terms (it favours equality) and less democratic from the political point of view because the Chinese political regime is authoritarian. Consequently, theoretically it is possible to argue that in order to democratize the US system from the social and economic point of view, the Americans could engage in redistributive policies – like those that President Obama wanted to implement which but began to be dismantled by Trump – while from a political angle the Chinese could improve its political regime by reforming its electoral and representative system. However, as in both cases these processes are moving within the scope of self-determination and non-intervention – in accordance with the principles of the UN Charter – these issues must be resolved freely by the American and Chinese peoples, without foreign interference and when they decide to proceed accordingly. Therefore, it is important is to acknowledge that the concept of democracy implies both process and structure, and that, despite being interconnected and interdependent, the political system (competitive elections, respect for fundamental freedoms, checks and balances), the economic system (redistribution of wealth, promotion of equality) and the social system (citizen participation, respect for human rights, indigenous, minorities', women's, cultural and religious rights) have different dynamics. Consequently, diachronically a democratic political regime can coexist with inequality (as in the US) and democratic social regimes can coexist with authoritarianism (as in China).[21]

Basically, we find ourselves facing once more the old political dichotomy between freedom and equality, present in the politics of all national states since the time of the French Revolution. Finding a balance – assessment or valuation – is very difficult to carry out without introducing axiological judgements that concern the value of *freedom* on the one hand, and the value of *equality* – not just before the law but in *economic and public policy practices* – on the other. And in passing, it is pertinent to realize that *fraternity*, the other great value of the French Revolution (closely linked with the principle of *solidarity* in social terms) is a fundamental part of sustainable development, which essentially consists of effective management of the economic and social dynamics, and takes into account the interests of future generations by conserving and respecting terrestrial ecosystems.[22]

In any case, if we want to reach some preliminary conclusions about democracy in the world it is clear that the Scandinavian countries are exemplary in making such balances between freedom and equality, and that most countries of the world – especially in Latin America – could have a lot to learn from them. A country like Guatemala, for example, with a per capita income of US $3,673, has a Gini coefficient of 55.1, which places it among the most unequal of the subcontinent, and this inequality brings, among other consequences (such as the enormous migratory flows to the US) a malnutrition index of 15.6%, which means that two and a half million of its inhabitants suffer or have suffered hunger in their lives. However, instead of being inspired by the Scandinavian model, a powerful segment of the Guatemalan ruling elites recently unleashed an aggressive ideological campaign against Norway and Sweden, accusing the governments of both countries of intervening in Guatemalan internal affairs through cooperation programmes directly aimed at financing civil society organizations or the UN commission against impunity in Guatemala (CICIG), which led Norway to close its embassy and the conservative, oligarchic and corrupt

---

[21] Therefore, if we analyse the situation of Hong Kong or Taiwan with this kind of methodological approach, it is clear that for the time being the democratic processes within China are not compatible but it is also important to accept that foreign interference is not the solution.

[22] For the great French writer and thinker of the nineteenth century, Alexis de Tocqueville, the "passion for equality" is one of the main reasons why people struggle for democracy. In the years that he wrote his masterworks, Tocqueville mainly studied the French Revolution and American democracy, explaining how the abolition of the privileges of nobles and aristocrats is fundamental to the establishment of formal equality before the law. Nowadays, it is possible to argue that with the same passion we could engage in the fight for equality – understood as a decrease in economic inequalities – and that is why less unequal countries – like the Scandinavians – are much more democratic than other developed countries (Tocqueville 2003, 2011).

Guatemalan regime to close its embassies in Oslo and Stockholm. I mention this fact not only as an example of extreme ideological conservatism – something detrimental even to the long-term interests of such elites – but also as a demonstration of the absurd inability of some rich elites in 'poor' countries to learn from those who could have something to teach them.[23]

However, in this era of globalization the concentration of wealth is not just the result of national policies, because at world level inequality is a consequence of hegemonic globalization and neocolonialism (Santos 2009). This phenomenon prevails in the world economy and it is scandalous to know that that big transnational corporations have deposits in tax havens of some US $20 trillion, apart from the enormous concentration of wealth that implies that just over 2,000 super-rich tycoons are the owners of another $7 trillion of the world's wealth (Sachs 2017). The weakness of the national governments that cannot stand up to these transnational corporations and the individuals owning some of them is easily understandable when we realize that the volume of capital and the income of these enterprises are usually greater than those of many states of the world.

For example, the famous Wal-Mart, which is the world's largest and most powerful corporation, has a financial size comparable to some national economies: it is just below Norway (with revenues of $486 billion), which is almost double the income of Pakistan ($247 billion) and superior to that of the Philippines ($285 billion), Vietnam ($186 billion), New Zealand ($188 billion), Malaysia ($327 billion) and Thailand ($374 billion), to quote some examples. Obviously, Wal-Mart is not the only one, so – just to get an idea – other big transnational corporations like Volkswagen ($269 billion), Royal Dutch Shell ($431 billion), Exxon Mobil ($383 billion), Apple ($183 billion), British Petroleum ($359 billion), Samsung ($196,000 billion), Toyota ($248,000) and the French Total ($212 billion) also have higher incomes than those of many countries in the world, including, of course, small Central American countries (Juniper 2016: 30–31).

Unquestionably, world economy is dynamic and things can change in relatively short periods of time. According to experts, if the global growth parameters are maintained, by the year 2050 the G7 will have decreased its share of world GDP and will be displaced by seven emerging powers: China, India, Russia, Brazil, Mexico, Indonesia and Turkey. However, figures and predictions aside, what I want to highlight here is how the mindset of unlimited growth continues to permeate the vision of those who really plan the economy – transnational corporations not governments – confirming what the Pope says in his Encyclical Letter not only about the technocratic paradigm but about the way it informs mainstream economics nowadays:

> The origin of many difficulties in today's world is above all the tendency, not always conscious, to constitute the methodology and objectives of techno-science in a paradigm

---

[23] And more recently (2018) Guatemalan authorities decided to put an end to the work of the UN *International Commission against Impunity in Guatemala* (CICIG). They were worried about investigations carried out by the CICIG concerning high-level officials and private sector personalities involved in organized crime, money-laundering and corruption. The figures cited about this problem (of the US, China, the Scandinavian countries and Guatemala) are taken from Juniper (2016: 72–73).

of understanding that conditions the lives of people and the functioning of society [which inhibits us] to recognize that the objects produced by the technique are not neutral, because they create a framework that ends up conditioning lifestyles and orientates social possibilities in line with the interests of certain power groups ... No one can think that it is possible to sustain a cultural paradigm and use technology as a mere instrument, because today the technocratic paradigm has become so dominant that it is very difficult to dispense with its resources and even harder to use them without being dominated by their logic ... The man who owns the technique knows that, in the end, it is directed neither to utility nor to welfare, but to domain and control ... The technocratic paradigm also tends to exercise its dominance over economics and politics. The economy assumes all technological development in terms of revenue, without paying attention to eventual negative consequences for the human being. Finance chokes the real economy. The lessons of the global financial crisis were not learned and the lessons of environmental deterioration are learned very slowly ... It is not a question of economic theories that perhaps nobody dares to defend today, but of their installation in the factual development of the economy ... the market by itself does not guarantee integral human development and social inclusion (Pope Francis, *Laudato Si'* 2015: 84–86).

What the Pontiff says in his Encyclical Letter shows with clarity that mainstream economics are ultimately part of the ideology of the supposed neutrality of technology, which should not make value judgements or judgements about the social consequences or the objectives that are pursued. This is an attitude that must be changed because, as Pope Francis (2015: 84) says, "the market by itself does not guarantee integral human development and social inclusion". In other words, since the market ideology is determining to a large extent the behaviour of human beings, especially the conduct of those who make decisions either at the level of governments or at the level of the executive offices of large transnational corporations, and since it is also evident that such decisions are made with the motivation for profit maximization directed by the market logic, and that those decisions are not guided by the welfare of human beings but by ideology (Habermas 1987),[24] the technocratic paradigm must be transcended and replaced with the holistic and cosmopolitan paradigm advocated in this book.

3.8   Consumerism and the Culture of Discarding

Consumerism has become a kind of social pathology that manifests itself through a way of thinking (a consequence of mental manipulation carried out by advertising and marketing techniques) according to which one must be buying merchandise to feel good and maintain the social status of the individual: from branded clothing to cell phones, cars, food and beverages or the type of restaurants and shopping malls that are frequented. This is the type of behaviour that encourages the consumer mentality, fostering the habit of discarding objects or clothing that are still useful

---

[24] I have already mentioned the position of the German school of critical theory regarding ideology. Nonetheless it is worth remembering that the seminal contribution regarding this matter was made by Habermas (1987) in his text (quoted in Chapter I) about "technology and science as ideology"., translated into Spanish by Manuel Jiménez Redondo and published by the Tecnos editorial house in Madrid in 1986 with the Spanish title *Ciencia y Técnica como 'Ideología'*.

just because they have gone out of fashion or because of a desire to have the latest technological innovation (as with mobile phones). This generates tons of discarded goods unnecessarily (large quantities of clothing and used cars in rich countries are sent for sale in poor countries), a phenomenon that allows both the manufacturers and the shopping malls to remain overcrowded by consumers buying unnecessary objects, especially at the times of year that tend to encourage sales, such as Christmas and other special days promoted with that purpose.

Obviously, consumerism and discarding are types of behaviour exhibited by the upper classes and the upper-middle classes throughout the world, not by the lower classes living in poverty. But sustainable development planners should be mindful that there is a risk involved in improving the economic situation of people living in poverty (as SDG1 indicates), because leaving poverty could take this new 'middle class' to consumerism, which could be a negative result of a State's social policies, especially from an environmental angle. That is why the new paradigm of sustainable development must be applied everywhere, taking as a parameter to evaluate its implementation the criteria of human needs not of income per capita, as will be seen later in this book. Another problem regarding poverty concerns the migratory flows provoked by the job market in industrialized countries, where the active population has been diminishing for demographic reasons. As a result, a considerable amount of remittances are keeping the economies of a good number of countries in the Global South afloat, but the long-term unsustainability of that kind of 'model' seems very clear, among other reasons because a significant part of the money from remittances is used in consumerism. It is important to become aware that, with globalization at the root of the mobility of both humans and merchandise, the only way to solve the excesses of trade exchange or migratory flows is to address globalization from a different perspective that favours the interests of people and human needs, not the interests of transnational corporations.[25]

But coming back to the policies addressed at the reduction of poverty, if, as a result of the implementation of SDGs and the redistribution of wealth via States' social policies, there is an increase in the middle classes throughout the world, then these segments of the population (with a daily expenditure of between 10 and 100 dollars),

---

[25] A report of the *International Fund for Agricultural Development* (IFAD) prepared by Pedro Vasconcelos and cited by Roshni Mujamdar in a newspaper article, states that remittances sent by migrant workers doubled in the last decade, increasing 51% from $296,000 million in 2007 to $445,000 million in 2016, which helped lift countless families from poverty throughout the world. The largest shipments of remittances come from immigrants in the United States, followed closely by Saudi Arabia and Russia. The money sent is used for health, education and food expenses, says the IFAD study. Although 85% of immigrants' income remains within the host country, the remaining 15% that is sent to their countries of origin can constitute 60% of household income in rural areas. Since one in seven people in the world is a sender or beneficiary of remittances, this means that about a billion inhabitants of the planet participate in these transactions in some way. Even during the 2008 financial crisis, remittance flows remained stable, and it can be said that the flow of money has exceeded the migratory flow, which grew by only 28% in the last decade. This means that there are approximately 800 million people around the world who depend on the monetary remittances of migrant workers, which number around 200 million worldwide. See at: http://www.ipsnoticias.net/2017/06/remesas-migrantes-suman-billones-la-economia-mundial/.

whose number is around 1.8 billion people worldwide in 2009 figures, could reach almost five billion in 2030 (Juniper 2016: 28–29). This implies that if these five billion people fall into habits of consumerism and the consequent culture of discarding, the whole planet will suffer the consequences because it will not have the carrying capacity for such a huge number of people.[26]

The United States, Canada, members of the European Union, Australia, New Zealand and Japan are already predominantly middle-class countries. However, if other countries such as India, China, Indonesia, Nigeria and Brazil, with their billions of inhabitants, started to develop large middle classes with consumer habits, it would be impossible for the planet to cope. In China alone, which, thanks to its planned economy with high annual growth rates has increased per capita income by more than 2,000%, there is already a middle class higher than the total US population. However, this middle class still maintains different consumption habits, which depend significantly more on its own internal market, as the external market is far more costly given the need to import luxury goods for the high purchasing power segment.[27] The danger, then, lies in the parallel rise of the two negative aspects of the improvement of living standards: consumerism and the culture of discarding, as Pope Francis calls it in his Encyclical Letter.[28]

The demand for industrial manufactured goods for the new middle classes (in any country of the world) would also have harmful consequences for the environment,

---

[26] The carrying capacity is a concept that refers to the maximum weight that the environment can support of a biological species living in a certain ecosystem or, for some, the "maximum load limit that an ecosystem can support before it cannot recover or self-regenerate". Therefore, it is defined as the maximum ecological load that a certain number of inhabitants can exert on the environment, the number of people that an environment can sustain indefinitely with resources such as food, habitat, water and other available needs, and there must exist a certain balance between inhabitants and the ecosystem.

[27] *Business Insider*, stated the following: "China had an urban population of 730 million people in 2015. So even if that figure doesn't change (and it will only grow), by 2022 over 550 million people in China will be considered middle class. That would make China's middle class alone big enough to be the third-most populous country in the world. According to McKinsey, in 2012, 54 per cent of China's urban households were considered 'mass middle' class, meaning they earned between US $9,000 and US $16,000 per year. But by 2022, thanks to a growing number of higher-paying high-tech and service industry jobs, 54 percent will be classified as 'upper middle' class – meaning they earn between US $16,000 and US $34,000 a year." See at: http://www.businessinsider.com/chinas-middle-class-is-exploding-2016-8.

[28] "[The] culture of discarding affects both excluded human beings and things that quickly become garbage. Note, for example, that most of the paper that is produced is wasted and not recycled. It is hard to recognize that the functioning of natural ecosystems is exemplary: plants synthesize nutrients that feed herbivores; these in turn feed the carnivorous beings, who provide important amounts of organic waste, which give rise to a new generation of vegetables. On the other hand, the industrial system, at the end of the production and consumption cycle, has not developed the capacity to absorb and reuse waste. It has not yet been possible to adopt a circular model of production that ensures resources for all and for future generations, which means limiting the use of non-renewable resources to the maximum, moderating consumption, maximizing the efficiency of use, reusing, and recycling. Addressing this issue would be a way to counteract the culture of discarding, which ends up affecting the entire planet, but we note that progress in this regard is still very scarce", in: *Laudato Si'* (2015: 20).

unless the development patterns are changed. I will observe here what could happen with two concrete cases: the new demand, artificially generated, for bottled drinking water and individual cars as a means of transport. Bottled water, a new product created by marketing experts – because tap water was previously used – is sold in plastic or glass containers and although the liquid used for this can contribute to the depletion of sources such as springs or underground aquifers, the main source of problems is the energy used to transport the containers (45% of their cost) and the bottling factory (50%). Additionally, bottling in plastic is particularly negative, because the discarded containers are increasingly contaminating lakes, rivers and oceans. According to some sources,[29] over 877 plastic bottles are discarded every second worldwide and, what is worse, it takes 700 years for a plastic bottle to disintegrate naturally, i.e. they are practically indestructible. Hence there is a need to manufacture another type of material for the packaging of liquids in general because, obviously, this type of container is not only used for bottling water.

As for individual cars, the increase in their manufacture and use has also been vertiginous, with the pernicious consequences ranging from the increase in the required metallic and other inputs (aluminium, iron, steel and electronic products, in addition to rubber for tyres and other materials that are used in each manufactured unit), to the number of vehicles circulating on streets and highways: 487 per thousand inhabitants in the European Union, 463 in Japan, 404 in the United States, 300 in South Korea, 259 in Russia, 147 in Brazil, 50 in China and 13 in India, to name just a few. Apart from manufacturing and the sheer number of cars, the use of individual vehicles as a means of transport causes other problems, ranging from the increase in atmospheric pollution caused by the carbon dioxide that automobiles emit due to the use of fossil fuels (gasoline and diesel – both from oil) to the inefficiency and decrease in productivity resulting from time lost in traffic congestion, and the need to build an adequate infrastructure for the vehicle fleet to the detriment of other priorities, including the loss of lives or damage in traffic accidents and the amount of 'garbage' generated by vehicles that are no longer used.

The two examples I have chosen are evidence of an indubitably unsustainable type of development that must be replaced by other alternatives, both for drinking water and for individual transportation. If Brazil, China or India were to move from the 147, 50 and 13 cars respectively for every thousand inhabitants of the present day, equating with the 400 vehicles of the big industrialized countries, this would be catastrophic for the environment. The planning of the SDGs, then, forces us to find alternative solutions, such as improving collective transport and increasing the use of electric or hydrogen cars, which implies a transformation of the automotive industry, something that is already being done but not with sufficient speed.[30]

---

[29] Juniper (2016: 89–90).

[30] "The market for purely electric vehicles is in its infancy. The Nissan Leaf was the first to become available in the U.S., with Ford, Toyota, and Honda rolling out models in 2011 and 2012. The Nissan Leaf sold 8,720 in its first 11 months. Nissan expects to sell over 10,000 of the Leaf within the first year of rollout. The Tesla Model S, a luxury BEV, received considerable attention including Motor Trend's 'Car of the Year' award in 2012. In the long term, Pike Research projects that BEVs will account for 0.8% of U.S. car sales by 2017.The market for PEVs and EREVs is more developed,

3.9   Toxic Waste Contamination and Garbage

When Pope Francis (2015) criticizes the "culture of discarding" he has in mind the trophic cycles of nature that have the virtue of recycling everything without producing waste. In the natural world there is no garbage; everything is recycled. It could also be said that this is clear proof that nature is much more intelligent than human beings, except for the fact that, as stressed before, we ought to be aware that humanity is part of nature (*Pachamama*), not different or separate from her, which means that we have to imitate nature with regard to producing or recycling waste. It is a profound mistake to divorce humankind from nature, as occurred during medieval times with religion and in our contemporary society due to positivism and neoliberalism. So the natural intelligence of human beings should now take us out of alienation and the quagmire in which we are living and return us to the communion with nature, as happened in the early days of human kind.

Although, thanks to the knowledge provided by environmental sciences, awareness of how to dispose of garbage is greater and better – especially in industrialized countries, as one might expect – there is still much to be done. At the beginning of the twentieth century the world produced half a million tons of solid waste every day, but by the year 2000 this amount had suffered a sixfold increase, and by the end of this century it is expected to have quadrupled again, to reach some 12 million tons a day without having yet adopted multilaterally agreed guidelines that prevent, for example, the gross contamination of the oceans with discarded plastic.

Obviously, in western countries there is a greater number of industrial waste and in developing countries there is more organic waste, but in any case the challenge is to adapt industrial production in such a way that as much waste as possible is either recycled or converted into organic fertilizer called *compost*. This is something which is already being done in some industrialized countries instead of burying or incinerating it, since burying gives rise to filtrations that can contaminate underground water deposits, and produces methane, a greenhouse gas. As for the incineration, this procedure contaminates the atmosphere. As the recycling of glass, metals, paper, cardboard and some types of plastic is also a better procedure for disposing of waste, it would be advisable to insist that the industry uses the best recyclable metals, such as aluminium and the new generation of recyclable plastics under study. The compost that comes from organic waste like food, plants or agricultural products can be also used as an energy source (biogas) for industrial production. According to data from the OECD, the pioneer countries in the use of biogas are Sweden (11%), Austria (45%), Italy (35%), the USA (8%), France (15%), Germany (18%), Spain (32%), Australia (70%), and Japan (7%). Regarding waste recycling, the OECD points out

---

but has yet to reach rapid deployment. Hybrids have been retrofitted for plug-in capability since they were introduced in the early 2000s. The Honda Insight was the first hybrid available in the United States. Since then, most of the major auto makers have introduced hybrid models in the U.S. The best-selling hybrid currently on the road is the Toyota Prius, which sold nearly one million units between 2000 and 2001." See at: http://www.iedconline.org/clientuploads/Downloads/edrp/IEDC_Electric_Vehicle_Industry.pdf.

that the United States recycles 24%, Sweden 32%, Austria 45%, France 18%, Finland 30%, Spain 10%, and the United Kingdom 18%, but Mexico – a sad example of the situation in developing countries – only 3% (Juniper 2016: 88–91). Hence, there are advances but much remains to be done, especially in developing countries.

Industrial waste is another serious problem derived from consumerism and hegemonic globalization. Paul Robbins, author of a book on political ecology, describes the recycling process of discarded industrial products, including new ones which have not been used and that are the result of overproduction, which are taken to Ghana in Africa to be recycled by workers, who, according to their own account, have great skill and knowledge because they apply their customary knowledge in the following way:

> It is hard not to notice, however, incredible technical inventiveness, ecological knowledge, and economic innovation on display here as well. Trucks of junk have been directed here by local team-leaders, who bid for access to shipping containers that make their way to the distant dockyards from China and the Americas. These teams together organize labour to disassemble and process the materials for sale to middlemen, whose massive industrial scales are positioned along the perimeter of the dumpsite, awaiting negotiations over prices of copper, lead, and steel. The men at work prying apart circuit boards and stripping components out of relict computers quickly sort materials that can be easily resold or refurbished from those that must be processed. They have a terrific grasp of the workings of the electronics, as well as the obsolescence of its components. The melting of lead is a delicate operation, conducted by people who can sift off materials for match-heads and purify the element to satisfy buyers. This is done with such efficiency, I am told, that the site can make a mountain of computers disappear in months or weeks. Livelihoods are being practised in this landscape, by people who sometimes lack a grade school education, but who possess far-ranging knowledge of markets, chemistry, and engineering. But one more thing is drawn to my attention: the radios are totally unused. As one worker pulls square angles of Styrofoam from their boxes and threads these along a length of twine, it becomes clear that these hundreds of music players have arrived on site encased in the very packaging in which they left their factory in China. This final fact changes the scene in an inexplicable way. Rather than the necessary outcome of contemporary consumer society and an unfortunate inevitability of modern life (someone 'has to' process waste after all!), the ingenious workers of Agbogbloshie appear as part of a bizarre engine that maintains a self-replicating worldwide system of over-production. Oceans of organic and inorganic material are drawn from the earth and flow into an enormous feeding machine that reforms them into myriad configurations (refrigerators, televisions, printers), devours energy in their transportation across the globe, and then summarily dumps them here, unused, in this deadly metabolic intestine of labour. There is Wonderland logic at work here that could only be considered comic if the human and environmental price was not so obviously high (Robbins 2012: 2–3).

Robbins also argues that the *Political Ecology* approach aims to show that ecological analysis can move from a method that emphasizes the destruction of the environment, highlighting human influence, to another type of methodology that focuses on the production of 'socio environments' and the way they are 'co-constituted' by different types of human and non-human actors (such as globalized markets). To cite specific cases, for Robbins the idea of East Africa (Kenya and Tanzania) that one is accustomed to imagining, thanks to cinema and the media, is that of wild nature, populated with animals such as lions, elephants, zebras, gazelles, antelopes, giraffes, leopards and hippos. Such an idea is pleasant and attractive, so that one thinks of the beautiful landscapes of the Serengeti National Park, Mount Kilimanjaro and Mount Kenya without relating them to human beings. However, it turns out that the ethnic group of the Masai, who for centuries were the owners of those lands, were stripped of them in order to turn them into attractive national parks for the tourism industry.

Nevertheless, the Masai still inhabit these areas, despite the fact that land extensions for agriculture have been reduced considerably. Thus, when one speaks of the danger in which nature finds itself, it tends to be assumed that this is due to the intrusion of the human being and, particularly, of the impoverished peasants of the region whose population has increased notably because, as noted in the section on demography, African fertility rates are still very high.

However, if you stop just observing what happens in the Serengeti savannahs and examine what happens in the ecosystem as a whole, you acquire a much more complete view of the complexity of what happens. For example, comparing the deterioration of the environment and animal life in Kenya with that of Tanzania, it is found to be much greater in the first country than the second. This is due to the fact that:

> [G]azing across the Serengeti-Mara ecosystem both in time and in space, habitat loss and wildlife decline appear more complex and more connected to the daily lives and routines of urban people in the developed world. Cross-border analysis shows that the decline in habitat and wildlife in Kenya is far higher than in Tanzania. Why? Rainfall, human population, and livestock numbers do not differ significantly. Rather, private holdings and investment in export cereal grains on the Kenyan side of the border have led to intensive cropping and the decline of habitat. These cereals are consumed around the world, as part of an increasingly globalized food economy. As Kenya is increasingly linked to these global markets and as pressure on local producers increases, habitat loss is accelerated. Less developed agricultural markets and less fully privatized land tenure systems in Tanzania mean less pressure on wildlife. The wildlife crisis in East Africa is more political and economic than demographic (Homewood et al. 2001). These facts undermine widely held apolitical views about ecological relations in one of the most high-profile wildlife habitats in the world. They also point to faulty assumptions about the nature of 'wild' Africa. Firstly, the image of a Serengeti without people is a fallacious one. The Masai people and their ancestors inhabited the Central Rift Valley for thousands of years before European contact, living in and around wildlife for generations. Indeed, their removal from wildlife park areas has led to violent conflicts (Collett 1987). More generally, the isolation of these places is also a mistaken perception. Export crops from Kenya, including tea and coffee in other parts of Kenya beyond the Central Rift Valley, continue to find their way to consumers in the first world, even as their global prices fall, constraining producers who must increase production, planting more often and over greater areas, further changing local ecological conditions. With three-quarters of the population in agriculture, economic margins for most Kenyans become tighter every year, and implications for habitat and wildlife more urgent. The migration of the wildebeest, and its concomitant implications for grasslands and lions, therefore, does not occur outside the influences of a broader political economy. Land tenure laws, which set the terms for land conversion and cash cropping, are made by the Kenyan and Tanzanian states. Commodity markets, which determine prices for Kenyan products and the ever-decreasing margins that drive decisions to cut trees or plant crops, are set on global markets. Money and pressure for wildlife enclosure, which fund the removal of native populations from the land, continue to come largely from multilateral institutions and first-world environmentalists. All of these spheres of activity are further arranged along linked axes of money, influence, and control. They are part of systems of power and influence that, unlike the imagined steady march of the population "explosion", are tractable to challenge and reform. They can be fixed (Robbins 2012: 12–13).

The difference between Robbins' contextual approach and the traditional way of dealing with environmental problems is what makes political ecology different from so-called 'apolitical' ecology. According to Robbins, you must first identify the broader (global) context of a problem before progressing to the local one. Therefore you must realize that ecosystems are loaded with power that is not reduced to the local sphere and that local actors are also linked to the global sphere. Therefore, they are not 'politically inert', which is why it is essential to be explicit about what needs to be done, through normative approaches, as ecosystems can be 'fixed' and are amenable to reform. Hence, instead of assuming the typical supposedly neutral ('apolitical') position of technocrats, who refrain from making value judgements, the political ecology approach must be adopted to solve environmental problems.

In another example, Robbins wonders what happens when coffee prices fall because of the 'fault' of global markets, thus considering the consequences for the farmers who depend on their exports, and what happens with the forest cover used in the coffee plantations. We should always be aware of the role of the communities and the social context. When the World Bank finances projects for the preservation of biodiversity, reforestation programmes, or the construction of infrastructure such as hydroelectric plants or roads, it must be clear what impact this will have on the biodiversity, as well as on the local population, which is why the local communities and the indigenous population should always be consulted, as established in Convention 169 of the ILO. In this regard it is interesting to note that the complex contextual methodology used by Robbins has also been used by well-known American social scientist Jared Diamond. For instance, the analysis concerning the causes of the collapse of entire societies such as the Anazasi Indians in the American South West, the Mayans in Guatemala and the Yucatan Peninsula, the Polynesians of Easter Island and the Norwegians in Greenland, as well as the environmental vicissitudes of certain contemporary societies in Rwanda, the Dominican Republic, Haiti, Papua New Guinea, China, Australia and Montana in the USA could be considered perfectly suited to the approaches of political ecology as Paul Robbins understands it.

Another ecological problem related to agriculture concerns the waste coming from the herbicides and pesticides that are used in large plantations of agricultural industrial products. These are the so-called *persistent organic pollutants* (POPs) that constitute a chemical cocktail that is not degraded or destroyed. By contaminating the food chains, it constitutes a serious danger to human and animal health. POPs were originally prepared as insecticides (DDT), or were the result of garbage incineration (dioxin), but DDT is by far the most dangerous as it spreads, being carried by rivers to lakes and large water reservoirs (dams). In this process, it is absorbed by aquatic algae (zooplankton), which are in turn ingested by small fish and later by larger fish, such as trout. Trout constitutes a sizable food source for bears, birds, and humans, so contaminating and poisoning these fish is particularly grave, as demonstrated by Jared Diamond in his research in the US state of Montana, as will be seen in the next chapter. And concerning DDT, although it is prohibited, it is necessary to maintain permanent surveillance to prevent clandestine use of it, something that has happened in many developing countries.

However, the biggest problem with waste disposal is the widespread use of plastics. Marine biology institutes estimate that some eight million tons of plastic enter the sea every year, because every second more than 200 kilos of garbage is carried to the oceans by the rivers. The elements of the garbage that are not biodegradable are deposited on the seabed (70%), while the rest floats on the surface of the water, harming marine life. Certain species such as turtles are at risk of extinction, not only because of the predatory action of humans but also because of the ingestion of plastics. Both WHO and UNEP have denounced this problem, but abandoning the use of plastic material (in supermarket bags, for example, or in the packaging of liquids) would require drastic action by all States (such as imposing taxes to discourage its production or prohibiting its use), which, as expected, is opposed by the petrochemical industry and oil producers.

3.10 Decarbonization and Renewable Energy

Burning fossil fuels (oil, coal and natural gas) is, to a large extent, one of the causes of the increase in atmospheric carbon dioxide (CO_2), which is causing the warming of the Earth's climate due to the greenhouse effect. Therefore, one of today's great environmental problems originates in the use of petroleum derivatives for almost 40% of the fuel needed to produce electricity around the planet, and almost as much from coal along with natural gas, leaving little more than 20% of our energy generated by renewable resources, such as wind, geothermal, hydroelectrical, tidal, nuclear, biomass or solar energy. Those figures have been rounded to represent everyone, as the variation between countries is remarkable (70% of the electricity generated in France, for example, comes from nuclear reactors) but what still remains to be done, in terms of substitution of energy sources, is a task of great importance.

In the field of renewable energies, wind and solar energy are the cleanest, but both have the disadvantage that the generation of energy is intermittent (it is dependent on the wind or the hours of the day when there is solar radiation), and the difficulties in storing it are considerable. However, the manufacturing costs of solar panels have been reduced, which is making solar energy quite competitive. Hydroelectric power plants are also a source of clean energy but their social and environmental impact is a disadvantage, and they also depend on the water supply, which can sometimes be reduced by droughts. Nuclear energy can also be considered clean from the point of view of carbon emissions, but the construction of nuclear power plants is very expensive and nuclear plants are also subject to the risk of accidents of human origin (as in the case of Chernobyl) or natural disasters, such as earthquakes or tsunamis (this was the case with the Fukushima nuclear reactor in Japan), not counting the difficulties in disposing of non-recyclable reactive waste. Even if it the proportion of this non-recyclable waste is very small – only 2% of the enriched uranium used as fuel – it has to be moved to special places and buried deeply, since the plutonium it contains, besides being dangerous from the point of view of radioactivity, is the raw material needed for the manufacture of nuclear weapons.

In terms of energy from biomass and fuels of vegetal origin, such as ethanol, it has the disadvantage of promoting deforestation, and the fact that sugar cane and basic grains are often much needed for human consumption. The hydraulic energy of the tides and coastal waves is very clean but the technologies are still in the initial phase so the cost is higher. Therefore, we are still using energy sources which require fossil fuels whose main advantage today is their considerable drop in prices (especially those of oil), but, as we have seen, their main disadvantage is that they are highly polluting. Furthermore, fossil-fuel-based energy sources are destined to be exhausted precisely because they are non-renewable. Hence, humanity is, in any case, obliged to adapt to their gradual replacement by sources of renewable energy. The fulfilment of the commitments assumed at COP 21 (the Paris Agreement) should facilitate the switch to renewables, but there is no doubt that clearer policies are required on the part of all governments.

These policies should consider charging a higher tax rate for gasoline and diesel, as well as higher taxes for the cars that use these fuels, all with the purpose of making investment in green technologies profitable both in terms of yielding new sources of energy, as in the manufacture of electric cars, and in terms of innovation in general, including the production of vehicles based on hydrogen engines.[31]

It should be noted that, among the innovative research on renewable energies I have not yet mentioned nuclear fusion, which unlike the nuclear fission that is currently used in normal reactors, is the type of energy from which the stars are nourished in the entire universe, including of course, our own solar star. Which is the equivalent of saying that if the nuclear fuel used by the reactors responsible for the generation of solar energy could be 'domesticated' with appropriate technology and dimensions, it is likely that this is truly going to be the energy of the future. For now, the multinational project ITER – installed in France – expects to have its first results by the year 2050. It is therefore a long-term project, and meanwhile the goals set by the 2030 sustainable development agenda are the best alternative.[32]

---

[31] An internet publication refers to hydrogen vehicles as follows: "With the recent arrival of the Honda Clarity, there are now three automakers offering cars powered by hydrogen fuel cells. The first was the Hyundai ix35 in early 2013, and then came the Toyota Mirai – and more than a dozen other automakers have fuel cell vehicles in development. Driving a hydrogen-powered car has some ups and downs. On the plus side, you get the green benefits of an EV without the range anxiety, because you can refill the car with more hydrogen. On the minus side, hydrogen refuelling stations are rare – at least for the moment. It's also challenging to obtain hydrogen in a way that's both green and efficient. Further, as with any new technology, there's some understandable hesitation to be among the first people to take the leap and commit to several years (at least) with a power source that may or may not work out. But keep reading and we'll give you enough background that you can make your own decision"; at: https://www.digitaltrends.com/cars/does-hydrogen-make-sense-as-an-automotive-fuel/.

[32] The most advanced project in nuclear fusion is the magnetic confinement or ITER (*International Thermonuclear Experimental Reactor*), a prototype based on the Tokamak concept. In addition to the reactor, this uses auxiliary systems but does not generate electricity, which is obviously the main purpose of all this great international effort in the field of renewable and clean energy. The European Union, Canada, the US, Japan and Russia are participating in this project. The objective is to determine the technical and economic feasibility of nuclear fusion by magnetic confinement for the generation of electrical energy, as a preliminary phase in the construction of a commercial

3.11 Food Security

Nearly 800 million people go hungry, and are undernourished in the world today, and one of the most important goals of sustainable development and the 2030 agenda is to end hunger in the world. People suffering from hunger are located in poor countries, especially in Africa, where droughts and the consequences of climate change seriously affect countries such as Ethiopia, with 32% of its population undernourished (31 million), Madagascar (33%, 8 million), Tanzania (32%, 16 million), and this list continues for a good part of African countries where the problem is greatest. In Asia, Pakistan, India and China have undernourished populations, as do North Korea, Sri Lanka, Mongolia and Tajikistan. In Latin America, Guatemala (already mentioned), Bolivia and Haiti are on the list. In the Middle East there are undernourished populations in Iraq and Yemen. In short, it is a global problem that is largely determined by phenomena such as droughts, but also by the degradation of the soil resulting from erosion provoked by wind and water, especially when there are floods or prolonged torrential rains, and by certain agricultural practices, such as excessive grazing, or cattle that leave the soil vulnerable to erosion by water, as well as deforestation. Furthermore, the issue of malnutrition is closely related to the failure of state policies which are supposed to ensure food for the nation's citizens and to combat both malnutrition and famine, but until the present have been absolutely incapable of solving the problem due to corruption and bad governance. There is also the political concept of 'food sovereignty', which refers to the right of each nation to define its own agricultural and food policies according to sustainable development and food security objectives.

Food sovereignty has a protectionist bias because part of the principle of food security means that at least basic foods such as corn, beans, rice, wheat and others must be produced in the country and not be imported. This inherent policy implies the protection of the domestic market against imported foods that are cheaper in the international market. Obviously, neoliberals who do not see problems – they are partisans of the policy of comparative advantages – will reject food sovereignty, as I will discuss hereafter. The origin of the concept is found in the World Food Summit organized by the *Food and Agriculture Organization* (FAO) in Rome in 1996, and it implies a rejection of free trade because it proposes to protect the local production of basic grains not only because of the elementary principle of food security but also to empower the rural population and improve their income. But the principle of food sovereignty goes beyond FAO's food security, which focuses on the mere availability of food, since it tries to have an impact on the way food is produced,

---

demonstration facility, although ITER is a technological project whose construction is estimated to take ten years and will require another twenty years to complete the investigations it has already started. Robotics, superconductivity, microwaves, accelerators and control systems stand out among the technologies used for its construction and subsequent operation and maintenance. As previously underscored in the ITER reactor now built, no electric power will be produced, only solutions to the problems that need to be solved to make future nuclear fusion reactors viable. This ambitious research project installed in France will have its first results in the long term: 2050; at: https://ene rgia-nuclear.net/que-es-la-energia-nuclear/fusion-nuclear.

giving preference to the local rather than the imported, among other reasons because bringing cheap food from abroad weakens both local production and local farmers, who see their income reduced and therefore their food security, which is also a major reason for migration.

A case in point is the industrialized corn producers in the centre and south of the US who have benefited from NAFTA, exporting cheap corn in large quantities to Mexico (and also to Central America, thanks to the free trade agreement signed with the Central American countries or CAFTA), where Wal-Mart deals with its distribution, harming all peasants who are unable to compete with the American agribusiness and consequently abandon the countryside and are forced to migrate to the US in search of employment. Bear in mind that the renegotiation of NAFTA could perhaps could be an opportunity to realize that it could be better for countries like Mexico to stop the importation of food from the US and promote national food security. The integral development plan that the President of Mexico is promoting for the less developed southern states could have food production as one of the main items, and if this policy operates it could also be the best way to reduce migratory flows to the United States.

The foregoing can easily be verified by taking the time to read the many studies and publications that have conducted evaluations of the first twenty years of NAFTA, which were completed in 2014, and although it is not the purpose of this text to provide an exhaustive bibliography on specific topics, I will survey part of what was written by Carlsen (2013), director of the programme for the Americas at the Center for International Policy, in an article published by the *New York Times* on 24 November that year:

NAFTA has cut a path of destruction through Mexico. Since the agreement went into force in 1994, the country's annual per capita growth flat-lined to an average of just 1.2 per cent – one of the lowest in the hemisphere. Its real wage has declined and unemployment is up. As heavily subsidized US corn and other staples poured into Mexico, producer prices dropped and small farmers found themselves unable to make a living. Some two million have been forced to leave their farms since NAFTA. At the same time, consumer food prices rose, notably the cost of the omnipresent tortilla. As a result, 20 million Mexicans live in 'food poverty'. Twenty-five percent of the population does not have access to basic food and one-fifth of Mexican children suffer from malnutrition. Transnational industrial corridors in rural areas have contaminated rivers and sickened the population and, typically, women bear the heaviest impact. Not all of Mexico's problems can be laid at NAFTA's doorstep. But many have a direct causal link. The agreement drastically restructured Mexico's economy and closed off other development paths by prohibiting protective tariffs, support for strategic sectors and financial controls. NAFTA's failure in Mexico has a direct impact on the United States. Although it has declined recently, jobless Mexicans migrated to the United States at an unprecedented rate of half a million a year after NAFTA. Workers in both countries lose when companies move, when companies threaten to move as leverage in negotiations, and when nations like Mexico lower labour rights and environmental enforcement to attract investment. Farmers lose when transnational corporations take over the land they supported their families on for generations. Consumers lose with the imposition of a food production model heavy on chemical use, corporate concentration, genetically modified seed and processed foods. Border communities lose when lower environmental standards for investors affect shared ecosystems. The increase in people living in poverty feeds organized crime recruitment and the breakdown of communities. Increased border activity facilitates smuggling arms and illegal substances (Carlsen 2013).[33]

What Carlsen described above is precisely what I alluded to when I said that, following Boaventura de Sousa Santos, we must differentiate between hegemonic globalization and non-hegemonic globalization., What has happened with NAFTA is a clear example of the former, since the workers (not the businessmen) in the two countries have lost out because of the famous agreement that was renegotiated on Trump's orders. Since 1994 half a million unemployed Mexicans have migrated to the United States annually, fleeing poverty and the inability to compete with producers of corn from the north as well as the reduction in labour and environmental benefits arranged to attract investments to Mexico, while in the United States workers are affected when employers threaten to close factories and move to Mexico where salaries are lower. That is exactly the kind of globalization that should be rejected from a 'non-hegemonic' perspective, precisely because Mexico is clearly not the hegemonic country.

In this same order of ideas, when the Executive Secretary of ECLAC, Bárcena (2018), was asked in an interview about the renegotiation of NAFTA and its possible termination, she said that since Mexico had bet everything on the US for twenty-three years, it was important "to change such a paradigm". Even finishing with NAFTA does not seem to be extremely serious, said Barcena, because, according to her, Mexico could decrease its GDP from 0.9 to 1% without major problems and the transition to an economy without FTA may be favourable for Mexico in the automobile industries. Automakers such as the Japanese Nissan would not leave Mexico since its installed capacity is very solid and the production of Japanese cars already has an outlet in an important niche of the market. But the Executive Secretary of ECLAC believes that it is particularly in the internal market that opportunities could be opened, because the end of NAFTA would give Mexico the opportunity to rescue the agricultural sector, thereby increasing its food sovereignty. As Barcena explains:

> Today Mexico is a net importer of food. 70% comes from the USA. Including corn, our great national pride. The country must return to being a food producer, and it has all the possibilities to do it: geography, technology … [It requires] a policy of support and financing, which has been the great sacrifice in the free trade agreement. There was a great dismantling of the banking that supported it and this has caused a great migration towards cities and towards the USA. The most productive workers in the US agricultural sector are Mexicans. Why can we not do the same here? Policies and added value fail … The salary remains the Achilles

---

[33] Carlsen's article can be found at: https://www.nytimes.com/roomfordebate/2013/11/24/what-weve-learned-from-nafta/under-nafta-mexico-suffered-and-the-united-states-felt-its-pain.   An interesting study on the first twenty years of NAFTA (1994–2014) with a number of comparative graphs about the performance of the Mexican economy compared to the rest of Latin America (some based on ECLAC studies) shows that Mexico did not fare well in indicators of growth, per capita income, or employment in addition to the social indicators of poverty already mentioned by Carlsen. The study (conducted by American researchers) points out, for example, that comparing poverty indicators with the region as a whole shows that Latin America reduced its poverty on average from 46 to 26% while Mexico went from 45.1 to 37.1%. Regarding these issues, see also Weisbrot et al. (2014), at: http://cepr.net/documents/nafta-20-years-2014-02.pdf.

heel. Mexico is one of the few countries in Latin America, or perhaps the only one, in which there have been no increases in recent decades. Jobs have been formalized and that is good news, but it is vital that income improves, which has been artificially contained. The salary is the product of a negotiation (between workers and patrons) and Mexico has dismantled the workers' organization and privileged the corporate unions ... The businessmen also agree [to the renegotiation of salaries]. We have to change the conversation between businesses, the State and society: the great pact of social equality of the Nordic countries was built in the 1940s under the agreement that whoever earned the most had to generate income for those who earned the least. Here that is not happening: the one who earns the most applies Adam Smith's maxim "all for me and nothing for others" ... We have to change the conversation, and in that change entrepreneurship is fundamental. In the case of Mexico, it is not realizing its relationship with insecurity. What do we want for the youth that falls into the crime or the drug trafficking that is productive here? ... You cannot continue to gain productivity by sacrificing the workers (Barcena 2018: 5).

Returning to the issue of food security in its global dimension, it should be remembered that since the mid-1990s the Rome Forum Action Plan – *Food for All, No Benefits for a Few* – insisted on the crucial role that civil society could and should play in implementing the commitments of the signatory governments of the declaration of the summit of 1996. Three other world events brought together social movements of civil society to deepen the concept of 'food sovereignty': The World Forum for Food Sovereignty in Havana met in August 2001, followed by the Civil Society Organizations Forum on the same subject in June 2002 (in parallel to the World Food Summit in Rome, in June 2002) and the International Forum on Food Sovereignty that met in the African Republic of Mali on February 2007.

In short, food sovereignty poses a framework for the governance of agricultural and food policies and incorporates a wide range of issues, such as land reform, the management of territory, local markets, biodiversity, autonomy, cooperation, debt, health, and other elements related to the ability to produce food locally. It covers policies which refer not only to localizing the control of production and markets, but also to promoting the fulfilment of the right to food (which is a crucial element of economic and social rights), as well as giving indigenous peoples and peasants in general access to and control over land, water, and genetic resources, including the promotion of environmentally sustainable forms of production.

Another problem closely linked to the food issue concerns the type of agrochemicals that some transnational corporations such as Monsanto are producing, given that they have been accused of contaminating soils through the use of these herbicides, which are then passed on to humans via food produced in the same soil. In fact, the use of glyphosate produced by this company as a broad spectrum herbicide has been subject to controversy in many countries, due to complaints by environmentalists regarding damage inflicted on human health, as well as serious contamination of the environment. In Colombia it was used to eradicate coca crops but its use ended up being prohibited for that purpose. In Guatemala, oil palm producers were accused of using it in their plantations, causing fish mortality rates to skyrocket in one of the most important rivers in the country, the Pasión River, which is a tributary of the Usumacinta, the border with Mexico. In Argentina, controversy over contamination with glyphosate used in soybean plantations that harm human health has been taken to court, and EU countries such as France and Italy have serious reservations about its use to the extent that there has been opposition to a controversial decision of the European Commission authorizing its use. As of July 2017 California included glyphosate in an official list of substances that can cause cancer, thus warning the purchasers and the manufacturers (Monsanto) against the herbicide brand 'Roundup' (Smith 2017: 25).

Last but not least, the inhabitants of the planet who find themselves in upper or middle classes must learn to moderate the consumption of foods of animal origin, be they from cattle, sheep, pigs, poultry or of marine origin. This is not just because the production of grains for animal consumption (such as corn) has a high cost and increases the carbon footprint, or because overgrazing and pasture growth for livestock consumption increases soil erosion, but because from the dietary point of view both overweight and excess protein are not good for human health, and although fish is a much healthier food than red meat, industrial fishing is depleting the planet's fish resources. Therefore, the vegetarian diets practised by some peoples of the world (such as Asian peoples who are Hindus or Buddhists) are recommended and worthy of being promoted by governments and health services, along with physical and spiritual exercises such as yoga and certain martial arts, since the health benefits that such practices entail are of great value and it is expedient to spread and increase their practice.

3.12 Water Resources

Water covers 70% of the surface of our planet but only 3% of it is fresh water suitable for human, animal and vegetable consumption, since most (97%) is salt water in the great oceans and seas. Of the remaining 3% of fresh water on the planet, 68.9% is stored in glaciers and ice on the planet, mainly in the Arctic (Greenland) and Antarctica or in the high mountain ranges, while 28.8% is found in underground deposits and aquifers. Consequently, only a tiny 0.3% is in liquid form accessible for use on the surface of rivers, lakes or swamps. So to be more aware of the importance of water, we must be clear that the vital liquid is only available in that tiny amount. Additionally, the melting of fresh water in glaciers and polar ice caps, the difficulties derived from the variability of the rainfall regime, and the impossibility of predicting climatic phenomena such as storms, hurricanes, floods or droughts make it is easy to realize that everything related to water is crucial for all the countries of the world and for the survival of the species that populate the planet. This is truly one of the greatest challenges for humanity in the twenty-first century, and finding an adequate solution at the level of multilateral negotiations and national policies for the management of water resources will depend on international peace and security, much more so than the solution that must be found for the abuse of other natural resources, such as fossil fuels like oil, gas and coal.

It is not the water we drink every day that constitutes the majority of human consumption. The largest water drain is hidden in food, manufactured products and energy. This is because, as with what has been called the carbon footprint,[34] there is also a water footprint, which means that in all the products we consume there is a certain amount of 'virtual water' that consists of the liquid that was used to produce the food we consume daily. The problem of the water footprint should be addressed, because it is precisely the water used in agriculture that constitutes the largest proportion of water used by humans, followed by industrial consumption and daily consumption (washing, drinking water, cooking, etc.) This 'virtual' water is exported, hidden in products, which are in turn consumed in the countries of destination, therefore benefiting the population of the country where the product is ultimately used.

Furthermore, it must be stressed that it is human activities that can seriously affect water resources via pollution, the over-exploitation of aquifers, the consequences of climate change, urbanization and deforestation. Pollution can contaminate surface water and groundwater with excess fertilizers and pesticides (the problems related to the use of glyphosate and DDT have already been mentioned). Also, badly managed agriculture can increase erosion, creating sediments which ultimately reach rivers and lakes, reducing their capacity to transport and store water. Poor road construction can also cause landslides that damage natural water courses, increasing sedimentation, as well as inappropriate wastewater discharges. Without proper treatment, they will contaminate rivers, lakes and groundwater, limiting their subsequent use.

In addition, pollution can damage water resources and aquatic ecosystems with the organic matter and the pathogenic organisms that the residual waters carry, since they transport residues of fertilizers and pesticides that have been used in agricultural work. There is also pollution from the air resulting from the use of heavy metals released by mining and industrial activities that, when precipitated on the ground, is called "acid rain". In sum, the overexploitation of water resources, both superficial and underground, has had catastrophic effects on many parts of the Earth. In Central Asia, the Aral Sea, for example, is almost completely dry due to the diversion of the Amu Daria and Sri Daria rivers for cotton irrigation in the time of the former USSR. Lake Chad in Africa is also a clear example of the overexploitation of water resources.

Overexploitation of groundwater is not as evident as that of lakes and rivers because there is less visual evidence and the effects of excessive extraction of ground-water (due to the opening of wells) take longer to be noticeable, although the pumping of aquifers has been increasing worldwide due to population growth and urbaniza-tion. In agriculture the benefits of irrigation with water extracted from the subsoil are ephemeral and end up translating into a decrease in the level of aquifers, in the drilling of deeper wells and sometimes in the depletion of underground sources.

---

[34] The carbon footprint is the amount of *carbon dioxide* (CO_2) emissions that come from certain products, activities or services. A citizen of an industrialized country has a footprint more than a hundred times greater than that of a peasant from India or a sub-Saharan country, because his consumption of food, transportation, heating, recreation, electricity, communications and products of industrial origin is within the activities that generate 10 tons of CO_2 per year. Lifestyles with higher consumption are becoming generalized as the middle class grows.

This means that the recharge capacity of underground aquifers must be taken into account because the benefits of their extraction are usually short-lived, while the negative consequences (reduction in water levels or depletion of resources) can be permanent. It is dangerous if their exploitation is prolonged for too long a period of time, as it will result in a non-renewable underground deposit. Moreover, the effects of climate change that have an impact on the planet's water resources are increasing the pressure on areas that already suffer from water shortages, droughts or desertification trends. The terrestrial and mountain glaciers are regressing more quickly in recent years as a result of global warming, which has also resulted in an increase in extreme weather phenomena, such as tropical storms, hurricanes and floods, and in droughts that not only affect food production but have also increased the occurrence and severity of forest fires, as in California, Oregon, Portugal, Catalonia and Australia. Such problems will continue to happen because droughts will be more frequent as a result of climate change.

3.13   Integrated Water Resources Management

All the water problems described in the previous subsection must be addressed and resolved at the national level, because even though the United Nations has convened different conferences and meetings on the water issue, ultimately it is on the member states that the obligation to take appropriate measures for the management of water resources falls. In that respect, national governments have been enacting their own legislation to appropriately address the challenge of managing water properly, or, as a bad example of the influence of neoliberalism in some extreme cases, they do not have legislation on that matter at all.[35] In this framework, the United Nations proposed the empirical concept of *Integrated Management of Water Resources* (IWRM), which is defined as "a process that promotes the coordinated management and development of water, soil and other related resources, with the purpose of maximizing economic outcomes and social welfare in an equitable manner without compromising the sustainability of vital ecosystems" (United Nations 2014). Although many of the elements of the concept had been used since the first conference on water (Mar del Plata 1977) it was not until after the Earth Summit in Rio (1992) – which approved Agenda 21 – that the concept of IWRM could be widely discussed and its use implemented in all countries.

Obviously, integrated management is indispensable because there are great differences in the availability of water both globally and locally. As deserts imply extreme scarcity, and tropical forests demonstrate the abundance of water, it is evident that the same kind of integrated management should not be applied in the Maghreb countries

---

[35] I refer to the case of Guatemala, a country where, even though the national constitution prescribes the obligation to enact a law on that matter, no normative has so far been approved by the Parliament. An interesting book on this topic has been produced by a researcher at the Institute of Investigations on State Policies at Rafael Landívar University (Padilla 2019).

of North Africa and in northern Chile. Nor should the water policies applicable to the warm tropical Amazon or Congo river basins and their tributaries be identical to those used in the cold rainforests of southern Chile, Canada or Russia, and still less should the policies that China must follow be applied on the great Tibetan plateau. Although sources of great rivers might seem similar, India and South East Asia – with the Ganges, the Brahmaputra and the Mekong – are different from the Anatolian highlands in Turkey, and consequently the water management policies for the Tigris and the Euphrates should not be like those for the rivers further east.

In addition, the water supply is also variable in terms of the rainfall regime. Although tropical countries have relatively well-established annual 'seasons' (the Indian monsoon, or the rainy season in tropical countries), there are no precise dates regarding the beginning and end, as the duration of rainfall is unpredictable. This implies a lack of reliability with regard to the resource, which represents a significant challenge for the planners of 'integrated management'.

Developed countries have overcome the natural variability by building infrastructures (dams, reservoirs and water storage tanks) to manage availability and ensure reliable supplies, thereby reducing the risks associated with dependence on unpredictable weather conditions. But these infrastructures are costly (such as the desalination of seawater that rich oil states can afford) and this often – as in the case of dams – involves a negative impact on the environment, as well as on local communities due to the flooding of crop lands and the obstruction of the free flow of water that a dam implies. Consequently, many States have realized that adequate IWRM means managing not only the supply but, above all, the growing demand for water resulting from demographic, economic and climatic pressures. For this reason, it has been necessary to implement measures of wastewater treatment, the recycling of water and demand management. In addition to the problems related to the integrated management of the available quantity of water, there are also problems with the quality of the liquid, because the pollution of springs, lakes and rivers is one of the main problems faced in the social sphere, as well as being a threat to the proper functioning of natural ecosystems.

Another problem, which I will examine next, concerns the effects of climate change on water resources, as it is in the most vulnerable countries and regions that the availability of water, both in quantity and quality, is being affected by the phenomenon. The shortage of rainfall or prolonged droughts, as well as the excess of rain and consequent floods caused by the increasingly frequent tropical storms (as occurred in Central America with hurricanes Mitch and Stan in recent decades), generate great difficulties for the integrated management of the resource, analogous to the difficulties caused by other factors, such as increases in demand which are the result of population growth, urbanization, and the expansion of agricultural and industrial borders that affect consumption patterns and production.

3.14   Climate Change

Without the atmosphere there would be no life on Earth, as it contains the air we breathe and protects us from solar radiation, playing a fundamental role in the water and rain cycle as well as in weather patterns. The thin layer of gases that surrounds our planet is composed mostly of nitrogen (78%) and oxygen (21%) with a lower percentage of other gases, such as carbon dioxide and methane, which are the main gases responsible for the warming that the Earth's climate has being experienced since the twentieth century. The increase in levels of carbon dioxide and methane has several contributing factors, including the burning of fossil fuels derived from petroleum, the use of coal to generate electricity and natural gas, the melting of Siberian and Canadian tundra permafrost, the use of fertilizers, herbicides and pesticides in agriculture, and the emission of chlorofluorocarbons, used in aerosols, air conditioning and refrigeration appliances (which together can be considered primarily responsible for the hole in the ozone layer), and so on. All these Green House Emissions (GHE) are the source of climate change and global warming, which is at the root of the increase in climatic hazards such as droughts, desertification, hurricanes, floods, heavy rain, heatwaves and dangerous sea-level rise.

3.15   The Ozone Layer and the Montreal Protocol

In the graphs below it is possible to see how ppm (parts per million of carbon dioxide, CO_2) have increased since 1975 in an absolutely explosive way compared with the normal average (as demonstrated in the graphs on the Great Acceleration in Figs. 3.2, 3.3 and 3.4). Between the end of the last glaciation (ice age) and the beginning of the geological period of the Holocene about 12,000 years ago, they remained between 260 and 280 ppm and now they have shot up to almost 400 ppm. The charts, produced by the US National Oceanic and Atmospheric Administration (NOAA 2017), show how CO_2, like nitrous oxide, methane and HCFC *hydrochlorofluorocarbons* (which replace CFCs), have also increased and display indicators showing, for instance, how methane has been released from the ice because global warming is melting the permafrost layer in the Siberian and Canadian tundra, and has also increased because of the burning of garbage, as mentioned on previous pages. Furthermore, nitrous oxide from the use of agricultural fertilizers and pesticides such as methylbromide, as well as the aforementioned chlorofluorocarbons (emitted by refrigeration appliances, freezers, sterilization equipment, aerosols and other products), have also had an impact on global warming or on the above-mentioned hole in the ozone layer, because all of them have a greenhouse effect. That is to say, they are those gases (carbon dioxide, methane, CFC) that, due to the fact that they continually reflect solar radiation and heat back into space, also contribute to the increase in the temperature of the Earth's surface. As is widely known, it is for these reasons that the *Inter-governmental Panel on Climate Change* (IPCC) has been demanding reductions in greenhouse gas (GHG) emissions, and that they were subject to the resolutions of the Paris Agreement of December 2015 within the framework of the intergovernmental Conference of the Parties (COP 21).

Chlorofluorocarbons (CFCs) are the only gases with demonstrable reductions in the level of emissions, and this has been achieved thanks to the signing of the Montreal Protocol of 1987,[36] a fact that demonstrates that it is possible to control pernicious emissions of anthropogenic origin, thereby avoiding an increase in the hole in the ozone layer in the stratosphere (at that altitude, the ozone layer constitutes a protective filter against solar radiation harmful to human health).[37] In the four graphics in Fig. 3.5 it is clear that the increase in GHG emissions from the period 1975–2015 in parts per million for carbon dioxide, and parts per billion for nitrous oxide and methane, has been constant and unstoppable, whereas gases altering the ozone layer have decreased, thanks to the 1989 international agreement to the Montreal Protocol.

However, on the bottom right we also can see that there has been an increase in *hydrochlorofluorocarbons* (HCFCs) belonging to a second generation of refrigerants whose use is allowed because, although they contain chlorine (which damages the

---

[36] The Montreal Protocol is a protocol arising from the Vienna Convention for the Protection of the Ozone Layer, designed to protect the ozone layer by reducing the production and consumption of numerous substances that have been studied and proven to react with the ozone layer. The Protocol addresses the causes that are believed to be responsible for the depletion of the ozone layer, with the objective of enabling it to make a full recovery. The agreement was negotiated in 1987 and entered into force on 1 January 1989. The first meeting of the parties was held in Helsinki in May 1989. Since then, the document has been revised several times, in 1990 (London), in 1991 (Nairobi), in 1992 (Copenhagen), in 1993 (Bangkok), in 1995 (Vienna), in 1997 (Montreal), and in 1999 (Beijing). It is believed that if all countries meet the proposed objectives of the treaty, the ozone layer could have recovered by the year 2050. Due to the high degree of acceptance and implementation that has been achieved, the treaty has been considered an exceptional example of international cooperation; at: https://en.wikipedia.org/wiki/Montreal_Protocol.

[37] The ozone layer was discovered in 1913 by French physicists Charles Fabry and Henri Buisson. Its properties were examined in detail by the British meteorologist G. M. B. Dobson, who developed a simple spectrophotometer that could be used to measure stratospheric ozone from the Earth's surface. Between 1928 and 1958 Dobson established a worldwide network of ozone monitoring stations, which continue to operate today. The Dobson unit, a unit for measuring the amount of ozone, was named in his honour. Ozone acts as a filter, or protective shield, against noxious, high-energy radiation that reaches the Earth, but it allows other rays to pass as the long wave ultraviolet, which thus reaches the surface. This ultraviolet radiation is what allows life to survive on the planet, since it is what enables photosynthesis to take place in the plant kingdom, which is at the base of the food pyramid. Apart from the ozone layer in the stratosphere, the 10% of the remaining ozone that is contained in the troposphere is dangerous for living beings because of its strong oxidizing character. High concentrations of this compound at surface level form the so-called photochemical smog. Ten per cent of this ozone is transported from the stratosphere and the rest is created by various mechanisms, such as the interaction between sunlight, hydrocarbons and nitrogen oxides, or by electrical storms that ionize the air and make it, very briefly, a good conductor of electricity (two consecutive lightning strikes sometimes follow approximately the same trajectory). In 2013, the dangerous exposure to the sun's rays without ozone protection reached the underwater world and caused the species that inhabit the Great Barrier Reef of Australia to suffer from skin cancer; at: https://en.wikipedia.org/wiki/Ozone_layer.

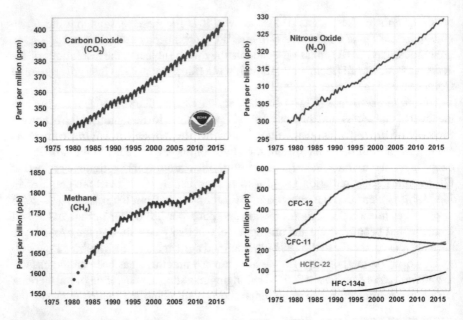

Fig. 3.5 Increase in GHG emissions compared with CFC and HFC emissions. *Source* https://www. esrl.noaa.gov/gmd/aggi/aggi.fig2.png

ozone layer), they also contain hydrogen, which makes them chemically less stable when going up into the atmosphere and consequently means that their depletion potential is very low (varying from 0.001 to 0.11) according to experts' data, which makes them less harmful to the ozone layer. This has allowed them to be used as substitutes for CFCs. Developing countries are allowed to continue using them until the year 2040, but the more advanced HFC-134a, which lacks chlorine, is currently used in refrigerants and air conditioning.[38]

In July 2016 in Vienna (Austria) the 3rd Extraordinary Meeting of the Parties to the Montreal Protocol made the decision to work on the amendment of the Montreal Protocol to reduce the production and consumption of *hydrofluorocarbons* (HFCs), because, as already pointed out, although they do not harm the ozone layer, they

---

[38] According to experts, Hydrofluorocarbons (HFCs) are considered the third generation of refrigerant gases, since they have been created to replace CFCs and HCFCs. At first, they were considered ecologically benign, because they do not damage the atmospheric ozone layer, but the presence of fluoride in their composition causes them to behave like a greenhouse gas when they are emitted, contributing to global warming. For this reason, they have to be subject to restrictions regarding their use to minimize their emissions. Their PAO (polyalphaolefin, a synthetic hydrocarbon) content is zero, but in general they have high Global Warming Potential (GWP) values, which implies a high influence on the global greenhouse effect. This means that in future all refrigeration and air-conditioning installations will be controlled by regulations related to the environment. Although HFCs represent only a small fraction of all greenhouse gases, they are growing rapidly in the atmosphere. The emission of these refrigerant gases could increase by almost twenty times in the next three decades if no measures are taken to reduce them (as can be seen in Fig. 3.4 of the US National Oceanic Administration Agency); at: https://www.esrl.noaa.gov/gmd/aggi/aggi.fig2.png.

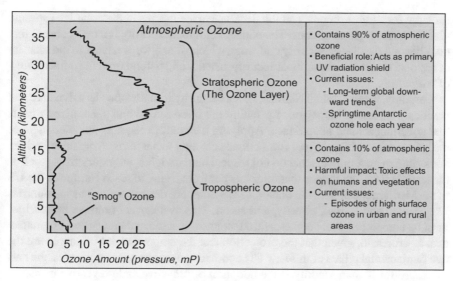

Fig. 3.6 Ozone layers in the stratosphere and the troposphere according to public domain information

have a more potent greenhouse effect than carbon dioxide, thus contributing to global warming. Therefore, even though the Montreal Protocol is a very successful environmental treaty – the production and importation of substances that harm the ozone layer has been reduced by up to 97% – it was necessary to renegotiate an amendment, which was signed in Kigali (Rwanda) in October 2016, in order to reduce HCFCs as well. The hole in the ozone layer is closely related to the problem of climate change, but because negotiations to mitigate it have been exemplary in their degree of success, it would be desirable for the COP 21 Paris Agreement to have the same results, despite the fact that the commitments agreed upon depend on the sovereign decisions of each member state, as there are no procedures to enforce them.

Figure 3.6 shows the distribution of ozone in the stratosphere (20–35 km above the Earth's surface) and the troposphere (1–15 km above the Earth's surface). The ozone of the stratosphere protects the planet from high-energy ultraviolet radiation, but allows the passage of long-wave radiation that in turn enables the photosynthesis of flora at the surface of the Earth. In contrast, the ozone in the troposphere is harmful to human health because it manifests as smog.

3.16 The Issue of Human Security

The problem of the environment is closely related to the problem of human security insofar as climate change has a large impact on phenomena such as hurricanes, cyclones of all kinds, droughts, unusual summer heatwaves that create the conditions for large forest fires, floods, and the decrease in ice masses in the polar ice caps and glaciers in the high mountains. The concept of human security is a new concept that goes beyond the classic security concerns, which were always related to the security of national States or international security, usually understood from the point of view of *realpolitik*.

Certainly, for the United Nations, human security is understood as a dynamic and practical regulatory framework for facing the intersectoral and generalized threats that governments and people face. Applying the concept requires an assessment of the human insecurities derived from climate change, social turmoil, organized crime, authoritarian governments that do not respect human rights, migratory risks (such as the terrible risks and threats that migrants confront trying to reach Europe or the US in the Mediterranean Sea, the Sahara or US Southern deserts, as well as in countries like Mexico and Libya), poverty, and so on. This evaluation of human insecurities must be comprehensive, people-centred, context-specific and prevention-orientated, and, as aforesaid, given that the protection and the empowerment of people are the two fundamental pillars of this new UN approach concerning security issues, the best safeguard for human security is the adoption of proactive and preventive measures in the face of these current and new threats: how to prevent forest fires or floods provoked by the heatwaves and tropical storms that have increased enormously due to climate change; how to prevent poverty that determines migratory flows; how to prevent violence triggered by juvenile gangs in poor countries; how to prevent the violation of human rights activated by migratory officials against migrants and refugees.

From that perspective it is clear that governmental measures to prevent human security being endangered must tackle the threats arising from issues like climate change as well as those arising from social insecurity, which in a great number of countries is the result of the absence of social public policies, lack of democracy, lack of transparency, and government corruption. Hence, we must act proactively and preventively in both directions. Concerning the environment, governments must have plans and implement actions to mitigate natural disasters or adapt to the inevitable, which could occur in the form of floods due to hurricanes and tropical storms, earthquakes and *tsunamis*, the increase in the frequency of phenomena such as 'el niño' in the oceans, as well as water stress manifested as droughts and other phenomena that require the instalment of an integrated system of water management in accordance with the respective UN SDGs. Concerning social policies, it is evident that the best way for governments to prevent social turmoil is also to establish a series of ongoing actions and public policy measures to fulfil the social pillar of the SDGs: the reduction of poverty, improved food security, investment in education and health, the empowerment of girls and women, diminishing social inequality, and so on.

In synthesis, an important achievement of the Convention on Climate Change is the production of a frame of reference for both human security and diplomatic negotiations which result in international treaties and conventions which not only focus on people but also seek to prevent the consequences of environmental phenomena like climate change which, if not addressed well in advance, will continue to cause future catastrophes, putting people's safety at serious risk. Hence, the strategies which should be used to address climate change are providing knowledge (through education) in order to empower people, and applying the mitigation and adaptation measures specified in the agreements of the COP 21. On the other hand, and concerning social issues, it is clear that freedom from fear (war, crime, violence, armed conflict) is related to freedom from want (poverty, disease, environmental deterioration) and that both freedom and equality, as well as sustainable development, are essential to guarantee positive peace and the conditions to live in dignity, with human rights protected and duly satisfied since they are key factors of human security.

3.17 Scientific Evidence of Climate Change

Given the fact that the Convention on Climate Change is based on the scientific knowledge provided by the reports of the IPCC,[39] one of the panel's greatest achievements has been to make everyone understand that climate change is a problem of enormous magnitude that requires urgent measures to mitigate its negative effects. As we have seen, climate change with the corresponding warming of the terrestrial climate is the result of the increase in greenhouse gas emissions (GHG), the biggest pollutant being carbon dioxide resulting from industry and the use of fossil fuels such as oil, coal and gas which are used as an energy source. The increase in temperature can clearly be seen in the following graph from the US government's National Oceanic and Atmospheric Administration (NOAA) (Fig. 3.7).

Regarding the increase in the global temperature, the graph in Fig. 3.8, produced by the NASA Goddard Institute of Space Studies, demonstrates the increase in global temperature over a period of more than a hundred years from 1880 to 2019.

All of this has had an impact on the polar ice caps in the Arctic, which between 1980 and 2015 decreased 13.3% per decade (averaged to the month of September, when the summer ends in the northern hemisphere) (Fig. 3.9).

The same has happened with the large masses of land ice in both Antarctica and Greenland, where there was an annual reduction of 127 gigatonnes of ice in Antarctica and 286 gigatonnes of ice in Greenland between 2002 and 2016, as can be seen in the following two graphics from NASA (and Public Domain Information) (Fig. 3.10).

---

[39] The Intergovernmental Panel on Climate Change was established by the *World Meteorological Organization* (WMO) and the *United Nations Environment Programme* (UNEP) in 1988. Its function is to analyse, in a comprehensive, objective, open and transparent manner, scientific information of technical and socio-economic relevance in order to understand the scientific elements of the risk posed by climate change caused by human activities, their possible repercussions and the possibilities of adaptation. Approximately 2,500 scientists and representatives of roughly 100 governments participate in this Panel (cf.: http://www.ipcc.ch/home_languages_main_spanish.htm#1).

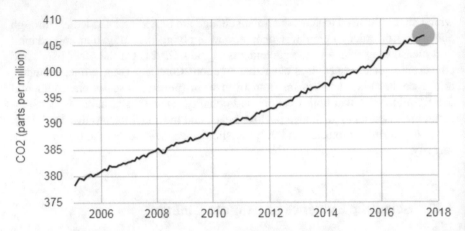

Fig. 3.7 Carbon dioxide: direct measurements, 2005–2017. *Source US National Oceanic and Atmospheric Administration* (NOAA). Information in the public domain

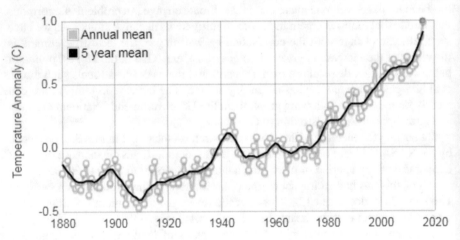

Fig. 3.8 Global temperature, global land-ocean temperature index. *Source* NASA Goddard Institute for Space Studies (NASA/GISS). Information in the public domain

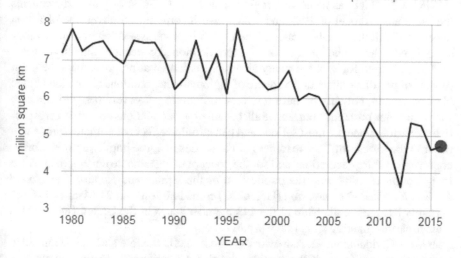

Fig. 3.9 Arctic sea ice. *Source* Satellite observations. *Credit* NSIDC/NASA

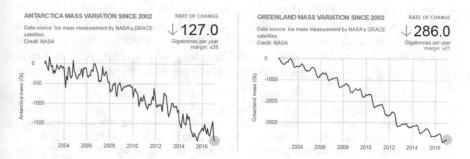

Fig. 3.10 Land ice. *Source* NASA's Grace satellites. *Credit* NASA (Public Domain Information)

3.18 The United Nations Framework Convention on Climate Change

The severity of climate change and global warming due to greenhouse gas emissions led to the signing of the commitments[40] of the United Nations Conference

---

[40] The commitments of the signatory States of the Kyoto Protocol "to promote sustainable development" are as follows: "(a) Apply and/or continue to develop policies and measures in accordance with their national circumstances, for example the following: (i) the promotion of energy efficiency in the relevant sectors of the national economy; (ii) the protection and improvement of sinks and deposits

on Climate Change in Paris in November and December 2015 (which was also
the 21st Conference of the Parties to the Kyoto Protocol on Climate Change, and
is widely known under the generic term of 'COP 21'). The COP 21 was organized
under the umbrella of the *United Nations Framework Convention on Climate Change*
(UNFCCC), which, as noted on previous pages, includes among its achievements
the Montreal Protocol of 1987 and which also incorporates a novel line of action
under which all the member states of the United Nations system recognize that they
are obliged to act in the interests of the aforementioned concept of human security.

The entry into force of the treaty also represented great progress from the point
of view of multilateralism in the sphere of diplomatic negotiations within the United
Nations system, not only because it is very difficult to get governments to agree on
such a complex problem[41] but, above all, because the negotiations concerned prospective scenarios, since the most serious and catastrophic effects of climate change will
occur in the long term. This readiness of the parties to agree implies a praiseworthy
concern for future generations and for the concept of sustainable development since
its inception in 1987 with the publication of the Brundtland Report. The *United
Nations Framework Convention* (UNFCCC) came into force on 21 March 1994 and
currently has the membership of the 193 member states of the United Nations plus
4 other parties, making a total of 197 parties.

It is also a document – a framework – whose text is flexible and can be amended
or developed over time if the parties consider it appropriate, so that humanity's
efforts in the face of climate change can adapt to new circumstances, reorientate
and be more effective. Furthermore, to the extent that it was negotiated and signed
by all the member states of the United Nations, it is also a very clear case of the
type of multilateral democratic diplomacy that should be practised among member
states of the UN because, as stated at the beginning of this book, there is a need
to 'democratize' the international system. This is precisely this kind of diplomacy

---

of greenhouse gases not controlled by the Montreal Protocol, taking into account their commitments
under the relevant international agreements on the environment; promotion of sustainable practices
in forest management, afforestation and reforestation; (iii) the promotion of sustainable agricultural
modalities in the light of climate change considerations; (iv) research, promotion, development and
increased use of new and renewable forms of energy, carbon dioxide sequestration technologies
and advanced and novel technologies that are environmentally sound; (v) progressive reduction or
gradual elimination of market deficiencies, tax incentives, tax and tariff exemptions and subsidies
that are contrary to the objective of the Convention in all greenhouse gas emitting sectors and application of market instruments; (vi) promotion of appropriate reforms in relevant sectors to promote
policies and measures that limit or reduce greenhouse gas emissions not controlled by the Montreal
Protocol; (vii) measures to limit and/or reduce emissions of greenhouse gases not controlled by the
Montreal Protocol in the transport sector; (viii) limitation and/or reduction of methane emissions
through their recovery and use in waste management as well as in the production, transport and
distribution of energy" (Art. 2 of the Kyoto Protocol to the UN Framework Convention on Climate
Change); at: http://unfccc.int/resource/docs/convkp/kpspan.pdf.

[41] The French internationalist Badie (2012) has criticized the parallel establishment of "oligarchic"
power groups that are not democratic in international relations and that practise a "democracy of
connivance". So the fact that the UNFCCC has been approved by 197 parties (a greater number than
the 193 member states of the UN) is of great importance, a true feat of democratic multilateralism
that must be regarded as a high point of contemporary international relations.

Table 3.5 List of the 12 countries with the highest *greenhouse emissions* (GHE)

Country	CO_2 emissions (kt) in 2015[a]	% CO_2 emissions per country	Emissions per capita (t) in 2015[b]	Emissions per country 2021
World	36,061,710	100		
China	10,641,789	29.51	7.7	2,806,634
United States	5,172,338	14.34	16.1	1,432,855
European Union	3,469,671	9.62	6.9	N/D
Indian	2,454,968	6.81	1.9	610,411
Russia	1,760,895	4.88	12.3	465,052
Japan	1,252,890	3.47	9.9	331,074
Germany	777,905	2.16	9.6	196,314
Iran	633,750	1.76	8.0	177,115
South Korea	617,285	1.71	12.3	160,119
Canada	555,401	1.54	15.5	146,494
Saudi Arabia	505,565	1.40	16.0	163,907
Indonesia	502,961	1.39	2.0	126,582

Source Own elaboration based on public domain information using data from 202

– one in which all countries of the world participate in the exercise of their political sovereignty – that I advocate. Conversely, the exclusive diplomatic clubs of the G7 or G20, that practise what Badie (2012) calls the "diplomacy of connivance" and make decisions which exclude the majority of member states of the United Nations, do not constitute a model of democracy in the international scene.

In addition, as it is clear that not all States have the same power and influence and that some are more powerful than others, it could be said that even in negotiations of this kind, there is an absence of democracy. However, in the large meetings of the multilateral system, at least the small and weak states can make coalitions and alliances to collectively represent their interests in a better way. This is what happened during the negotiations when it was decided that countries who are major polluters must consequently contribute the most funds, so that the Green Fund is in a position to finance the mitigation and adaptation actions to climate change needed by the small and most vulnerable countries. The graph below presents the list of the twelve countries that most pollute the atmosphere with greenhouse gas emissions (GHG) updated to the year 2015.

As can be seen in Table 3.5, although China is ahead of the United States in terms of emissions per country, the same is not true in terms of emissions per capita, which, as noted earlier, refers to the individual carbon footprint and is determined by lifestyle, customs, and, in general terms, the consumption of people.[42] Chinese

---

[42] Bear in mind that carbon footprint is one of the simplest ways to measure the impact or the mark that a person leaves on the planet in their daily lives. It is a count of the emissions of carbon dioxide (CO_2) which are released into the atmosphere due to their daily activities or the commercialization of a product. Therefore, the carbon footprint is the measure of the impact caused by human activities

peasants, workers, and middle-class individuals do not have the same consumption habits as their counterparts in the United States. They do not travel in 8-cylinder suburban SUVs, nor do they consume beef as part of their daily diet, or have a large-screen television at home. This explains why China, with more than 1,300 million inhabitants, has a per capita carbon footprint (7.7) that is less than half the average American per capita carbon footprint (16.1) even though the United States' population of 320 million inhabitants is a quarter the size of China's, and why India, with almost as many inhabitants as China, has a per capita carbon footprint as low as 1.9.

It is also interesting to note that the 500 million inhabitants of the countries of the European Union have a relatively low carbon footprint (6.9), a fact which is probably attributable to the greater ecological culture of European citizens. Conversely, a lower ecological culture determines that a country which is rich thanks to oil – Saudi Arabia, with only 30 million inhabitants – has a very high carbon footprint, equal to that of the USA (16.0), and the same can be said of countries such as South Korea (12.3), Iran (8.0) and Japan (9.9). Two countries which are not densely populated – Canada, with 35.8 million inhabitants, and Russia, with 144.1 million inhabitants – have surprisingly high per capita footprints (15.5 and 12.3 respectively), although this is probably due to the need to use fuels that emit GHG for heating during the long, cold winters of both countries, whose northern borders are in direct contact with the Arctic Ocean, which is why both countries suffer very low winter temperatures. Another explanation could be that both Canada and Russia are less efficient than, for instance, the Scandinavian countries, which have managed to be much more efficient in terms of public transport, buildings insulation and infrastructure in general, and thus have lower GHG and consequently do not appear among the countries with the highest GHG emissions.

3.19   Implications of Climate Change for Latin America

A study by ECLAC, the IDB and the WWF (Vergara et al. 2014) has identified the impacts of climate change, pointing out, among other things, that the warming of the terrestrial climate will cause a reduction in soil moisture in the region's agriculture, alterations to precipitation patterns, floods in coastal areas, major frequency of the *El Niño* maritime phenomenon in the Pacific Ocean, greater frequency and intensity of extreme weather events (such as heatwaves, droughts and tropical storms), an increase in tropical diseases, water level fluctuations due to the alterations in the rainfall

---

in the environment and is determined by the amount of GHG emissions produced, measured in units of the carbon dioxide equivalent. Obviously, manufactured products also leave a carbon footprint (analogous to the water footprint already mentioned), to establish it, an analysis must be made that covers the life cycle of a product (from the acquisition of raw materials to its management as waste). The carbon footprint of a product should be put on the label so that consumers with an ecological conscience can better decide which foods or products to buy based on the pollution generated as a result of the processes through which it has passed.

regime, and the contraction of the humid tropical forest, affecting the biodiversity and stability of ecosystems, to name just a few effects.

Given the vulnerability of the region, as well as the type of territorial distribution of its population and its dependence on fragile natural resources, it is expected that by the middle of the twenty-first century, if the temperature rise reaches 2 °C, it will be enough to significantly reduce the coral reefs of the Caribbean Sea. There is also the risk that glaciers located below 5,000 m in the mountains and volcanoes of the Andean region will disappear. Much of the Amazon Basin could become savannah if deforestation continues, which would seriously affect the yield of basic grain crops, and the flooding of coastal areas could increase, with the consequent rise in tropical diseases. Likewise, the destabilization of the hydrological cycle in important watersheds is expected, as well as the already mentioned intensification of extreme weather phenomena such as hurricanes, droughts and tropical storms.

With regard to carbon dioxide emissions, the study indicates that the countries with the highest GHG emissions are Brazil, Mexico, Venezuela and Argentina, which together account for almost 70%. Most of the GHGs come from agriculture (32%) as well as livestock, forestry, land use changes and expansion of the agricultural frontier (31%), while the energy sector – mainly transport – contributes 31%, the industrial sector is responsible for 2.3%, and the disposal of garbage and waste accounts for 2.9%. All this presents a frightening panorama in front of which the measures suggested by ECLAC do not seem to be up to the task. The remedies proposed include improvements to the efficiency of productive processes, the reorganization of public transport, more efficient use of energy, improvements to the disposal of waste through biological treatment or composting, and the use of *clean development mechanisms* (CDM) in accordance with the regulations of the Kyoto Protocol.[43] CDMs provide an opportunity for developing countries to mobilize additional funds for investment in the field of renewable energies and increased energy efficiency. It is therefore important for each country to have an institutional structure that allows CDM projects to be approved via framework agreements which define priority projects.

---

[43] "The *Clean Development Mechanism* (CDM) is one of the three mechanisms established in the Kyoto Protocol to facilitate the implementation of projects to reduce greenhouse gas emissions by developing country parties (parties not included in Annex I) in cooperation with developed countries (Annex I). The CDM is defined in Article 12 of the Protocol and aims, on the one hand, to help the countries that are parties to Annex I to comply with their targets regarding the limitation and reduction of GHG emissions, and on the other to foster sustainable development in countries that do not belong to this Annex. The mechanism allows parties not included in the aforementioned annex to benefit from project activities that result in certified emission reductions. The parties included in Annex I may use the certified reductions in emissions resulting from these activities in projects designed to comply with part of their emission limitation and reduction commitments." Be aware that these certificates have been criticized a lot because they are market mechanisms, since the credits resulting from GHG emission reductions are commercialized and used to fulfil the party's reduction commitments. However, between this and nothing, it is often considered better to have the CDM.

It must be noted that a relatively high percentage of such projects finance renewable energies or are related to methane and biomass in garbage dumps, including biofuels from agricultural waste, such as bagasse from sugarcane and palm oil.

3.20 Central American Vulnerability

According to an ECLAC report on historical climatological trends in Central America, it is already possible to detect an upward trend of between 0.6 and 0.76 °C in the average annual temperature in the last three decades. In contrast, the rainfall patterns during those same three decades compared to the period between 1950 and 1979 indicate that there is a slight downward trend in El Salvador, Guatemala and Honduras, while the pattern in Costa Rica and Nicaragua has remained relatively stable, and Panama and Belize register a slight increase. However, the report asserts that the most striking characteristic of annual rainfall, from a historical point of view, is its high inter- and intra-annual volatility and its geographical variability, given that in recent years the region has suffered from high climatic variability, as well as multiple extreme climatic events such as droughts, hurricanes and the so-called phenomenon of the *Niño-Southern Oscillation (ENSO)* that had two extremely intense manifestations in 1982–1983 and 1997–1998. Because of this it is possible to say that:

> The socio-economic vulnerabilities of Central America are exacerbated by its geo climatic location on a narrow isthmus that serves as a bridge between two continents, located between two oceanic systems, the Pacific and the Atlantic, with their corresponding climatic processes. Given that climate-dependent factors are significant contributions to economic activities such as agriculture, changes in climate will increasingly affect the economic evolution of the region during this century. In addition, the region contains valuable assets to be preserved for its contribution to the development of current and future generations, such as its abundant biodiversity ecosystems, forests, corals and mangroves, among others, providers of multiple services to the population. These ecosystems are dwindling and some are already severely degraded by the current unsustainable pattern of development, and will be further affected by climate change. The still relatively young population and the cultural, ethnic, linguistic and lifestyle diversity of the region are treasures that require investment to develop their capacities. The local and indigenous peoples' knowledge must be valued. The evaluation of the economic impact of climate change is subject to intense debate, whose development uses various methods and techniques (ECLAC 2010).

In a projection exercise, the ECLAC study considers several scenarios regarding the agricultural sector, water resources, biodiversity and increased intensity of hurricanes, storms and floods. According to the authors of the research, the cost of the impact on biodiversity would grow exponentially from the year 2050, with greater indirect agricultural costs. Extreme events would also show a tendency to increase from the year 2050, which implies that an increase in temperature will result in a greater intensity of this type of phenomenon, with higher costs for countries, Therefore, in a scenario of increasing emissions and global inaction, the impacts of climate change in Central America are significant and growing, confirming the truism that international asymmetry means that "…the developed countries that have polluted the most suffer less impact and have the resources to adapt. On the other hand, the countries that have contributed the least to the problem suffer the greatest impacts and have the least resilience" (ECLAC 2010: 36).

For these reasons, it is also confirmed that the costs of the impacts in a scenario of global inaction, particularly for the countries that are large GHG emitters, would be higher than those in a scenario with an equitable and inclusive international agreement that could significantly reduce emissions. As this study was published in 2010, the aforementioned issue pointed towards the concretion of the Paris Agreement, which ECLAC expected to be an accord with shared but differentiated responsibilities among the countries, which would help the most exposed countries to take adaptation and mitigation measures in a framework of sustainable development (in a manner similar that agreed in relation to the Green Fund).

Consequently, ECLAC study concludes by pertinently pointing out that:

Central American societies need to become daring managers of their water resources, ensuring their sustainable and efficient use for the benefit of the population and the production process. Ensuring food security in the face of climate change, particularly that of basic grains, and moving towards more sustainable agriculture is a great challenge, but a necessary one, in order to protect the poor population, both small producers and urban consumers. The protection of natural ecosystems and their biodiversity, including forests, mountains, river systems, and coastal-marine zones, including corals and mangroves, is vital to maintain the multiple services they provide to the human population and other living beings. An essential element of adaptation to climate change and the transition to low carbon economies is technological change, understood as access to modern technologies and the rescue of knowledge and traditional and local technologies, particularly of indigenous peoples and peasant communities. The region has developed a serious dependence on imported and highly polluting fossil fuel sources. The transition to an energy matrix based on local renewable sources would bring multiple benefits, improve energy security, save foreign exchange and reduce the negative impacts of fossil fuels on human health and GHG emissions (ECLAC 2010: 121).

Apart from the importance of carrying out the recommendations of the aforementioned study in Central America, it is indisputably true that the ecological challenge in this epoch of the Anthropocene is closely linked to the development model that governments have adopted for establishing their public policies, and it is clear that the said model is the neoliberal one. That has not changed in Central America, with the possible exception of Costa Rica. Hence, if growth is privileged as a criterion for guiding public policies, it is clear that the priorities will be set for the purpose of promoting such growth, which may be contrary to the achievement of the social and environmental *Sustainable Development Goals* (SDG).

Therefore, it is essential for policies to be set in accordance with the SDGs, because otherwise actions may be taken which are detrimental to natural resources or the interests of rural communities. For example, if what matters is to increase the productivity of the soil in order to increase economic growth, and for this purpose the use of fertilizers, herbicides and pesticides is required, concern for the environmental impact of these agrochemical inputs will not have any impact on the ultimate decision to use fertilizers. Similarly, if the priority is to promote the 'competitiveness' of a country by offering foreign investors low wages – i.e. underpaid national workers

– while simultaneously reducing taxes for the same investors (or eliminating tariffs in free trade zones) under the pretext of fiscal incentives to attract foreign capital and giving priority to infrastructure construction in order to facilitate mining or oil extraction etc., it is evident that in those circumstances governments will not worry about the redistribution of income via social investment policies. If a river has to be diverted to irrigate private plantations, or glyphosate is used in palm plantations to the detriment of the needs of the local community, this will also be done with impunity. That is why it is fundamental for the guidance framework for governments' policies in the underdeveloped countries of the Global South to become the sustainable development paradigm. Otherwise, the good intentions expressed in government plans or in the subscription to international commitments like those of the UN 2030 Agenda will just remain another rhetorical exercise, while poverty, lack of sustainable development, inequality and the concentration of wealth, including huge migratory flows of poor and desperate people towards the rich countries of the world, will continue.

3.21   Neoliberalism, Violence and Political Crises

The *Sustainable Development Goals* (SDG) and the ecological paradigm are not only supposed to have a holistic, inter and transdisciplinary vision of knowledge but also full awareness of the interdependence and interconnection that exists between politics, socio-economic problems and the ecological question. Both global governance and national governance depend on the proper functioning of the economy and the resolution of social and environmental problems. Establishing a harmonious relationship between sustainable development, job creation and environmental protection is quite a difficult task, especially since the warnings of the 1970s Club of Rome report that regarded the 'zero growth' policy (Meadows et al. 1972) as the only way to escape the possible exhaustion of raw materials and consequent ecological collapse, and also since the Brundtland Report, *Our Common Future* (1987).[44]

---

[44] The conclusion of the Club of Rome report (1972) was that if the current increase in world population, industrialization, pollution, food production and the exploitation of natural resources remained unchanged, the absolute limits of growth on Earth would be reached during the next hundred years (before 2072). The main thesis of the report sustains that in a limited planet the dynamics of exponential growth of population and industrialized production is not sustainable, since the Earth itself limits the growth, as the natural resources are not all renewable, arable land is finite and the capacity of the ecosystem to absorb pollution and garbage has limits. Thus, exceeding the exploitation of natural resources will lead to their depletion, followed by a collapse in agricultural and industrial production which will drastically affect the human population. For this reason, 'zero growth' or a steady state is considered a sensible way to avoid the collapse of the species, stopping the exponential growth of the economy and the population so that natural resources are not depleted by economic growth. The authors of the report believe that that is the only way to achieve an ecological stability that remains sustainable in the long term. Global equilibrium, then, should be based on the satisfaction of human needs, an approach which corresponds to the human development paradigm subsequently elaborated by Manfred Max Neef, Antonio Elizalde and Martin Hopenhayn and by

However, it is not only rich industrialized countries and the rich and middle classes that produce garbage, pollution and *greenhouse gas* (GHG) emissions. Even poor people cause pollution due to hunger and underdevelopment (the burning of wood as a fuel for cooking in poor countries, for example). Since all countries have committed themselves to the eradication of poverty by 2030 and, in accordance with SDG 8, governments must promote "sustained, inclusive and sustainable economic growth, full and productive employment and decent work for all", what means will be used to avoid the clash between the linear conception of development with the cyclical functioning of natural ecosystems?

This is a debate that I will focus on in the next chapter, keeping in mind the idea that in order to be sustainable, growth must be compatible with natural ecosystems and must simultaneously be aimed at the satisfaction of the social needs of the present without jeopardizing those of future generations. It is clear that a lot of obstacles and difficulties will be confronted, particularly in the poor countries of the Global South. Examples include the prevalence of fossil fuels as a source of energy; the continuity of export-orientated policies based on commodities and the exploitation of raw materials via mining and drilling (the so called *extractivist industries* that in countries plagued by corruption – like Guatemala – just embezzle oil and minerals); the accumulation of waste; the contamination of soil, water and the atmosphere; and unemployment, which also means an increase in migratory flows. Therefore, the sixteenth SDG, addressed at promoting peaceful and inclusive societies, and at facilitating access to justice through effective institutions and the rule of law, will be one of the fundamental parameters which might offer some degree of assurance that the 2030 commitments will be fulfilled.

3.22 Summary: Is the Breakdown of *Homo Sapiens* the Central Issue of the Anthropocene?

In relation to the starting point of the novel concept of the Anthropocene and the immensity of the geological time scale measured in billions of years, which places the human species in its infinitesimally small temporal dimension of the 12,000 years that have elapsed since the beginning of the Holocene, it is worth remembering some ideas expressed by the great French anthropologist and philosopher Bruno Latour in a paper delivered at the French Collège in London in November 2011:

> The fourth and last aspect that I want to describe is without doubt depressing. The disconnection that I have analysed here is built on the very idea of an immense threat against which we would be reacting slowly and to which we would be unable to adjust. That is the spring with which the trap was set. Of course, faced with such a threatening trap, the most reasonable of us react with the very plausible argument that apocalyptic prophecies are as old as humans. And it is true, for example, that my generation survived the threat of the nuclear holocaust – something that Gunther Anders analysed with great skill in terms very similar to those

the United Nations Development Programme (UNDP), which I will discuss in the next chapter, as well as the UN Report on Sustainable Development (Brundtland 1987).

used today by the prophets of the Last Judgement – and we are still here. In the same way, historians of the environment could argue that the warning about an Earth in the process of death is as old as the so-called Industrial Revolution. Undoubtedly, there seems to be licence for a greater dose of healthy scepticism when one reads, for example, that Dürer, the great Dürer himself, prepared his soul for the end of the world – projected for the year 1500 – while investing a fortune in printing his beautiful and expensive engravings of the Apocalypse in anticipation of a considerable profit. Those comforting thoughts offer reassurance against the madness of the prophecies of the Day of Judgement. Yes, yes, yes. Unless the opposite happens, what we are witnessing today is another consequence of playing the liar for too long (Latour 2017: 9)

As readers will have noticed, Latour's remarks are related to the central question in the conclusion of this chapter, because if the possible collapse of *Homo sapiens* could be the result of global warming (or of still unknown viruses in new forthcoming pandemics) as part of the sixth extinction process already underway, then humanity is indeed facing the crucial task of taking seriously what science says in its prospective long-term research, otherwise the entire human species runs the risk of collapsing and disappearing. Hence the deep meaning of the Anthropocene lies in the fact that the *Great Acceleration* of the middle of the twentieth century and the deepening of globalization after the end of the Cold War have already placed us in this new human-made geological epoch, even though only about 12,000 years have elapsed since the beginning of the Holocene. Consequently, if this tendency persists, it could be a serious threat to our species, against which we are reacting too slowly without showing signs of the ability to adapt, as stressed by Latour.

Indeed, in reply to the sceptics who remind us that this is not the first time that such apocalyptic predictions – the Day of Judgement and the threat of a nuclear conflagration, to use Latour's examples – have been made but not taken place, the fact is that there is no certainty that this ecological apocalypse will not materialize, with the aggravating circumstance that we would be walking towards it with flagrant disregard for scientific knowledge, as occurred in 2020 with the almost apocalyptic hecatomb provoked by a virus, despite the fact that credible medical forecasts based on former regional epidemics (Ebola, SARS, H1N1 etc.), as well as movies like Steven Soderbergh's film *Contagion*, have warned governments about the possibility of a world pandemic of colossal dimensions. Many believed it would not happen, but finally it has come to pass. For that reason, it is better to remember the fable of the lying pastor – as Latour says – and do everything in our power to prevent the foreseeable catastrophe, since the wolves of nuclear war and climate change continue to lurk in the shadows.

Therefore, for the good reason that what is really at stake is nothing less than the survival of humankind, we simply cannot stand by without making decisions that translate into at least the fulfilment of the commitments agreed at COP 21 in Paris to reduce *greenhouse gas* (GHG) emissions in order to stop global warming rising more than 1.5°. Additionally, the commitments of the SDGs and the 2030 Agenda should be upheld as a *minimum*, because the *maximum* could be the "retreat of sustainable development", as Lovelock (2006) says in his book about the "revenge of Gaia" that displays the vivid and eloquent subtitle *Why the Earth is Fighting Back and How We Can Still Save Humanity*. In other words, the aim is to avoid the final collapse of

Homo sapiens. This implies, for instance, making a serious commitment to renewable energy (including nuclear fusion, as Lovelock suggests and as mentioned above in the footnote about ITER) as a substitute for the energy currently derived from oil, gas and coal, thereby decarbonizing energy; and drastically reducing GHG emissions; but also – and overall – rejecting neoliberalism as the ideology that informs the current capitalist mode of production that prevails at world level.

In other words, sustainable development is not possible unless we stop worshipping 'markets' as if they were 'gods' entitled to determine economic public policies all over the world. If any collateral benefit can be derived from the drastic measures taken by all governments of the world to mitigate the coronavirus pandemic of 2020/21 and reduce the mortality rate, it is that the Keynesian approach to economics, whereby governments play a major role in economic management and the regulation of markets and transnational corporations such as pharmaceutical, oil and gas giants – thereby putting an end to the *laissez-faire, laissez-passer* neoliberal doctrine – will not be questioned. Consequently governments will have fewer problems applying the sustainable development principles and guidelines expounded by Sachs (2015) which are addressed at regulating the private sector in a way which is compatible with social dynamics and terrestrial ecosystems, even though this is not an easy task. It requires the linear perspective of development to be combined with the circular movement of ecological cycles. It is a complex and difficult endeavour and quite a difficult challenge for the next generations of politicians and rulers.

In the same vein, it is important to stress that, just as there is no doubt that the signing of the UN 2030 Agenda is a triumph of multilateralism, if any lesson can be learned from the coronavirus pandemic, it is not just that the world must abandon the doctrine of neoliberalism, but also that multilateral international cooperation – and the strengthening of the United Nations instead of the absurd confrontation with WHO unchained by the former US president – and cosmopolitanism – not nationalism – are the only ways to face to the approaching threats of natural disasters and catastrophes that climate change will certainly be unleashing everywhere.

Last but not least, social and human development must be promoted, because to address those threats the world need transnational and cosmopolitan citizens who embrace ethical values capable of being translated into new normative guidelines destined to end social exclusion, reduce poverty, promote decent employment, improve education and health systems, empower women and girls, draw attention to food security at national level, promote renewable energy, integrate the management of water and natural resources, protect biodiversity, and reform national political systems in order to deepen democracy and respect for human rights. To recapitulate, humanity needs nature but nature does not need humanity. Thus, social preparedness must include human and social development as well as the deepening of democracy, which essentially means increasing cosmopolitan consciousness and responsibility for the planet. The obsolete doctrine of "laissez faire, laissez passer" must be abandoned, as well as the misconception of material wealth as the quintessence of happiness, which is not only erroneous but usually leads to misfortune and unhappiness. Consequently, if the collapse of humanity is the central issue of the *Anthropocene*, global civil society must react straightaway with the lucidity, responsibility and intelligence that such a huge and vast challenge requires.

References

Angus, Ian, 2016: *Facing the Anthropocene: Fossil Capitalism and the Crisis of the Earth System* (New York: Monthly Review Press).

Badie, Bertrand, 2012: *Diplomacy of Connivance* (New York: Palgrave Macmillan).

Barié, Cletus Gregor, 2017: "Nuevas Narrativas Constitucionales en Bolivia y Ecuador: El Buen Vivir y los Derechos de la Naturaleza", in: *Política Internacional* (Guatemala City: Academia Diplomática): 48–67.

Bárcena, Alicia, 2018: "Executive Secretary of ECLAC, interviewed by Ignacio Fariza in Mexico City", in: *El País* (Spain).

Boff, Leonardo, 2018: *Etica de la Naturaleza: ¿Cómo Cuidar la Casa Común?* (Guatemala City: Cooperación Alemana [GIZ]).

Brauch, Hans Günter, 2019: *Peace Ecology in the Anthropocene*, in: Brauch, Hans Günter; Oswald Spring, Úrsula; Collins, Andrew E.; Serrano Oswald, Serena Eréndira (Eds.): *Climate Change, Disasters, Sustainability Transition and Peace in the Anthropocene* (Mosbach: Springer).

Brundtland, Gro Harlem; et al., 1987: *Our Common Future* (New York: United Nations).

Capra, Fritjof, 1975: *The Tao of Physics* (Berkeley: Shambala).

Capra, Fritjof, 1982: *The Turning Point* (New York: Simon & Schuster).

Capra, Fritjof, 1996: *The Web of Life* (New York: Anchor Books).

Chardin, Teilhard de, 1965: *El Fenómeno Humano* (Madrid: Taurus Ediciones).

Crutzen, Paul J.; Stoermer, Eugene, 2000: "The Anthropocene", in: *Global Change Newsletter*, 41 (International Geosphere-Biosphere Programme [IGBP]/International Council of Science).

Dalai Lama, 2005: *The Universe in a Single Atom: The Convergence of Science and Spirituality* (New York: Broadway Books).

Feenberg, Andrew, 1999: *Questioning Technology* (London/New York: Routledge).

Fukuyama, Francis, 1992, *The End of History and the Last Man* (London: Penguin).

Gudynas, Eduardo, 2017: "Ecología Politica de la Naturaleza en las Constituciones de Bolivia y Ecuador", in: *Política Internacional* (Guatemala City: Academia Diplomática).

Habermas, Jürgen, 1987: *The Philosophical Discourse on Modernity* (Cambridge, MA: MIT Press).

Harari, Youval Noah, 2014: *De Animales a Dioses: Breve Historia de la Humanidad* (Mexico City: Penguin Random House/Grupo Editorial).

Harari, Youval Noah, 2016: *Homo Deus: Una Breve Historia del Mañana* (Barcelona: Penguin Random House).

Heidegger, Martin, 1977: *The Question Concerning Technology and other Essays* (New York/London: Garland Publishing Inc.).

Juniper, Tony, 2016: *What's Really Happening To Our Planet? The Facts Simply Explained* (New York: Penguin Random House).

Laszlo, Ervin, 2006: *Science and the Reenchantment of the Cosmos: The Rise of the Integral Vision of Reality* (Rochester: Vermont Inner Traditions).

Latour, Bruno, 2015: *Face a Gaia: Huit Conférences sur le Nouveau Régime Climatique* (Paris: La Découverte).

Latour, Bruno, 2017: "Waiting for Gaia: Conference paper delivered at the French Institute in London, November 2011, on the occasion of the launch of the Political Science in Arts and Politics Programme (SPEAP)", in: *Política Internacional*, 3 (Guatemala City: Academia Diplomática).

Lovelock, James, 1979: *A New Look at Life on Earth* (Oxford: Oxford University Press).

Lovelock, James, 2006: *The Revenge of Gaia: Why the Earth is Fighting Back and We can Still Save Humanity* (London: Oxford University Press).

Matul, Daniel; Cabrera, Edgar, 2007: *La Cosmovisión Maya*, 2 vols. (Guatemala City: Amanuense Editorial).

Meadows, Donella D.H.; Meadows; Dennis, D.L.; Randers, Jorgen; Behrens, William; et al., 1972: *Los Límites del Crecimiento: Informe al Club de Roma sobre el Predicamento de la Humanidad* (Mexico City: Fondo de Cultura Económica).

Morin, Edgar, 1999: *Relier les Connaissances: Le Défi du XXIe Siècle* (Paris: Seuil).

Padilla, Luis Alberto, 2018: "Human Rights and Radical Democracy", in: Oswald Spring, Úrsula; Serrano Oswald, Serena Eréndira (Eds.): *Risk, Violence, Security and Peace in Latin America: 40 Years of the Latin American Council on Peace Research* (Cham: Springer International Publishing).

Padilla Vassaux, Diego, 2019: *Política del Agua en Guatemala: Una Radiografía Crítica del Estado* (Guatemala City: Universidad Rafael Landivar, Instituto de Investigación y proyección sobre el Estado (ISE)/Editorial Cara Parens).

Ricard, Matthieu; Trinh Xuan Thuan, 2000: *L'Infini dans la Paume de la Main: Le Moine et l'Astrophysicien* (Paris: Fayard).

Robbins, Paul, 2012: *Political Ecology: A Critical Introduction* (Oxford: Wiley-Blackwell).

Ruddiman, William, 2005: "How Did Humans First Alter Global Climate?", in: *Scientific American*, 292, 3: 46–53.

Russell, Peter, 2007: *The Global Brain: Speculations on the Evolutionary Leap to Planetary Consciousness* (London: Routledge & Kegan Paul).

Sachs, Jeffrey, 2015: *The Age of Sustainable Development* (New York: Columbia University Press).

Santos, Boaventura de Sousa, 2009: "Un Discurso sobre las Ciencias", in: *Una Epistemología del Sur* (Mexico City: CLACSO/Siglo XXI Editores).

Smith, Scott, 2017: ICEFI y CIG Difieren por Impacto de Industrias Extractivas, in: *Diario La Hora* (July): 31.

Tocqueville, Alexis de, 2003: *Democracy in America* (Cambridge: Cambridge University Press).

Tocqueville, Alexis, 2011: *The Ancien Régime and the French Revolution* (Cambridge: Cambridge University Press).

Todd, Emmanuel, 2002: *Après l'Empire: Essai sur la Décomposition du Système Americain* (Paris: Gallimard).

Vernadsky, Vladimir Ivanovich, 1997: *The Biosphere* (Madrid: Fundación Argentaria).

Wallace, Allan, 2003: *Buddhism & Science* (New York: Columbia University Press).

Wilber, Ken, 2007: *Integral Spirituality* (Boston/London: Integral Books).

Zalasiewicz Jan; Williams, Marc; et al., 2008: "Are We Now Living in the Anthropocene?", in: *GSA Today*, 18, 2 (February): 4–8.

Internet Links

Arias Maldonado, Manuel, 2017: "Antropoceno: El Fin de la Naturaleza", in: *Revista de Libros* (February); at: http://www.revistadelibros.com/articulos/antropoceno-el-fin-de-la-naturaleza.

Business Insider, 2016.

Carlsen, Laura, 2013: https://www.nytimes.com/roomfordebate/2013/11/24/what-weve-learned-from-nafta/under-nafta-mexico-suffered-and-the-united-states-felt-its-pain.

Clean Development Mechanism (CDM).

Economic Commission for Latin America and the Caribbean (ECLAC/CEPAL), 2010: *La Economía del Cambio Climático en Centroamérica: Síntesis*, Santiago de Chile; at: http://repositorio.cepal.org/bitstream/handle/11362/35228/1/lcmexl978e.pdf.

Geological Society of America: Time Scale; at: https://www.geosociety.org/documents/gsa/timesc ale/timescl.pdf.

GHG emissions, 2015. Based on data from the United Nations Statistics Division.

Hydrogen Vehicles, 2017; at: https://www.digitaltrends.com/cars/does-hydrogen-make-sense-as-an-automotive-fuel/.

Instituto Nacional de Estadísticas (INE), Guatemala, at: https://www.ine.gob.gt/index.php/estadisti cas/tema-indicadores.

International Fund for Agricultural Development (IFAD), 2017: Majumdar, Roshni: "Remesas de Migrantes Suman Billones a la Economía Mundial", 2017, at: http://www.ipsnoticias.net/2017/06/remesas-migrantes-suman-billones-la-economia-mundial/.

International Geosphere Biosphere Programme (IGBP), 2015: *Science for a Sustainable Planet*; at: http://www.igbp.net/news/pressreleases/pressreleases/planetarydashboardshowsgrea taccelerationinhumanactivitysince1950.5.950c2fa1495db7081eb42.html.

International Monetary Fund (IMF).

International Thermonuclear Experimental Reactor (ITER); at https://energia-nuclear.net/que-es-la-energia-nuclear/fusion-nuclear.

Juniper, T., 2016: "What's Really Happening To Our Planet: Gross Domestic Product of the World Economy according to the World Bank"; at: https://www.weforum.org/agenda/2017/03/worlds-biggest-economies-in-2017/.

Liu, Jianguo; Dietz, Thomas; Carpenter, Stephen R.; Alberti, Marina; Folke, Carl; Moran, Emilio; Pell, Alice N.; Deadman, Peter; Kratz, Timothy, Lubchenco, Jane; Ostrom, Elinor, Ouyang, Zhiyun; Provencher, William; Redman, Charles L.; Schneider, Stephen H.; Taylor, William W., 2007: "Complexity of Coupled Human and Natural Systems", 2007; in: *Science*, 317, 14 (September): 1513–1516; at: https://science.sciencemag.org/content/317/5844/1513.full.

Montreal Protocol on Ozone Layer, 1989; at: https://es.wikipedia.org/wiki/Protocolo_de_Montreal.

National Oceanic and Atmospheric Administration (NOAA), 2019: "The NOAA Annual Greenhouse Gas Index"; at: https://www.esrl.noaa.gov/gmd/aggi/aggi.fig2.png and https://www.esrl.noaa.gov/gmd/aggi/aggi.html.

Ozone Layer: https://es.wikipedia.org/wiki/Capa_de_ozono.

Sachs, Jeffrey, 2015: "The Age of Sustainable Development", Lecture at the International Institute for Applied Systems Analysis (Laxenburg: IIASA).

Steffen, Will W.; Broadgate, Wendy W.; Deutsch, Lisa L.; Gaffney, Owen O.; Ludwig, Cornelia C., 2004: "The Trajectory of the Anthropocene: The Great Acceleration", in: *The Anthropocene Review*, 2, 1: 81–98.

UN Framework Convention on Climate Change, at: https://unfccc.int/process-and-meetings/the-convention/what-is-the-united-nations-framework-convention-on-climate-change.

United Nations: "Decenio Internacional para la Acción: El Agua Fuente de Vida", at: http://www.un.org/spanish/waterforlifedecade/iwrm.shtml.

Vergara, Walter; Rios, Ana; Galindo, Luis; Gutman, Pablo; Isbell, Paul; Suding, Paul; Samaniego, José Luis, 2014: "El Desafío Climático y de Desarrollo en América Latina y el Caribe: Opciones para un Desarrollo Resiliente al Clima y Bajo en Carbono" (Washington DC: CEPAL-BID-WWF); at: https://publications.iadb.org/publications/spanish/document/El-desaf%C3%ADo-clim%C3%A1tico-y-de-desarrollo-en-Am%C3%A9rica-Latina-y-el-Caribe-Opciones-para-un-desarrollo-resiliente-al-clima-y-bajo-en-carbono.pdf.

Wake, David; Vredemburg, Vance T., 2008: "Are We in the Midst of the Sixth Mass Extinction?", in: *Proceedings of the National Academy of Sciences*, 105, Supplement 1 (August): 11,466–11,473; https://doi.org/10.1073/pnas.0801921105; at: http://www.pnas.org/content/105/Supplement_1/11466.

Weisbrot, Mark; Lefebvre, Stephan; Sammut, Joseph, 2014: "Did NAFTA Help Mexico? An Assessment After 20 Years" (Washington, DC: Center for Economic and Policy Research, February); at: http://cepr.net/documents/nafta-20-years-2014-02.pdf.

Chapter 4
Sustainable Development or Sustainable Systems?

> *Economics is not a science that we can apply universally like chemistry. Economics is not even a general science of human society, such as anthropology, which attempts to study all forms of culture. The data of economics come mainly from accounting and bookkeeping because the models of economics apply mainly in the function of accountants and bookkeepers. Instead of saying that every society has an economic base, we should say that every culture has an ecological context. The general science of humanity's interaction with the Earth and other living systems is ecology as bionomics, not economics. Culture is the overall survival strategy for Homo sapiens, its ecological niche. Economic society is a special form of culture.*
> Richards (2017: 25)

> *Following the exponential increase in the global production and consumption of fossil fuels from 1800 onwards and especially since the accelerated global industrialization after World War II, a 'silent transition' in geological time has occurred, which is why Crutzen (2002) announced that "we are now in the Anthropocene". The impact of this new social construction of reality is not yet reflected in most publications in the social sciences, political science, international relations or security, peace, environment and development studies. It is not yet well understood in the global political discourse, which fails to recognize that now "we are the threat", and that we are members of the human species that has for the first time directly interfered in the Earth System.*
> Brauch (2016: 29)

4.1 Introduction

The risk of human extinction posed by the Anthropocene is a consequence of the predatory features of neoliberal capitalism, characterized by the constant production

© Springer Nature Switzerland AG 2021
L.-A. Padilla, *Sustainable Development in the Anthropocene*,
The Anthropocene: Politik—Economics—Society—Science 29,
https://doi.org/10.1007/978-3-030-80399-5_4

of goods irrespective of the real needs of society,[1] just for the sake of the accumulation of capital, a fact that provokes consumerism and the squandering of produce, food and all kinds of agricultural and industrial outputs. Thus, the current mode of production must be changed if we don't want to run the risk of extinction. Nevertheless, a world revolution of the type that Communists attempted during the nineteenth and twentieth centuries does not seem to be a viable way out of this crisis. The type of *Great Transition* (Raskin 2002) proposed by the Global Scenario Group could be a solution, but such a transformation requires time, the scarcity of which is not exactly any help, unless the 2020/21 pandemic triggers all nations to support the international cooperation needed to transform capitalism at world level, a scenario that does not loom large on the horizon. Therefore, a reform of the capitalist system to carry out as a minimum the SDGs of the UN 2030 Agenda is necessary and unavoidable if sustainable development is to be implemented. The latter has been the best alternative for restructuring capitalism since the end of the 1980s, but unfortunately the predominance of the neoliberal paradigm has obstructed the reformist agenda. In the following pages I will outline the historical difficulties and problems of sustainable development since its origins, with the aim of demonstrating that the contradictions that exist between a circular economy and a linear one are real and deserve attention and study. To achieve that goal, I will review the proposals made at the end of the twentieth century concerning human and sustainable development, the UN proposals of Agenda 2030, and certain normative intentions to change the model – such as the cultural and community development intentions expressed by the constitutions of Ecuador and Bolivia – and compare all these cases with historical examples of failed and collapsed societies of the past and with successful experiments of sustainable systems described by well-known researchers like the US social scientist Jared Diamond. In short, it is possible to say that sustainability refers to the conservation of Earth ecosystems, given its cyclical nature, and concerns the maintenance of balance in the survival of any species, hence it has a close relationship with biodiversity in the field of natural sciences. However, it also relates to social and environmental sciences, since this field of knowledge seeks to sustain human actions over time without exhausting resources or harming the environment, which implies the ability of a society to make responsible use of such resources without exceeding the carrying capacity of the planet or putting them at risk for future generations. Sustainability refers to the ability to remain, to durability, resilience and endurance, and to last over time, whereas sustainable development is the process by which the balance between the socio-political, ecological and economic factors essential to guide individual actions and public policies is achieved in order to satisfy human needs.

---

[1] "It's funny how the economy is about to collapse because people are only buying what they need." Tale circulating in the social media during the confinement of the coronavirus pandemic.

4.2 Historical Background of the Sustainable Development

The concept of sustainable development was introduced into the lexicon of the international community in 1987 when the famous Report of the Brundtland Commission defined it as a development that meets the needs of the present generation without compromising the ability of future generations to satisfy their own needs (Brundtland 1987). This clearly contains an implicit reference to social structure (evident in the reference to present and future generations), to the human actors (implicit in the reference to their needs), and to the social processes in a historical-ecological context, since the present must not compromise the future – all within the framework of culture, because the capacity to satisfy needs is determined by society and education. It is therefore a holistic, integral concept that simultaneously covers the economy and the environment, individual subjectivity (needs) and the intersubjective (culture), all within the framework of a prospective social dynamic (the present generation in relation with future ones). The concept of 'sustainable development' is therefore a *tour de force* which expresses with so few words such complex and innovative ideas compared with the traditional definition that reduces development to the economic, ignoring its human and sociocultural dimensions. Hence, sustainable development must be understood as a concept that alludes not to the parts of a whole, but to the whole itself in such an intrinsically interconnected manner that any analysis of its discrete parts will fail. Therefore it is essentially a synthetic concept, not an analytical one.

The idea of sustainability, then, is part of the prospective sciences[2] that enable us to think and make plans for the long term, visualizing a future where economic, environmental, social, cultural and human considerations maintain a relative balance in the search for a better quality of life, rather than things being done for purely economic purposes. Hence it is necessary to differentiate between sustainability, which concerns the natural ecosystems, and sustainable development, which concerns long-term objectives such as those posed by the 2030 agenda of the United Nations, and which in particular refers to the means to achieve objectives and goals, such as good governance, agricultural and industrial production methods that pollute as little as possible and recycle most of their inputs to reduce waste, and scientific research, innovation and new technologies to reduce the harmful effects of human activity on the environment.

In any case, it seems clear that we must distinguish between sustainability as a development process which involves the transition from a lower stage to another that is considered superior, and sustainability as a system that works in a circular manner, which recycles its components and therefore has durability, but does not transit from one point to another in a linear way but returns to the starting point like all natural cycles. This question will be dealt with in this chapter and in Chap. 6, which suggests

---

[2] The term 'prospective', of French origin (Gaston Berger, Michel Godet), is defined as the science that studies social evolution, thereby making it possible to forecast the future. Gaston Berger founded the journal *Prospective* in 1957, while Michel Godet's website, "La Prospective" urges people to think and act differently; see at: en.laprospective.fr.

certain research paths that could provide further insights regarding solutions to these problems.

In the meantime, returning to the concept of sustainable development, I would like to reiterate that the novelty of the Brundtland Commission's proposal lies in linking the satisfaction of the human needs of present generations with those of future generations, all within the framework of the ecological dimension of international relations. It is clear that a satisfactory relationship between people and their environment depends, to a large extent, on natural resources (water, land, oxygen, forests, and biodiversity in general) being conserved in such a way that the actions of human beings do not endanger the carrying capacity of the planet, that is to say, without putting at risk or compromising the possibilities that the future generations we will be able to make use of these natural resources too (Brundtland 1987).

This idea allows us to relate two central concepts: need, which refers to the fact that sustainable development seeks the satisfaction of human needs and in this sense is tantamount to human development and sustainability; and the self-sufficiency, autonomy and independence from external factors of any given society. Thus, the natural cycles of the Earth (the rotation and translation of the planet, the rain regimes, the four seasons, the marine and atmospheric currents) are by definition indefinitely sustainable, because although everything ultimately depends on the sun, both the sun and the planet we inhabit are part of the solar system, which is at the very origin of life. Therefore, the sun is not an external factor but a consubstantial phenomenon of the Earth, and it is not by chance that one of the most important renewable energy sources that is promoted everywhere today to replace the non-renewable ones (like fossil fuels) is solar energy. Therefore, the main characteristic of ecosystems is their sustainability, precisely because they are cyclical, autonomous and self-sufficient. However, since natural sustainability is not linear but circular, the concept of development cannot be applied to it.

Satisfying the basic needs of the population is obviously the human dimension of the concept of sustainable development, while the ecological dimension concerns the limits that are placed on development by both technology and social organization, as well as by the carrying capacity of the *planetary boundaries* – in other words, by the natural ecosystems that are not linear but circular. This means that sustainability must be evaluated not only in relation to the human needs of both present and future generations but also with regard to the needs of natural ecosystems or *Mother Nature* (called *Pachamama* by South American indigenous people). Therefore, even though the SDGs do not oppose growth,[3] they do impose a change in quantity (growth must be 'sustainable', i.e. limited by the planet's livelihood capacity) and quality (it must be socially inclusive, which implies the intervention of the State to redistribute its benefits through tax mechanisms and social policies). Hence I maintain that the satisfaction of the social needs of the present should not put at risk those of the future, and at the same time we must protect the natural ecosystems or, in other words, be aware of the needs of *Mother Nature*.

---

[3] Goal 8 concerns growth and refers to the promotion of a "sustained, inclusive and sustainable economic growth, full and productive employment and decent work for all".

4.3   The Planetary Boundaries

We have seen that both our planet and the sun are part of the solar system, and even if the energy provided by the sun comes from space, it is also part of our planet because otherwise life cannot exist. No life is possible without the sun. So, the true planetary boundary is not extra-terrestrial space, but rather the limits of firm ground and the resources (including the oceans) that the human species has at its disposal to subsist and reproduce. The planetary boundaries, then, are those limits defined by Jeffrey Sachs, the intellectual father of the SDGs, (as Director of the Earth Institute and advisor to the UN Secretary General), who has called the limits of human action on the planet "planetary boundaries" because neither the world's economy nor the population can surpass the limits imposed by the finitude – the carrying capacity – of the planet itself. This raises the fundamental question of whether it is possible to reconcile environmental sustainability with economic growth: are there sufficient resources – water, air, land, forests and food – to sustain the growing world population? Is the improvement in our living standards doomed to be lost when natural resources come to an end? These concerns increase as climate change, soil degradation, water scarcity and loss of biodiversity worsen. According to Sachs, the planetary boundaries are anthropogenic climate change, the acidification of the oceans, the loss of ozone in the stratosphere, the pollution caused by nitrogen and phosphorus used by agricultural fertilizers, the overuse of water resources, the excessive use of land suitable for agriculture, the loss of biodiversity, the pollution of the air we breathe by aerosol particles from the burning of fossil fuels, and population growth. Thus, if the capitalist system mode of production surpasses those limits – as seen in the iconic graphs of the Great Acceleration in the previous chapter – a catastrophic future awaits humanity.

The increase in the temperature of the Earth's surface is, as we know, caused by the increase in the carbon dioxide particles produced by burning fossil fuels such as coal, oil and natural gas, as well as deforestation. The changes in the use of soil – for agriculture and livestock rather than forests containing carbon dioxide deposits – contribute to the amount of carbon dioxide in the atmosphere. A rise in temperature of more than 2 °C by the end of this century could end up causing major climatic catastrophes, such as an increase in ocean levels due to the melting of the polar ice caps as well as the frequency of large tropical storms, droughts, floods, winters or extremely rigorous summers, to name just a few risks.

The acidification of the oceans is also the result of greenhouse gases, because when carbon dioxide dissolves in the oceans, it is transformed into carbonic acid, which is harmful to marine life, including corals, crustaceans, fish, lobster and other species. As for the hole in the ozone layer that filters out ultraviolet solar radiation, good results have been achieved so far by reducing it through controlling the emissions of *chlorofluorocarbons* (CFCs) used in refrigeration appliances etc., in accordance with the terms of the Montreal protocol, so we can expect the phenomenon to remain under control.

Another boundary is atmospheric pollution engendered by the increase in nitrogen and phosphorus due to the use of fertilizers in agriculture. This continues to be a serious problem because it is leading to the growth of algae, which – when decomposed – produce bacteria that decrease aquatic oxygen, thereby contributing to the decrease in fish, lacustrine and marine species, as is happening in more than a hundred rivers and lakes around the world.

Excessive consumption and use of fresh water is also a planetary boundary, because although human consumption is relatively small (10% compared to the 70% of water resources that are destined for agriculture and the other 20% that is used in industry), overuse of water is leading to increasing scarcity, especially since underground aquifers are failing to recover with rainfall, and the amount of water used for industrial, mining and irrigation activities in agriculture and plantations continues to increase.

The use of land for agriculture and livestock is another important boundary, not only because of deforestation (which increases *greenhouse gases* [GHG]) but because the habitat of other species is destroyed, which leads to the loss of biodiversity, which in turn harms the proper functioning of ecosystems, the productivity of crops, human health and the very survival of the species. Hence, if governments don't undertake world multilateral policies in order to prevent the increase of deforestation across the globe, this path could be leading humankind towards the sixth mass extinction, this time caused by our own species, since growth and the capitalist neoliberal model could be accelerating the loss of other forms of life, contrary to what happened in the geological past of the Earth, since the five preceding mass extinctions occurred because of natural causes and phenomena.[4]

Another 'planetary boundary' concerns the population itself, because its growth cannot be unlimited either, as infinite growth would risk exhausting the capacity of the planet to sustain *Homo sapiens* as a species. It is clear that the best solution to the demographic explosion is to decrease poverty (the number 1 SDG) because it favours social mobility, and the middle classes spontaneously reduce their birth rates, which is partly the result of the empowerment of women,[5] who, thanks to reproductive health programmes are able to plan their pregnancies, and partly due to the fact that the children of the middle class are not obliged to work. The problem resides in the fact that the pace at which this is happening is not fast enough. The reduction in the growth rates of the population required to affect change in the poorest regions

---

[4] The occurrence of the mass extinctions is evident from the geological history of the planet, which is measured in millions of years. The first, caused by huge glaciation, took place about 440 million years ago, and the second was the result of another glaciation that occurred at the end of the Devonian period, about 300 million years ago. The third mass extinction, which occurred about 255 million years ago at the end of the Permian period, was provoked by enormous volcanic eruptions. The fourth mass extinction resulted from a 'comet shower' at the end of the Triassic period. As for the fifth mass extinction, palaeontologists attribute it to the impact of a meteorite that caused climate change of such magnitude that it killed the dinosaurs 65 million years ago, at the end of the Cretaceous period.

[5] SDG 5 refers to gender issues and aims to achieve gender equality and empower all women and girls.

of sub-Saharan Africa, the Middle East, and some South East Asian countries, and for those parts of the population that suffer social exclusion such as peasants and indigenous people in Latin America and other regions of the world, is much higher and will not be achieved in a short period of time.

Therefore, it is essential to implement complementary policies that are part of the objective concerning reproductive health and the empowerment of women and girls, so that humanity moves in the direction of maintaining a population level in accordance with the carrying capacity of the Earth (environmental sustainability) as well as conserving and increasing natural resources, promoting the reorientation of technology, and adopting measures to control risks. In this manner coherence will be given to economic policy in relation to demographic and environmental issues. Consequently, since the UN boundary for the end on the century is the forecast of 10.9 billion people, urgent action must be undertaken to accelerate the accomplishment of SDG number 1, i.e. the end of hunger and poverty, and, as highlighted above, the empowerment of women.

4.4 The Origins of Human and Sustainable Development as UN Paradigms

Any development worthy of the description 'sustainable' implies the existence of a democratic political system that ensures the effective participation of citizens in decision-making to help redistribute surpluses in an equitable manner while simultaneously negotiating solutions to the inevitable conflicts that result from social inequalities.[6] Democracy is also essential to ensure that the productive apparatus respects the obligation to preserve the ecological basis of development and does not go against the ability of governments to sustain and preserve natural resources. Furthermore, in order to reduce carbon emissions, democratic governments must promote the search for new technological alternatives – mainly in the field of renewable energy – while establishing intrinsically flexible administrative procedures which are capable of self-reform and correction while promoting sustainable trade and financing patterns.

It has been more than four decades since the United Nations promoted research that focused on responding to the means to satisfy basic human needs without transgressing the outer limits of the biosphere, that is, within the framework of the resolutions of the first United Nations Conference on the Human Environment (Stockholm, 1972) that were published in the book *Another Development: Approaches and Strategies* (Nerfin 1977). In this collective work, papers on different topics written by authors like Johan Galtung, Marc Nerfin, Fernando Henrique Cardoso, Rodolfo Stavenhagen, Bolívar Lamounier, Cinthia de Alcántara, Paul Singer, and Sergio Bitar

---

[6] Bear in mind here SDG 16 on the issue of good governance, which establishes the intention to: "Promote peaceful and inclusive societies, for sustainable development, provide access to justice for all and build effective, accountable and inclusive institutions at all levels".

presented an innovative vision of development which is not based on growth and capi-talist modernization. Since then, the narrow concept that considered development to be synonymous with economic growth and industrialization[7] is not the gener-ally accepted paradigm in academia. A good number of Latin American scholars – including the Norwegian Johan Galtung, who was living in Chile at the time – were among the pioneers of the new paradigm of sustainable development. The kind of "other development" which these Latin American pioneers were talking about is one that must be orientated towards satisfying a more comprehensive concep-tion of human needs, i.e. one which encompasses material, economic, social, and cultural and spiritual needs, as can be seen in Neef and Elizalde's matrix of needs and satisfactions reproduced in Table 4.1.

This comprehensive understanding of development is a matter of *human develop-ment* and not simply economic development measured by GDP. It began to be used by the United Nations in the 1990s with the annual publication of human develop-ment reports that introduced social categories such as health, education and political freedom as parameters for assessing development in this broad sense. This concept of human development is based on the ideas in the seminal text *Development on a Human Scale: An Option for the Future*, by Neef et al. (1989), which gave continuity to the paradigmatic line of research initiated in the previous decade by academics such as Stavenhagen (1981, 1990, 2013), Cardoso (1969, 2006), Singer (1980, 2002), Galtung (2003a, 2003b, 2004), and Nerfin (1978), who were among the authors of the innovative book *Towards Another Development: Approaches and Strategies* (1978). Hence, *Development of Human Scale Development* was the result of a collective effort that crystallized in a theoretical systematization of theses that in 1978 outlined only its more general features. It was designed to give precision and coherence to such ideas, and to establish the framework for a new approach to the theory of devel-opment that was not reduced to a "mere cosmetic arrangement of a paradigm in

---

[7] The concept of development without qualifiers was always understood to be the rough equivalent of economic growth, industrialization, trade and GDP increases, and modernization in general terms (Rostow 1962). In the European, Marxist or socialist tradition (Marx 2010) the concept of development is not reduced to its economic dimension because a permanent preoccupation with social welfare – such as improving education, health and other services provided by the State – exists in the socialist movement, but the concept continues to be mainly evolutionary and linear, which is at variance with the idea of sustainability as a systemic form comparable to the way that natural ecosystems function, which are cyclical. Thus, the concept of a *circular economy* (EU 2020; Raworth 2017; Esposito et al. 2018) that is *not linear* emerged as an approach that is complementary to that of sustainable development. Obviously, the idea of a type of development that involves both the classic linear economic development and the non-linear ecologist policy proposal that asserts that the environment must not be changed by human agency and therefore needs to be protected and preserved is contradictory. Therefore, a sustainable approach that does not imply any 'development' (and could even adhere to policies of degrowth), and that entails the conservation of natural ecosystems – as the Club of Rome (1972) and the Brundtland Report (1987) have both suggested – pose a quite difficult predicament. But that is precisely the kind of dilemma which needs be solved in the foreseeable future.

Table 4.1 Matrix of needs and satisfiers

Needs according to ontological and axiological categories	*To be*: Individual subjective dimension, spirituality, affectivity (being)	*To have:* Objective and material dimension, economic and social aspect (having)	*To do*: Intersubjective and cultural dimension (doing)	*Be*: environment, ecological environment (interacting)
Subsistence	Physical health, mental health, balance, solidarity adaptability	Food, housing, work	Feed, procreate, rest, work	Vital environment Social environment
Protection	Attention, adaptability, autonomy, balance, solidarity	Insurance system, social security, savings, human rights legislation, family work	Cooperate, prevent, plan, cure, defend	Social contour, vital contour, residence (housing)
Affection	Self-esteem, solidarity, respect, tolerance, generosity, receptivity, passion, will, sensuality, humour	Friendships, relationships, family, pets, plants, gardens	Making love, caressing, expressing emotions, sharing, caring, cultivating, appreciating	Privacy, intimacy, home, meeting spaces
Understanding	Consciousness, criticism, curiosity, astonishment, discipline, intuition, rationality	Literature, books, teachers, method, educational policies, communication	Investigate, study, experiment, educate, analyse, mediate, interpret	Habits of participative interaction, schools, universities, academies, groups, communities, family
Participation	Adaptability, receptivity, solidarity, conviction, dedication, respect, passion, humour	Rights, responsibilities, obligations, attributions, work	Affiliating, cooperating, proposing, sharing, disagreeing, accepting, dialoguing, agreeing, giving opinions	Areas of participatory interaction: party, clubs, associations, churches, communities, neighbourhoods

(continued)

Table 4.1 (continued)

Needs according to ontological and axiological categories	*To be*: Individual subjective dimension, spirituality, affectivity (being)	*To have:* Objective and material dimension, economic and social aspect (having)	*To do*: Intersubjective and cultural dimension (doing)	*Be*: environment, ecological environment (interacting)
Leisure	Curiosity, will, intuition, imagination, rationality, autonomy, inventiveness, curiosity	Games, shows, parties, rest	Wander, abstain, dream, long for, fantasize, evoke, relax, have fun, play	Privacy, intimacy, meeting spaces, free time, environment, landscapes
Creation	Passion, will, intuition, imagination, rationality, autonomy, inventiveness, curiosity	Abilities, skills, methods, work	Work, invent, devise, compose, design, interpret	Areas of production and feedback: workshops, seminars, groups, audiences, spaces of expression
Identity	Belonging to a social or ethnic group, coherence, difference, self-esteem, assertiveness	Symbols, language, habits, customs, reference groups, sexuality, values, norms, history, work	Behave, integrate, confront, define, know, recognize, update, create	Socio-rhythms, everyday environments, areas of belonging, maturational stages
Freedom	Autonomy, self-esteem, will, passion, openness, determination, audacity, rebellion, tolerance	Equality of rights, national and international legal protection systems	Disagree, choose, differentiate, disobey, meditate	Temporal/spatial plasticity

Source The matrix was originally published in a special issue of the Swedish publication *Development Dialogue* (Uppsala: Dag Hammarskjöld Foundation/Santiago: CEPAUR – Development Alternatives Centre, 1986) and was designed by the Chilean scholars Manfred Max Neef, Martin Hopenhayn and Antonio Elizalde. The version used in this book is taken from a reprint published under the title *Human Scale Development: An Option for the Future* (1989: 33). It is used with the authorization of the authors both here and in one of my other books, *Paz y Conflicto en el Siglo XXI: Teoria de las Relaciones Internacionales* (2009: 244–245) crisis", as the authors of the text say, but instead was a genuine effort to transform and substantially modify existing thinking (Neef et al. 1989: 12–13).[8]

Indeed, in a recent book on sustainable development, the German researcher Göpel reproduces the same matrix in her criticism of mainstream economics. To the extent that it is based on economic growth and the individual search for profit, mainstream economics can be seen as responsible for the multiple obstacles and difficulties that sustainable development has suffered in all parts of the world. For Göpel, a change of mentality and the adoption of a new holistic paradigm that effectively seeks the full satisfaction of human needs in all their complexity are indispensable tools. Not only the most elementary and primary needs such as subsistence, protection and shelter, but also affection, creativity, freedom, political participation, leisure, and understanding, must be properly articulated within sustainable development processes (Göpel 2016: 64–65).

But it is not enough for development to be merely orientated towards needs. For human development to be sustainable, it must also be endogenous in nature, i.e. it must evolve within each country, community or region according to that society's own path rather than being based on a general model supposedly valid for all societies, as postulated by the 'developmentalist' and Marxist doctrines of the 1950s and 1960s. As already stated, sustainability also entails rejection of the linear conception of development, according to which development is a regimented and universal model consisting of successive stages that each country inevitably has to go through – a view which is paradoxically characteristic of both the American-inspired Rostow type of 'developmentalism' and Marxism, which exerted a lot of influence in Latin America in those years.[9]

---

[8] The book *Development on a Human Scale: An Option for the Future* by Manfred Max Neef, Antonio Elizalde and Martin Hopenhayn is the source of the concept of human development introduced by the UNDP in its reports at the beginning of the 1990s. This seminal masterwork was the result of multiple research projects, reflections and discussions during seminars carried out in Brazil, Chile and Sweden which brought together Latin American scholars from different latitudes and disciplines, among them Hugo Zemelman (Argentina); Jesús Martínez (Colombia); Jorge Dandler (Bolivia); Jorge Jatobá (Brazil); Felipe Herrera (Chile); Rocío Grediaga (Mexico); Franz Hinkelammert (Costa Rica); Manfred Max Neef, Luis Weinstein, Martin Hopenhayn, and Antonio Elizalde (Chile); and the Swedish scholar Sven Hamrell, Director of the sponsoring entity, the Dag Hammarskjold Foundation of Sweden.

[9] *The Stages of Economic Growth* (Rostow 1962) is the most important book with regard to this topic. As for the concept of development in general terms, it is well-known that Marxism postulated the historical (and linear) progressive transit of the slave and feudal modes of production to capitalism. The latter is a stage forcibly previous to socialism and communism, which – according to historical materialism – are going to be the next stages in the evolution of the economic infrastructure of society. Thus promoting the development from backward agrarian structures to modern capitalist forms of production is a positive task for communist parties, since it implies the enlargement of working classes, according to Marx in his classic book, *Das Kapital* (1867). Other authors, such as the Brazilians Cardoso (1969) and Enzo Faletto, who proposed the *Theory of Dependence* that was very influential in Latin America, are also among the pioneers of this change in thinking, although Cardoso, who was elected President of Brazil in the 1990s, could not do much to change the social inequalities and lack of human development in Brazil. President Luiz Inacio Lula da Silva and the Workers Party made a more successful attempt to improve social and human development and

In addition, it should be understood that non-linear sustainability is ecologically sound, i.e. the resources of the biosphere should be rationally used with full awareness of the potential of local ecosystems and the need to use appropriate and environmentally sound technologies. Sustainability also offers societies the highest degree of autonomy possible, all within the framework of self-reliance, although this does not mean the rejection of interdependence as a characteristic phenomenon of the present interconnected and globalized world. Hence, it is not a matter of proclaiming autarchy[10] (a term that refers to the isolation and economic self-sufficiency of a society) but autonomy, to underline the fact that each country must rely primarily on its own strengths and resources in terms of the energies of its inhabitants and its natural and cultural environment. Self-reliance, which is another way of describing autonomy, should be regarded as a way to reduce centralization and promote administrative and political decentralization for local levels and regions within a country. However, this does not mean going against interdependence; it merely opposes those forms of globalization that seek to impose the hegemony of big capital over small local economies. In any case, it is crucial to remember that the concept of self-reliance is rooted at local level, in the practice of each community.

Another aspect of human development is that it involves the structural transformation of socio-political relationships, which are essential to guarantee satisfaction of the needs of participation, identity and freedom. They are further needed to favour the changes of mentality that facilitate the awareness necessary to create the conditions required for self-administration, freedom, and the strengthening of civil society. In other words, for there to be participation in decision-making by citizens, deepening democracy by making it participatory in rural communities, municipalities and regions until it reaches national level requires the establishment of effective and imaginative mechanisms (such as the use of social networks, improving electoral procedures to elect political representatives, and improving social audit procedures to combat corruption) in order to encourage citizens and the population in general to have greater and better involvement in the national policy of each country. Therefore increasing citizen participation is a crucial factor of democratic deepening and a pre-requisite for development.

Thus considerable effort is required to establish a new paradigm for development theory that goes beyond 'developmentalism', be it neoliberal, Marxist or technocratic, as Pope Francis has called it, since all these lines of thought have been shown to have a narrow and limited character. The insufficiencies of developmentalism have given rise to the appearance of a new perspective which has been reinforced by the papal encyclical *Laudato Si'* discussed in the previous chapter. The new trend is

---

diminish inequalities in Brazil, but the reaction of the conservative establishment was to promote a *legalfare* that promoted the 'technical' *coup d'état* that impeached President Dilma Roussef in 2016 and placed former President Lula in jail, preventing him from running for office in 2018, a fact that allowed the far-right populist Jair Bolsonaro to became President in 2019, stopping the process of *human scale development* in Brazil.

[10] As was the case – to a certain extent – with the countries of the Soviet Bloc before the fall of the Berlin Wall and present-day North Korea, since they pretended to have no links with the capitalist world.

characterized by its orientation towards people, i.e. towards real and concrete human beings. In addition, the new trend is committed to questioning the values that guide economic activity (as Göpel does when referring to the importance of a change in mentality), which entails abandoning the criterion of profit as the only reason for business activity and growth as a fundamental criterion for the evaluation of economic development. In this line of thought it is interesting to quote a recent document from the *Congregation for the Doctrine of the Faith* of the Catholic Church, which states the following:

> Any progress of the economic system cannot be considered as such if it is measured only with parameters of quantity and effectiveness in obtaining benefits, as it must also be evaluated based on the quality of life that it produces and the social extension of well-being that spreads, as well-being cannot be limited to its material aspects … Therefore, well-being must be evaluated with criteria much wider than the gross domestic product (GDP) of a country, taking into account other parameters, such as security, health, the growth of 'human capital', the quality of social life and work. The benefit must always be sought, but never at all costs, nor as a single reference for economic action. Here the importance of parameters that humanize, form cultures and mentalities in which gratuity – that is, the discovery and exercise of the true and just as intrinsic goods – becomes the norm of measurement and where profit and solidarity are not antagonistic (Ladaria et al. 2018: 4).

Orientating development towards the person is therefore a chief characteristic of this new paradigm that seeks to focus on the importance of the real and actual human being. This supposes that the satisfaction of human needs should be the central criterion in development policies, whether at the level of macro state policies (of the ministries of development or planning agencies), international organizations and international cooperation agencies, or in the area of the micro-policies of *non-governmental organizations* (NGOs).

This systemic perspective is holistic and transdisciplinary in nature, which means that it is based on an integral methodology for the formulation of policies aimed at generating increasing levels of self-reliance that engender sustainability in the planning, management and execution of projects. This new paradigm also seeks to establish organic articulations of human beings with nature, technology, industrialization and all kind of processes, including the coordination of global processes with local policies. The foregoing means that the concepts of human needs, self-reliance and organic articulations are fundamental pillars of the new paradigmatic vision that must be constructed with the aim of transforming the role of the person from the object to the subject of development.

Democratizing development could also be the emblematic leitmotiv of sustainable development because the transition from the status of 'object' to 'subject' can only be achieved as the processes of citizen participation deepen. This 'deepening' is understood here as an increase in participatory democracy and not merely in the formal aspects like competitive elections or political parties, which are necessary but not sufficient on their own. Participatory democracy also implies greater citizen awareness, less paternalism and fewer guidelines imposed from 'above', and greater conscious involvement of citizens in the decision-making of all matters that directly concern them through referendums or direct democracy.[11] In short, sustainable human development that contains this new paradigmatic vision of development also entails transforming the international system in order to reach sustainability in the midst of a dense network of new local economic micro-orders that are based on social solidarity and collective cooperation and not on individual greed and desire for personal enrichment.

4.5   The Satisfaction of Human Needs as the Main Purpose of Human Development

If the satisfaction of human needs is going to be the true guiding criterion of policies, plans, projects, programmes and specific government actions, a new way of interpreting and understanding reality is essential. The three basic postulates of this new interpretation are: (1) Development refers to people and not capital or material things. (2) Development should not be measured by economic growth but by indices that reflect the improvement in people's quality of life. (3) Quality of life depends on the opportunities that people have to meet their basic human needs.

What are these fundamental human needs? One of the most important contributions to the theory of human needs is undoubtedly Abraham Maslow's Hierarchy of Human Needs,[12] which inspired the works of the Chileans Neef, Elizalde and Hopenhayn. It is important to realize that these ideas are linked to an integral vision of the human being, since the adequate satisfiers of vital needs, such as subsistence (food, housing, work) and protection (social security, citizen security, health systems) are provided mainly by infrastructure or public services, but in this perspective some values (such as freedom and creativity), socio-political categories (like identity and participation) and political systems (like democracy) are transformed into needs, which means that they are embedded in a person's subjectivity.

An example of a 'concept-value' that becomes a necessity in this new perspective is freedom. Since the French Revolution, political science has dealt with freedom as

---

[11] The fact that Switzerland's political system embraces direct democracy based on referendums demonstrates that the deepening or 'radicalization' of democracy is perfectly feasible.

[12] The scale of needs is described as a five-level pyramid. Higher needs are only met when lower needs have been met, such as the need to breathe, hydrate, feed, sleep (rest), eliminate bodily waste, avoid pain, maintain body temperature, etc. In addition, there are security and protection needs (physical security, health, work, housing); social needs derived from the relationship function (friendship, partner, colleagues and family), social acceptance and esteem (recognition); and the need for personal self-realization. Although inspired by Maslow's works, the Chilean social scientists Neef et al. (1986) criticize the hierarchization of his paradigm of human development, according to which needs are few, finite, classifiable and universal. They propose instead a system of nine needs with four forms of realization: subsistence, protection, affection, understanding, participation, creation, recreation, identity and freedom, through being, having, doing and relating. They also assert that needs and satisfiers (the means to satisfy needs) vary according to person and culture. However, Spanish translations of Maslow's works, such as *Visiones de Futuro* and *La Persona Autorrealizada*, published by Kairos Editorial in Barcelona, for sure influenced Neef et al.

an essential value of liberal democratic political theory and human rights, and that is why the notion of 'fundamental freedoms' appeared, but in the perspective of Neef, Hopenhayn and Elizalde freedom is also considered a *necessity*, which means that it is not just a matter of the objective organization of the subjective dimension. In this respect the concept of need concerns both the physiology and the psychology of each person (including the rational and emotional dimensions), which means that freedom is no longer exclusively regarded as a *value* of the political system, but also as a psychological need in the subjective dimension of each individual.

The same happens with several other needs, such as the need for cultural identity, which is located in the field of ethnic groups (indigenous people) or social minorities, immigrants, refugees, feminist groups, and so on; the need for democratic participation, which is grounded in the political sphere; the need to express oneself creatively through art or any aesthetic or cultural manifestation; and the need for emotional well-being, which includes everything related to love, friendship and sexual preferences, which are considered affective needs. There are also needs in the intellectual field, such as those of understanding and knowledge, and in the physiological and psychological sphere, where the needs of leisure, relaxation and recreation are located.

In brief, some needs can be placed in the field of the physiology and psychology of the human being, while others can be regarded as part of the intellectual and rational dimension – e.g. the needs of understanding and knowledge, as well as those of a political nature: democratic participation and everything which has to do with individual human rights, such as fundamental freedoms (of assembly, association, human mobility, migration), free speech, freedom of conscience and of religious creed.

Consequently, according to this perspective, freedom is not only a value and individual right that must be respected by the State, but at the same time (a remarkable novelty of the human development paradigm) it constitutes a human **need** so that, for instance, the respect for fundamental freedoms can be also a parameter to be used in its assessment, which means that the concept of sustainability must include the political commitments of SDG 16 concerning the rule of law and good governance. The matrix of needs and satisfiers by Neef, Elizalde and Hopenhayn is summarized in Table 4.1.

Accordingly, all human needs can be divided into the basic categories of being (individual), doing, having and being (locational), which then allude to the subjective and objective dimensions, as well as the ecological and socio-political environment of people closely linked to each other in such a manner that it is capable of satisfying the higher needs of affection, understanding, participation, freedom, identity, and creation, which are as important as the basic needs of subsistence, security and protection. Therefore, concepts such as identity should not only be understood as an important social and anthropological category but also as a necessity, which implies that the full realization of them should be promoted through policies appropriate to ethnic plurality and cultural diversity. As a result, if all UN member states are guided by these principles and norms, which are part of the universal declaration of human rights, we can rightly say that the fulfilment of human needs is a benchmark for evaluating sustainable development.

As for other important categories such as *political participation*, if we view it as an aspect of the human need to participate, which is essential to consolidate and deepen the processes of democratization, and the need to understand, which is of vital importance for both the educational system and scientific activity, it is of great consequence for the collective formation of a social critical conscience as well as for individuals to acquire an attitude of tolerance and behaviour open to dialogue and negotiation in conflicts or violent socio-political contexts. Thus this holistic conception is important not only for assessing human and sustainable development but also for encouraging the materialization of cosmopolitan consciousness on a planetary scale.

4.6   Cultural and Community Development

Cultural and community development is the kind of sustainable development that seeks to expand and consolidate the areas of one's own culture, especially in indigenous communities in countries with this kind of population. For instance, the 'counter-hegemonic globalization' based on the *ecology of knowledge* that the Portuguese social scientist Boaventura de Sousa Santos advocates in his books could be accomplished by strengthening the ability of a culturally differentiated community to make autonomous decisions and guide its own development in the exercise of an equitable and appropriate form of power. This means that a linguistic and cultural community or ethnic group can become a political-administrative unit with authority over its own territory, and with decision-making capacity in the areas that constitute its own development within the framework of a growing process of autonomy and self-management.[13]

The positive contribution that indigenous populations can make to the protection of the environment deserves special consideration, as the issue of land remains crucial. Modernization puts pressure on indigenous territories and this can affect the economy, the habitat and the social, religious and cultural systems of indigenous people. This is a problematic issue, as these communities are depositaries of a vast accumulation of traditional knowledge and experience, which is what Boaventura de Sousa Santos has called the *ecology of knowledge* (Santos 2009), the disappearance of which would be a loss for any national society that could learn a lot from indigenous people, especially in terms of skills based on tradition to manage complex ecosystems in a sustainable way.

An example of the above is reported by Santos himself, given what happened on the island of Bali (Indonesia) during the 'green revolution' of the 1960s, when Western

---

[13] This type of governance structure could be considered tantamount to the autonomous regions or *Länder* in developed countries like Spain and Germany. Certainly the latter can be a source of comparison and inspiration.

'experts' insisted on introducing a new irrigation system for rice crops, replacing the traditional one even though it had worked efficiently for over a thousand years in accordance with traditional knowledge. Hence, according to Santos:

> The traditional irrigation systems were based on ancestral and religious knowledge and were used by the priests of a Hindu-Buddhist temple dedicated to Dewi-Danu, the god of the lake. These systems were replaced precisely because they were considered to be based on magic and superstition, the 'rice cult', as they were disparagingly called. It happened that this replacement had disastrous results in the rice fields: the crops declined more than 50%. The results were tremendously disastrous to the point that the irrigation systems had to be abandoned and the traditional system restored (Lansing 1987, 1991; Lansing and Kremer 1993). This case also illustrates the importance of the precautionary principle in dealing with the question of the possible complementarity or contradiction between different types of knowledge. In the case of the irrigation systems of Bali, the incompatibility budgets between two systems of knowledge (the religious and the scientific), both concerning the same intervention (irrigating rice fields), resulted from an incorrect evaluation based on the abstract superiority of scientific knowledge. Thirty years after the disastrous technical-scientific intervention, computer-generated models, at the time an area of the new sciences, showed that the water maintenance sequences used by the Dewi-Danu divinity priests were more efficient than any other conceivable system, be it scientific or otherwise (Santos 2009: 186–191).

As can be seen from what has been described, it is a terrible irony that modernization penetrates tropical forests, deserts and other relatively isolated environments (as has also happened to the *Inuit* people of the Canadian Arctic and the *Sami* in Scandinavia), destroying cultures that have proven themselves capable of surviving in this environment for centuries. Therefore, the starting point for a fair and humane policy towards indigenous groups is the recognition and protection of their traditional rights to land and other resources that sustain their way of life according to custom, which means that although customary law usually does not fit within the official legal system of a country, the government of those states must recognize these legal systems that are based not only on customary law but on ancestral knowledge, since its proper functioning is crucial to maintain harmony with nature and ecological awareness characteristic of traditional ways of life.

Therefore, the recognition of indigenous rights must go hand in hand with measures designed to protect local institutions that ensure the responsible use of resources. This recognition should give local communities active participation in the decisions on the manner in which the resources of the regions where they live are used. The protection of indigenous rights and indigenous people must be accompanied by appropriate measures to increase the welfare of the community in ways that are consistent with their lifestyles. For example, profits from activities such as handicrafts can be increased with the introduction of market mechanisms as long as these ensure a fair price for the products, but also through measures to conserve and increase the base resources and their productivity. Promotion policies that have an impact on the lives of indigenous communities and indigenous peoples must also understand how to differentiate between artificial conservation measures which are necessary, and those that are not be required by the people and may lead to the destruction of their traditional ways of life.[14]

Health policy measures are another example, as traditional health practices can complement official medicine, and should thus be preserved. Also, if deficiencies in nutrition can be improved by introducing new food crops, this should be encouraged through public policies to supplement the current food sources or replace them when necessary, but all in agreement and after thorough deliberation.[15]

Another issue of particular importance concerns the obligation to carry out consultations with indigenous communities when the State authorizes the construction of hydroelectric plants, infrastructure or the installation of mining companies because of the requirements of ILO Convention 169. The fact that these communities are small settlements often makes them more vulnerable, and their own marginalization is a consequence of state policies that refuse to take into account the human, cultural, social and ecological aspects inherent in the sustainable development model. These preventive policies of the conflicts and the rejection of the local communities must precede the new projects that open certain rural areas to so-called 'economic development'. In addition, negotiations must be initiated with local communities so that they derive concrete and substantial benefits from such projects, which go beyond the salaries of the local inhabitants who are employed by transnational corporations, since otherwise there is a risk of popular rejection and even of violent action against such projects.

One should be aware that native languages are part of the cultural heritage of any nation, given that language is much more than a means of daily communication, as it is a means of transmitting culture and identity. The dangers posed by the aforementioned type of investment can also have an impact on the cultural diversity of a country. These native languages run the risk of disappearing through the onslaught of modernity and, in certain cases, because of the imposition of an official language.

4.7   The New Constitutionalism of Bolivia and Ecuador

During the first decade of this century left-wing political parties won elections and have made significant changes to the constitutions of Bolivia and Ecuador, which largely resulted from the presence of important nuclei of indigenous population within their respective national societies. It ought to be noted that the new constitutions of both countries express a deepening or radicalization of democracy (in the

---

[14] For instance, an artificial conservationist measure could be the endeavour to maintain corn crops in coastal communities when artisanal fishing projects should be promoted; but in another instance a very different situation might result from the purchase of imported corn in highland communities if it contributes to the destruction of traditional ways of life that are based on maize production that is additionally an important component of food security.

[15] For example, *quinoa* is a nutritious food highly popular among the indigenous populations of the South American highlands, especially in Peru and Bolivia, and is now being introduced in Central American countries.

terms that have been described in previous pages). Because of the important role played by these population groups in the electoral triumphs of Evo Morales and Rafael Correa, they have been able to re-found – at the constitutional normative level – their respective national states in an inclusive manner, thus expressing a new social pact that is manifested in the new constitutional regulations. I will here examine briefly what constitutes those remarkable characteristics of the constitutions of both countries.

According to Barié (2017), the concepts of *buen vivir* and *Madre Tierra* or *Pachamama* are the most novel aspects of the new constitutions of both Bolivia and Ecuador. Regarding the concept of *buen vivir* (good living), it is evident that all good living can be understood as the satisfaction of human needs, because 'living well' and 'good living' are quite clearly terms that refer to a rich life in which both the subsistence and protection needs (decent work, adequate housing, food, medical and educational services) as well as the affective ones (living with the family, community, loved ones, being respected and appreciated) and those relating to knowledge (understanding), cultural identity, freedom and political participation, creativity, leisure and spirituality in general are fully met. However, what I want to highlight in the case of the indigenous people of the Andean region (Quechua and Aymara) is that these needs are understood from the point of view of their respective *cosmovisions* and under their corresponding terminology in the native languages, so although 'good living' is a subjective notion not quantifiable or definable in objective terms, it is clear from its name that the concept is closely related to human values and needs. This explains why it is appropriate to place it not only in the sphere of the aspirations of the nation as a whole, but also in the field of human development, thus enabling it to acquire the precision that can be obtained from United Nations Human Development Indices.

A review of the main changes reveals, for instance, that in the Constitution of Bolivia the concept of "good living" appears in the preamble, which refers to the fact that the Bolivian state is based on and guided by new principles and values, such as "sovereignty, solidarity and equity in the distribution and redistribution of the social product", and above all is based on the "search of living well". This is one of several new ethical-moral principles, which are presented in the Aymara language as *sum qamaña*, very similar to the Quechua term of *sumac kawsay*. With this concept, the purpose is to make the idea of material well-being compatible with both social peace and support and mutual solidarity among the people. Consequently, a person who lives well (*qamiri sum*) is not the one who is rich but the one who shares:

> The *qamiri sum* happens to be the one who lives and coexists well, because he is welcomed by everyone and knows how to welcome and collaborate with everyone, regardless of whether you have little or much. In a certain way it can no longer be given individually but only in and with a larger social group … Living well turns out to be a kind of meta-value (to which other more common values must be subordinated, such as equality, inclusion and social equity). Even the educational system and the new economic model must be guided by the principle of living well (Barié 2017: 57).

Although the economic model is not defined by the new constitution, Barié points out that it is a flexible model that Bolivians are called upon to continually seek, based on that quest to live well, which is roughly equivalent to improving the quality of life. Furthermore, they are asked to give priority to the collective versus the individual interest, because "the social and community economy will complement the individual interest with the collective good living" (Barié 2017: 58).

In the new Ecuadorian constitution, good living has different connotations, Barié points out, although it appears in the context of the re-founding of the State: "We, the sovereign people of Ecuador […] decided to build a new form of citizen coexistence, in diversity and harmony with nature, to achieve good living, *sumak kawsay*" (Barié 2017: 59). The concept of good living is also associated with popular wisdom and the ancestral cosmovision. Incidentally, *sumac kawsay* has been part of the historical claims of the *Confederation of Indigenous Nationalities of Ecuador* (CONAIE) since its foundation, and has also been proposed by CONAIE as a criticism of capital accumulation because, as Barié argues, "the objective and the principles of the economy should not be about profitability, but human well-being, living well: the *sumak kawsay*. The economy is only a tool at the service of the community" (Barié 2017: 61).

For this reason, the Constituent Assembly proposed the concept of *sumac kawsay* as an ethical principle, based on reciprocity and promoted by indigenous communities so that the 'rights of good living' are located within economic, social and cultural rights, such as the right to water, food, education, physical culture, work, social security, healthy environments and others that support good living. It should be noted that the emphasis on good living in the Ecuadorian Constitution is broader than the Bolivian, since it includes two complete chapters on inclusion and equity as well as on biodiversity and natural resources, both divided into sections on different topics.

Another aspect of the Ecuadorian constitution highlighted by Barié is that, following criticism of the predominance of the market typical of neoliberal 'mainstream economics', one of the main axes of this concept of good living is, according to the CONAIE, the creation of a Social and Solidarity Economy that can to a large extent be compared with the same concept proposed by the Chilean-American academic Howard Richards, which I will examine in the next chapter. This system recognizes the human being as the subject and end of the economy, its main goal being to guarantee the production and reproduction of the material and immaterial conditions that make good living possible (Barié 2017: 64).

The constitution also establishes different forms of organization of production, favouring State forms and modalities that "ensure the good living of the population", including public debt and the environment. Development is defined as the whole sustainable dynamic of economic, political, socio-cultural and environmental systems in order to achieve good living, with the State having the obligation to plan development at a national level and in a sustainable manner, as well as to organize the redistribution of resources for access to good living. Consequently, within the framework of a government that promotes development, natural wealth must benefit people, including all communities and ethnic groups, so that they can live in a good manner and simultaneously respect the right to a healthy environment that must be ecologically balanced in order to ensure good living (Barié 2017: 60).

The other novelty in both constitutions is that they give rights to nature, personified as Mother Earth or *Pachamama*. Normally legal theory only grants rights to human beings, therefore such innovations are truly revolutionary from the point of view of constitutional law. The preamble of the Bolivian constitution has a clearly spiritual tone, stating that Bolivians populate the 'sacred *Mother Earth*' in order to fulfil the mandate of the people 'with the strength of our *Pachamama*'. However, the constitution also refers to industrialization as one of the main objectives of the Bolivian State, which must promote the responsible and planned use of natural resources, including industrialization and the conservation of the environment (Barié 2017: 66). Other precepts give the State ownership of natural resources while simultaneously designating it the "guardian of nature, the environment, and biodiversity". However, even though the constitution reiterates that natural resources can be used for industrialization, these purposes must be achieved without damaging the environment, because, as Barié points out:

> The industrialization of natural resources to overcome the dependence on the export of raw materials and achieve a productive base economy, within the framework of sustainable development [must be done] in harmony with nature [because] all organizations must protect the environment, and the principles of harmony with nature, the defence of biodiversity and the prohibition of private appropriation should guide international relations and negotiations (Barié 2017: 67).

Nonetheless, it is interesting to note some differences between the constitutional tenets of the two countries, since, as highlighted by the Uruguayan anthropologist Gudynas (2017), the Ecuadorian constitution has a *biocentric dimension* that is not present in the Bolivian one:

> In the Ecuadorian Constitution, nature becomes a subject of rights and therefore there is an admission that it possesses intrinsic values … [and] in the recognition of those rights it expresses a 'biocentric' perspective different from the 'anthropocentric' one, where Nature is valued for the utility or benefit that it holds … Biocentrism defends intrinsic values as independent of the utility of the non-human world for human uses and purposes … The Bolivian situation is very different, since conventional wording prevails in the constitutional text, referring to environmental rights under the umbrella of third-generation citizenship rights. But the position diverges even more with calls to 'industrialize' natural resources. This developmental mandate is unusual and reproduces a utilitarian view of Nature (Gudynas 2017: 140–141).

As for the indigenous people, the constitution assures them of the right to the exclusive use and exploitation of renewable natural resources and the definition of their development according to their cultural criteria and principles of harmonious coexistence with nature. It is notable that Bolivia has developed the question of the rights of Mother Earth in special legislation such as the *Law on the Rights of Mother Earth* and even more notable that an Ombudsman of Mother Earth has been established in the tenets of the framework of the Law of Mother Earth and Integral Development.

In Ecuador, *Mother Earth* is the subject of a special chapter in the constitution, and, like individuals and groups, is considered to have rights. Consequently, *Pachamama*, where life is reproduced and performed, has the right to be fully respected in terms of its existence and the maintenance and regeneration of its life cycles, structure, functions and evolutionary processes, as stated in the constitutional text. Every person, community, town or nationality may demand from the public authority the fulfilment of the rights of nature, including, as an innovation of great significance, the right to the repair of any ecological damage suffered by human action. The constitution establishes nature's right to restoration, and states that this right to restoration is independent of the obligation of the State and natural or legal persons to compensate individuals and groups who depend on the affected natural systems. The cycle of life that includes human beings as well as nature must be respected and preserved.

As in Bolivia, in Ecuador the State is both the guarantor of these rights and responsible for exploiting strategic natural resources, so it reserves for itself the right to administer, regulate, control and manage the strategic sectors within which energy is located in all its forms: telecommunications, non-renewable natural resources, the transportation and refining of hydrocarbons, biodiversity, genetic heritage, the radio spectrum, water, and all other forms determined by law. The constitution also states that non-renewable natural resources belong to the state patrimony and that exercising protection over the environment and the joint responsibility of the citizens in its preservation will be articulated through a decentralized national system of environmental management in charge of the advocacy of the environment and nature.

As for the people and social groups, although they have the right to live in a healthy environment and in harmony with nature, they also have the obligation to protect the rights of nature because "These are the duties and responsibilities of Ecuadorian women and men. Ecuadorians [...] respect the rights of nature, preserve a healthy environment and use natural resources in a rational, sustainable and sustainable way" (Barié 2017: 66). Consequently the State's vigilance with regard to preserving the rights of nature must be combined with continuous observation and citizen monitoring, since "every person, community, town or nationality may demand from the public authority the fulfilment of the rights of nature" (Barié 2017: 67).

Exports must also be guided by a criterion of environmental responsibility and respect for nature, regulated when the latter is affected by public debt or by the operation of productive companies. This restricts certain types of inputs, since Ecuador is a country which has been declared free from transgenic crops and seeds – a unique achievement in the region. Briefly, the State is responsible for planning the social development regime for the realization of good living and *sumac kawsay*. Although both Evo Morales and Rafael Correa have had problems with indigenous movements during their respective administrations, it seems that no major problems concerning the new constitutional tenets have occurred. And, in passing, with regard to the philosophical issues raised by Eduardo Gudynas about the biocentrism of Ecuadorian constitutional law, there is a significant relationship between Lovelock's *Gaia* theory, deep ecology and biocentrism.[16]

---

[16] Recall this the author of Gaia Theory, which regards the planet as a living entity and highlights the difference between 'biocentrism' (focused on life) and 'anthropocentrism' (focused on man). For Gudynas, the difference between the two constitutions is that "the texts diverge radically. In the Bolivian Constitution the industrialization of natural resources is a goal, while in the Ecuadorian

In any case, it is clear that this constitutional law refoundation of both states means above all a legal transformation carried out in the framework of a socio-cultural confrontation, because it seeks to establish new symbols, ideologies, customs and subjectivities with the aim of forming a new cultural hegemony. In the foreseeable future this means the possibility of a collective change of mentality or *mindshift*, as Göpel calls it, which can lead to the consolidation of a social movement in which ethnic groups, Afro-descendants, women, and peasants have the opportunity to participate in alliance with other groups and social classes. The latter assumes that the re-foundation of the State is, above all, a civil demand and as such "requires an inter-cultural dialogue that mobilizes different cultural universes and different concepts of time and space [within the framework of a] minimal convergence of very different political wills, historically formed predominantly by cultural shock instead of by cultural dialogue, or by the ignorance of the other rather than by recognition of it" (Santos 2010: 82–84).

In addition, because of its scope, the re-foundation of the State is aimed at changing not only the institutional political structures but also the articulations and relations between the different economic systems, social relations and culture. Furthermore, when it comes to culture, it must be borne in mind that while for the indigenous movement's allies the re-foundation is about creating something new, for indigenous people what the State wants to be re-founded is rooted in ways that preceded the so-called Spanish 'conquest' — ways that, despite repression, managed to survive fragmentarily and diluted in the poorest and most remote regions. Also, according to the Portuguese scholar, when they exist, these people manifest only at the local level (Santos 2010: 83). Hence the re-foundation of the State is a long-term project which is inevitably based on trial and error. The reformist measures may also be inspired by social-democratic models such as those in Europe, although there is no doubt that this *transformative constitutionalism* is a formidable achievement which is making changes promoted 'from below' by the subaltern classes, and establishing a new political regime of intercultural democracy in which transformative constitutionalism has given rise to:

> one of the instances (perhaps the most decisive) of the use of hegemonic instruments because the modern constitutions are often said to be sheets of paper to symbolize the practical fragility of the guarantees they consecrate and, in reality, the Latin American subcontinent has dramatically lived the distance that separates what the Anglo-Saxons call the *law-in-books* and the *law-in-action*. This can also happen with transformational constitutionalism and its anti-hegemonic character, because settling on the strength of social mobilizations that fight the hegemonic visions and manage democratically compels a vision against hegemony that does not necessarily protect it from that possibility. Hegemonic institutions are the expression of the inertia of classes and hegemonic ideas. They are social relations and therefore also fields of dispute…Thus, any fracture in the mobilization can reverse the oppositional content of the constitutional norms or empty its practical effectiveness (Santos 2010: 95–96).

---

case Nature is presented for the first time as a subject with rights. Despite its positive aspects in other fields, the Bolivian text ends up reproducing the attachment of modernity to progress, while the Ecuadorian option allows a rupture with that perspective" (Gudynas 2017: 141).

4.8   Sustainable Development as a UN Paradigm

Many people remember the famous phrase of Pope Paul VI in his Encyclical *Populorum Progressio* (1967)[17] about development as "the new name of peace", but after what has been said on previous pages about the nature and history of the concept of sustainable development, and its relationship with both the environmental sciences and the theory of human needs, including cultural and community development and the *cosmovision* of indigenous people, it should be clear that *sustainable peace* requires sustainable development, among other reasons because the type of peace that goes beyond the absence of war – positive peace, as Johan Galtung calls it – requires the satisfaction of human needs to sustain itself.[18]

Hence, to be sustainable, peace must be based on respect for natural ecosystems as well as human development. Consequently, ecological peace is also sustainable peace, as Brauch (2016) and Gómez Camacho (2017) argue, because the sustainability of peace is linked to policies which are both environmentally and socially appropriate. Another approach of particular importance is that of Jeffrey Sachs (2015), who argues that sustainable development essentially consists of the proper management of four intersecting spheres: terrestrial ecosystems, social dynamics, the techno-economic sphere, and governance. If this tetralogy is not managed properly, the result translates into social instability, violent conflicts, increased crime and other expressions of the social, economic, ecological and political crises that afflict the contemporary world.

Consequently, guaranteeing peace would require at least the kind of 'solidarity development' of which Paul VI speaks in his Encyclical, but not just 'development' without qualifiers, because that is the path that until now has led the world on a track of unlimited growth, the concentration of wealth (and its corollaries in terms of increased inequality and poverty), crime, transnational terrorism, the ecological

---

[17] "Development is the new name of peace", said Paul VI in his encyclical *Populorum Progressio*, which deals with the theme of the integral development of human beings and proposes solidarity development. The importance of solidarity action is presented by the Pope as "the passage from less human conditions of life to more human conditions of life", which means not only an improvement in living conditions (education, health, housing, work) but also the promotion of spiritual values, such as respect for the dignity of others. Regarding solidarity development, it is considered that people's "obligations are rooted in human and supernatural fraternity and are presented under a triple aspect: duty of solidarity, in the aid that rich nations must provide to developing countries; duty of social justice, straightening defective business relationships between strong and weak peoples; duty of universal charity, for the promotion of a more humane world for all, where all have to give and receive, without the progress of the one being an obstacle for the development of the others." Development, then, responds to a demand for justice on a global scale and, understood in this way, should guarantee peace on the entire planet, which is why it is said to be "the new name of peace" (Pope Paul VI 1967).

[18] Galtung (1981) is a pioneer of peace research and speaks of an important difference between positive peace and negative peace. The first is a consequence of development, of the full satisfaction of human needs and, in short, of the absence of structural violence. In that sense, peace is closely related to sustainable development because it ends with structural violence. Negative peace, on the other hand, concerns geopolitics and *realpolitik* and refers solely to the absence of inter-state wars or internal armed conflicts.

crisis arising from anthropogenic causes, and other negative manifestations outlined in the previous chapter. Therefore, if those processes were to be described in any way, they would have to be called 'bad development' or unsustainable development.

Indeed, the main social problems brought about by the economic development promoted by hegemonic globalization (Santos 2009) are derived from growing inequality both between and within countries. In rich countries reformed capitalism has allowed the formation of societies in which the majority of the population is middle class. By contrast, in the countries of the Global South the increase in poverty has caused four phenomena of global magnitude: (1) the so-called population explosion; (2) the increase in international migration from the Global South to the Global North, or, more accurately, from poor to rich countries; (3) the increase in armed conflicts of all kinds, including terrorism and transnational organized crime; (4) climate change resulting from global warming, which in turn is the result of the increase in greenhouse gas (GHG) emissions caused by the exaggerated use of fossil fuels and in general by industrialization, whose mark on the Earth's surface has already, according to Paul Crutzen, placed the planet in the new Anthropocene geological epoch whose characteristics were analysed in the previous chapter.

These four negative consequences of hegemonic globalization can be opposed by a globalized counter-hegemonic social movement (Santos) or by striving for a *Great Transition* (Raskin),[19] but in the meantime a more prosaic task concerns reformist governments: implementing the commitments of the UN 2030 Agenda as an absolute necessity. As stated in a previous chapter, in 2015 all United Nations member states made a commitment to accomplish the *Sustainable Development Goals* (SDGs). To reaffirm their importance, it is timely to remember certain essential theoretical issues through a holistic and comprehensive approach. Thus, returning to planetary boundaries and the demographic issue, we must be aware that despite the fact that in all countries of the world social development policies in the field of health and education have reduced the fertility rate (a matter already discussed in Chap. 3), this is still far from sufficient, especially in Africa, the Arab countries and the poor countries of the Global South, including, of course, those in Latin America.

---

[19] Paul Raskin and the Global Scenario Group (GSG) are convinced that a new paradigm and a great planetary transition to a different model of development based on what indigenous peoples have called *sumak kawsay* or *buen vivir* (good living) is possible, but in that case it will be necessary to go beyond policy reform scenarios such as the SDGs and the UN 2030 Agenda. The GSG argues that: "Policy Reform is the realm of necessity – it seeks to minimize environmental and social disruption, while the quality of life remains unexamined. The new sustainability paradigm transcends reform to ask anew the question that Socrates posed long ago: how shall we live? This is the Great Transitions path, the realm of desirability. The new paradigm would revise the concept of progress. Much of human history was dominated by the struggle for survival under harsh and meagre conditions. Only in the long journey from tool-making to modern technology did human want gradually give way to plenty. Progress meant solving the economic problem of scarcity. Now that problem has been – or rather, could be – solved. The precondition for a new paradigm is the historic possibility of a post-scarcity world where all enjoy a decent standard of living. On that foundation, the quest for material things can abate. The vision of a better life can turn to non-material dimensions of fulfilment – the quality of life, the quality of human solidarity and the quality of the Earth. With Keynes (1972), we can dream of a time when 'we shall once more value ends above means and prefer the good to the useful'" (Raskin 2002: 41).

Hence, in order to reduce the fertility rate to an average of two children or fewer (the rate that currently characterizes the developed countries and the former Communist Bloc, including China) and avoid the vertiginous jumps that saw the population leap to a billion at the beginning of the nineteenth century, 3 billion half a century ago (1959) and more than 7.8 billion at the time of writing, we must continue to enhance the aforementioned development policies, with the SDGs in mind and particularly the empowerment of girls and women with the aim of reducing gender inequality.

However, according to the French demographer Todd (2002), the demographic transition (the reduction in fertility rates) and the modernization that development entails also requires a mind shift, which largely explains the social turbulence that exists in countries where this phenomenon has been occurring. In any case, what is undoubtedly true, and what all the demographic figures prove, is that the change in social status (vertical mobility or rise to the middle classes) determines that children are seen as people who should not work during their childhood and school years. Their parents begin to understand that resources should be invested in their education, health and well-being, which, together with the empowerment of women and reproductive health education, contributes to a reduction in fertility rates and a decrease in population growth. Thus, population reduction can be obtained through sustainable development policies, and especially through ending poverty and hunger, improving the health system, as well as through education and promoting lifelong learning opportunities and the empowerment of all women and girls with the aim of achieving gender equality.

Consequently, as the era of globalization is characterized by global interconnectivity, decreasing demographic growth and promoting social development and the vertical mobility of workers undeniably has a direct impact on the migratory phenomenon. The migration of unaccompanied minors and the waves of mass migration in 2018 from Central American countries to the United States could be resolved in the foreseeable future by getting people out of poverty through investment and the creation of employment, as the Mexican President López Obrador has suggested with his 'integral' development plan proposed to the White House in 2019 after the clash of the two governments regarding the implementation of the right of asylum international norms and Trump's threat to impose tariffs on Mexican exports.[20] And undoubtedly the Mexican President was right, because social investment and any integral plan must consider *decent work* as defined by ILO transnational standards, including appropriate remuneration, so that international migratory flows can be reduced. However, these proposals – the importance of which nobody denies – entail public policies designed to stimulate the establishment of internal markets in poor countries, whereas their economies are based on low-salary industries (the so called 'maquilas') and on the export of commodities aimed at external markets, which

---

[20] Trump wanted to obtain from Mexico the "Third Secure Country" status foresee by the UN Convention for Refugees of 1951 but Mexico refused. In July 2019 a weak and submissive Guatemalan President started negotiations with the White House towards the same end but was stopped by the Guatemalan Constitutional Court. Trump wrathfully threatened in a Tweet to impose tariffs on Guatemalan migrant worker remittances, which amount to 11.2% of the country's GDP (US $10,000 billion annually).

contradicts the unofficial – but very real – neoliberal policy ordained by the ruling landlords and oligarchic elites of those countries ('laissez faire/laissez passer'), who encourage production for external markets and the export of low-salary workers to industrialized countries in order to keep the economies of these 'poor' countries afloat with their monetary remittances.[21]

As for organized transnational crime and drug trafficking, they are undoubtedly nourished by the prohibition of the free movement of both people (which causes people trafficking) and substances (which leads to drug trafficking), which is why the reform of both the drugs and the migration policies in the multilateral UN arena could be a way to solve this kind of global scourge.[22] Therefore, a new regulatory policy multilaterally negotiated in the United Nations (such as the Global Compact for International Migration signed in Morocco in 2018), addressing both the migratory flows and the transit and consumption of prohibited substances, could contribute decisively to the solution of these global plagues of transnational organized crime nourished by prohibition policies.

Concerning climate change, the increase in terrestrial temperatures and extreme weather events does not only harm harvests and diminish food security but also constitutes a source of migratory flows.[23] Therefore, the SDGs oblige all countries to use oceans, seas and marine resources sustainably, to protect, restore and promote the sustainable use of terrestrial ecosystems, to adopt effective sustainable forest management, combat desertification, halt and reverse land degradation, and curb the loss of biological diversity (objectives 13, 14 and 15). Such measures constitute, more

---

[21] For instance, in the case of Guatemala, the US $10,000 billion of remittances sent annually by migrant workers in the US are not only more than Guatemala's GDP of 10%, but also exceed the 9.2% fiscal charge of 2018, which is so low because rich people hire lawyers who are experts in tax evasion. Thanks to the control that they exert on parliaments through corruption, it is extremely difficult for governments to make tax reforms.

[22] Ultimately, the whole problem of illegal drug trafficking is related to the undue intervention of states in the sphere of personal freedom when deciding to prohibit the consumption of certain substances (narcotics) because they are harmful to health on account of their addictive effects. The reason why other substances with pernicious and addictive effects (from alcohol to tobacco and even salt and sugar since diabetes and high blood pressure are endemic diseases) are not equally forbidden has more to do with the interests of global agribusiness than human health. Furthermore, if everything that has harmful effects on social welfare and human health were to be subject to prohibition, the production, sale and 'consumption' of small weapons rather than narcotics should be in first place, because although drugs do harm individual health, they are not designed to be used to hurt or kill other people.

[23] There is a general consensus that the illegal US invasion of Iraq in 2003 that led to the dismantling of Saddam Hussein's army (composed mainly of Sunni officers and troops) contributed decisively to the emergence of terrorism in that country, especially when the Shia majority won the elections and installed a Shia government in Baghdad. It also seems evident that the engagement of certain Western countries in policies addressed at overthrowing the Syrian dictator Bashar el Assad has worsened the situation rather than improving it – as happened in Libya when Gaddafi was overthrown. Westerners also tend to forget the vulnerability of the entire Middle East region to climate change. The historic region of Mesopotamia, corresponding to large areas of modern-day Iraq, Kuwait, Turkey and Syria, is similar to Egypt because both rely heavily on rivers – the Nile and the two great rivers that descend from Turkish Anatolia, the Tigris and the Euphrates – therefore the increase in droughts and desertification is also a cause of violence and migratory flows.

or less, the most effective way to help curb the migratory flows that are so worrying to industrialized countries. And, of course, in spite of its limitations, the COP 21 Paris Agreement is essential to reduce not only the harmful effects of climate change in the near future but also international migration, as well as enforce the sustainable consumption and production patterns requested by the SDGs and Agenda 2030.

4.9   Jared Diamond: Unsustainable Systems and Sustainability

The reasons why some societies and civilizations withstand adversity, enduring difficult times with resilience and success, whereas others fail and collapse (some completely) can be better understood when applying the holistic analysis of Jeffrey Sachs's previously mentioned four spheres of sustainable development: governance, techno-economic factors, social dynamics and natural ecosystems. The contradiction between unsustainable systems and sustainability is key to understanding the problems faced by societies that have collapsed in the past, as explained by Diamond's (2007) description of the way the Mayas in today's Yucatan and Guatemala, the Norwegian settlers in Greenland between the ninth and fourteenth centuries, the pre-Columbian Anazasi indigenous peoples of the south-west United States, and the Polynesian peoples in South Pacific islands such as Easter Island (Pascua or Rapa-nui), Pitcairn Island, Henderson Island and Mangareva failed and collapsed in a catastrophic manner in the not so distant past.

Diamond's book also refers to contemporary societies, such as the US state of Montana, the Dominican Republic in the Caribbean, Haiti, Rwanda, Japan, China, Australia, the highlands of Papua New Guinea, and the island of Tikopia in the South Pacific Ocean as interesting cases of sustainability, and consequently as successful examples of adaptation to the environment and endurance over long periods of time.

Nevertheless, as previously stated, the concept of sustainability makes it important to differentiate between circular and linear processes. Indeed, the term 'development' intrinsically implies a *linear* vision of progress, from a lower level to a higher one. Hence this concept cannot be applied in a case like that of Tikopia with 3,000 years of sustainability. Similarly, the idea of *conservation* (not 'development') is what characterizes the way that the forested areas of countries like Papua New Guinea, the Dominican Republic, Japan and Germany are preserved. All these constitute good examples of circular process, where natural ecosystems perform in harmony with social dynamics and economy thanks to good governance.

The resilience of human societies, including their historical sustainability (or failure), always depends on multiple factors, such as environmental deterioration (overgrazing, erosion, deforestation, industrialization, mining), bad governance, violence and war, cultural or religious causes, and so on. For instance, climate change, such as the sudden cooling of the Earth's climate in Greenland or the droughts in Yucatan and the North American South-West, played a fundamental role in the collapse of those societies, but social factors like conflicts with hostile neighbours or religious beliefs (the construction of temples, the use of ornaments and the enhancement of places of worship) prevailing over social needs must also be taken into account. Similarly, having friendly neighbours with whom you can trade and cooperate is another important factor in the success and stability of societies. In other words, it is always a holistic and complex set of factors and variables that explains the failure or achievements of a civilization and culture.

The widely varying examples of both current and past societies presented by Diamond range from the state of Montana in the United States, where the environmental deterioration caused by mining and forestry companies led to such activities being regulated and substantially modified in such a way that the economy is now mainly orientated towards attracting wealthy investors who wish to acquire luxury residences, enticed by the mountain landscape and sports for rich people (such as golf and trout fishing in lakes and rivers), to the prevailing situations in societies such as the small island of Tikopia in the South Pacific Ocean, the Dominican Republic, New Guinea and Japan. Even more dramatic cases are included, involving the collapse of entire human groups that disappeared collectively in a mysterious and brutal way: Norwegian settlers in Greenland between the ninth and fourteenth centuries; the classic Mayan civilization in the Yucatan peninsula and the Guatemalan Petén between the sixth and twelfth centuries; the Polynesian colonizers of the islands of Easter, Henderson and Pitcairn in the Pacific Ocean (ninth century to the sixteenth century); and the Anasazi natives of the North American South West (seventh to twelfth centuries). The demise of all these communities is attributable in part to the mismanagement of their relationship with the environment, internal violence and the inability to resolve conflicts within the group, as well as socio-political and cultural factors, which played a fundamental role in the case of the Norwegians in Greenland, the Polynesians in Easter Island, and the Mayans in Petén and Yucatan.

Therefore the fifth set of factors, which concerns the responses of political power and society to environmental problems, is of great relevance, as it involves the cultural aspects that often determine the short-term behaviour of chiefs, priests and nobility, or, in modern times, executives of transnational corporations, democratically elected rulers, dictators, and those belonging to oligarchic groups.

In any case, it seems important to Diamond to underline the fact that there are parallels between different cultures and historical epochs. In Mayan times the use of wood for the ornamental coating (stucco) and construction of the temples which constituted symbols of the rulers' power was more important than the care and conservation of the forest. This prioritization led to the issue of properly maintaining irrigated food crops, such as corn, due to the resurgence of periodic droughts resulting from forest depredation, which caused a water shortage of such magnitude that it led to innumerable conflicts between rival cities, given the importance of control over water resources. Diamond specifically refers in his book to the Mayan city of Copan and the way in which inappropriate management of corn fields on deforested hillsides surrounding the city led to the aggravation of soil erosion as well as more frequent droughts. In addition, the constant wars resulting from conflicts with neighbouring cities (in the case of Copán with the neighbouring city of Quiriguá located in the

current department of Izabal, Guatemala) eventually led to the collapse or abandonment of the city by its population, as well as Mayan cities of the classic period such
as Tikal, Uaxactun, Piedras Negras, Aguateca, Ceibal, Dos Pilas, Yaxha and others
that were located in what is now the Petén department in Guatemala (Fig. 4.1).

As for the Norwegian settlers in Greenland, for cultural reasons they brought
livestock (cattle, sheep and goats) to Greenland, and in order to build stables and
pastures they devastated the natural forest of the island without adapting to the most
appropriate type of food (fish, whale and seal meat, which the Inuit natives ate). As
they were trading with Norway, they were preoccupied with obtaining decorative and
ceremonial objects and special materials (crucifixes of gold and silver, consecrated
wine glasses, bells for churches) and exchanging polar bear skins and ivory walrus
tusks that were hunted beyond the Arctic Circle with great difficulty, so they neglected
to obtain timber which they truly needed, together with metal utensils and iron.

Fig. 4.1 The Great Jaguar pyramid in Tikal, Guatemala. *Source* Eduardo Sacayon (Guatemalan
photographer). Photograph reproduced with photographer's authorization

According to Diamond, many innovations could have helped to avoid the collapse of the whole community and improved the material conditions of the Norwegian settlers in Greenland, such as importing more iron and fewer luxury goods, spending more time exploring places where they could have obtained iron and wood, and adopting the Inuit's customs, especially concerning food (eating fish and seal meat instead of cattle), boat design and hunting techniques. But Diamond thinks that those innovations could have threatened the power, prestige and limited interests of the ruling elite:

> In the closely controlled and interdependent society of Norwegian Greenland, the leaders occupied a position from which they prevented other members of the community from testing this type of innovation. Thus, the structure of the Norwegian Greenland society produced a conflict between the short-term interests of those who held power and the long-term interests of society as a whole. Much of what chiefs and clerics appreciated proved to be ultimately harmful to society. So, the values (cultural, religious) of that society were both the basis of its strength and its weakness. The Norwegians of Greenland did manage to create a unique form of European society and survive for 450 years being the most remote outpost in Europe. We, the current Americans, should not rush to label them as failures, since their society survived in Greenland longer than our English-speaking society has survived so far in North America. At the last moment, however, the leaders discovered that they had no followers. The last right they had preserved for themselves was the privilege of being the last to die of hunger (Diamond 2007: 364–365).

Ultimately, the decisive factor in the collapse of the Norwegian settlement in Greenland (about 5,000 people living on 250 farms) was the inability to adapt to new environmental conditions that in turn originated in factors of a cultural and political dimension (the decisions made by the leaders). As Diamond emphasizes in his book, the Scandinavians could have copied the autochthonous Inuit people (still living in Greenland) by adopting their techniques of making and sailing light canoes (*kayaks*), using harpoons to hunt seals and whales, and eating them, instead of cattle and sheep (which died when the ice caused by climate change covered most of the grasslands and prevented the possibility of grazing). Instead, despite the difficult circumstances, Norwegians refused to modify their eating habits, which included the refusal to eat fish. This has been proven by investigating garbage dumps which contained no evidence of fish bones or food of marine origin (Fig. 4.2).

Something similar happened on Easter Island, where the heads of the ruling clans forced the townspeople to dedicate a large part of their lives to the manufacture of the monumental heads (*mohai*) which were a symbol of their power. These had to be carved from basalt, and then transported with great effort after being placed on wooden platforms that were rolled over logs with ropes made from material from the felled trees. These large heads had to be installed in special ceremonial sites, located on the high parts of the island at a considerable distance from the stone quarries, and all this work was done without regard for the forest depredation that was causing less and less rain, with the consequent pernicious effects on agriculture and human food.

Other factors also contributed to the almost complete collapse of this society. Occupation of the island began in the year 900, but by the sixteenth century there were very few inhabitants left and by 1877 there were just 111. One of the circumstances

Fig. 4.2 The ruins of Hvalsey Church in the former Norse settlement of Greenland (ninth to fourteenth centuries AD), which collapsed entirely due to deforestation, climate change and the inability to adapt culturally to the new conditions of living after the drop of temperatures

which led to the decline of the population was the impossibility of communicating or trading with islands or neighbouring lands (the Chilean coast is 3,700 km away, and the closest islands – Pitcairn and Henderson, whose human settlements also collapsed, leaving them completely uninhabited – are 2,100 km away). With no active volcanoes nearby, and the island's remoteness preventing it from receiving volcanic ash transported by the wind to fertilize the land, it did not have karst soils, and it had very little fresh water provided by springs and wells (with a rainfall of only 1,200 mm per year). Political and cultural factors also played an important role in the decline and collapse of Easter Island's society, as Diamond says:

> The isolation of Easter makes it the clearest example of a society that destroyed itself by exploiting its resources … after the collapse of Easter (there are) only two main sets of factors: the environmental impact of the human being, especially the deforestation and the elimination of bird populations, and the political, social and religious factors behind these impacts, such as the impossibility of migration as an escape valve due to Easter's isolation, its dedication to the construction of statues for the reasons analysed and the competition between clans and chiefs that drove the increasing construction of statues, which in turn required more wood, more ropes and more food. The isolation of the Easter Islanders surely explains why it seemed to me that their collapse, more than that of any other pre-industrial society, obsesses my readers and students. The parallels between Easter Island and the modern world as a whole are chillingly obvious. Thanks to globalization and international trade, to air flights and to the Internet, today all the countries of the Earth share resources and affect each other,

just as the dozen Easter clans did. The Polynesian Easter Island was as isolated in the Pacific Ocean as Earth is today in space. When the inhabitants of Easter Island found themselves in difficulties, there was nowhere they could flee to or that they could turn to for help, nor can we, the modern Earthlings, have recourse to any other place if our problems become acute. Those are the reasons why people see in the collapse of the Easter Island society a metaphor, the worst possible scenario, of what the future may be giving us (Diamond 2007: 164–165).

The above quote reminded me of Christopher Nolan's splendid film *Interstellar*, a film as high-quality as Stanley Kubrick's classic film *2001: A Space Odyssey*, in which a group of scientists from a Planet Earth on the brink of collapse for ecological reasons in the not-so-distant future are faced with the need to abandon it. Technology based on quantum physics and the Theory of Relativity was used to build a spacecraft capable of transporting a group of pioneer astronauts to several planets located in a solar system beyond ours (which, as we know, does not have any planets which are capable of housing human life, apart from the Earth). However, an interstellar journey of that nature is currently impossible, given the inability of existing technology to exceed the speed of light. Nevertheless, science fiction aside, technology (which the inhabitants of Easter Island did not have at their disposal) could be the lifeline for a humanity (and planet) on the verge of collapse, but for that to work, it is necessary to heed what science tells us, particularly the sciences that study the environment, and not to ignore what science teaches (Fig. 4.3).

However, so far, the evidence is not enough to inspire great optimism about what the future holds, not only because in 2016 a relative majority of voters in the most powerful country in the world elected a character who ignores science (though his successor does not), nor because of the disturbing findings of Pope Francis in his

Fig. 4.3 Mohai on Easter Island (Rapa Nui, Chile), a land that was completely deforested by the inhabitants of the island in order to build these symbolic statues of power

Encyclical *Laudato Si'*, but because mainstream economics (to use Göpel's term for the dominant economic paradigm) has led the world economy and globalization to acquire cultural and political influence analogous to that of the elite inhabitants of Easter Island, the Norwegian settlers of Greenland, and the Mayan priests of that civilization's Classic Period. Today, most of the great magnates and oligarchs of the planet – just 1% of the world's population (the rentiers, as Piketty calls them) – maintain sumptuary consumption habits of scandalous proportions, just like their employees (the super-paid executives or CEOs of large transnational corporations), with the difference that instead of building pyramids, exchanging goods obtained with great difficulty for ornamental and religious articles or raising large heads as a symbol of power, they build gigantic mansions on private islands and waste their income on luxury goods.

Even worse, the capitalist system itself operates in such a way that the purpose of production (the well-being of people) is no longer the value that guides it. It is all for the sake of a productivity or competitiveness that does not seek to fulfil social needs, but is concerned exclusively with increasing profits and the accumulation of capital. What is of interest to transnational corporations is that the technocrats at their service should remain committed to the constant renewal of products, from mobile phones to computers, software, and electronic devices of all types, including automobiles. The commitment to this is required to be so great that concern for the environmental impact disappears, among other reasons because they do not worry about the damage that industrial waste causes to the environment. The interests of big capital are always very short term: to retain the decision-making power within companies, to stay competitive, and to accumulate capital to satisfy the shareholders, who, in turn, usually waste their sumptuous income on sumptuary consumption without caring about the sustainability of the 'model' in the long term, and still less about its long-term sustainability, i.e. about the conservation of the environment and terrestrial ecosystems in the framework of human needs, understood according to the matrix of Neef, Elizalde and Hopenhayn reproduced earlier in this chapter.

The most concerning aspect is that this type of attitude and behaviour (along with the consumerist ideology that justifies it), is rooted not only in the rich classes but also in the middle classes – including the middle classes of developing countries across the world – something largely due to the demonstration effect that is transmitted by advertising and the media in general. Diamond cites, as a paradigmatic case of this 'contagion' of consumerism social pathology, the custom of discarding in the Dominican Republic. This example can be applied to most middle-class sectors that have been emerging in the whole world, as it demonstrates that failure to take the appropriate measures could lead to an apocalypse that will not take the form of a devastating planetary earthquake, a new ice age or a gigantic tsunami like the catastrophes in movies, but will simply be that "we will all die buried by garbage", as a Dominican citizen told the distinguished Californian professor.[24]

---

[24] Although Diamond refers positively to the Dominican Republic's policy of forest conservation due to the establishment of ecological reserves and national parks (owing to environmental figures and organizations that had the support of the dictator Joaquin Balaguer), it is interesting to quote what

However, in the same book Diamond offers some glimmers of hope. In the US state of Montana, despite the fact that mining companies were forced to retreat and that both agriculture and livestock have decreased due to land being repurposed for recreational activities for wealthy people who do not reside in the state, the important point is that environmental protection now takes precedence over mining, agricultural and industrial activities. Furthermore, in a notable example of 'social responsibility', the transnational company Stillwater Mining, owner of a platinum and palladium mine, came to an understanding with environmental groups (including a trout protection NGO named *Trout Unlimited*), and also reached agreements with the local community regarding employment, education, electricity and citizen services in exchange for ecologists and citizens not opposing the operations of the mining company. A similar scenario occurred when the transnational corporation BP (*British Petroleum*) bought a large copper mine formerly owned by Anaconda, and rejected the previous mine-owner's traditional policies of refusing to clean up pollution, financing local support groups favourable to its interests, declaring bankruptcy and other similar measures. Instead, BP made a commitment to carry out clean-up work without shirking its responsibilities towards the environment.

In other meaningful cases, which are significant because they involve transnational corporations, Diamond cites the policies of companies such as the Chevron oil company in Papua New Guinea in the Kutubu oilfield that functions as a *de facto* national park. He further cites Chevron's understanding with the Norwegian government in the North Sea; Rio Tinto in California; the American Dupont in titanium farms in the rutile-rich sands of Australian beaches; the proposed mining and sustainable development project in the United States (MMSD: *Mining Minerals and Sustainable Development*); and the experiences in the field of logging and the timber industry with the forest certification system established by the *Forest Stewardship Council* (FSC), to which companies such as Home Depot, Columbia Forest Products, the famous IKEA of Sweden, B & K of Great Britain, Anderson Corporation, Collins Pine, Lowe's, Svea Skog and many others belong. This eco-labelling

---

he wrote about the pernicious effects of consumerism (the human impact per capita or individual ecological footprint) on the generation of waste pollution (garbage), because the same phenomenon occurs in the rest of Latin America. Naturally, this includes Guatemala and the Central American countries: "More serious than the demographic growth of the country is the rapid increase in human impact per capita (with this concept … I mean the average consumption of resources and waste production per person … The overall impact of a society is equal to the per capita impact multiplied by the number of inhabitants). The overseas trips of the Dominicans, the visits made to the country by tourists and television, make the population fully aware of the higher standard of living in Puerto Rico and the United States. Everywhere there are billboards advertising consumer goods, and at all the important junctions in the cities I saw street vendors selling mobile phones and compact discs. The country is giving itself more and more to a consumption that at present is not supported by the economy or the resources of the Dominican Republic itself, and that depends in part on the income that Dominicans who work abroad send to their homes. All those people who buy huge quantities of consumer goods are, of course, generating huge amounts of waste that collapse the municipal waste disposal networks. You can see how garbage accumulates in streams, along roads, on city streets and in the countryside. A Dominican told me: 'Here the apocalypse will not take the form of an earthquake or a hurricane, but the one that the world has been buried by garbage'" (Diamond 2007: 459).

experiment (based on certificates) has also been adopted by the fishing industry under a similar name, the *Marine Stewardship Council* (MSC), in which a range of companies from the *World Wildlife Fund* to companies such as Unilever, Marks & Spencer, Safeway, Young's Bluecrest Seafood Co., Sainsbury's, Safeway, Whole Foods, Migros, France Foods and several others are included. This is all based on the idea that while a company can maximize its benefits in the short term, it is in its long-term interest to adopt corporate social responsibility policies, not just to avoid causing harm to the environment and the population, and to ensure the adequate conservation of the forests and the fishing resources that provide them with their economic activity, but also to project a responsible image for the purposes of their commercial relations throughout the world.

Alongside the protection and forest conservation policies of the Scandinavian and European countries in general, among which Germany stands out, and also the conservationist policies that were put in place in countries such as Japan (since the Tokugawa dynasty in the sixteenth century),[25] some developing countries should also be highlighted. The Dominican Republic is a particularly good example of a country which has embraced sustainable forest development. A third of its territory has been declared protected areas and there are seventy-four national parks. This occurred as a result of the happy coincidence between the interests of environmental groups and personalities and the support they received from Joaquín Balaguer, a dominant figure in Dominican history during the last four decades of the twentieth century. In contrast, unipersonal dictatorships, structural violence, and the disastrous policy of forest depredation in Haiti, the country with which the Dominican Republic shares the island, led the Haitians into a situation of conflict and violence that forced the United Nations to intervene in order to avoid the total collapse of the society and its State.

Another positive example worth mentioning is the case of the tiny island Tikopia in the South Pacific, an area of just 6 km^2 with a population of some 1,200 inhabitants, which has remained that size for almost three millennia. It constitutes a remarkable example of ecological sustainability, among other reasons because all the islanders participate in, and are conscious of, the decisions that have to be made (it is therefore a 'communitarian democracy', as Santos would call it). For the sake of its survival over the past 3,000 years, its inhabitants have had to resolve problems ranging from feeding the community to the growth of the population, their relationship with their environment, and, of course, the organization of human interactions within that social

---

[25] The conservationist policies applied by Japanese rulers to the country's forests and national resources can be compared with the fact that in spite of being the world's third-largest economy (after the US and China), Japan remains a very conservative country with an almost 'zero growth' economy that, from the perspective of reductionist neoliberal 'mainstream economics', could have been regarded as an economic crisis. However, not only is there no such crisis, but Japan's situation demonstrates that economic growth is not really needed to make an economy both sustainable and in harmony with ecosystems. Japan's average rate of economic growth is actually only 1.7%, according to the World Bank, and since 1992 it has been very low (i.e. less than 1%) or between 1% and 2% (except for the year 2000, when it was 2.8%), excluding the years of Wall Street's financial crisis when the country had red numbers (minus 5.4% in 2008 and minus 0.1% in 2009) and very low ones afterwards (an average of less than 1%).

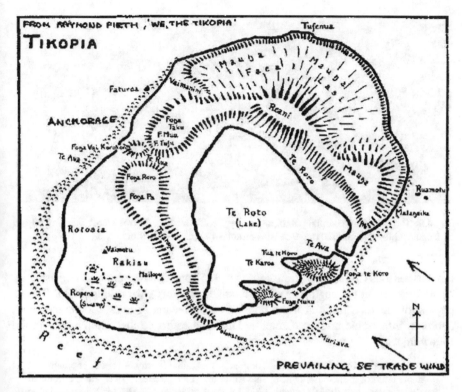

Fig. 4.4 Tikopia. *Source* http://tikopia.co.uk/map-tikopia.html

microcosm. This all had to be achieved in a peaceful way, since, as previously noted, the violence that is generated within a community is often one of the most likely causes of its collapse and disappearance (Figs. 4.4 and 4.5).

All this was adequately resolved by the inhabitants of Tikopia, which is not a minor achievement, given the millenary sustainability of the system and the fact that on other islands of the region it was not possible to organize similar systems, which is why the collapse of communities on the islands of Pitcairn and Henderson was absolute (when the Westerners discovered them they were uninhabited) and on others partial, but devastating all the same (as in Mangareva and Easter Island). Producing food for over a thousand people in a territory of 6 km^2 while preventing the population from growing above a number that it is not possible to maintain autarkically with local resources was, and still is, a formidable achievement. There is no room here to relay all the details of the case (see instead Diamond 2007: 378–387), but it should be noted that, apart from favourable objective factors – high rainfall, moderate latitude, and location in an area where the winds carry volcanic ash to fertilize the soil – the subjective factors are fundamental, as these are relevant to the democratic social organization and involved the procedures which were put in place to control population growth. Although these measures are no longer allowed today for

Fig. 4.5 The island of Tikopia, photographed by Geoff Mackley. *Source* Geoff Mackley https://chrismielost.blogspot.com/2016/04/en-islas-remotas-tikopia-la-isla.html

obvious reasons (the arrival of Europeans in 1857 and the conversion to Christianity of the islanders at the beginning of the twentieth century), for millennia the Tikopia population decided to control the population by means of natural contraceptives, abortion, infanticide and even suicide. Diamond refers to the Tikopia case in the following way:

> The other prerequisite for occupying Tikopia in a sustainable way is for the population to be stable and not to increase. During his visit in the years 1928 and 1929, Firth accounted for the population of the island with 1,278 inhabitants. From 1929 to 1952 the population increased at a rate of 1.4% per year, which is a modest growth rate, certainly lower than that of the generations after the first occupation of Tikopia about three thousand years ago. Most of these seven methods to keep the population constant are no longer practised. The British colonial government of the Solomon Islands prohibited crossings by sea and war, while Christian missions preached against abortion, infanticide and suicide. As a result, the population of Tikopia went from 1,278 inhabitants in 1929 to 1,723 inhabitants in 1952, when two devastating cyclones destroyed half of the Tikopia harvests in a period of thirteen months and caused widespread famine. The British colonial government of the Solomon Islands responded to the most immediate crisis by sending food, and then tackled the long-term problem by allowing or encouraging the inhabitants of Tikopia to alleviate their overpopulation by settling in the less populated Solomon Islands. Currently, the heads of Tikopia limit the number of inhabitants allowed to reside on the island to 1,115 people, a figure that approximates the size of the population that was maintained traditionally through infanticide, suicide and other means which are unacceptable today … Virtually the entire island is managed to produce food continuously and sustainably, rather than through the slash and burn agriculture prevalent in many other Pacific islands. The inhabitants of Tikopia use almost all the vegetable species of the island for one thing or another: even the grass is used as much in the orchards, and the wild trees are used as food sources in times of famine … (Thus) the inhabitants of Tikopia had been practically self-sufficient in their small speck of micro-managed land for three millennia. At present, the inhabitants of Tikopia are divided into four clans, each of which is led by a hereditary chief who concentrates more power than a great non-hereditary man from the highlands of New Guinea. However, the evolution of subsistence in Tikopia is better described by the metaphor of bottom-up management than by top-down management (Diamond 2007: 385–387).

Consequently, aspects of a political nature and governance, such as the organization of the political power of clan leaders, the participation of the community in collective decision-making, and the cooperation to cope with adversity, are as important for the sustainability of the model as aspects of economic order such sustainable agriculture and fish farming or the absence of deforestation, and they complement each other to maintain a mode of production and life that has been successful for millennia. It is particularly fortunate that the colonial presence of Western culture and religion (of the British in this case) has not brought pernicious consequences. The introduction of prohibitions on abortion, suicide and infanticide as a method of controlling population growth was expected, but, as explained in Diamond's book, the formula for emigrating to the Solomon Islands seems to have given good results without causing greater upheavals to the islanders. In any case, Tikopia is an interesting example of a sustainable process, though not precisely of 'sustainable development', since the concept of development has connotations of progress or transit from one stage to another that is supposed to be 'better' or 'superior', whereas Tikopia's economy is circular, in the sense of resilience and endurance rather than progress. Thus, the social dynamics and mode of production, as well as the island's way of governance, have clearly demonstrated its sustainability. In short, Tikopia is a striking example of a durable process with positive results in terms of its socio-political and economic organization through three millennia, which also demonstrates the way a sustainable system functions.

Diamond also mentions other cases, some successful – such as the management of casuarina forests in the New Guinea highlands by local natives, and the *shogun* decisions of the Tokugawa dynasty in sixteenth-century Japan that resulted in an adequate conservationist policy for its forests to date – and others catastrophic – such as the overpopulation in Rwanda (according to Diamond one of the causes of the 1994 genocide), and the introduction of new species (e.g. foxes and rabbits) to an unsuitable habitat, as happened in Australia, when British colonists attempted to reproduce the cultural patterns of their homeland (the fox hunt; rabbits were introduced as a source of lean meat). However, I will not continue to describe such cases in detail, as it is enough to indicate roughly that they are examples of Diamond's main thesis, namely that some societies somehow choose or 'decide' to fail when they establish non-sustainable systems, while others are successful in their relations with ecosystems.

4.10   Summary

This last topic leads to an issue that is worth putting forward as a point of debate because I do not have a definitive position on this matter. Indeed, as indicated in the initial pages, development is a process that involves the transition from a stage C or B that is considered inferior to another A, that is considered superior, which implies a linear vision of progress. While it is easy to see that cyclical processes are not linear but circular,[26] it is evident that in certain cases and specific circumstances, such as that of Tikopia, which has maintained a population of 1,200 inhabitants of the island, it is not a precisely a process of sustainable development but a sustainable cyclical system, because even when the islanders were confronted with changes resulting from the population increase due to the introduction of prohibitions stemming from western colonization (abortion, infanticide and ritual suicide were prohibited), they were resolved through the decision of the British colonial authority to move the surplus population (about 500 people) to other islands in the region, so there was no 'progress' because the number of islanders remained the same. Moreover, it seems evident that the Tikopians have been happy or have had a 'good life' (*buen vivir*), in their own way, throughout their millenary history, and that even today they probably live much better, in their self-sufficient isolation and autarchy, than a good number of Earthlings who travel in cars and purchase what is necessary to survive in shopping malls and supermarkets, let alone the marginalized ones who endure a difficult subsistence in poverty.

The conservation of forests in countries such as Japan, Germany, the Scandinavian countries, Papua New Guinea, southern Chile, and the Dominican Republic can also be seen as the establishment of those sustainable (or sustained) cyclical systems that are the result of the environmentally aware decisions of their respective societies and governments. Such decisions have resulted in the successful management of their relations with the environment in a way that was not catastrophic, unlike the cases of the ancient Norwegian settlers in Greenland, the Mayans in Guatemala, the Easter Islanders and the current-day Haitians – all collapsed societies. It is worth remembering that nature's own conservation systems are circular, not linear. For instance, trees function in a cyclical (circular) way that sustains timber production while the regeneration of the forest occurs naturally. Hence forests in protected areas can be preserved in a rational and appropriate manor which allows some of the timber to be harvested sustainably. Of course, governments have to fight pests, forest fires, illegal predators and other problems of various types, but, again, it is about maintaining the system, not pushing it forward. This means that conservation, not growth, is the criterion by which the performance of the system is judged – its permanence in time – and this is the kind of approach that sustainable development needs.

Could such a system be established at the planetary level? Is conservation compatible with the capitalist mode of production? Are social systems sustainable without economic growth? It is easy to understand the complexity of these questions if

---

[26] For example, natural biogeochemical cycles concern the movement of elements (nitrogen, oxygen, hydrogen, calcium, sodium, sulphur, phosphorus, potassium, carbon etc.) between living beings and the environment (atmosphere, biomass and aquatic systems) through a series of production and decomposition processes. In the biosphere, as a living envelope of the Earth or global ecosystem, matter is limited, so recycling it is a key factor in the maintenance of life on the entire planet, otherwise nutrients would run out and life would disappear. Natural cycles are, therefore, circular phenomena, not progressive or linear. This occurs, for instance, when the carbon dioxide exhaled by animals or produced by decaying organic matter is absorbed by plants, which then produce oxygen by means of photosynthesis, and the cycle repeats, circularly, not progressively.

we consider the fact that at least one twentieth of the world's population lives in poverty, lacks food, decent work, housing, health and educational services, appropriate infrastructure and so on. Therefore, if the Sustainable Development Goals are to be achieved, progress and a certain degree of growth are needed, as clearly stated in SDG 8, but implementing them in a form compatible with the preservation of ecosystems is quite difficult, and maybe only science and technology can help humanity respond to the challenges satisfactorily.

What is undoubtedly evident is that if development is to be sustainable in the long term, we must avoid falling into the trap of unlimited growth, consumerism, the "culture of discarding" and the generation of polluting waste that is making the planet sick with plastics and non-recyclable industrial waste. In addition, we must adhere to the commitments to reduce *greenhouse gas* (GHG) emissions, agreed upon at the COP 21 in Paris, which puts us face to face with the dilemma of how to deal with a mode of production that is fundamentally based on the accumulation of capital, economic growth, individual enrichment and little or no care for the environment. This system is a product of modernity, which means that it is part of the Cartesian and modern delusion that nature is at the service of man and that humankind consequently has the right, regardless of whether this is of divine origin or a rational decision of governments, to exploit the planet and its natural resources to satisfy its appetites and cravings.

In the next chapter I will explore some of the proposed macroeconomic solutions for making capitalism compatible with the social and environmental spheres, such as taxes on big capital (Piketty), a surplus recycling mechanism that will realize the ideas of John Maynard Keynes (Varoufakis), and redistributing wealth so that governments have more resources to invest in health, education, housing and work sources, according to the human needs in each country. At the microeconomic level I will explore some ideas like those implemented in the Kingdom of Bhutan in the Himalayas and some initiatives of the business community, like the concept of 'social responsibility' put into practice by some of the largest transnational corporations, such as Chevron, Shell and Rio Tinto, including some companies that participate in the eco-labelling system (acknowledged by such respectable and suitable councils as the FSC and the MSC) for the timber and fishing industry worldwide, as well as some other experiments, such as the cities in transition and the commons movement described by Göpel (2016).

However, all of the above requires a change to the predominant paradigm in both the economic and the political spheres of the international system because the current crisis is affecting both the Westphalian order and capitalist hegemonic globalization. The narrow conception of sovereignty that has awoken dormant nationalist feelings all over the world must be overcome, as must the policies of 'de-territorializing' practised by transnational corporations, which primarily obey the interests of their shareholders without caring for the needs of their workers, local communities or governments because they can 'raise anchors' at any time and take their investment elsewhere. Hence, transnational corporations must understand that constantly relocating their investments must be regulated by multilateral agreements, or globalization could be stopped in the future by angry neonational movements on both sides of the political spectrum. Exactly the same can be said about the common practice of tax evasion through fiscal paradises all over the world.

In other words, since not all companies have the good will to proceed according to the code of conduct that indicates corporate social responsibility or transnational cosmopolitan citizenship,[27] it is essential for national governments to initiate negotiations at multilateral fora in order to reach agreements that can be implemented as global public policies capable of imposing social responsibility on transnational corporations. Governments should also abandon the diplomacy of connivance and embrace instead a diplomacy centred on the solidarity needed to accomplish the sustainable development goals defined in the UN Agenda 2030.

References

Barié, Cletus Gregor, 2017: "Nuevas Narrativas Constitucionales en Bolivia y Ecuador: El Buen Vivir y los Derechos de la Naturaleza", in: *Política Internacional* (Guatemala City: Academia Diplomática), 3: 48–67.

Brauch, Hans Günter; Oswald Spring, Úrsula; Bennett, Juliet; Serrano Oswald, Serena Eréndira, 2016: *Addressing Global Environmental Challenges from a Peace Ecology Perspective* (Cham: Springer International Publishing).

Cardoso, Fernando Henrique, 1969: *Dependencia y Desarrollo en América Latina* (Mexico City: Siglo XXI Editores S.A.).

Cardoso, Fernando Henrique, 2006: *The Need for Global Governance* (Washington: Library of Congress).

Castoriadis, Cornelius, 1977: *L'Institution Imaginaire de la Société* [*The Imaginary Institution of Society*] (Paris: Seuil).

Delanty, Gerard; Mota, Aurea, 2018: "Governing the Anthropocene: Agency, Governance, Knowledge", in: *Política Internacional* (Guatemala City: Academia Diplomática), 5: 83–110.

Diamond, Jared, 2007: *Colapso: Por Qué Unas Sociedades Perduran y Otras Desaparecen* (Mexico City: Random House Mondadori).

Galtung, Johan, 1981: "Specific Contribution of Irenology to the Study of Violence", in: *Violence and its Causes* (Paris: UNESCO).

Galtung, Johan 2003a: *Paz por Medios Pacíficos: Paz y Conflicto, Desarrollo y Civilización* (Bilbao: Bakeaz).

Galtung, Johan 2003b: *Violencia Cultural* (Gernika-Lumo: Gernika Gogoratuz).

Galtung, Johan, 2004: *Trascender & Transformar: Una Introducción a la Resolución de Conflictos* (Mexico City: M&S Editores).

Gómez, Juan José, 2017: "Paz Sostenible: Un Nuevo Paradigma para el Trabajo de Naciones Unidas", in: *Política Internacional* (Guatemala City: Academia Diplomática), 4: 117–126.

Göpel, Maja, 2016: *The Great Mindshift: How a New Economic Paradigm and Sustainability Transformation Go Hand in Hand* (Cham: Springer International Publishing).

Gudynas, Eduardo, 2017: "Ecología Política de la Naturaleza en las Constituciones de Bolivia y Ecuador", in: *Política Internacional* (Guatemala City: Academia Diplomática), 4 (July–December): 136–143.

---

[27] There is already a 'worldcentric' cosmopolitan consciousness in global civil society but, unfortunately, this type of world citizenry that cares about the planet and has a cosmopolitan ideology is still a minority at world level. In my last chapter I will discuss some of the ideas of authors such as Jürgen Habermas, David Held, Daniele Archibugi, Ulrich Beck, Richard Falk, and Walter Mignolo on the kind of cosmopolitanism that the subtitle of this book advocates.

Harari, Yuval Noah, 2014: *Sapiens: From Animal into Gods: A Brief History of Humankind* (New York, NY: Penguin Random House).

Ladaria, Luis; Cardinal Tukson, Peter; et al., 2018: *Oeconomicae et Pecuniariae Quaestiones: Considerations for an Ethical Discernment on Some Aspects of the Current Economic and Financial System* (Rome: Press Office of the Holy See).

Marx, Karl, 2010: *El Capital: Crítica de la Economía Política: Antología* (Madrid: Alianza Editorial).

Meadows, Donella; Meadows, Dennis; Randers, Jorgen; Behrens, William, 1972: *The Limits of Growth* (New York, Universe Books).

Neef, Manfred Max; Antonio, Elizalde; et al., 1986: *Development on a Human Scale: An Option for the Future*, in: *Development Dialogue* (Uppsala: Dag Hammarskjold Foundation).

Nerfin, Marc; et al., 1977: *Another Development, Approaches and Strategies* (Uppsala: Dag Hammarskjold Foundation).

Padilla, Luis Alberto, 2009: *Paz y Conflicto en el Siglo XXI: Teoría de las Relaciones Internacionales* (Guatemala City: IRIPAZ).

Raskin, Paul; et al., 2002: *Great Transition: The Promise and Lure of the Times Ahead: A Report of the Global Scenario Group (GSG)* (Stockholm: Stockholm Environment Institute; Boston: Tellus Institute).

Rostow, Walt W., 1962: *The Stages of Economic Growth* (Mexico City: Fondo de Cultura Económica [FCE]).

Searle, John R., 1997: *La Construcción de la Realidad Social* (Madrid: Ediciones Paidós Ibérica).

Sousa Santos, Boaventura, 2009: *Una Epistemología del Sur: La Reinvención del Conocimiento y la Emancipación Social* (Mexico City: Siglo XXI Editores; Buenos Aires: CLACSO coediciones).

Sousa Santos, Boaventura, 2010: *Refundación del Estado en América Latina: Perspectivas desde una Epistemología del Sur* (Bogotá: Universidad de los Andes; Mexico City: Siglo XXI Editores).

World Commission on Environment and Development, 1987: *Our Common Future* (New York: Oxford University Press).

Internet Links

Esposito, Mark; Tse, Terence; Soufani, Khaled: *Introducing a Circular Economy: New Thinking with New Managerial and Policy Implications*, 13 March 2018; at: https://doi.org/10.1177/000 8125618764691.

European Union: *Circular Economy Action Plan: The European Green Deal*; at: https://ec.europa.eu/environment/circular-economy/.

Göpel, Maja: "Resilience, Community Action and Societal Transformation", in: Henfrey, Thomas; Maschkowski, Gesa; Penha Lopez, Gil (Eds.): *People, Place, Practice, Power, Politics & Possibility in Transition* (21 June 2017): 120; at: https://www.amazon.com/Resilience-Community-Action-Societal-Transformation/dp/1856232972.

Gudynas, Eduardo 2011: *Ecología Política de la Naturaleza en las Constituciones de Bolivia y Ecuador*.

Pope Paul VI, 1967: *Encyclical Letter Populorum Progressio*; at: http://w2.vatican.va/content/paul-vi/en/encyclicals/documents/hf_p-vi_enc_26031967_populorum.html.

Chapter 5
Alternative Paths Towards Post-capitalism or a Renewed Democratic Socialism

Every age generates a constellation of values coherent with its social arrangements. The modernist ethos once rose in concert with incipient exigencies but has now become out of sync with twenty-first-century realities. Modernity's canon of perpetual progress gains little purchase in a time of thwarted expectations and existential apprehension. An international order based on the Westphalian model of inviolable state sovereignty clashes with global interdependence and the very idea of Earthland. The destabilization of the biosphere debunks the idolatry of markets, the myth of perpetual economic growth, and the fetish of consumerism. Corrosive inequality and hollowed-out communities sap allegiance to dog-eat-dog capitalism.
Raskin (2002: 25)

The first part of sustainable development – the analytical part – is to understand the interconnections of the economy, society, environment and politics. The second part of our sustainable development agenda is to do something about the challenge we face to implement the SDGs and achieve them. Our global goal should be to find a global route, made up of local and national routes, in which the world promotes inclusive and sustainable economic development, which combines economic, social and environmental objectives. This can only be achieved if a quarter of the target – the good governance of both governments and the corporate sector – is also achieved.
Sachs (2015: 13)

In the decades to come, old and alternative paradigms will struggle to fill the shoes of what could become the Second Enlightenment. Our task is to fill the reservoir of social and cultural inventions with ideas, norms, principles, and values that sustain a free perspective of mercantilism towards human needs, nature and money, based on the natural and social sciences of the 21st century which include many non-quantifiable variables. These will provide an alternative meaning, legitimacy, and options of practice for all those involved in the intense political struggles for the transformation of sustainable development. This is what The Great Change of Mind means.
Göpel (2016: 39)

© Springer Nature Switzerland AG 2021
L.-A. Padilla, *Sustainable Development in the Anthropocene*,
The Anthropocene: Politik—Economics—Society—Science 29,
https://doi.org/10.1007/978-3-030-80399-5_5

> *Capitalism will never be subverted; it is not made for that.*
> *Capitalism will be sucked down, so to speak, by the alternatives*
> *that will appear in all parts of the world. And maybe, because*
> *there is not enough planet for capitalism.*
> Latour (2015: 218)
> *As soon as the technological change of exponential progression*
> *is extended (by the cascade effect), from silicon chips to food,*
> *clothing, transport systems and health, the cost of the*
> *production by the labour force will decrease dramatically. At*
> *that time, the economic problem that has defined human history*
> *to this day will become minuscule or will simply disappear.*
> *Those that will probably catch our attention by then will be*
> *problems related to sustainability in the economy and, beyond*
> *this, those with the competition between the different models of*
> *human life (or post-capitalism).*
> Mason (2016: 165)

5.1   Introduction

In the foreseeable future wild capitalism (neoliberalism) will be over because the unlimited growth and the monstrous inequalities that this economic model produces across the whole world is absolutely incompatible with planetary boundaries and the carrying capacity of the planet we inhabit. That is the main lesson of the Anthropocene. So what kind of model will replace neoliberalism? In this chapter I summarise some alternatives formulated by social scientists and movements, and refer to the Kingdom of Bhutan, which happens to be a real-life example, not based on any model but on the philosophical principles of Buddhism, which informs social practice, thinking and customs in that country. From my perspective, the name of the model to be established is unimportant; what matters is that neoliberalism must be terminated, otherwise humanity risks extinction. On the other hand, as seen on previous pages, IR theory has several dimensions in accordance with the different components of the subsystems that are part of the international system. The economic subsystem is a fundamental component of them, since the production of goods, services and trade is an indispensable part of satisfying human needs.

However, the modern capitalist system brought about the idea that the main purpose of an economy is personal enrichment and the accumulation of capital. Thus, according to this ideology, the owners of any business must prioritize growth and gain over the satisfaction of human needs. The well-being of people and the conservation of ecosystems are put in second and third places after the objectives of production and economic activity. The result of this mode of thinking is not just the alienation of the working class and the divorce between human beings and nature, but the Earth's reaction against *Homo sapiens*. Consequently, in the future capitalism will be "sucked down" by "the alternatives that will appear in all parts of the world", as Latour argues (2015: 218), because the political order is always interwoven with the natural order. The current pandemic demonstrates how the economic system has played a major role in producing this 'Revenge of *Gaia*', as Lovelock (2007) argues, or 'punishment of *Medea*', as Ward (2009) would say. Even worse, because nature and culture are interconnected (Chardin 1955), capitalism is to blame for the outbreak of coronavirus, consequently:

> In a global economy that increasingly intervenes in the ecosystem, it is not surprising that new viruses emerge and then migrate from one side of the world to the other at lightning speed. There is nothing at all 'natural' about that. The speed at which the virus is spreading is driven by economic globalization, and the asymmetry by which it does this is guided by socio-economic inequality. This also applies to the way viruses such as COVID-19 enter our society. The first thing to understand, as head of the Global Virome Project Dennis Caroll recently suggested, is that when it comes to these kind of viruses, "Whatever future threats we're going to face already exist; they are currently circulating in wildlife." In recent years, we have seen the demand for wildlife on the food market rise. With this marketization of wildlife, we are getting in closer contact with millennia-old ecosystems that were previously closed off from interaction with humans. On the basis of that argument, biologist Rob Wallace, author of *Big Farms Make Big Flu*, concludes in a recent interview that by way of deforestation and through the marketization of wildlife "Many of those new pathogens previously held in check by long-evolved forest ecologies are being sprung free, threatening the whole world." It is this catastrophic entanglement of global capitalism and eco-colonialism that has brought us both the urgent epidemiological threat of the coronavirus and global climate destruction (Ieven/Overwijk 2020: 1–6).

Another important consequence of the pandemic is that the virus must not be regarded as an external force because it is too simplistic to see it as a sort of 'intruder' threatening humanity from the 'outside' (as Trump did when finger-pointing at China and the WHO) when, in fact, it is the result of the marketization of wildlife and a global economy that is increasingly intervening in the ecosystems, as Ieven and Owerwijk argue. The collective self-isolation and social distancing that people all over the world are, according to Ieven/Overwijk (2020), "willingly adhering to" (though, in many cases, it would be more accurate to say "grudgingly complying with") have, according to them, "all the formal characteristics of a general strike" and consequently there is a "shimmer of utopia" in these confinement policies. Compared with the 1968 Paris riot, when students demanded the abolition of wage slavery (but instead precipitated a reduction in stable jobs and an increase in precarity), the prevailing neoliberalism continues to exclude workers from that kind of 'utopian shimmer' in such a way that workers must suffer either income loss or the danger of infection. However, this economic interruption that allegedly looks like a "general strike" paradoxically coincides with the kind of public policies that climate change militants have been demanding for a long time: an end, for instance, to mass tourism, mining and extractivism in order to stop global warming, and the degrowth of the economy. Thus, as work nowadays has become a matter of life and death, it is possible to understand why labour is both utterly political and ecological. Therefore, extending this logic to the wider threat of ecological collapse could be the turning point to break with:

[B]usiness as usual and the neoliberal politics of containment. We need an outbreak of politics. The corona pandemic is an immense human and social tragedy; but in spite of, or maybe because of this, it should also be a real turning point. We should start by thinking differently about the relationship between society and the ecosystem. We simply can no longer afford to see that ecosystem as a pure 'outside' that stands over and against society; an 'elsewhere' that we can endlessly exploit, expropriate and exhaust. The coronavirus outbreak requires a different attitude. As Bruno Latour made clear, our ecosystem is, in all its complexity, a political actor that is as much a part of our society as the average citizen. Culture and nature are not opposed to each other, but irrevocably intertwined. Just as the coronavirus is part of society, the way we organize our communal life and the way we work is inextricably linked to a much more comprehensive ecosystem. This is the great lesson to be learned from this ecological catastrophe. So if we want to escape neoliberal containment, we must use our democratic and political power to connect these two themes: our relationship to the ecosystem and the organization of life and work within it. It points to the large-scale transformation of our society often captured in the rallying cry for a Green New Deal. No bail-outs for big business, but a bail-out for the people and the planet. This requires large public investments to make our societies climate neutral, implementing policies of degrowth, and a different social relation to our environment. Naturally, the benefits and burdens must be shared fairly, as we realize an alternative organization of work (Ieven/Overwijk 2020: 1–6).

Thus, there is a view that this crisis must not result in governments again bailing out big business (as happened in 2008 with the *Wall Street* financial crisis) but instead provide a bail-out for the people of the planet, a *basic social income* or participatory socialism, as Yanis Varoufakis and Thomas Piketty are proposing. However, this objective can be reached only if global civil society is able to mobilize social forces in order to establish an ecological socialism centred on life (biocentric), as the Andean indigenous peoples have proposed in the new constitutions of Bolivia and Ecuador based on the rights of Mother Earth (*Pachamama*) and the concept of good living (*buen vivir*) as a substitute for personal gain and the accumulation of capital. In this chapter I will explore some of these possibilities, starting with the paradigmatic change needed to put an end to the structural violence generated by neoliberalism to ensure the *positive peace* which goes beyond the simple absence of war and promotes, if not degrowth as suggested by Ieven and Overwijk, at least sustainable development and ecological socialism ('peace ecology') as a way out of the current coronavirus pandemic crisis.

5.2   The Need for a Change of Paradigm

Capitalism has a starting point that goes back to the sixteenth century, but its global dominance did not become effective until the nineteenth century. This process is described skilfully by Polanyi (1957) in his work on the *Great Transformation* of European culture, when the ideology of corporate profit and individual enrichment was introduced as the driving force of economic activity.

Ideas about individual enrichment and the accumulation of capital create a mentality or mindset whose origins lie in the paradigmatic change that occurred during the modern age, when the scientific revolution took place, putting an end to the medieval world-view of ecclesiastical scholasticism dominated by the Catholic Church. It is well known that this religious vision prevailed for more than a millennium – the Middle Ages – but in the sixteenth century it began to be questioned by Descartes via the philosophical process of methodic doubt, and by the subsequent boom of English empiricism (Hume, Locke) as well as German rationalism (Leibniz, Kant, Hegel). These ideas were somehow transformed into the ideology of economic liberalism (Bentham, Ricardo, Smith), the basis of what Marx called the "capitalist mode of production", which became the dominant paradigm of economic science, i.e. the *mainstream economics* of nowadays, to use the terminology of Maja Göpel, a former head of the Berlin section of the Wuppertal Institute and former Secretary-General of the German *Advisory Council on Global Change* (WBGU, short for Wissenschaftlicher Beirat der Bundesregierung Globale Umweltveränderungen). According to Göpel, it is necessary to initiate a new scientific revolution like those that occurred in the past, as described by thinkers such as Karl Polanyi and Thomas Kuhn. These paradigmatic shifts occur when there are significant changes over time, and there is no doubt that humanity is facing this type of situation during the present time of *Great Transition* alluded to by Paul Raskin and the Global Steering Group (GSG). Among other reasons, this is because a change in paradigm is essential if the United Nations commitment to sustainable development made by all governments of the world is to be fulfilled. Göpel argues:

> The term 'paradigm shift' originates from the philosophy of science and usually references Thomas Kuhn as the original thinker in this context. In his 1962 book *The Structure of Scientific Revolutions* he wanted to describe a change in the thought patterns and basic assumptions with which scientific analyses are addressed. In scientific terms, paradigms comprise assumptions that are epistemological (what can we know?), ontological (what can be said to exist and how do we group it?), and methodological (which guiding framework for solving a problem is suitable?). In the context of world-views many add axiological aspects (which values are adopted?). Depending on how these are defined, one and the same event will be interpreted very differently. Kuhn examined how the standard definitions of these assumptions determine which questions will be raised when assessing a certain issue, how they will be raised, what will be observed and how these results will be interpreted. Usually, competing paradigms hold different assumptions and therefore one and the same event will be analysed differently and proposed solutions to the same problem will vary significantly, depending on assumptions about actor behaviour, processes of development and system characteristics (Göpel 2017: 120).

Therefore, what we know about the economy is mediated by what is thought to be the main purpose of this human activity, namely individual enrichment, which is what really matters for average individuals who want to accumulate capital within the framework of the market economy. The opposite sustainable development paradigm, which would regard the satisfaction of human needs as the purpose of the economy, must take into account the concrete situations of societies that are generally stratified and unequal. Hence work, production and economic exchanges do not occur in any ideal market, and require legal and political regulations to try to achieve an elusive and difficult common good for the majority of the population. Such a paradigm, with a humanistic purpose, does not guide human actions in practice, as it does not correspond to the prevailing ideas, which consist of extreme individualism and selfishness. Thus it is problematic when solutions like the SDGs are proposed, based on a paradigm that is not pursued in practice because decision-makers are under the influence of the ideology of personal enrichment and neoliberalism.

For these reasons, we need to change the economic paradigm, since sustainable development forces us to think about future generations as well as those existing now. In addition, we need to take into account the needs of the entire society and not only those of the classes or sectors that exercise economic power. However, this requires the world-view (or 'cosmovision', as indigenous people call it) to be modified. As long as social and ecological issues are subordinated or reduced to the economic dimension, as is currently the case, there will be neither sustainable development (linear progress forward) nor systemic sustainability (cyclical reproduction or conservation), because reducing sustainability to one of its parameters or pillars leaves it incapacitated and without full support for the issue as a whole. According to Sachs (2015), sustainable development entails social dynamics, techno-economic factors, government, and ecosystems. When only techno-economic factors prevail, the remaining three pillars (socio-cultural, ecological, and government) are overshadowed by the economic dimension that prevails over the rest, triggering an imbalance in sustainability. It is essential, then, to relocate the system on its four pillars and articulate the four dimensions of development – social, natural, economic, and political – so that they work in harmony to meet human needs. This clearly requires a new paradigm that is integral and holistic to replace the partial and reductionist one that has prevailed until now.

However, governments and business elites worldwide do not seem to realize the urgency of this change of paradigm, and have so far been operating as if only minor changes and adjustments are required both in economic thinking and in the functioning of the economy itself. This trend prevails despite the financial crisis which the European Union has suffered since 2008, 'infected' by the infamous sub-prime crisis provoked by Wall Street the same year, which is unquestionably the root of problems in the EU, such as *Brexit*, the rise of right-wing neo-nationalist movements, and the xenophobic anti-immigration trends which have acquired considerable political presence in Hungary, Poland and Slovakia in Eastern Europe and even in the old democracies of Germany, France, Italy, the Netherlands, and Scandinavia. To this list of critical issues must be added the global ecological crisis that is the consequence of a form of growth which does not respect planetary boundaries and which is clearly unsustainable – as Pope Francis has reminded us in the encyclical *Laudato Si'*[1] – but it is also causing a climate change of such magnitude that the survival of the human species may be at risk.

Therefore, if the crisis is rooted in the predominance of the economic sphere over the social, ecological and political spheres, humanity is at the crucial moment to make the type of *mind-shift* (Göpel 2016) required to fulfil at least the global commitments of the United Nations' 2030 Agenda, even if the GSG's *Great Transition* is not yet possible. It is necessary to realize that the sustainable development paradigm has been

---

[1] In my opinion, concepts such as the 'dominant economic paradigm' and 'mainstream economics' are equivalent to the 'technocratic paradigm' used by Pope Francis in his Encyclical and the GSG's 'market forces' paradigm, as all of them allude to the predominance of a reductionist 'economic' mentality that places emphasis on growth, individual gain, consumerism, industrialization, and capital accumulation as the main purpose of development, ignoring its social, ecological, cultural and political dimensions.

operating in an unbalanced and incomplete manner (Sachs 2015), with only a single point of support (the economic pillar), for at least thirty or forty years, depending on whether we take as a starting point the Brundtland Report (1987) or the Club of Rome's (1976) report on zero growth. This is the main reason for establishing an appropriate balance through a change of paradigm.

In order to reorientate development towards sustainable development, it is essential to adopt the aforementioned novel thinking or new mentality, which is open to a *social imaginary*[2] capable of providing the requisite theoretical basis and political space. Even if they are initiated by means of small 'niches' or pioneering activities, solutions which transform the way in which we see things in the medium and long term must be put into practice. Until the sustainable development agenda is transformed through new ways of thinking ('software'), the social and environmental variables will continue to be subordinated to the dominant economic thought (quantification, the ideology of commodification at the service of unlimited growth). Hence, Göpel argues that:

> a transformational sustainable development agenda needs new 'software' that opens up the imaginary and thus political space for radically different development solutions and systems. And I feel we might be at a turning point: the first 40 years of sustainable development agenda left the economic paradigm widely unchallenged. Instead of integrating economic, environmental and social dimensions of development – as mandated by the Brundtland Report defining sustainable development – social and environmental concerns have been inserted into an economic way of seeing and therefore governing the world. As a result, quantification and marketization in the service of endless 'growth' has become the dominant mode of organizing ever more areas of life. Diversified governance solutions have been homogenized to fit in with this paradigm (Göpel 2016: 5).

In other words, if we keep thinking in terms of the subordination of sociocultural and ecological systems to the economy, instead of properly articulating the four dimensions of techno-economic factors, natural ecosystems, social dynamics and

---

[2] The power of a *social imaginary* (and the scientific concept of it) has been studied through the lenses of psychology (Freud and Lacan), political science (Castoriadis), and sociology (Taylor) among others. In the work of Lacan, although the term *imaginary* fundamentally means 'fictive', it is not synonymous with 'fictional' or 'unreal' because imaginary identifications can have very real effects. The term was developed in political science by the French scholar Cornelius Castoriadis in his work on the *imaginary institution of societies*. He asserts that societies, including their laws and institutions, are founded upon an imaginary world-view expressed through various creation myths which explain how the world came to be and how it is sustained. Thus, according to Castoriadis, *imaginaries* are the main source of culture and religion: Ancient Greeks and Romans believed in an imaginary in which the world stems from chaos, while in the imaginary of Muslims, Jews and Christians the world stems from the will of God. Thus, Romans and Greeks constructed a political system where the laws were established in harmony with the will of citizens, while monotheist religions developed a theocratic system according to which people must enforce the will of God. Hence, while not constituting a reality, the social imaginary is nevertheless a social force inasmuch as it constitutes the system of meanings that rules a given social structure. In consequence, these imaginaries can be understood as historical constructs defined by the interactions of subjects in society or by *intersubjectivity*, as Jürgen Habermas calls it. In that sense, the imaginary is not necessarily 'real', as it is an *imagined* concept of a particular social subject. Nonetheless, the *imaginary* can be quite real in the sense that it is precisely this 'imagined reality' that explains the behaviour of individuals and social groups and collectivities (Castoriadis 1975).

government action (as Jeffrey Sachs's diagram reproduced ahead in Fig. 6.1 illustrates), we are not walking in the right direction. Hence the importance of integrating or 'recoupling' the economy with nature and society after the previous 'decoupling' that allowed the bond between nature and the economy to cease to be fundamental. It is necessary to free ourselves from the traditional ideological parameters by giving another purpose to the agenda:

> The radical repurposing agenda could be summarized as recoupling economic processes with human well-being and nature's laws by making the economic dimension the one that needs changing. Given the structural reality of today's path dependencies, the foremost strategy for successive change in this direction – the incremental strategies that can achieve it – is double-decoupling: 1. Decouple the production of goods and services from unsustainable, wasteful or uncaring treatment of humans, nature and animals (do better). 2. Decouple the satisfaction of human needs from the imperative to deliver ever more economic output (do well). The latter has been given much less attention because the world-view informed by the mainstream economic paradigm cannot even countenance it (Göpel 2016: 10).

Consequently, according to Göpel, in order to make economic processes compatible with human welfare and nature's ecosystems and establish a new purpose for economic activity which is radically different from traditional mainstream economics since it will be based on the sustainable development paradigm (*radical repurposing agenda*), we must proceed through *incremental strategies* of change (*double decoupling*) which increase action gradually. Those strategies have a procedure based on two concepts or key ideas: *do (things) better*, which means separating or decoupling the production of goods and services from unsustainable modalities (plastics, fossil fuels, coal) and processes which currently do not care for people or animals and natural realms (and which also do not recycle their products, thereby increasing waste to the detriment of the environment); and *do (things) well*, which points to the necessary decoupling or separation of the satisfaction of human needs from the omnipresent economic imperative of growth, because, among other reasons, it is evident that many needs – such as affection, leisure, creativity and freedom – have absolutely nothing to do with the sphere of the economic, as highlighted in the Neef matrix of human needs reproduced in the previous chapter.

In addition, the dimension of *doing things well* has been the one that has received much less attention since the dominant mindset is strictly about economic growth and individual enrichment and says nothing about concepts such as good living (the *buen vivir* of indigenous people) and even less about satisfying leisure, affection, creativity, freedom, cultural and ethnic/national identity needs. These kinds of values and perspectives are absolutely out of the 'mainstream economics', which "cannot even countenance" them, as Göpel underlines. Another important issue that can be deduced from both concepts and ideas (*do better* and *do well*) is that they could become an instrument or guide to evaluate development processes and determine their sustainability.[3] It is evident, for example, that conserving nature (such as a forest, biodiversity or fishing resources of the ocean) does not mean not using it

---

[3] In many developing countries there are numerous conflicts, and there is a permanent debate between those who support investments in mining sites or hydroelectric plants and the local communities that oppose such projects. Evaluating whether such projects are appropriate (or not) to the

at all, but neither is it acceptable for natural resources to be squandered on luxury consumption, the accumulation of capital, or individual enrichment. In other words, a balance must be found between the *use value* and conservation.

It should be added that Göpel's proposals coincide with and are related to the concepts of *buen vivir* (good living or *sumak kawsay*) and *Pachamama* of the indigenous people explained in the previous chapter. According to Barié, these concepts are of paramount importance since they have been recognized in the new constitutions of Bolivia and Ecuador. Thus, for the German expert, the new constitution of Bolivia is extremely innovative in its acceptance of the idea of good living, which:

> [O]riginally comes from the Aymara expression *suma qamaña*, similar to *sumac kawsay* in Quechua. It combines the idea of material well-being with peaceful social coexistence and mutual support and solidarity. The one who lives well, the *qamiri sum*, is not the rich but the one who shares: The *sum qamiri* happens to be the one who lives and coexists well, because he is welcomed by all and knows how to welcome and collaborate with all that have little or much. In a certain way it cannot be given individually but only in and with a larger social group. Living well then turns out to be a kind of meta-value (to which other more common values must be subordinated, such as equality, inclusion and social equity). Even the educational system and the new economic model must be guided by the principle of good living. The constitutional text also leaves some contextual clues about (the meaning of) living well, especially in Title I on the economic organization of the State, although it does not offer a precise definition: it does not seem to be a rigid and static concept, but something flexible that Bolivians are called to continually search for – "the quest of good living", which is characterized by having multiple dimensions and is linked to improving the quality of life. It also seems to express and contain collective interests, complementary counterbalances to individual interests, particularly in the framework of a plural economy: "The social and community economy will complement the individual interests with (the idea of) living for the collective good" (Barié 2017: 51–52).

The idea that living well – or good living – is not getting rich but sharing, being welcomed by everyone and being aware that living well essentially consists of knowing how to welcome and collaborate with everyone, whether they have little or much, has a lot to do with the satisfaction of non-economic needs, such as affection, identity and participation. This demonstrates the alignment between the perspectives of Western academics like Göpel (doing better, doing well) and those of the indigenous people of the South American highlands. The same can be said of the new Ecuadorian constitution, even though it is distinguished from the Bolivian constitution, according to Eduardo Gudynas, by its biocentric nature (as stated in the previous chapter), as the rights of nature (*Pachamama*) are granted to serve nature itself and not the interests of the State, as is the case for Bolivia. Consequently, for Gudynas:

> The new Ecuadorian Constitution presents a large number of articles which directly or indirectly refer to environmental issues. The basic framework includes a section on 'rights of nature', together with another section on 'living well' rights (including norms for a healthy environment), which are understood as part of the relationships between a regime of development and living well. The terms *nature* and *Pachamama* are presented on the same level, and defined as 'where life is reproduced and carried out' (Article 72). This is a novel formulation. However, it is not less valid to use both the term *Pachamama* and Nature, since the

requirements of sustainable development, from the perspective of the alternative paradigm described here, could help to resolve such contentious points of the debate.

former is anchored in the cosmovision of indigenous peoples, and the latter is typical of the European cultural heritage. Also, concepts such as ecosystem or environment come from western culture, and leave aside the visions of the original peoples (Gudynas 2017: 138).

There are other authors who have referred to the problem resulting from the need for a paradigm shift, or the replacement of dominant thinking in the economic field, such as the Spanish scholar Gonzalo Vitón García for whom:

> In short, many of the logics of free market capitalism are incompatible with a model of sustainable development. Perhaps the solution is to move to a de-growth model, at least of the most developed economies, that would allow a lower global consumption of resources and therefore a recovery of the Earth. What is clear is that if we want to save the planet Earth we need more planning and administration, not only of resources, but of societies in general. Leaving the solutions in the hands of the free market has shown us in recent decades that it does not work when it comes to caring for the Earth. The public … needs to be aware of this fact, not just to control a certain part of the activities of the great powers that seek growth while minimizing economic costs, even if that means maximizing environmental costs, but … so that we can change many of our uses and customs. We are not going to change the global suicidal dynamic with small daily gestures; structural in depth changes are needed. However, small everyday gestures are equally important, because they must be part of a larger change, a change of mentality (Viton-García 2017: 100).

It is interesting to note how people from different parts of the world – Germany, Ecuador, Bolivia and Spain – agree on the need for a change in the dominant economic paradigm. In addition, most experts, including those who have highlighted the constitutional changes in Andean countries with a majority of indigenous population – also concur when referring to the fact that sustainable development requires a shift towards this new mentality or '*cosmovision*', as the indigenous people call it. The following pages will outline the approach of experts in macro-economics who criticize capitalism as a dominant system.[4]

5.3 Alternatives for Overcoming Neoliberal Capitalism

Economic development and growth are the main components of the capitalist economic system, and since the seminal work of Karl Marx everybody knows that capitalism is prone to periodic crises. It is also widely known that the social reform policies introduced by the European social democratic parties at the beginning of the twentieth century, the economic reforms (the *new deal*) launched by the Roosevelt administration in the wake of the major crisis of 1929 and later reinforced by the measures of economic reform proposed by Keynes, and the enormous profits made

---

[4] It is important to mention the fact that neither constitutional amendment has been applied in an appropriate manner. In Ecuador the return to neoliberal economic policies under the rule of President Lenin Moreno provoked violent indigenous protests in 2019 that obliged authorities to repeal their decisions. In Bolivia, in spite of the good results of some important economic reforms, President Evo Morales was forced to resign in 2019 after public protests against his intention to remain in office via election fraud.

by American industrialists thanks to the demand for industrial products generated by the Second World War were the factors that enabled the 1929 crisis to be decisively overcome.

There has been intense debate among experts about both the oil crisis of the 1970s that led Washington to abandon the gold standard during that decade and the Wall Street financial crisis of 2008, which had disastrous repercussions in the European Union. However, the criticisms of scholars like Stiglitz (2005, 2008) and Krugman (2012) refer mainly to the economic policies adopted by the US government rather than the structural root causes of the crisis. This chapter will present the views of authors who study the underlying structural problems of capitalism.

5.3.1 Yanis Varoufakis

According to Varoufakis, all human societies have cemented their success thanks to a double mechanism of the production of surpluses and public acceptance of how they are distributed, which gives legitimacy to the system. In the past – for instance, under the feudal regime – the production of surpluses and their distribution were quite transparent, since the part each individual received depended on a transparent mechanism understood by everybody. This situation began to change when capitalism was implanted and the market extended its reign to the countryside and artisanal workshops. Both land and labour ceased to be inputs and became merchandise whose nature became opaque, as did production processes in the hands of distant capital investors over which workers lacked influence, as they were treated like gears in a vast foreign machine. Varoufakis describes the new social dynamics in the following way:

> The process recalls an underground, almost ironic, conspiracy between the paradox of success and the paradox of prophecy: the creation of growth and wealth requires the use of machinery, the development of new technologies and the intensification of labour productivity. Market societies thrive when commodification, financialization and technological innovation are booming. The more rationalized and mechanized production becomes, the less human contribution to its existence, and the cheaper it will be. But then, the more product is squeezed from a given amount of human creative input, the lower the unit value of the product. If mobile phones and all kinds of gadgets are becoming cheaper, it is because their production is becoming more and more automated, almost without the implication of human work. Hence, profit margins decrease. When they fall below a certain threshold, the first bankruptcies occur. Like light snowflakes at first, in the end their fall triggers an avalanche. Then the *crisis* begins. Once they have society trapped in their iron stocks, the gremlins of the system (the labour and monetary markets) refuse to allow it to escape before humanity has paid a high price in the form of a lost generation. In short, as long as human labour resists total commodification, society can produce value; but only under circumstances that also produce crises, and sometimes also a crisis like the one of 1929 or in fact, the one of 2008 (Varoufakis 2015: 82).

Varoufakis argues that in 1929 the greatest crisis of world capitalism was resolved by the *new deal*, which is equivalent to saying that it was thanks to the Keynesian policies applied by the Roosevelt administration. However, it is also well known that the Second World War played a fundamental role in revitalizing the economy because the war effort required weapons, food, equipment and all kinds of supplies for the fighting troops in Europe and the Pacific. Production and employment were invigorated, and payment was often made in gold, enabling the United States to emerge as the great economic power of the post-war period.

In this context, John Maynard Keynes, head of the British delegation at the Bretton Woods conference, was responsible for the global plan that resolved the crisis by designing an institutional framework to prevent large-scale depressions in future. Keynes was well aware that capitalism could no longer be managed effectively at a national level; rather, it had to be done on a global scale. Thanks to this framework, the Bretton Woods institutions (the IMF and the World Bank) were established. Because the US held most of the world's gold reserves, the US dollar (backed by a fixed exchange rate of $35 per ounce of gold) was made the 'reserve currency', and, in accordance with Keynes's proposal, a *global surplus recycling mechanism* (GSRM) was launched to prevent the accumulation of systematic surpluses in some countries and persistent deficits in others.

Varoufakis regrets that some of Keynes's proposals regarding this institutional and global surplus control mechanism were discarded, even though the dollar was converted into the world's reserve currency. Dollarization facilitated the boom in American exports to a Europe and Japan devastated by war and enabled the US economy to overcome its commercial and budgetary deficits, as well as rise above crises like the financial one unchained by Wall Street in 2008, because the banks were rescued by huge amounts of paper money printed by the Treasury Department. The dollarized world economy explains why the US domestic economy is not affected by trade and budgetary deficits. Thanks to its status as the world's banker, the US economy can manage those deficits – notwithstanding Trump's complaints that they had obliged him to unleash a tariffs war with China – even if this US monetary policy is against all the dogmas of the prevailing neoliberal ideology.

But going back to the post-war period, it is widely known that the Marshall Plan played a decisive role in the economic recovery of Europe (together with the *European Coal and Steel Community* that began European integration thanks to the cooperation of the two former enemies France and Germany), while in the Far East the Korean War helped to restart the Japanese economy. The financial support provided by the United States to aid the European recovery enabled Germany to become an economic giant again. Varoufakis points out that this was remarkable because never before had a victor helped a society it had recently defeated to increase its power in the long term (Varoufakis 2015: 114).

This clever set of Keynesian policies applied by the United States led to the 'golden age' of the 1960s, which lasted until the huge deficits accumulated by the American treasury during the Vietnam War triggered the crisis of the 1970s. That crisis forced the dollar to be delinked from gold, which led to the floating exchange rates than now exist and the collapse of Keynes's original plan at the Bretton Woods conference, which was the global mechanism of surplus regulation to avoid the periodic crises of capitalism. To get out of the crisis of the 1970s, the President of the Federal Reserve,

Paul Volcker, set out to attract capital from all over the world by raising interest rates from 6% in 1971 to 11% in 1979. Under the Reagan administration, interest rates reached 21.5% in the 1980s. The process that ensued is described by Varoufakis in the following manner:

> Thus, began a new phase. Now the United States could mock with impunity a growing trade deficit, while the new Reagan administration could also finance the huge expansion of its defence budget, and defend its gigantic tax cuts for the wealthiest sectors of the US. The ideology of the 1980s of the supply economy, the mythical trickle-down effect, the imprudent tax reductions, the prevalence of greed as a form of virtue, etc. all these things were mere manifestations of the new 'exorbitant privilege' of America: the opportunities to expand its double deficit were practically unlimited and arose courtesy of the influx of capital from the rest of the world. American hegemony had taken a new turn. The reign of the global Minotaur had dawned (Varoufakis 2015: 140).

Incidentally, it should be noted that Varoufakis's findings coincide with the viewpoint of Todd (2002), who believes that American capitalism has been in crisis since the dollar became the world's reserve currency and the Federal Reserve launched its policy of printing bills to cover the enormous US trade and budgetary deficits. Todd's book has prophetic appraisals of what is happening now:

> Economic scholasticism perceives, describes, and invents a perfectly symmetrical ideal world in which each nation occupies an equivalent place and works for the common good. This theory whose seeds were isolated by Smith and Ricardo, is currently cultivated and produced by 80% by the large American universities. It constitutes, together with music and cinema, one of the largest cultural exports of the United States. This degree of adaptation to reality is also of Hollywood type: weak. [The theory] loses its flexibility and remains silent when it comes to explaining the problematic fact that globalization is not organized by a principle of symmetry but of asymmetry. The world produces more and more for the United States to consume. There is no balance between imports and exports. The autonomous and super-productive nation of the immediate post-war period has become the nucleus of a system with a vocation to consume, not to produce. The list of US trade deficits is impressive because it covers all the major countries in the world. We list those of the years 2001: 83 billion deficit with China, 68 with Japan, 60 with the European Union of which 29 with Germany, 13 with Italy and 10 with France; 30 billion deficit with Mexico, 13 with Korea. Even Israel, Russia and Ukraine have surpluses in their exchanges with the US of 4.5, 3.5 and 0.5 billion dollars respectively (Todd 2002: 79–81).

And those are the figures for the year 2001, when Todd was writing his book. In 2016, the deficit figure in the US trade balance was already 502.3, according to Tim Worstall of *Forbes Magazine*.[5] Now it is substantially higher. Todd asserts that if the

---

[5] According to Worstall, "The American trade deficit with the rest of the world rose to $502.3 billion last year, that's in 2016 … The US trade deficit narrowed slightly in December, but the improvement was not enough to keep the deficit for the entire year from rising to the highest level since 2012. The deficit in December fell 3.2% to $44.2 billion, the Commerce Department reported Tuesday. Again, in exports of commercial aircraft, heavy machinery and cars offset to rise in imports. For the whole year, the deficit rose 0.4% to $502.3 billion, the highest annual imbalance since 2012. The US trade gap was the largest since 2012. The last time the country ran a surplus was in the mid-1970s when Gerald Ford was president. The deficit with Mexico, the biggest target of President Donald Trump's wrath, rose 4.2% to $63.2 billion in 2016 to mark a five-year high, according to government data … The economy grew at 3.5% pace in the third quarter".

US commercial deficit were to be correlated with just the industrial sector (not the GDP as a whole), we would arrive at the incredible result that 10% of its industrial production is not covered by the export of products manufactured in the United States. The social repercussions of this problem explain the electoral behaviour of the unemployed Rust Belt workers of the American North East and other old white people in similar circumstances.[6]

This phenomenon occurs, as also noted by Worstall, despite the increase in exports of commercial aviation, heavy machinery and automobiles, which confirms Todd's ideas in the sense that the industrial sector's tendency to contract remains, despite the fact that the US continues to have a relatively good performance in the export of state-of-the-art goods (aeronautics, IT, medical equipment). The world trade indices indicate that China now has a 33% share, while the US share has been reduced to 14%.

In aeronautics alone, in 2003 Airbus produced as many aeroplanes as Boeing, so the speed at which the industrial deficit occurred is impressive, as Todd points out:

> On the eve of the Depression of 1929, 44.5% of world industrial production was in the USA, versus 11.6% in Germany, 9.3% in Great Britain, 7% in France; 4.6% in the USSR, 3.2% in Italy and 2.4% in Japan. Seventy years later, the American industrial product is a little lower than that of the European Union and hardly greater than that of Japan. This fall in economic power is not offset by the activity of American transnational corporations. Since 1998 the profits repatriated to the US are lower than those of foreign firms that have settled in the US and repatriated profits to their respective countries (Todd 2002: 79–81).[7]

In such a situation, it was only a matter of time before the outbreak of the financial crisis that led to the bankruptcy of Lehman Brothers on Wall Street and was triggered in 2008 by sub-prime mortgages. According to Varoufakis, it was Paul Volcker (the same man who managed to get capital around the world invested in Wall Street in the 1970s, thanks to the rise in interest rates) who saw it coming. In an article published in the *Washington Post* in 2005 he pointed out that what was keeping the US economy afloat was the massive influx of capital from the rest of the world ($2 billion a day). Although by that time interest rates had fallen considerably, this was absolutely untenable for banks, which decided to convert these funds into toxic money by lending them without sufficient guarantees in the form of sub-prime mortgages. Hence those financial policies triggered the crisis, which was only stopped when the US federal government intervened with all its economic power to rescue banks and institutions from bankruptcy, as occurred with the insurer AIG and the three major automotive companies of Detroit (GM, Ford and Chrysler).

---

[6] Those unemployed workers have aggravated the problem of narcotics consumption due to the increased use of medicines with opiate content in pain medication, legally prescribed by medical doctors. In other words, the increase of addicts is linked to mental illnesses that in turn are derived from the frustration, pessimism and negative attitude which are side-effects of the kind of unemployment caused by transnational corporations when they move abroad seeking low-wage workers.

[7] This happens because the profits which are not repatriated are deposited in tax havens around the world. According to the statement of Jeffrey Sachs in the high-level ECOSOC segment at the United Nations (held in New York in October 2016), these deposits amount to the incredible sum of US $20 trillion.

Afterwards, the American government, acting in accordance with neoliberal ideology, refused to do what would have been the logical conclusion of its policy of financial rescue with public funds, i.e. put the firms generously rescued with federal funds (taxpayers' money) under state control. Neither did Washington acquire any power over a Wall Street that continued to practise the same policies, as evidenced by the transmission of the crisis to Europe, Japan and other countries. At the same time, a White House that paradoxically demonstrated itself to be powerless against Wall Street, despite the bank rescue and the change from the Bush to the Obama administration, was unable to even consider negotiating the "surplus recycling mechanism" that – according to Varoufakis – could have been a more effective way out of the crisis. So neither the G7 (at that time still G8) nor the newborn G20 knew anything about those alternatives.

However, even though Wall Street maintains its power inside the United States, Varoufakis asserts that it is impossible for it to recover the global economic power it had before the 2008 crisis because the two major deficits of the US economy (budget and trade balance) are still gigantic, and the US is no longer in a position to continue buying mountains of net import goods and balancing them with a similar volume of capital flows, a fact which means that "for the first time since the Second World War, the US has lost its capacity to recycle the surplus of the planet. Without an alternative mechanism to carry out this recycling, the ability of the US (and the world) to recover is severely limited" (Varoufakis 2015: 306). This means that if, in these conditions, the United States wants to continue to be a hegemonic world power in the sphere of capitalism, it should understand that its power must be rejuvenated not by extracting more from its 'subjects', but by: "investing in its capacity to generate surpluses. To extract something from his subjects the hegemonic power must dominate the art of giving them something in return. In order to maintain power, he has to strengthen his surpluses, but for that he must redirect large parts of them among his subordinates" (Varoufakis 2015: 306).

However, if the power behind the throne in the United States continues to be the financial sector, it does not seem likely that the banks will assume the corresponding social responsibility. The former Greek Finance Minister asserts that current capitalism is no longer predominantly industrialized but financialized, and that banks are the real economic power. Thus, according to Varoufakis, bankers have established a sort of 'bankocracy' that rules de facto instead of using democratic procedures, a fact evident in the EU, where the opposition of bankers made it impossible to solve the Greek financial crisis. Varoufakis thinks that the capitalization of Wall Street is now too shallow to attract the tsunami of foreign capital that kept the United States in good shape, so if its banks can no longer recycle the world's surplus alone, it will reach a state of "irreversible degeneration":

> Global capitalism will not regain the lost balance if central banks focus on price stability, and the task of rebalancing the world economy is left to the magical machinations of supply and demand. This is the most threatening error of libertarians. The stability of global capitalism, but also of the regional one, requires a Global Mechanism of Recycling of Surpluses, a mechanism that the markets, however globalized or however free they could be and well that they could work, cannot provide (Varoufakis 2015: 325).

So if banks are the new rulers of the capitalist system and have governmental support as a result of the neoliberal policies launched by Reagan and Thatcher in the 1980s, the collapse of the welfare state is perfectly explicable. Varoufakis argues that the welfare state 'civilized' and 'stabilized' capitalism in those years, so the neoliberal wild capitalism prevailing again in current times must once more be put under state control, especially if the working class is no longer socially insured and their wages stagnated since the introduction of the so-called 'Washington Consensus'. Consequently, to solve the current crisis (aggravated by the pandemic of 2020/21) the neoliberal narrative that maintains that wealth is created by the private sector must be discarded because, in fact, wealth is created by society (taxpayers who finance research and development) and appropriated by the owners of capital. This is demonstrated by simple facts: the internet is the result of research financed by the state but was appropriated by Google or Microsoft, the components of cellular phones and computers are also the result of state-financed research appropriated by Apple, the vaccines developed with public funding are appropriated by pharmaceutical corporations, and so on. Hence, Varoufakis affirms that the best way to stabilize and 'civilize' the system again would be the establishment of a *universal basic income* that could also help to solve the problem of the huge unemployment created by the pandemic crisis and prevent the same problem in the foreseeable future, when joblessness will increase due to automatization and robotization. Thus the need for a social safety net is not just a matter of social justice but of keeping the economy afloat by subsidizing demand.

5.3.2   Thomas Piketty

In his book about capitalism in the twenty-first century, Piketty (2014) deals mainly with the factors that drive the dynamics of capital accumulation and the distribution of wealth. In addition, he explores the evolution, from the eighteenth century to the present, of the phenomenon of inequality, and explains the reasons for the tendency towards the concentration of wealth in more than twenty countries. In his book, which is almost 700 pages long, Piketty shows that although the growth and diffusion of knowledge have helped to avoid inequality on the apocalyptic scale predicted by Marx, it has not been possible to modify the deep structures of capitalism, because the main factor that determines the phenomenon of inequality is the tendency of capital gains to exceed the rate of growth. This has led to the extreme inequalities that now threaten and undermine the European democratic system, and has generated the rise of neo-nationalist movements and parties (Badié/Foucher 2017), confirming the existence of a tension between the political system and the economic system that is endangering the process of integration and the EU itself.

Incidentally, this phenomenon had already been analysed by Jürgen Habermas in the 1980s in his classic study on communicative action, which pointed out that "the capitalist dynamic proper to the economic system can only be preserved to the degree in which the production process is decoupled from orientations towards values of use" (Habermas 1988: 481). Following the thinking of Claus Offe, he added that the tension between capitalism and democracy is the logic of social and economic integration facing each other

> ...with the differentiation and privatization of production and with its socialization and politicization. These two strategies intersect and paralyse each other. As a consequence, the system is constantly confronted with the dilemma of having to abstract and, nevertheless, not be able to do without the normative regulations of the action and the references of sense of the subjects ... This paradox also finds expression in the fact that when the [political] parties get power, they have to maintain the confidence of private investors and at the same time the trust of the masses (Habermas 1988: 488–489).

In this we can clearly see the origin of the dilemma of the European ruling political classes today, in terms of preserving the confidence of investors while maintaining the neoliberal policies of austerity, whereas, according to Habermas, to preserve the trust of *the masses* the same politicians should return to the policies of the welfare state or the social economy of market – the kind of policy which has never excited the almighty Finance Minister of Germany, to say the least.

However, returning to the work of Piketty, it must be emphasized that this approach could be regarded as more social than economic, despite starting from an economic analysis (the tendency of 'capital gains' to exceed the growth rate). This is because the profits of capital go into the pockets of the owners of capital (the small global elite of the super-rich); in other words, this tiny group appropriates a vast quantity of surpluses to the detriment of national economies, exceeding the growth rate and affecting national political and fiscal decisions. Hence its effect has become more social than economic.

Indeed, Piketty argues that the concentration of wealth in these rentier elites of owners of capital has in turn led to an exaggerated increase in social inequalities, a phenomenon that is undermining and threatening democracy itself among the member countries of the European Union. Given that the structure of social inequality is the central problem of twenty-first-century capitalism, Piketty dedicates his book to the study of it in several special chapters which deal with issues such as inequality and the concentration of wealth; the inequality of income at work; inequality in the ownership of capital; merit and inheritance in the long term; and inequality on a global scale.

According to Piketty, the market economy based on private property needs to be regulated by the State, otherwise, without rules and a legal order, very much in Habermas' analytical line, "powerful forces of divergence ... threaten democratic societies and the social justice values on which they are based" (Piketty 2014: 571). Piketty also emphasizes the fact that the main destabilizing force of capitalism is that the private rate of capital gains can be considerably higher than the income growth of wage earners, small and medium entrepreneurs and the rest of society. This means that wealth accumulated in the past grows faster than production and wages. Such inequality therefore expresses "a fundamental logical contradiction", which inevitably leads the owners of capital to become *rentiers* with increasing dominance over those who rely on employment or their workforce. This is how a situation that favours luxury consumption and small market niches of luxury goods and services for these elites is generated, to the detriment of productive investment and employment.

The fact is that once capital works, it reproduces much faster than production, and the consequences in the long term lead to a brutal concentration of wealth so that on a global scale the problem becomes enormous and without simple solutions. During the twentieth century, two World Wars were necessary to reduce the excessive profits of capital, somewhat creating the illusion that the fundamental structural contradiction of capitalism had been overcome. With the current average of profits (4–5%) wars will hopefully not become the central phenomenon of the twenty-first century, as they were during the twentieth century (Piketty 2013: 572). Consequently, for wealth not to continue concentrating excessively, Piketty proposes creating an annual tax on capital in order to stop this "spiral of infinite inequality". The rates suggested by Piketty are in the order of 0.1–0.5% for fortunes under 1 million Euros; 1% for capital of between 1 and 5 million Euros; 2% for the scale of 5–10 million and between 5 and 10% for fortunes of between several hundred and several thousand million Euros:

> This would stop the unlimited growth of inequality in the distribution of global wealth that is expanding at a rate that cannot be sustained in the long term, and that should worry even the most fervent self-regulated market champions. Historical experience also shows that these immense inequalities in [the distribution of] wealth have very little to do with entrepreneurship and are not useful for promoting growth … The difficulty is that this solution, a progressive tax on capital, requires a high level of international cooperation and regional political integration. It is not within the reach of nation-states where social commitments are negotiated. Many worry that moving in the direction of greater cooperation and political integration, as seen within the European Union, only subverts the achievements already made … Although the risk is real, I do not see any genuine alternative: if we are to recover the control over capitalism we have to bet everything on democracy – and in Europe, democracy on a European scale. Larger political communities such as the United States and China may have a broader range of options, but for small European countries, which will soon appear smaller in relation to the global economy, retreating to national borders can only lead to worse frustrations and dislikes than those that already exist. The nation-state is still the appropriate level to modernize any number of social and fiscal policies and develop new forms of governance as well as shared ownership between public and private, which is one of the greatest challenges for the remainder of the century. But only regional political integration can lead to an effective regulation of the globalized patrimonial capitalism of the twenty-first century (Piketty 2013: 572–573).

The previous lines were written by Piketty in 2013, the year of the first edition of his book in French. It is worrisome that since then the situation has worsened with the unfortunate vote for *Brexit* by British people exasperated by EU regulations; Trump's election by disenchanted American unemployed workers; the huge flow of refugees resulting from the wars in the Middle East, especially Syria; terrorism menacing European cities; the racism and xenophobia that these last phenomena have aroused throughout Europe; and, of course, the COVID-19 pandemic in 2020/21. To all this should be added the populist neo-nationalism in spite of Biden's election, the damage caused by former. Trump administration, unchaining a new cold war against China, threatening Iran, Venezuela and North Korea, intimidating Mexico and the Latin American countries with his racism and xenophobia, distancing the US from its EU and NATO allies, and so on. Thus the picture could not be grimmer now in relation to the scenarios outlined by Piketty in the conclusions of his book. In any case, it is clear that there is no way out other than betting on democracy and the reform of capitalism at macro-economic level. At least in the European case, it is from their own institutions that Brussels must make appropriate decisions – such as the circular economy and the *European Green Deal* – to impose effective regulations on this "globalized patrimonial capitalism" which characterize these early decades of the twenty-first century.

In any case, the prospect of new concord in the European leadership and the possible reconstitution of the Paris-Berlin axis looms on the horizon thanks to the crisis provoked by the pandemic. Thus in a recent interview Piketty (2020) stated that it is now necessary to change the economic system because even though, after the fall of communism, changing the system was no longer fashionable, the pandemic has again put this issue on the agenda. It is clear that a different economic system is absolutely necessary if the EU really wants to solve both the social inequality crisis and the climate change threat produced by what Piketty calls 'hyper-capitalism'. He is in favour of a participatory socialist system based on reforms such as sharing power with workers within big corporations and enterprises in order to give workers the right to vote on strategies; the establishment of a basic income; and the "permanent circulation" of wealth through fiscal reform, including a progressive tax on patrimony and heritage in order to change the current situation where in France and Italy, for instance, 50% of the population owns just 5% of the real estate, financial and professional assets whereas the richest 10% in those countries own 50% of the assets and another 1% owns 25% of the wealth.

5.3.3 Richard Wolff

According to the American scholar Richard Wolff, capitalism and democracy are in permanent contradiction, given the fact that corporations and enterprises are ruled by CEOs and managers who answer to the owners of capital, not to society and citizens. Consequently, radical reform of the economic system is needed to democratize the economic system by creating, for instance, producers' cooperatives, or democratically managed enterprises that might be an alternative which generates a new, genuinely socialist mode of production which is fully consistent with the ultimate rationale underlying Marx's theoretical approach in *Das Kapital*. Furthermore the proposition that firms should be run by the workers on their own was endorsed in the past by famous and well-known liberal thinkers such as John Stuart Mill and John Dewey – one of the prominent philosophers of the twentieth century – as well as Marxists like Karl Korsch, Wright Mills and Antonio Gramsci. Besides, the stark contrast between the innermost driving force behind capitalism and generally recognized ethical values, such as Christian ethics that extol virtues such as benevolence and care for our fellow-beings, condemns greed and discourages the accumulation of wealth, so there is nothing to be admired in individuals whose actions are solely guided by the personal profit motive, rather than the duty to take care of their fellow-beings, i.e. the behaviour of people operating in capitalistic systems of interpersonal ties does not proceed from democracy, but from the development of capitalism. Thus the establishment of a new system founded on solidarity is an absolute necessity, given that in capitalism "all the higher bonds of love and solidarity are dissolved: from the bonds of craftsmen's guilds and social castes to those of religion and the family" (Gramsci 1994: 134). In a new democratic economic system where a cooperative or enterprise is ruled by workers' councils, "limits are placed on the sway of capital in the workplace" and people gain autonomy and freedom as members of collectives. Hence socializing production also involves humanizing the workplace, making it is possible to say that citizens are better forged in the factory than in the electoral district. This also explains why it is important to press for the permanent inclusion of workers in the strategies and bargaining agendas of all trade unions. Furthermore, if the supreme value of life is not *to have* but *to be*, being oneself and enjoying life will be facilitated by a socialist order founded on being, rather than having, and this will be the result of the active involvement of workers in economic life as free citizens.

5.3.4   Ian Angus

In his book *Facing the Anthropocene: Fossil Capitalism and the Crisis of the Earth System* (2016) Ian Angus underscores the depth of the environmental crisis, situating the onset of climate change in geological and historical eras that are now converging to mark the end of the ecological conditions that allowed *Homo sapiens* to prosper. Angus describes the rapidity with which the threshold has been reached, revealing how our species – along with many others, because we are in the midst of the sixth mass extinction – sits on a precipice while the dominant classes tighten their grip on the structures that have brought us to this point and to the emergence of the Anthropocene as the classification that best defines the current geological epoch in which the human impact upon the Earth – including the new substances that human industry has devised (such as plastics) – has brought environmental changes of a magnitude comparable to that of all the previous great geological transitions.

As stated in previous chapters, what the Anthropocene replaces is the Holocene – that unique atmospheric and climatic configuration which, with its relative stability and predictability over eleven millennia, has been the scene of all the achievements of human civilization. It is extraordinary that the end of the Holocene has come so fast, but the suddenness is not unprecedented. Relatively sudden climatic change between epochs – abrupt yet long-term warming or cooling within the span of a few years – has happen before. While the build-up of destabilizing factors may be gradual and barely perceptible, their cumulative effects – sped up by feedback loops and tipping-points – collide into our lives without warning, in the heightened severity of climate events like floods, earthquakes, droughts, tropical storms, hurricanes, cyclones and so on.

Such is the progression of geological time. The historical timescale is incomparably shorter but structurally similar, in the sense that each epoch of class-rule likewise entails a gradual build-up of aggravating factors, culminating in some variety of breakdown or explosion – chaos or revolution. The historical side of the present drama is the accelerating extraction of fossil fuels, in the form of carbon and hydrocarbons (including methane) and their processing either via simple combustion (as energy sources) or via industrial chemistry (to produce synthetic substances, from plastics to fertilizers and pesticides). This whole petrochemical complex, full of gasoline vehicles, aeroplanes and highways, grew exponentially in the decades following World War II (as a result of the *Great Acceleration*, as explained in Chap. 3), and this phenomenon was the 'golden age' of capitalism, although in the subsequent period globalization brought about deindustrialization, massive energy-waste, the privatization of public property, and the delocalization of work and factories etc., which is what lay at the core of the epochal environmental breakdown. Angus uses the phrase 'fossil capitalism' to evoke the historical links between capitalism and the reliance on fossil fuels, noting, for instance, that nineteenth-century mill-owners switched from water power to coal power not because coal was cheaper or more reliable than water-mills, but rather "because it gave the factory-owners better access to and control over labour" (Angus 2016: 129). The tight symbiosis between capital and fossil fuels has persisted, albeit in different forms: "Global proven oil reserves were worth about $50 trillion. No capitalist would willingly exchange that for the chance to sell green electricity twenty years from now" (Angus 2016: 173).

Big corporations' persistent veto of ecological rescue, expressed through indirect means ('proxies' like the usual practice of US government obstruction at UN environmental summits), is also addressed by Angus. Although it is true that environmental collapse ultimately threatens everyone, the struggle against such collapse could pit global civil society against the great powers – decision-makers who, viewing the forthcoming breakdown as a 'security' threat, calculate that they can save themselves and their fortunes by increasing their agenda of power even further: "they are prepared to bring the world down to protect their power" (Angus 2016: 216).

Calamity has already struck for many among the poorer populations, whether from floods, droughts, heatwaves, or food shortages. Shortages and droughts provoke internal violent conflict – as in the Middle East – which in turn generates further scarcity alongside the violence, triggering massive migratory flows that then strain the resource-base in other countries, like Turkey and several European countries. The dialectic of scarcity feeds and is fed by terrorism and military intervention in countries like Afghanistan, Iraq, Syria, Palestine and Yemen, and the consequences are terrorist attacks against civilians in both Muslim and predominantly non-Muslim countries.

In the last two chapters Angus outlines the main points of his eco-socialist alternative, underlining the importance of replacing carbon-based fuels (including biofuels) with clean sources of power under community control (wind, geothermal, wave and solar power), promoting collective transport instead of individual vehicles, promoting food sovereignty, drastically reducing greenhouse emissions, and so on. He stresses the main features of a social movement that must be an ecological and majority counterpower, pluralist and open to different views within what he calls "the green left", and overall able to constantly extend the analysis and programme in the light of changing political circumstances and scientific knowledge. It should be internationalist and anti-imperialist, capable of overcoming all kinds of environmental struggles – large and small – and unite the eco-socialist consciousness in order to create an ecological civilization.

5.3.5 Howard Richards

Richards (2017) is an American social scientist who spent several years working and living in Chile. He argues that the determining factor of social injustice and what he calls "the march towards the decline of the biosphere" is the need to maintain favourable conditions for the accumulation of capital. To counteract this, an *economy of social solidarity* (ESS) is needed, particularly within the small and medium economic business sector. According to Richards, most businesses (small and medium-sized enterprises owned individually or by families) do not accumulate capital, because their profits are barely sufficient to cover the drop in the value of money due to inflation. The profit that is obtained is used to ensure the survival of the business and the provision of a legal and satisfactory income to live off. And that – not personal enrichment – is the reality for the small or medium entrepreneurs.

Therefore the most important playing field of the economy is situated in these small businesses sectors for the good reason that they are the main generators of employment.[8] This category includes enterprises owned by a family or individual for whom work serves mainly to live, not to amass money and goods. Hence, the income of people in the informal or social solidarity economy is usually small, similar to those people who obtain their main economic income from waged labour, administration or professional services. However, it is necessary to differentiate between small and medium enterprises that work within the framework of the dominant economic paradigm and those that could work within the framework of the sustainable development paradigm, according to the ideas of Richards. For him, it is possible to build a new theory of the economy based on the ancient origins of the concept of solidarity and a holistic vision:

> Various socio-political disciplines and perspectives such as cooperativism, socialism, grassroot Christianity, ecology, fair trade and responsible consumption, popular education and food sovereignty, among others, have existed for at least 250 years [and] found expression in the utopian socialists who referred to an economy controlled by the workers and not by the state and the capitalists. We could summarize the great principles of the social solidarity economy in that it does not have a permanent and growing logic of profit and income; a tendency to reinvest profits in better working conditions; a democratic management of the

---

[8] This is corroborated by the official statistics of almost all the countries that place the largest job generators in the SME sector. There are also small businesses in the informal sector of the economy which generate employment in poor countries, but obviously, precisely because they are informal, statistical registration is lacking since they do not pay taxes or social security.

company; different ways of working, in which work and life are reconciled; and respect for the environment, although this is a more recent orientation (Richards 2017: 8).

According to Richards, the principles of the social solidarity economy could also have been inscribed within postmodern philosophy, lessening the obstacles to change that obstructed the revolutions and reforms of capitalism attempted during the twentieth century. Furthermore, it could be inscribed within postmodernism because it incorporates elements of the traditional wisdom of peasant communities and indigenous peoples, emphasizing that this type of knowledge constitutes an epistemological rupture with respect to modern science,[9] since one of its fundamental principles is to share the surplus. In that sense, Richards agrees with Enrique Dussel, for whom overcoming capitalism means rising above modernity and accepting traditional popular wisdom or what Sousa Santos calls *ecology of knowledge*. The social doctrine of the Church also has aspects that coincide with these ideas, as do pre-Columbian indigenous economies, where they "recover the values of archaic societies such as fraternal solidarity and the right of each individual to feel appreciated as a son in their home … The incorporation of archaic norms in the ESS establishes another epistemology, another way of doing science, but in no way means renouncing science and embracing superstition" (Richards 2017: 10).

In summary, Richards's *Economy of Social Solidarity* (ESS) coincides not just with views like those of Sousa Santos but also has parallels with technical and sophisticated macro-economic approaches, such as those of Varoufakis about the surpluses recycling mechanism that constitutes his key criterion to determine whether capitalism could be reformed. In the same manner that global-scale capitalism should be reformed to diminish inequality by sharing surpluses among the poor and establishing, for instance, a global tax on *financial transactions* (FTT) to finance the universal basic income advocated by Varoufakis, the ESS model deems that surplus recycling must take place at a local and community level. This is because it is in this physical location that economic surplus can be shared instead of accumulated for personal enrichment. Individuals who hoard their surplus cease to be responsible members of the community – or society – becoming *rentiers*, as Piketty similarly says in his critical historical analysis of the huge concentration of wealth and inequalities of twenty-first-century capitalism. Richards's views could also be compared with those of Morin (2011), who, in an interesting book about 'humankind policies', refers to the cooperative movement, micro loans, fair trade, demography, indigenous peoples, water, and environmental experiences in different countries, including France (Morin 2011).

---

[9] According to Richards, the French term for epistemological rupture – *coupure épistémologique* – was originally proposed by Gastón Bachelard, and alludes to a theory of knowledge preferable to that of Thomas Kuhn's concept of paradigm shift because "as Kuhn clarifies in the second edition of *The Structure of Scientific Revolutions*, a paradigm is always a concrete example of what science is and should be according to a determined scientific community. We propose, on the contrary, that the economy of social solidarity (ESS) is and should not be a concrete example but an invitation to exercise an infinite creativity in the perfection of a great variety of material practices" (Richards 2017: 3).

Furthermore, for Richards, this economy of social solidarity could also be considered a confluence of doctrines in traditional ethics, although he recognizes that a more precise investigation would be needed to establish whether or not it has a relationship with "modern economic sciences as they are the Ricardian income, the surplus value, the rent without social function as analysed by Thomas Piketty and by the Nobel Prize winner Joseph Stiglitz, just to name a few, the theories of the quasi-rent and situation rent of Alfred Marshall, and to be more specific, the copper revenues in Chile, analysed by the former Chilean Minister of Economy, Jorge Leiva" (Richards 2017: 7). In conclusion, Richards's reference to the *economy of social solidarity* (ESS) is tantamount to Latouche economy in the micro-economic field, thanks to its epistemological rupture with mainstream economics. In that sense, it could be properly inscribed within the framework of sustainable development, seen from the stance of the incremental transformations that Maja Göpel proposed, not only because it exists and works in social practice itself, but also because there are undoubted links that can be established with other theoretical perspectives. Examples of these other perspectives include, but are not limited to, the points of view of Boaventura de Sousa Santos, Yanis Varoufakis and Edgar Morin, the standpoint of the people of a kingdom such as Bhutan in the Himalayas, and the principles of the Andean indigenous people's world-view (*cosmovision*) of good living (*buen vivir* or *sumak kawsay*), which are part of the new constitutional norms of Bolivia and Ecuador and a good example of how the Bolivian and Ecuadorian indigenous peoples have, for the first time in history, achieved a social pact with the traditional ruling elites.

5.3.6   John Holloway

John Holloway, an Irish professor resident in Mexico, where he works at the University of Puebla, published an interesting article, "Change the World Without Taking Power: The Meaning of Revolution Today" (2002), in which he maintains that the struggle of the left against capitalism failed throughout the twentieth century, as did half the political struggles in both their revolutionary and reformist aspects, due to an erroneous conception of the nature of political power, in which the State was supposedly an instrument or tool in the hands of the "dominant classes". Indeed, for Holloway – and this is something that had already been raised in the 1980s by Claude Lefort in France – the State is a nodal point of power relations and therefore constitutes a political space where class struggle is manifested as the different interests of various sectors of society, including economic and ethnic groups, religions and, in general terms, all the components of a nation state. The State cannot and should not be viewed as an 'instrument' or kind of tool which a political party or movement can use to achieve its political goals when it reaches power through the electoral process, revolution or a *coup d'état*. This is because, regardless of the nature of the political system (democracy, dictatorship, monarchy, authoritarian state), political power is a rapport or correlation between different actors, such as political parties, economic corporations and associations, civil society groups, non-governmental organizations, ethnic groups, social movements and so on. Consequently, for Holloway, it is wrong to think that it is necessary to seize state power to promote change. It is more important for social movements which become aware of their situation to empower and promote changes they consider necessary from below. According to Holloway, who has written extensively about the Zapatista social movement:

> The paradigm of the state, i.e. the assumption that winning state power is central to radical change, dominated the revolutionary experience during most of the twentieth century: not only the experience of the Soviet Union and of China, but also the numerous national and guerrilla liberation movements of the Sixties and Seventies. If the state paradigm was the vehicle of hope for much of the century, it became increasingly the executioner of hope as the century progressed. The apparent impossibility of the revolution at the beginning of the twenty-first century reflects, in reality, the historical failure of a particular concept of revolution: the one that identifies it with the control of the State. Both approaches, the 'reformist' and the 'revolutionary', have failed completely to meet the expectations of their enthusiastic supporters. The 'communist' governments of the Soviet Union, China and elsewhere certainly increased the levels of material security and decreased social inequalities in the territories of the states they controlled (at least temporarily), but did little to create a self-determined society or to promote the reign of freedom that has always been central to the communist aspiration. In the case of social democratic or reformist governments, the situation is not better: although some have achieved increases in material security, their performance in practice has differed very little from that of governments that are openly in favour of capitalism, and most social democratic parties have long abandoned any pretence of being the bearers of radical social reform (Holloway 2002: 16).

Although Holloway has not put forward an alternative proposal – such as Richards's economy of social solidarity or Mason's post-capitalism – his sympathies for the Zapatistas and his experience in Mexico have led him to express his ideas about what should be understood by the concept of changing the world without taking power. When asked in an interview how to break with the logic of capitalism, he said:

> Actually, the question of the rupture is central. We want to break with the logic of capitalism. And we want to do it in thousands of different ways. We are going to create spaces where we are not going to reproduce the logic of capital, where we are going to do something else, have other types of relationships, develop activities that make sense for us. So, the problem of the question of the revolution, instead of how we take power, is: how do we break with the logic of capital? For me, the most obvious example is that of the Zapatistas. From the moment we cross into their territory there is a sign that says "Here the people rule and the government obeys". It is about the creation of a space with another logic. But if we begin to think about it, we see that it's not just the Zapatistas. There are also community radios, social centres, communities and autonomous municipalities, in short, a series of ruptures that are not necessarily territorial, but may also refer to an activity, such as student protests against the introduction of market logic in education. This type of revolution can also occur in the relationship between husband and wife, or with children, treating love as an attempt to create a relationship in which the capitalist logic of profit and merchandise is not accepted. The only way to think about the revolution is in terms of those spaces or moments that can be thought of as cracks in the social fabric of capitalist domination. The only way to think about the revolution is in terms of creation, multiplication and expansion of these cracks, because they move, they are dynamic (Holloway 2015: 27).

When the interviewer asked again about the concrete results that the movement has produced after so many years of existence and struggle, Holloway answered:

First, it is clear that the Zapatista uprising…changed the world for us who live in Mexico, but also for the left throughout the world. First, simply because they got up when there seemed to be no more room for getting up, but also for the fact of rethinking the whole question of the meaning of the revolution… There is a change in the Zapatista subjectivity after 2001, in the sense that after the San Andrés Accords in 1996, the campaign and the Zapatista movement aimed at the acceptance and implementation of these agreements (but)…after they came to Mexico City and it became clear that the government was not going to implement the agreements, they said, after a silence: "Let's be the ones who are going to do it." It seems to me that there is a very important change, because they left behind the policy of demands. Since 2001 they practically said, we are not going to ask for anything, we are not going to demand anything, it does not make sense to demand anything from the government, we learned that, we are the ones that are going to make the changes, then we are going to take responsibility, we are going to implement our system of education, health, etc. That means a rethinking of the Zapatista movement, in the way I understand it: the centre is no longer in command; instead the communities are, because they are the ones that are implementing these changes. And that creates the impression that they really are not doing much, because they are simply doing it, they are no longer demanding, they are no longer issuing communiqués as before. But it seems to me that, at the same time, the actual process of transformation is quite profound, with all the difficulties of the world (Holloway 2015: 28).

In short, Holloway's ideas of changing the world without taking power and without a political party ('vanguard of the revolution') are quite innovative and coincide with Santos's theory of the 'rear-guard', according to which intellectuals accompany the social movement and think with the leaders without pretending to be intellectually superior or trying to command them, hence the dialogue with the popular leadership feeds theoretical formulations, not the other way around. Sustainable development – which implies the management of social dynamics, natural ecosystems and economic production – can, of course, function at this micro-level space of governance, but field research is indispensable to find out what kind of changes have been made by the Zapatistas in the territories where they operate in the Mexican state of Chiapas.

5.4 Circular Economy

As we have seen before, and independently of the neoliberal model which refers mainly to the relationship between the economy and the political system, over the last 150 years, the industrial economy has been dominated by a one-way model of production and consumption in which goods are manufactured from raw materials, sold, used, and then incinerated or discarded as waste. In the face of a rising global population and the associated growing resource consumption and negative environmental impacts, it becomes increasingly apparent that business as usual is not an option for a sustainable future. While the concept of a circular economy has been discussed since the 1970s, switching from the current linear model of economy to a circular one has recently attracted increased attention from major global companies and policy-makers like the Davos World Economic Forum. Furthermore, the *Ellen MacArthur Foundation* (EMF) and the McKinsey Company published a report which evaluates the potential benefits of replacing the existing economic paradigm with a circular economy paradigm (Ellen MacArthur Foundation 2012: 5).

The concept of a circular economy cannot be traced back to one single date or author, but one of the main exponents of the school is the ecological economist Kenneth Boulding. In a paper entitled *"The Economics of the Coming Spaceship Earth"* he argued that a circular economic system is a prerequisite for maintaining the sustainability of human life on Earth (Boulding 1966: 8–10) because the planet can be compared with a spaceship which is "without unlimited reservoirs of anything, either for extraction or for pollution, and in which, therefore, man must find his place in a cyclical ecological system which is capable of continuous reproduction of material form even though it cannot escape having inputs of energy" (Boulding 1966: 8). Boulding describes the traditional economic model as an open system in which nature is perceived as limitless because apparently no limit exists on the capacity of the planet to supply or receive energy and material flows, which is clearly an erroneous perception because the traditional *linear economy* is based on the ideology of growth and economic development and is characterized by its huge environmental impact (like the one described in the iconic graphs of the Great Acceleration in Chap. 3) and by the enormous amounts of pollution, waste, garbage and careless and exploitative behaviour. For Boulding, this kind of economy is built around a flawed understanding of nature in the long run, which is why a new model (what he calls the "spaceman economy") is necessary. Boulding's model regards the planet as a closed system (with practically no exchanges of matter with the universe except for solar energy), where the economy and the environment are characterized by a circular relationship where everything is input into everything else.

Consequently, *environmental economics* – part of the circular economy – have emerged in opposition to the misperception that the economic system and the environment are unrelated, when the environment actually encompasses everything, and is made up of ecosystems and interrelationships between all living species and non-living or abiotic structures. Because simple economic models have ignored the economy-environment interrelationship, a shift from the traditional linear or open-ended economic system to a circular one is needed. The circular model is based on the fact that there is extensive interdependence between the economy and the environment and on the perception that human economic activity is causing unacceptable environmental changes. This is why the circular economy paradigm emerged in opposition to the erroneous idea that the industrial system can be separated (or isolated) from the environment.

Thus the core idea of the circular economy is the redesign of industrial processes as specific ecosystems within the biosphere. Accordingly, the concept of environmental economics relies on a systemic, comprehensive and integrated analysis of an industrial system and all its components within its environment, considering them as a joint ecosystem. This approach aims to identify how the industrial system works, how flows of material and energy (the *industrial metabolism*) are regulated, and how they interact with the biosphere. The analysis of the industrial metabolism is then used as a basis for optimizing the cycle of industrial materials (from raw materials to finished product and the disposal of waste), taking inspiration from the natural ecosystems function. The socio-economic and environmental benefits of environmental economics are that productivity increases with fewer resources and fewer impacts on the biosphere. Furthermore, the circular economy applied to industry addresses issues of pollution and environment by looking at the energy and material flows and the industrial system as a whole. Industrial ecology hence challenges the competitive neoliberal dogma by promoting collaboration between distinct entities to ensure efficient resource management through, for instance, the creation of eco-industrial networks.

Therefore, industrial economy (IE) represents a way for companies to better exploit their products and resources (including their waste) more efficiently and thus more profitably. Erkman (2001: 533–534) defines four key principles which guide the reorganization of our industrial ecosystem toward a sustainable industrial society. Waste and by-products must be systematically valorized; traditional recycling is just one of many material recovery strategies. Erkman calls for the creation of eco-industrial networks: smart networks of resources and waste in which the residues of one company can become the inputs of another industrial process. So-called *industrial symbiosis* is probably the best-known example of a working *industrial ecosystem*. For example, the waste and *by-products* from one industrial process can serve as raw materials for another, thereby reducing the use of raw materials, waste and pollution. In another example, the loss caused by dispersion must be minimized: products and services must be designed to minimize dispersion or at least eliminate its harmful effects on the environment and health. Also, the economy must be *dematerialized*, i.e. the goal of production should be to minimize *total material flows* while ensuring that equivalent or *higher levels of services* are provided. Within the context of industrial economy, one of the best ways to dematerialize the economy is to evolve towards a *service-orientated society* in which *the use* of products is prioritized over *selling* them (autonomous vehicles, for example, but also the use of hired bikes as a mean of transport in big cities like Paris and London). The *value of use* then becomes more important than the *exchange value* (as it is called in Marxist theory) – two notions that have also been discussed by Mason and Varoufakis. While recent advances in information and communication technologies (ICT) have also contributed to the "dematerialization of the economy" (Mason 2016), the real impact on resource consumption might be ambiguous, due to the well-known 'rebound effect' (for example, the increasing use of electric vehicles requires more electricity, which in some countries is still mainly generated from fossil-based sources).

Of course, the circular economy postulates that energy must not rely on fossil fuels and hydrocarbons. A large part of its literature deals with the development and application of various methods for measuring the environmental impacts of industrial production and human consumption, making lifecycle assessments, and describing practical cases of products, materials and eco-industrial practices. On the other hand, while the economic and environmental benefits of the ecosystem approach for the design of industrial systems have already been highlighted, its practical implementation is still limited because current technology is often inadequate to support this approach, and inter-enterprises synergies are not naturally created. Therefore, the intervention of the State through publicly-founded projects is needed to eliminate another neoliberal dogma. Hence, the circular economy builds on the analysis and optimization of industrial systems at a micro-level, scaling it up to an economy-wide system in which products and processes are redesigned to maximize the value of resources through the economy. Accordingly, the circular economy promotes the transition from open to closed cycles of materials and energy, thus leading to less wasteful industrial processes through reuse (second-hand markets), reman-ufacturing, upgrading, and recycling using reprocessing activities which recover secondary materials like paper or plastics.

But the more important issue regarding the circular economy is that since the European Union has officially adopted it, the results of this decision will be tested in the coming years. Indeed, on 11 March 2020 in a press release of the Euro-pean Commission under the heading "Changing how we produce and consume: New Circular Economy Action Plan shows the way to a climate-neutral, competitive economy of empowered consumers", the EU Commission stated that the EU had adopted the Circular Economy Action Plan as one of the main parts of the *Euro-pean Green Deal*, Europe's new agenda for sustainable development. Establishing measures along the entire lifecycle of products, the new Action Plan aims to make the European economy fit for a "green future", the press release says, "strengthening the competitiveness of the Union while protecting the environment and giving new rights to consumers. Building on the work done since 2015, the new Plan focuses on the design and production for a circular economy, with the aim of ensuring that the resources used are kept in the EU economy for as long as possible. The plan and the initiatives therein will be developed with the close involvement of the business and stakeholder community" (European Commission 2020). The bulletin also mentioned that the executive Vice-President for the *European Green Deal*, Frans Timmermans, said:

> To achieve climate-neutrality by 2050, to preserve our natural environment, and to strengthen our economic competitiveness, requires a fully circular economy. Today, our economy is still mostly linear, with only 12% of secondary materials and resources being brought back into the economy. Many products break down too easily, cannot be reused, repaired or recycled, or are made for single use only. There is a huge potential to be exploited both for businesses and consumers. With today's plan we launch action to transform the way products are made and empower consumers to make sustainable choices for their own benefit and that of the environment (EC 2020: 2).

The bulletin also quotes the Commissioner for the Environment, Oceans and Fisheries, Virginijus Sinkevičius, who added:

> We only have one Planet Earth, and yet by 2050 we will be consuming as if we had three. The new Plan will make circularity the mainstream in our lives and speed up the green transition of our economy. We offer decisive action to change the top of the sustainability chain – product design. Future-orientated actions will create business and job opportunities, give new rights to European consumers, harness innovation and digitalization and, just like nature, make sure that nothing is wasted (EC 2020: 3).

Hence, it is possible to say that, thanks to the regional integration process that makes it evident that the European Union is genuinely committed to the SDGs and the UN 2030 Agenda, the transition towards a circular economy is already underway, and that the process includes businesses, consumers and public authorities committed to this sustainable model. As the Commission has stated, the EU will make sure that the circular economy transition will deliver opportunities for all, "leaving no one behind" (EC 2020: 8).

5.5   The UN 2030 Agenda and the SDGs

The UN 2030 Agenda is not a revolutionary one. Since its inception at the end of the twentieth century, SD was intended to reform capitalism but neoliberalism hampered and blocked its implementation. Thus the 17 SDGs, to which the 193 member states of the United Nations committed in 2015, are a renewal of that old proposal and of the *millennium development goals* (MDM) of the year 2000. The novelty is that they have four major 'pillars' (social, economic, environmental and political) or four major spheres – social dynamics, techno-economic factors, terrestrial ecosystems, and government policies – and, of course, since all 193 member states of the United Nations have approved the 17 SDGs, they are all obliged to make an effort to achieve them. However, it is still not clear how the coronavirus crisis will affect the ability of all countries to reaching those goals.[10]

The social pillar has main six social goals. SDGs 1 and 2 are to end poverty and hunger – i.e. achieve food security and nutrition – through sustainable agriculture; SDG 3 refers to healthy lives and well-being; SDG 4 concerns inclusive and equitable quality education and lifelong learning opportunities for all; SDG 5 aims to "achieve gender equality and empower all women and girls"; and SDG 10 concerns the reduction of inequalities both within countries and between them.

The SDGs in the economic sphere are closely linked to the environment and concern water, energy, economic growth, infrastructure and industry, urbanization, consumption and production in the search to: "ensure availability and sustainable management of water and sanitation for all" (SDG 6); "ensure access to affordable, reliable, sustainable and modern energy for all" (SDG 7); "promote sustained, inclusive and sustainable economic growth, full and productive employment and decent

---

[10] To give an idea of the 169 targets assigned to each SDG and the indicators that will allow the UN to verify their accomplishment, they include reducing poverty by half; implementing social protection systems and measures; ending hunger and ensuring access to healthy, nutritious and sufficient food; ending all forms of malnutrition; doubling agricultural productivity and the income of food producers; maintaining the genetic diversity of seeds; increasing investments; correcting and preventing trade restrictions and distortions in world markets by eliminating subsidies; reducing the maternal mortality rate; ending avoidable deaths of newborns and children under five years of age; ending the AIDS, tuberculosis and malaria epidemics; guaranteeing universal access to health services; ensuring that all children have free and equal quality primary and secondary education; eliminating gender disparities; building and adapting school facilities; and increasing the scholarships available for developing countries.

work for all" (SDG 8); "build resilient infrastructure, promote inclusive and sustainable industrialization and foster innovation" (SDG 9); "make cities and human settlements inclusive, safe, resilient and sustainable" (SDG 11); and "ensure sustainable consumption and production patterns" (SDG 12).

The third pillar is that of the environment. SDG 13 refers to the adoption of urgent measures to combat climate change and its effects; SDG 14 refers to the conservation and sustainable use of oceans, seas and marine resources; SDG 15 seeks to protect, restore and promote the sustainable use of terrestrial ecosystems, sustainably manage the forest, combat desertification, stop and reverse land degradation, and stop the loss of biodiversity. Finally, the SDGs of good governance and international cooperation are included in SDG 16, which concerns the promotion of peaceful and inclusive societies for sustainable development, the facilitation of access to justice for all, and building effective and inclusive accountability at all levels. In a broad interpretation this means that it is about consolidating good governance, democracy and the rule of law, while the last SDG, 17, is to strengthen the means of implementation and revitalize the World Alliance for Sustainable Development.

The implementation of the SDGs is a central issue of great difficulty and complexity. For instance, social development cannot be obtained via a magic trickle-down of wealth from financial markets to the unemployed. State intervention is necessary to correct "market failures". It is easy to imagine the obstacles and difficulties that the implementation of the SDGs involves and that range from the social sphere (poverty reduction, food security, health and education, gender equality) to the economy (energy, sustainable growth, urbanization, infrastructure, sustainable production and consumption, decent work) and the ecosystems (water and sanitation, climate change, oceans, seas and marine resources, terrestrial ecosystems, sustainable management of forests, stopping desertification, land degradation and loss of biodiversity) as they reach public policies. It seems to require the quadrature of the circle, among other reasons because the sphere of the environment is not linear (ecosystems are cyclical and therefore circular) and the intervention of the State in economy is (or used to be) the nemesis of the oligarchy, as demonstrated by Valdez (2015) in a book about the ideology of the ruling elites in Guatemala.

Furthermore, public policies are not exactly characterized by their qualities of 'good governance'. The fact that legislation and the State itself are at the service of corporative interests has been highlighted in a recent investigation concerning the problem of water resources in Guatemala (Padilla 2018). Indeed, it was in Guatemala that an agreement with the UN to investigate and punish corruption was abandoned when it was discovered that both government officials and private sector important personalities were involved, forcing CICIG to leave the country.[11]

---

[11] CICIG was the *International Commission against Impunity in Guatemala*, an UN body established in 2007 at the request of the Guatemalan government to assist the general prosecutor of the Ministry of Justice (MP) to investigate the most significant cases of impunity. The work of the Commission went well until it started to prosecute high-level officials (a former president and a vice president of the republic were forced to resign and are currently in jail; the former President Morales was also investigated). In 2018 the government announced that the Commission agreement would not be renewed and prohibited the Commissioner, the distinguished jurist and former Colombian judge

5.5.1   Central America

As we have seen in Chap. 2, since the 1990s, when the Central American integration process was relaunched and SICA was established, an "alliance for sustainable development" was established in the sub-region. However, twenty years later, social indicators continue to be quite low except for Costa Rica and Panama. In a comparison of public expenditure in Central American countries (as a percentage of GDP) regarding social variables such as social security, housing, health and education, 17 and 18% are invested by Panama and Costa Rica, while Honduras and Nicaragua do not reach 12%, Guatemala barely exceeds 6% and El Salvador's figure is even lower (ICEFI 2016).[12] During the decade 1997–2006, the average social spending in both Costa Rica and Panama remained between 10 and 12% (with a tendency to rise in Panama), while, again, Guatemala and El Salvador (with averages below 2%) were behind and Honduras and Nicaragua were in an intermediate location with averages of between 2 and 4%. These four countries always appear at a notorious disadvantage with respect to Costa Rica and Panama, which each have a lower population and – in the case of Panama – higher foreign exchange earnings thanks to income from the canal and the fact that Panama City has been transformed into a financial centre. As for Costa Rica, its better performance can be explained by the democratic system, less social inequality, and the capacity of the government to cope with its population's social demands. Concerning Nicaragua, ruled by a former leftist guerrilla commander who made a pact with the oligarchy to establish an authoritarian regime, the fact that it is a less populated country and the remains of social policies of the Sandinista revolutionary times could explain the differences, which is more or less also the case with Honduras (Fig. 5.1).

As the experts of the *Central American Institute of Fiscal Studies* (ICEFI) point out in relation to this problem:

> In Central America, the amount and manner in which public social spending is distributed continues to show deficiencies in meeting the social needs of the most vulnerable population. As in the rest of Latin America, this has led to slow progress in alleviating poverty and reducing inequality. On the one hand, this expenditure is insufficient and is administered under budgetary constraints caused by low taxation rates. On the other hand, the structure of public expenditure must adapt permanently to emerging social needs, without first satisfying existing ones (ECLAC 2007) … the simple average of public social spending in the region was equivalent to 9.5% of GDP, while for 2006 it reached 11.7%. Only Costa Rica and Panama have a public social expenditure higher than this average, 17.1% and 18.2% of GDP respectively. Despite the increase observed in Guatemala and El Salvador, both countries continue to be the ones that dedicate fewer resources to social spending (Carrera 2008: 16).

The ECLAC (2011) reports on social spending in Latin America reveal that among the countries that allocate the most funds to social spending in their budgets are Cuba,

---

Iván Velásquez, from entering the country. The General Attorney and Prosecutor Telma Aldana (also a former president of the Supreme Court) was dismissed and is currently in exile because she presented her candidacy to the presidency of the republic in the 2019 elections.

[12] However, it seems that a change of government facilitated more social investment in El Salvador and by 2018 Salvadorans had improved their indicators concerning reduction of poverty.

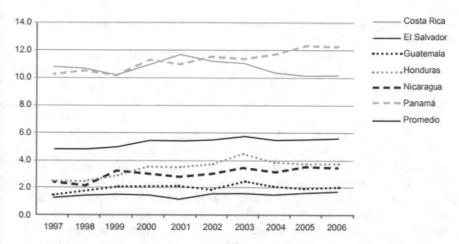

Fig. 5.1 Central American social expenditure on health and social security as a percentage of GDP, 1997–2006. *Source* ICEFI with data from ECLAC United Nations

Brazil, Uruguay, Chile, and Costa Rica, while Guatemala and Honduras are among those which allocate the least. Therefore, concerning the Millennium Development Goals (MDGs), Guatemala lagged the most and in some indicators it even reversed, as in the case of extreme poverty, which increased from 18.1 to 23.4%. As a former Guatemalan diplomat and UN official explains:

> Regarding the mortality rate of children under five, progress was made: from 80.6 in 1990 it was reduced to 31.0 in 2013 for every thousand children born alive, but we are still the third highest in the entire continent, and we are the third highest in maternal mortality also. The proportion of our children under one year vaccinated against measles is only 85%, which puts us third in the queue. Guatemala did not reach the goal in proportion to the population that uses improved sanitation services, where we are also third parties at the end of the list. The UNDP Human Development Index shows how Guatemala has taken steps backwards. It is worth noting Guatemala's limited progress in the MDGs period with respect to the objectives and goals of health in general, and sexual and reproductive health in particular, expressed for example in the still very high incidence of maternal mortality, is a significant pending issue, which is considered a priority in the 2030 Agenda. This is added to the worrying increases in teen pregnancy rates associated with poorer households. As a corollary, another important fact: our population is the fourth lowest internet user in Latin America and the Caribbean, which makes us less competitive, informed and innovative (Mulet 2016: 153–154).

The points made in the previous paragraph are also the consequence of the fact that, regardless of the ideology, conservative or progressive, of current governments or the endemic scarcity of resources given the low tax burden and tax evasion, the low social investment of governments responds to the proliferation, throughout Latin America, of the criminal social pathology which corruption has become. This is another cause of the lack of effectiveness of public spending on social matters, since corruption diverts large amounts of money – specifically, public funds – into the pockets of unscrupulous officials and entrepreneurs whose criminal activity has common traits throughout the world.[13]

Thus, it turns out that the corruption of high-level State officials and transnational corporations is not a phenomenon exclusive to the Brazilian Odebrecht corporation in Latin America, but also occurs in the United States itself with the Koch brothers, the owners of 20th Century Studios and large oil industries which pay campaign bills, buy judges and congressmen, and currently manage public policies from federal departments and agencies in charge of the environment, foreign trade or energy, as Sachs denounces. This makes it clear that a sector of the economic system is increasingly involved in corrupt activities in a way that has led to the establishment of special research entities – such as *Global Financial Integrity* – which are designed to detect illegal financial flows, underbilling practices and price adulteration with a view to tax evasion, money laundering, bank secrecy, tax havens, cardboard companies, transnational organized crime, financing terrorism, and so on – and this is without even mentioning the problem of narco-trafficking that would require an entire book to be addressed properly. What I want to stress is that capitalism as the dominant mode of production on a global scale has so many linkages with criminal organizations in the fields of prohibited drugs, money laundering, illegal trafficking in people and armaments, and all sort of analogous activities that, as Manuel Castells argued at the *end of the millennium*, the "flexible connection of these criminal activities in international networks constitutes an essential feature of the new global economy" (Castells 1998: 167).[14]

For example, the graph below illustrates the way in which illegal financial flows from developing countries have increased considerably in less than a decade.

---

[13] Corruption is the true cancer of capitalism and has been increasing worldwide, facilitated by the phenomenon of globalization that has proliferated as transnational organized crime for several years. Thus, there are corporations such as the Brazilian firm Odebrecht that would have bribed more than a thousand high officials to obtain large contracts for public works up to an amount of US $2,200 million. The case of the so-called *Panama Papers* uncovered the huge corruption and money laundering network existing on a world scale and, as mentioned before, in Guatemala both the former President and Vice-President were forced to resign in 2015, and both the former Peruvian and Spanish presidents Kuscinsky and Rajoy resigned because of corruption charges in 2018.

[14] In the 1990s Manuel Castells asserted that there was a perverse connection between governments and organized crime as follows: "Crime is old as humankind. But global crime, the networking of powerful criminal organizations and their associates in shared activities throughout the planet, is a new phenomenon that profoundly affects international and national economies, politics, security and, ultimately, societies at large … The economies and politics of many countries … cannot be understood without considering the dynamics of criminal networks presented in their daily workings. The flexible connection of these criminal activities in international networks constitutes an essential feature of the new global economy, and of the social/political dynamics of the Information Age. There is a general acknowledgment of the importance and reality of this phenomenon, and a lot of evidence from well documented journalists' reports and the conferences of international organizations. Yet the phenomenon is largely ignored by social scientists, when it comes to understanding economies and societies, with the argument that data are not truly reliable, and that sensationalism taints interpretation. I take exception to these views. If a phenomenon is recognized as a fundamental dimension of our societies, indeed of the new globalized system, we must use whatever evidence is available to explore the connection between these criminal activities and societies and economies at large" (Castells 1998: 167–168).

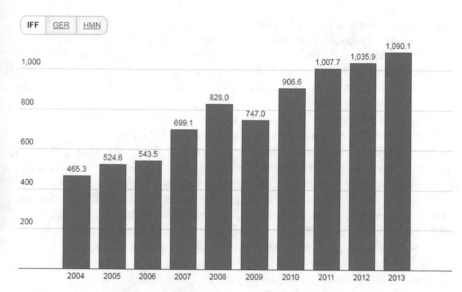

Fig. 5.2 Illicit financial flows from developing countries, 2004–2013. *Source* Global Financial Integrity: http://www.gfintegrity.org/issue/trade-misinvoicing/

However, if the nature of the economic system itself is putting democracy and the rule of law at risk, if growth as it is understood from the point of view of mainstream economics is unsustainable, we must ask ourselves in what way we should then understand the sustained growth of the SDGs. Is it enough to ask that growth be socially inclusive to give it a different character? (Fig. 5.2).

It could be argued that 'sustained growth' is a contradictory concept because a sustainable system must be based on recycling in a circular economy, not on unlimited growth and the increase of *gross domestic product* (GDP) or *per capita income* criteria. Business policies and market techniques designed to make consumers believe (via advertising) that happiness consists of shopping at malls that sell the latest fashion novelties, designs and technical gadgets to keep industry running for the sake of personal gain, the accumulation of capital and an increase in shareholders' revenues is absolutely wrong and misleading. These are goals that do not concern the satisfaction of human needs, that harm the environment, and that have nothing to do with the SDGs of the UN Agenda 2030 and even less with well-being (*buen vivir*) or individual happiness.[15] Instead, they are influenced by business obsolescence plans (industrial products designed to have an average durability of a short enough

---

[15] The increase in drug addiction and violence in high schools in the United States, and, indeed, the whole problem of drug trafficking is linked to the unhappiness of people and the psycho-social pathologies related to the phenomenon. Even terrorism, racism and xenophobia can be regarded as endemic social illnesses generated by structural violence. Thus the treatment of these social diseases should be the priority of governments' public policies, not the repression and punishment of addicts. Addiction and youth violence are public health problems, not issues of national or international 'security'.

span to ensure demand for – and sales of – the new products that are periodically released on to the market), thus promoting unsustainability in the long term, and the waste of unsold products, the increase in toxic waste and the *culture of discarding* harshly criticized by Pope Francis in his Encyclical *Laudato Si'*, not only for being unsustainable but also for promoting consumerism, the psychopathological features of which are increasingly prevalent in human behaviour.

Furthermore, and not just because of the 2020/21 COVID-19 economic crisis, postulating, as a goal of 'sustainable development', a GDP annual growth of 7% for the 'least developed countries' is of dubious benefit. Apart from such a goal being unrealistic, the main point of debate is the category of growth itself, because, as repeatedly stated, if the carrying capacity of the planet is not taken into account – not just because of the obvious waste and garbage generated by industrial production but because of the transgression of the planetary boundaries (Sachs 2015) – humankind itself is at risk of disappearing in the wake of the sixth mass extinction. So we need a paradigm shift which forsakes the concept of growth if we really want to comply with the SDGs in the matters of climate change, the protection of the oceans and terrestrial ecosystems, and stopping the loss of biodiversity. Tackling climate change means decarbonizing energy by replacing fossil fuels with renewable sources. Protecting marine life and oceans means reducing the tons of plastic thrown into them. Sustainably managing forests, combating desertification and soil degradation, and halting the loss of biodiversity means reducing deforestation and the use of chemical fertilizers in agriculture, and completely stopping the use of persistent pollutants (POPs), among other urgent measures.

As a consequence, these are the greatest challenges involved in fulfilling the 2030 Agenda, otherwise the possibility of a total breakdown of human civilization, as happened with the classic Maya civilization in Guatemala, the Norwegian settlers of Greenland and the inhabitants of Easter Island, as described in Diamond's book *Collapse* (2005), is not an implausible phenomenon. In addition, precisely because the debate about the compatibility of sustainable human development[16] with an economic system (which sees nature as an inexhaustible source of resources) is not

---

[16] The terms 'human development' and 'sustainable development' have been used separately in the context of the United Nations and the academic world in general. However, both concepts are part of the new paradigm proposed by the United Nations that has its origin in both the theory of human needs and respect for nature and the planet we inhabit (*Gaia*) as a vital source of all that exists, on which humanity depends and which we cannot help ourselves to as if it were an inexhaustible supply of resources, on pain of exposing *Homo sapiens* to the risk of its own disappearance as a species. In this sense it is interesting to quote the famous French philosopher and anthropologist Bruno Latour, who in an interview answered the question about how we should understand the Gaia concept: "Gaia is an entity of Greek mythology that was reinvented scientifically by James Lovelock. As a mythological figure she has a complex and cruel history. She is a goddess of many facets and metamorphoses, which describes her character and personality very well. And Lovelock turned to her to imagine a way to pay attention to the planet not as something inert but as a finished, complete form. Gaia means Earth, system, planet, but it is not a holistic definition, it is rather a set of connections between human action and natural action. Human history and planetary history come together in a process that I call 'geo-history'. Our predecessors never imagined that we would have to take the entire planet, with its geological ages, as part of our history" (Latour 2015).

resolved, it is logical to ask to what extent the global elites will be able to overcome the capitalism crisis provoked by the 2020/21 pandemic.

As noted before, the island of Tikopia is the only successful example of a self-sustaining, self-sufficient system which has existed for 3000 years. Naturally, it is not replicable anywhere else in the world because the seven billion inhabitants of the Earth cannot be compared with the 1200 people living in Tikopia – and nor can Tikopia be compared with the diverse, interdependent, interconnected and complex global system of the present time. Therefore, it is clear that the solution lies in the sustainability of a new planetary system that – if the *Great Transition* envisioned by Paul Raskin comes to life – will have to be constructed in the decades to come. However, we have to bear in mind that if, as Bruno Latour says, capitalism cannot be *subverted*, it could be *sucked down* because clearly there is "not enough planet for capitalism", as he argues. And, of course, as happens with the Tikopians, humanity is not in a position to leave our small 'island-planet' in the middle of the Universe, at least not with the existing technology of the twenty-first century.[17]

Consequently, later in this chapter I will briefly explore some of the alternatives to the so-called mainstream economics of the current 'model' of capitalism, and depict some of them according to their main features. Such alternatives have slowly appeared in various countries in both the Global South and the Global North, and they could corroborate Latour's premonitions about the transformation of the system by global civil society.

5.5.2 The Tribulations of World Capitalism

What can we ourselves do here and now in the face of the described tribulations of world capitalism and the need to promote this kind of *Great Transformation*, as Karl Polanyi called the rise of the liberal model in the eighteenth century. The 2020/21 pandemic crisis is not exactly a good context for promoting sustainable development, and it is evident that the UN SDGs are at great risk of failure, to say the least. The only positive outcome of such a catastrophe is that it will be quite difficult to return to *business as usual*, so – in theory – it could be easier for global civil society to get rid of neoliberalism. It seems that the conservative political sectors all over the world will no longer be able to question either the Keynesian policies applied everywhere or

---

[17] This would explain why a recurring theme in science fiction movies is the exploration of other planets for the purpose of extra-terrestrial colonization or leaving Earth after an ecological catastrophe, as depicted in Christopher Nolan's magnificent film *Interstellar*. In addition, in the same Chilean interview, Latour adds the following: "Be that as it may, capitalism has no future and has nothing to do with the future. It is a definition of what happens now and above all of what happened in the past. I think that everywhere…every human being is trying to find an alternative to capitalism and to get out of the apocalyptic vision of the left of the twentieth century, which still maintains the idea of subverting capitalism … Capitalism will never be subverted; it is not made for that. Capitalism will be sucked down, so to speak, by the alternatives that will appear in all parts of the world. Because maybe there is not enough planet for capitalism …" (Latour 2015).

the need to be prepared to face pandemics through state intervention in the economy and the reform of public health systems, especially in countries like the United States and Brazil, where the incompetence and irresponsibility of populist neoliberal right-wing rulers have led their respective countries to the worst situation in terms of the loss of human life and contagion.

It is also evident that life must be prioritized over the economy because with sick workers or consumers no market can function. The return to 'normality' will be impossible without a vaccine, but obviously the patents regime is an obstacle. Consequently, any vaccine must be free and put under the control of the World Health Organization not of a pharmaceutical transnational corporation. And a state basic income must provide for the needs of the huge numbers of people who have lost their employment, including the small and medium enterprises that went bankrupt. Besides food security is such an important issue that governments will have to protect peasant economy not agroindustry. The rich will have to pay for all these changes and tax reform must assure that states will have the financial means to face the cost of the crisis. Neither GDP nor economic growth must be the paramount criteria to determine 'development' performance. Instead a new index – like the one suggested by Joseph Stiglitz or the UNDP Human Development Index – which includes the satisfaction of human needs (the 'good living' or *sumac kawsay* of the Andean indigenous people) as well as respect for ecosystems and nature (Mother Earth or the *Pachamama*) must be created.

And, of course, the threat of climate change and global warming needs to be addressed seriously by government officials (as Greta Thumberg has been demanding) and a multilateral agreement to impose a tax to financial transactions is indispensable as well as sanctions to those governments that do not meet the targets to diminish GHE. The tax on financial transactions is important because the trillions of dollars invested by the banks in oil, petrochemicals, agroindustry, automobiles, aeronautics, armaments and so on must compensate for the polluting 'externalities' that harm the environment and human and animal health. A world tax on the most polluting industries and carbon emissions could also be agreed multilaterally to finance the UN and organizations like WHO, UNEP, WMO, ILO, UNICEF, etc. The financial sector of world economy must share the costs of the climate change mitigation measures, vaccines, universal basic income and so on because most environmental problems all over the world are the result of the domination exerted by the world transnational oligarchy of the super-rich. As Úrsula Oswald Spring – former president of the International Peace Research Association – explains, there is a

> crucial nexus between water and energy; water and biodiversity; water and food; energy and food; food and biodiversity; and energy and biodiversity in the current political arena of neo-liberalism. Key national figures act as part of a transnational oligarchy. Their productive and service systems are integrated within multinational enterprises and interlinked with global financial flows, the international homogenization of culture, fashion, military expenditure, and the arms trade. Their interaction maintains business-as-usual and concentrates wealth in the hands of a small oligarchy. They exercise pressure on governments through lobbying, bribes, and support for political campaigns. Through television, films, and social networks they influence global society, and through propaganda and fashion they promote

a consumerist behaviour that benefits their business. International organizations – the International Monetary Fund, the World Bank, and the World Trade Organization – support this global neo-liberal model. In the international political arena, no powerful institution exists to negotiate global governance. In the security area, the five permanent members of the Security Council of the United Nations (the United States, China, Russia, the United Kingdom and France) may veto decisions that challenge their geopolitical interests. This global arena and its actors promote a policy of business-as-usual, causing human insecurity, especially that of the most vulnerable (UNEP 2014). The multiple nexuses between water, food, energy, and biodiversity security, because of this focus on state-centred concepts of security, seem to reinforce this model of reference, but at the same time are affecting issues of gender and environmental security. The global economic, political, ideological, and military lobby is responsible for numerous obstructions in the UN against achieving universal legally binding agreements on the three conventions negotiated at Rio de Janeiro in 1992 on climate change, UNFCCC, desertification UNCCD, and biodiversity CBD (Oswald Spring 2018: 135–136).

The main concern of this transnational neoliberal oligarchic group is to maintain *business-as-usual*, exercising pressure on governments through lobbying, corruption financial support for political campaigns, and so on. These kinds of policies determine human and planetary *insecurity* instead of security because the agenda of political actors within the global economy (such as the financial sector) with regard to energy, water, food and biodiversity means they are financing fossil fuel industries, petrochemicals, fracking and biofuels instead of the good renewable energy sources like geothermal, tidal, hydro, solar and wind power. Nor have these powerful banks ever invested in organic agriculture or renewable energy, as they usually put their money into the oil and automotive industries, armaments, fertilizers, pesticides and the agroindustry controlled by mega corporations like Monsanto, Dupont and Union Carbide. Thus both the world financial sector and the mega corporations in the field of fossil fuels, biofuels, carbon, agroindustry, armaments, transports etc. should pay for the policies addressed at reducing GHE gases via a tax on financial transactions and carbon emissions and stop talking about 'externalities' as their ideological justification.[18]

---

[18] A *Financial Transaction Tax* (FTT) is already implemented by some governments (Belgium, Colombia, Finland, France, Greece, India, Italy, Japan, Peru, Poland, Singapore, Sweden, Switzerland, Taiwan, the UK and the US) but as a national contribution. The idea of a *"Global Tobin Tax"* was originally suggested by John Maynard Keynes himself and afterwards reviewed by James Tobin, a US Nobel Prize Winner in economics, and proposed again in the year 2000 by the 'pro-Tobin tax' French NGO ATTAC. The proposed tax could have been used to fund national sustainable development projects in the face of increasing income disparity and social inequity and inequalities. The *Tobin Tax* could represent a rare opportunity to capture the enormous wealth of an untaxed sector and redirect it towards the public good on a world scale. Conservative estimates show the tax could yield from $150 to $300 billion annually and the revenues raised could be used for the sustainable development objectives (SDGs) of the UN 2030 Agenda. Another possible tax could be the *Robin Hood Tax* proposed on February 2010 by a coalition of 50 global civil society organizations that launched a campaign for a tax on global financial transactions that could have affected a wide range of asset classes, including the purchase and sale of stocks, bonds, commodities, unit trusts, mutual funds, and derivatives such as futures and options. Another possibility for global taxation imposed on financial institutions and TNCs is the French proposal for a *G20 Financial Transaction Tax* (G20 FTT) that was suggested in 2008 to raise revenue to reduce global poverty and spur global economic growth, but it failed to be approved at the Cannes meeting of that year. Apropos, Bill Gates

Nonetheless, it is precisely for this reason that the implementation of the commitments adopted within the framework of the 2030 Agenda of the United Nations is going to face difficulties and not just because of the coronavirus crisis. For example, the tenth SDG seeks to diminish inequality between nations, terminating the dependency and *de facto* limited sovereignty of peripheral and underdeveloped countries. Nonetheless it will be extremely difficult to implement the sustainable development paradigm if the neoliberal economic model is not abandoned. Regarding this issue, Göpel (2016) suggest a "radical repurposing agenda" in order to prioritize social welfare and the conservation of natural ecosystems as absolutely indispensable for fulfilling the commitments of the 2030 Agenda. In another example, SDG 8 endorses the idea of promoting "sustained, inclusive and sustainable economic growth, full and productive employment and decent work for all", which essentially means that decent work – wages according to the cost of living – must be prioritized over growth. But the neoliberal mantra explains that "comparative advantages" to attract *foreign direct investment* (FDI) essentially consist of paying low wages. Obviously, workers who do not accept this neoliberal dogma decide to migrate to the US precisely because, despite exploitation, they are better paid, as demonstrated by remittances. Thus, if governments really want to implement SDG 8 they must facilitate social inclusion, which means they will have to question neoliberalism and develop the capacity to deal effectively with the change in the so-called *Washington Consensus*. Without serious negotiations with the private sector the attainment of the said goal would not be possible. Consequently, it is clear that decent work and "sustained growth" are contradictory, indicating the need to elect progressive leaders to negotiate with the business elite of each country. It is also evident that even if the 2020/21 pandemic favours a change of perspective among oligarchic groups – taking into account that

---

is among the supporters of a G20 FTT, arguing that even a small tax of 10 basis points on equities and 2 basis points on bonds could generate about $48 billion from G20 member states or $9 billion if only adopted by larger European countries. Concerning the carbon tax, the Carbon Tax Center says on its website: "A carbon tax is a fee imposed on the burning of carbon-based fuels (coal, oil, gas). More to the point, a carbon tax is the core policy for reducing and eventually eliminating the use of fossil fuels whose combustion is destabilizing and destroying our climate. A carbon tax is a way – the *only* way, really – to have users of carbon fuels pay for the climate damage caused by releasing carbon dioxide into the atmosphere. If set high enough, it becomes a powerful monetary disincentive that motivates switches to clean energy across the economy, simply by making it more economically rewarding to move to non-carbon fuels and energy efficiency. Carbon chemistry is potent but also simple. The amount of CO_2 released in burning any fossil fuel is strictly proportional to the fuel's carbon content. This allows the carbon tax to be levied 'upstream' on the fuel itself when it is extracted from the ground or imported into the US, which vastly simplifies its administration. The energy essence of every fossil fuel is its carbon and hydrogen atoms. Oxidizing (combusting) those atoms releases their heat energy but also converts carbon to carbon dioxide. Natural gas, with a high ratio of hydrogen to carbon, is the least carbon-intensive fuel, while coal is the most. The CO_2 released from burning these fuels rises into the upper atmosphere and remains resident there – typically for around a century – trapping heat re-radiated from Earth's surface and causing global warming and other harmful climate change. The carbon content of every fossil fuel, from anthracite and lignite coal to heating oil and natural gas, is precisely known. A carbon tax obeys these proportions, taxing coal more heavily than petroleum products, and much more than natural gas. This makes a carbon tax simple to document and measure" (Carbon Tax Center 2020).

sustainable development is in the long-term interests of rich people – short-term economic interests could also influence the decision. Consequently, a possible way out of the dilemma could be pressure from the international community. In other words, if rich countries of the Global North really want to stop migration from the 'poor' countries of the Global South, they need to understand that conservative oligarchies must be convinced that their long-term interests must prevail over economic growth and personal gain.

Thus it seems quite clear that sustainable development supposes a displacement of the neoliberal ideology from its current hegemonic position of mainstream economics and that the criteria that should guide both national and global public policies require cosmopolitan global governance based on the well-being of both the people and the planet. Hence, neither unlimited growth nor consumerism should be regarded as suitable parameters for the assessment of sustainable development. Instead the principles of *doing things better* and *doing things well* suggested by Göpel (2016) should be used. *Doing things better* means discontinuing the production of non-recyclable goods, either because they are harmful to the environment or to stop the excess of waste and garbage. *Doing things well* means decoupling production from the imperatives of growth, because if the satisfaction of human needs and respect for natural ecosystems are the paramount objectives of sustainable development, they must prevail over growth, and new parameters (such as Buthan's 'gross national happiness') must be found to assess progress and social well-being.

Sustainable Development Goals will not be achieved unless governments seriously start using the criteria of human needs to guide their policies and disregard the quantitative indicators of economic growth and per capita income. In another example, the sixth SDG concerns every government's obligation to "ensure availability and sustainable management of water and sanitation for all". However, water resources are mainly being used for agroindustry, mining and industrial production, not for drinking and sanitation, with terrible consequences for health (such as child malnutrition and the impossibility of washing hands in these pandemic times). Therefore, the priorities of poor countries ruled by neoliberal oligarchies must be changed to meet the basic needs of the population through the use of water for sanitation and drinking purposes instead of prioritizing the interests of the extractive sector and big agroindustry plantation owners.[19]

---

[19] In Guatemala, to give concrete examples, on two occasions (concerning the OXEC hydroelectric company and the San Rafael mining company) the highest courts ruled in favour of the local communities when they protested against the breach of international standards, according to ILO Convention 169, which requires prior consultation with the local population. However, in OXEC's case, the pressure of the business elite led the Constitutional Court to reverse a previous ruling in favour of the community, while the case of Minera San Rafael, whose operations are suspended due to legal issues related to the issue of the query, is still not resolved. On the other hand, and bearing in mind what was mentioned earlier regarding the need for social policies to be financed by taxes in the framework of the redistributive obligation of the State (and the reduction of inequality as requested by SDG 10), it is interesting to note that in the face of the private sector's argument that alleges the risk of job losses, a reduction in State tax revenues and discouragement of foreign investment when such cases are brought to the courts, Lourdes Molina, a researcher from the ICEFI (*Central American Institute for Fiscal Studies*), said in a press interview that the contributions

What is at stake is both human development and ecosystems sustainability, which echo the issue concerning the rights of Mother Earth (*Pachamama*) that have been recognized by the new constitutions of Bolivia and Ecuador and grant individuals the right to act on behalf of the Pachamama.[20] The recognition of the traditional philosophy of indigenous people in both countries has also promoted an ecological ethic in which the care of nature includes the assumption of full respect for ecosystems and biodiversity, as Pope Francis demands in his Encyclical Letter *Laudato Si'*. Consequently, environmental regulations (concerning sewage treatment plants, prohibition of plastic bags, etc.) are urgently needed, especially in countries of the Global South.[21] Hence, neoliberalism is absolutely incompatible with sustainable development, and in order to achieve the *new normality* required all over the world after the pandemic full respect for nature and environmental balance (which depends on cultural vision and ecological culture) also require constitutional and legislative reforms to provide the legal basis needed to apply public policies of management and protect natural resources and ecosystems.[22] Hydroelectricity is an important instrument for decarbonizing energy and reducing GHE emissions, but the construction of

---

of the extractive sector (mining) to the Guatemalan economy during the period 2005–2016 were only 0.07% of GDP according to data from the central bank (*Banco de Guatemala*), so for every one hundred 'quetzales' – the national Guatemalan currency – produced in the country, only 0.70 (cents) corresponded to the extractive industry. In terms of employment, the IGSS (the Guatemalan social security institute) indicates that extractive mining generated only 4,800 jobs in 2016 and in fiscal terms the tax burden of the sector only represented between 0.85 and 5.6% – well below the national average – while in relation to the budget its contribution is about Q.400 million annually (about US \$55 million). Royalties are only between 0.3 and 0.7% or Q.67.4 million (about US \$9 million) which was just 0.10% of the 2016 state budget. All this is according to data from the Superintendence of Tax Administration (SAT). Given such low figures, the best public policy for the country could be to declare a moratorium on extractive activities. Concerning water, Guatemala simply does not have a system for regulating the use of water resources (Padilla 2018).

[20] I am perfectly aware that according to traditional legal theory, only persons can have subjective rights. However, in the new constitutions of Bolivia and Ecuador, Mother Earth itself (the 'Pachamama') has been declared a holder of rights, as described on previous pages. The new constitution of Bolivia has five articles and the Ecuadorean one has twenty-seven articles regarding this issue. In the case of Bolivia, the concept has a spiritual meaning and is rooted in indigenous traditions: "We populate this sacred Mother Earth with different faces [...] Fulfilling the mandate of our peoples, with the strength of our Pachamama and thank God, we re-found Bolivia." According to Barié (2018), the concept of *Pachamama* as a benevolent deity and fertility goddess justifies the belief in Ecuador that the deity deserves a special chapter in the Constitution and is considered a subject of rights because "nature or Pachamama, where life is reproduced and carried out, has the right to be fully respected [in terms of] its existence and the maintenance and regeneration of its lifecycles, structure, functions and evolutionary processes. Every person, community, town and nationality may demand from the public authority the fulfilment of the rights of nature" (Barié 2018: 54–55).

[21] An example of this is the Motagua River in Guatemala, because the amount of plastic garbage and other solid waste that it drags into the Caribbean Sea is of such magnitude that it has provoked protests from Honduras. The Pacific Ocean is also suffering from the drainage of the hyper-contaminated Lake Amatitlán near Guatemala City.

[22] Incidentally, and as the cases of the Motagua River and Lake Amatitlán were mentioned in the previous footnote, it is important to underline the fact that Guatemala is probably the only country in Latin America with the dubious distinction of lacking a legal water standard due to the powerful

such facilities must be done in accordance with the needs of the communities that depend on those water resources, and the same policy must guide the installation of other sources of green energy, such as wind, geothermal energy, solar power and biomass. However, it is important to remember that some international environmental NGOs (such as the IUCN and *The Nature Conservancy*) and authors (e.g. Lovelock) have argued that human development is contrary to environmental conservation and protection policies regardless of the nature of the economic model.

Thus alternatives for economic 'de-growth' suggest deeper changes to the theory, as proposed by authors like Rockstrom (2009) and Latouche (2009, 2012). According to these ideas, humanity has already severely breached the limits of three fundamental ecological 'frontiers': climate change, loss of biodiversity, and interference with the nitrogen cycle due to the atmospheric fixation by industry and agriculture, the exaggerated burning of fossil fuels, and the pollution of rivers and coastal areas. Therefore, as nature is subsidizing economic growth (the so-called 'externalities' of economists), unlimited growth is incompatible with sustainability, and, as Latouche argues, development is just a slogan to conceal what he calls the 'Westernization' of the world. Consequently, what is needed is a complete transformation of the economy to promote degrowth. Latour's criticism of economic neoliberal orthodoxy condemns the encompassing reductionism of everything to economy, utilitarianism and concepts like development, efficiency, growth and sustainable development itself, because humanity must organize degrowth "to survive or last", which – in light of the global crisis provoked by the coronavirus pandemic – seems a premonitory vision of what was going to happen in 2020. His criticism of what he calls "grand society" refuted by the "informal sector" deserves to be quoted extensively:

> The normal response to announcements that modernity's grand society is collapsing is a sceptical one. Hopes placed on the performance of the barefoot friars who live in the city margins or in the countryside of another age would seem to be fragile indeed. Even the best successes in the informal sector or the achievements of some bush peasant groups look a little derisory beside the opulence of jet-age society capable of conquering space and manipulating nature. ... So it becomes inconceivable, once exposed to this true path, not to follow it – unless, of course, you are excluded from it. ... Considered in these terms, the informal is a serious threat and a paradox for modernity. Understandably enough, its existence was for a long time denied outright by Western analyst – it was trivialized as being the mere vestige of a disappearing past. But finally it has had to be tolerated in the face of evidence that it refuses to go away. ... It seems that grand society, whose own rationality decrees all other societies to be irrational and which acts the self-fulfilling prophet by exterminating them, has ended up by creating and nourishing a new irrational which, refusing to be abolished, must be recognized as incarnating an *other* reason. The proliferation of the informal thus scandalizes Western reason. History itself starts to lose sense. If several social forms, even all forms, can coexist, and some of them that we thought that were outdated can come back into use, then neither evolution nor progress can be said definitely to exist! ... Judged unfit to play a role in the techno-economic machine, these outcasts survive by reactivating solidarity networks and reinventing a lost mode of social interaction. The network of reciprocity and the logic which permits their persistence and proliferation derive from a *re-embedding* of the economic within the social (Latouche 1993: 48–49).

---

opposition of those who benefit from the absence of regulations and the prevailing neoliberal model in the economy: *laissez-faire, laissez-passer* (Padilla 2018).

Hence, according to Latouche, this informal nebulous patchwork of hangovers from the past and borrowings from modern techniques constitutes a fundamental challenge to the exclusive pretensions of "grand society" because if the hypothesis of a separation of the economic dimension from the social sphere is refuted in agreement with Karl Popper's epistemology of the presence of economic informality at the social level, the neoliberal assertion that capitalism is a unique mode of production and the only way to reach progress and development is false. This is why Latouche says that the reactivation of solidarity networks ("a lost mode of social interaction") is a way of re-embedding the economic within the social, thereby ending the erroneous neoliberal idea that the economic level (international financial markets) functions independently from society and the State. Certainly, this analysis can also be applied to the peasant indigenous communities that produce food in the highlands of Guatemala and in sub-Saharan countries. The subsistence lifestyle of this 'traditional' economy substantiates methodological holism and the fact that both the transdisciplinary approach of Morin and Nicolescu and the *ecology of knowledge* perspective of Santos (2010) are on the right path. Life and objects of the material world are interconnected and interdependent, which explains why, from a cosmopolitan philosophical perspective, holism and transdisciplinarity are better equipped to deal with the complexity of reality, the "nature of things" and the cosmos.

5.5.3   Are We in the Midst of the Sixth Mass Extinction?

Latouche's ideas about degrowth and his criticism of development as a hoax, and Lovelock's omen about the "revenge of Gaia" and proposal to abandon the idea of sustainable development because for him any 'development' is unsustainable due to climate change and the impossibility of stopping global warming, have influenced the title of this book: is the Anthropocene the signal that our species is in the middle of the sixth mass extinction?

Palaeontologists like Wake/Vredemburg (2007) opened a debate about the mass extinctions that have occurred in the geological history of the planet, all of which are related to the risk of the disappearance of the human species if its current economic behaviour continues to transgress the planetary boundaries. The planet existed without humans for millions of years. The first *Homo sapiens* date from only about 200,000 years ago and the introduction of agriculture, which initiated the substantial changes caused by man on the environment, took place approximately 12,000 years ago, which was the beginning of the Holocene. Before these interferences by mankind, the changes in matters of biodiversity (extinctions included) can be considered 'natural'. This prompts the question whether humanity is nothing more than a mere 'chapter' (the Anthropocene chapter?) in the geological history of the planet, because even though cultural evolution could have replaced biological evolution in humans, natural selection is still shaping our biology in response to environmental change. Therefore, humanity's current state of evolution is clearly not necessarily the last word in evolutionary terms. Furthermore, there is no guarantee that we, humans, will still participate in the long-term geological evolution of the planet.

On the other hand, the phenomenon of mass extinction has led the paleontologist Peter Ward (2009) to propose the *Medea hypothesis*, which contrasts with Lovelock's optimistic *Gaia vision*. Ward argues that evolution and the process of change in nature, which is not linear but cyclical (as noted previously in this book), imply that the mass extinctions that have happened throughout the millions of years of Earth's geological history will continue to happen, therefore at any moment the planet could react like Medea killing her own children and make *Homo sapiens* disappear from Earth.

Ward's Medea hypothesis stands in opposition to Lovelock's theory of Gaia and the idea that the most important characteristic of the planet as a living entity is its tendency to optimize the conditions of terrestrial life. Ward argues that if the mass extinctions that Planet Earth has had in its geological past are cyclical – as natural processes tend to be – the cyclical ruptures of planetary equilibrium are also natural. This conception differs from the Darwinian theory of a linear and progressive evolution and disagrees with the philosophy of the palaeontologist and theologist Teilhard de Chardin, who regards the *omega point* as the culmination of the evolution of the spirit manifested in the geological history of the planet. Likewise Ward does not share the pantheistic optimism apparent in James Lovelock's palaeontological debate on how the deep interior of our planet works. Ward contends that Gaia can convert herself into a sort of Medea goddess who does not worry about optimizing life on the Earth. His pessimistic vision includes our species, which is hence doomed to extinction.

However, although, according to the most widely accepted hypothesis, the last mass extinction that caused the disappearance of the dinosaurs some 65 million ago was the result of an external agent (the impact of a meteorite) that provoked a devastating greenhouse effect which led to cataclysmic climate change, according to Ward the previous five mass extinctions were clearly endogenous. Examples of these phenomena include the oxygenation of the atmosphere caused by the proliferation of cyanobacteria that in turn generated photosynthesis, the appearance of the plant kingdom and all the aerobic beings that breathe oxygen, and the temperature drop that caused the glaciations and ice ages of the planet, all of which were the result of natural cycles. In that line of thought, the sixth mass extinction is being caused by the loss of biodiversity as a result of the impact of human activity on the planet, which is causing humanity to assume the role of Medea, a heartless mother who, according to the Greek myth, killed her own children in revenge for her husband's infidelity.

Of course, this hypothetical extinction of our species in the foreseeable future could eventually be accelerated by human action, hence the term 'Anthropocene' for a phenomenon that cannot exclusively be related to the natural cycles in the geological history of the planet. In an interview published in *New York Magazine*, Ward acknowledged that he is in favour of decisive action to combat global warming, underlining the danger posed by the polar thaw and the threats related to the increase in ocean levels that would require the construction of a massive and enormously costly infrastructure to impede flooding in order to safeguard coastal cities. He additionally held that the rise in ocean temperatures will cause the disappearance of marine species due to the reduction of oxygen in the water, and that global warming could also make life in the tropics unviable and jeopardize human life in regions like Australia and the Maghreb, where desertification is leaving Tunisia – one of the breadbaskets of Rome in ancient times – almost without the production of wheat. But one of the worst forecasts of Ward is that desertification will advance in such a way that in the twenty-first century violent conflicts and wars will be fought over water and food (Ward 2017).

From my point of view, this analogy with the tragic Greek Medea myth equates to a fatalistic philosophical vision that denies human freedom since – according to Ward – our species is condemned to suffer an ineluctable destiny. I do not share Ward's determinist vision, although I admit that the debate opened by his hypothesis is valid, especially regarding the crisis of the 2020/21 pandemic, because the predatory behaviour of human beings towards the planet and Mother Nature could lead to our extinction if we don't react promptly and adequately to threats like climate change and pandemics provoked by malign viruses – the coronavirus is just one more of a long chain of lethal microbes. This again indicates that humankind must embrace a mind-shift by disowning neoliberalism, the main culprit of the disarray and disasters created by an ideological economic doctrine that does not care about health and life (human security) or the environment and planetary boundaries (global security). I outlined in the previous chapter how, according to social scientist Jared Diamond, some civilizations (the Mayas in Yucatan and Guatemala) and settlements (the Norwegians in Greenland, the Polynesians in Easter Island) collapsed totally after failing to adapt to environmental strains after climate change triggered by cultural and collective comportment, while other civilizations and cultures (the Japanese of the Tokugawa dynasty, the natives of the highlands of Papua New Guinea, the natives of the tiny island of Tikopia in the South Pacific Ocean) succeeded in avoiding disastrous climate change by respecting nature (forests and the typical tropical vegetation) and taking appropriate decisions in time, thanks to the forethought and wisdom of their respective ruling elites. This essentially means that I believe in the capacity of humankind to overcome the challenges of destiny using intelligence, knowledge and freedom. Thus, if Gaia can conquer Medea our species is not doomed and hopeless. But I cannot make any predictions; it all depends on our behaviour from now on.

So, regarding sustainable development when our species is in the midst of a sixth mass extinction accelerated by human agency, we must ask ourselves how can we manage to activate the change in collective and individual mindsets that good governance requires and what kind of policies must be implemented by states in order to get rid of the neoliberal "mainstream economics" that have prevented governments fully complying with SDG commitments. We must also assess the options for rescuing the UN 2030 Agenda in the wake of the COVID-19 pandemic. On the following pages I will review some of the micro-economics experiments and policies carried out in different countries and regions following the publication of Maja Göpel's book *The Great Mind-Shift* (2016), in which she asserts that a change of paradigm can be obtained through what she calls "radical incremental transformations" and "small niche-innovations", including methods like the 'good matrix' and balance sheets; the experiment in transition towns for resilient local solutions; the 'beyond GDP measures community'; and the kingdom of Bhutan in the Himalayas, that 'new stark utopia'.

5.6   Maja Göpel and Renewable Energy in Germany: A Change of Paradigm?

It is now time to refer to some of the more positive experiences that could indicate a change of paradigm, such as the situation in Denmark, where appropriate policies made it possible for local communities to launch numerous renewable energy projects, and the most notable case of Germany, which has exceeded its official goal of obtaining at least 45% of its energy from renewable sources by 2020. This change of paradigm in energy matters is of such magnitude that a substantial transformation in the thinking of citizens and the private sector has already occurred – a *mind-shift* described by Göpel (2016: 25) which demonstrates that it is feasible to change the dominant paradigm in the field of *socio-technical systems* (STS). This turn means that a multilevel perspective and its *incremental radical transformations* can also be applied to other spheres of human activities, such as the environment, production, trade, and so on. This is not a minor issue, since Germany is not just an economic power but also the most influential country within the European Union. Despite being highly dependent on the importation of hydrocarbons, it fostered a group of visionary entrepreneurs (who could met with opposition through the lobbying of conservatives) in order to promote the ecological niches indispensable for transforming the energy matrix by replacing fossil fuels with renewables. Göpel describes the process by which a clever mechanism effectively created a return on investment security that convinced banks and even risk-averse investors to lend small businesses, farmers and citizens money to install renewable energy technology, thereby creating a market for renewable energy even though the sector had previously been dominated by an oligopoly of powerful companies:

> Since these older business models had rendered the transaction costs of switching to renewable energy solutions prohibitively pricey in the past, no pioneering movement had been possible. The Renewable Energy Law hedged the risks of a plethora of new, decentralized energy producers and unleashed the competitive activity of many *small- and medium-sized enterprises* (SMEs) active in technology development. By the mid-2000s the tipping point into the acceleration or navigation phase had been reached, and pioneering activities had become mainstream considerations. Fossil energy suppliers now felt threatened and tried to fight the regulation at all levels, e.g., attempting to make the EU declare feed-in tariffs incompatible with energy market integrations. But the renewable energy sector grew very quickly, created many jobs in rural areas with high unemployment and turned Germany into an international technology leader that inspired other countries. Thus, an environmental issue had found technology solutions and became solidly economic when it served the export interests of the German economy and found wide, bottom-up citizen support. The share of renewable energy in the electricity mix increased steadily, debunking the strongly spun narrative that renewable energy systems were technologically unfeasible – although concerns about black-outs remain. However, a new narrative was established in which a transformation of the

energy sector was both possible and in progress, drawing in many new participants (Göpel 2016: 25).

In other words, the strategy of a gradual radical transformation worked: it started from a pre-development phase, moved towards the take-off phase, and reached a point of maturation (tipping point) that led to a third phase of acceleration, until finally, the *socio-technical system* (STS) stabilized at a new level. This same process can be seen in a multi-level perspective, demonstrating that changes can be promoted from small innovative niches which, once gathered in a design that brings them together, can take advantage of windows of opportunity to lead to a new regime that exerts influence on the upper level of the 'socio-technical landscape'. That is the level at which the ideology, the 'mindset', and the predominant way of seeing things in society is located and, despite belonging to an exogenous context, will react downwards so that the preferences of markets and consumers are properly articulated within public policies, culture, industry, science and technology in order to provide feedback through expectations and social networks to the small innovators who started the whole process. Figure 5.3 clearly shows how, starting from innovative actions at micro level (niches), influence is exerted on science, public policies, market patterns, social practices and the same technology at an intermediate level, thereby affecting the higher level, which is the field of 'meta-narratives' where world-views ('cosmovision') and paradigms are located. This all provides feedback to terrestrial ecosystems, which benefit from the change in the energy matrix by completing the

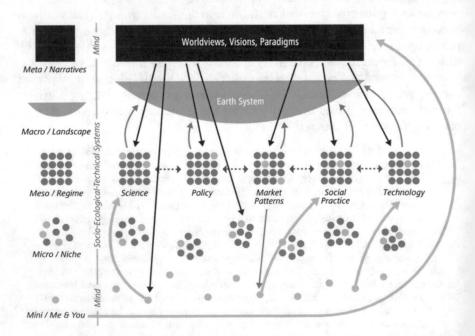

Fig. 5.3 Mindsets in the multilevel perspective on transformation. *Source* Göpel's book (2016: 47). Reproduced with Göpel's permission

cycle and, in the example provided, is consistent with moving from the predominance of non-renewable energy to renewable energy.

It is also important to point out that the *mind-shift* necessary to promote sustainable development coincides with the ideas of Sachs (2015), who argues that reconciling economic growth with ecological realities requires a profound change of economic model. It also aligns with the views of Klein (2015) and Wolff (2012), who are in favour of models in which local companies are managed democratically in the form of cooperatives or communal enterprises to produce food, energy, crafts and goods of all kinds in a renewed version of socialism.

5.7   Other Experiences and Economic Alternatives

5.7.1   The Matrix of the 'Common Good'

The first pioneering initiative worth describing comes from the business world, and originates from the well-known concept of corporate social responsibility, based on the idea of finding a balance between responsibility towards the community and free enterprise because neither works in isolation: individuals need to cooperate to "flourish and create wealth", while the community needs individual entrepreneurship to maintain its ability to adapt and diversify the goods and services that are produced. According to the proposal of Christian Felber, author of *The Economy for the Common Good* and the main leader of the movement, the key to sustainability lies in connecting the private sector to the old Christian idea of the *common good*, hence the name, because if a company is conducted in an appropriate and honest way it undoubtedly contributes to the welfare of all, i.e. operates for the common good.

The movement proposes new rules for the game, one of which is to avoid regarding the *externalization* of social or environmental costs (use of water or forest resources, polluting greenhouse gases) as a 'competitive advantage'. Naturally, either all entrepreneurs agree to internalize these costs (paying special taxes for emitting GHG or for the use of water, for example), or competitors who do not internalize them have greater benefits thanks to these subsidies of nature. Hence new rules are required and these must be accepted by all on the basis of social responsibility, or at least by a majority if the internalization responds to a legal obligation. Consequently, these new rules with internalized costs (formerly subsidized by nature) must be reflected in balance sheets which no longer measure economic well-being by indicators of *exchange-value* but instead by those of *use-value*.

Indeed, the common good movement has proposed that GDP be replaced at *macro-economic level* by the new indicator *product of the common good*, which should be reflected at *micro-economic level* with a new type of balance sheet for the common good (*Common Good Balance Sheet* or CGBC). Since 2010 this new indicator has been used by a pioneering group of 70 European businessmen who by the middle of 2015 had increased to 1811, along with 232 clubs and more than 6,000 individual members. As for the indicators of the common good, there are five categories – human dignity, solidarity, social justice, ecological sustainability, and democratic decision-making and transparency – which can be applied to groups or *stakeholders* such as suppliers, financial creditors, employees and co-owners, partners, and clientele. Moreover there are indicators of *bad behaviour* connected with the violation of labour standards established by the ILO and the OECD, such as human rights, labour protection, environmental standards, tax evasion, the secret use of payments to lobbyists, subsidiaries, prices below cost, and dumping.

The initiative of the common good encourages communities and local governments to support this type of business with consumer loyalty and public recognition, and even to make their own evaluations in the so-called "common good regions" which already exist in 45 countries, including Austria, Germany, Switzerland, Italy, Spain, Portugal, the UK, Greece and the United States. The vision of the movement is to have a global reach, which means fostering self-reliance and promoting changes to overarching regime structures in such a way that the 'normal' hegemonic mentality guided by interest in economic gains and profits is transformed, along with the way of doing business, introducing, for instance, Göpel's ideas about *doing things well* and *doing things better* (Göpel 2016: 112–126).

5.7.2 Communities in Transition: Resilience in Local Solutions

Originating in the United Kingdom, the *Transition Towns* movement has already spread across Europe, seeking to promote a reflexive relocation promoted by *resilient* communities. Göpel tells us that *resilience* should be understood as a systemic characteristic, according to which systems capable of resisting crisis situations endure them with bounciness and flexibility, thus overcoming and recovering from them sooner in a way similar to those communities where services are restored rapidly after natural catastrophes like earthquakes, floods, volcanic eruptions, hurricanes and tropical storms. Therefore, this resilience has a dynamic quality that allows a system to learn, create and redesign essential processes for its proper functioning, thanks to its capacity for self-organization and restoration, as occurs with the *autopoiesis* of nature.[23]

---

[23] The French social scientist Serge Latouche argues that the informal sector is resilient. This feature is understood as a systemic characteristic which does not just apply to resilient communities after natural catastrophes but also to autopoietic 'common good' communities, transition towns, the informal sector solidarity networks in developing countries, the social solidarity economy groups of Howard Richards, and even the Zapatista social movement described above (Richards: 239–243). Indeed, communities around the world which resist and overcome the 2020/21 pandemic will do so as a result of their resilience and autopoietic capacity. It is appropriate to note what Maja Göpel affirms in her quotation above, because what must be pursued is not the resilience of the economic system but that of the ecological system and its biophysical qualities which can be 'irreversibly

One of the leaders of the movement, Rob Hopkins, also adds a socio-political component to the definition of resilience when he says that all resilient communities are sustainable because they are structured around a variety of subsistence sources, have a modular structure that gives them autonomy and flexibility in front of external systems, and have firm feedback loops. In *Transition Towns*, economic activity is not the fundamental parameter which guides development, but a process subordinated to the satisfaction of human needs, which must also be carried out without contradicting the natural laws of terrestrial ecosystems. As the economy is not the main focus, moving away from economies of scale (producing in larger quantities to reduce prices) is viewed positively, which not only means that growth is not the primary concern but also that a certain decrease in growth as a consequence of manufacturing better-quality, long-lasting goods is acceptable, with the advantage of reducing the risk of interruptions to supplies when there are transport difficulties or when companies have financial problems. Each transition or transformation of a transition town is expected to take twenty years, but the movement's website also includes a map with registered initiatives from all over the world (479 official initiatives in 43 countries in 2015) which can be located through an interactive map that from a multilevel perspective forms perfect niches ready to build coalitions for regime change, as Göpel (2016) remarks.

It is important to stress that the movement's vision is in line with the idea that the purpose of production is to satisfy human needs, and it is interesting to note that the movement's training workshops use the matrix of Manfred Max Neef to discuss various strategies to achieve a better quality of life, forsaking the criterion of individual enrichment as an objective. Thus issues such as community welfare, meaningful and satisfying work, friendship between members of the community, and sharing experiences are central aspects of the labour relationship. As Maja Göpel points out in the *Great Mindshift*:

> We can clearly see how many of these principles fly in the face of mainstream models. Actors are explicitly requested to change their way of thinking and being and to share instead of compete. The processes of providing energy and food are intentionally made less efficient so that they become more resilient. The economic system is analysed as a subset of SES's that can and should fundamentally change if it hurts the balance of ecological reproduction circuits. It is therefore not the resilience of the economic system that successful development strategies should pursue but that of the ecological systems with biophysical qualities that can be irreversibly altered. This point is important because otherwise 'resilience', as a concept, might be used in as undifferentiated a fashion as 'efficiency'. Resource efficiency is basically always desirable, but making efficiency a core value in and of itself, for all processes, is taking it too far. The same holds true for resilience, as the financial system shows nicely. From a sustainability perspective it is utterly damaging, but its immaterial qualities mean that very quickly many new financial products and instruments emerge if others are ruled out. It thus shows a great self-organizing capacity and has bounced back from the 2008 crisis without significantly changing its functions, structure, identity or internal connections. Instead, the financial system would require a decrease of resilience so that transformation could take place. Thus, the goal of improved resilience can only be of added value for sustainability

---

altered' by the predatory behaviour of capitalism. The financial system showed such a resilient and self-organizing capacity after the 2008 crisis that it is possible to think that less resilience would have been better because then governments would have been able to regulate the financial markets.

when the purpose and setup of the system in question is one in line with what sustainable development on the macro scale requires (Göpel 2016: 19).

Apart from the substantive issues related to the fact that economic actors are explicitly asked to change their way of being and thinking and learn to share instead of competing, what Göpel says about the concept of 'resilience' is really fundamental, because it applies to sustainable development. It is therefore essential to be aware that all four spheres – economic, social, environmental, and governance – need to be taken into account to strike the right balance, because otherwise resilience may be inconvenient. For example, the financial system needs to be less resilient, not more, as noted by Göpel for the reason that the way in which the banking system 'recovered' after the sub-prime crisis of 2008 in the United States (thanks to the funds granted by the US federal government) was incredibly rapid but unfortunately continues to cause havoc both in the US and Europe, as pointed out previously when Varoufakis's ideas were outlined. Because the Obama administration refused to significantly change the functions and structure of the financial system and allowed Wall Street to remain the power behind the throne, European countries – especially Greece and Italy – are still struggling with the financial crisis, another fact which demonstrates that the said system requires less resilience, as Göpel and other authors have argued,[24] to give governments the leverage to make changes in the public interest.

An additional important aspect of these communities in transition is the concept of empowerment, autonomy and self-reliance that is the essence of their existence, enabling communities of all types and sizes to adapt to a mission which is intended to "inspire, promote, connect support and empower communities while adopting and adapting to the transition model on their journey to rebuild the resilience essential to drastically reduce greenhouse gas emissions" (Göpel 2016: 128). This essence is based on principles which emphasize being positive (conducting campaigns 'for'

---

[24] These authors include Paul Mason, who – coincidentally – refers to the inconvenience of having a resilient financial system, as argued by Göpel. For Mason:

When the global financial system collapsed in 2008 it did not take long to discover the immediate cause of such a debacle. The culprits were the hidden debts in products known as 'structured investment vehicles' (subprime) valued at deceptively unrealistic prices and the network of companies not regulated and domiciled in known tax havens, from the moment the system began to implode, as the 'shadow banking system'. Then, when the judicial proceedings began, we were able to verify the scale of all the criminal activity that led to the crisis and how normal it had seemed to us … Seven years later the system has been stabilized. By increasing the public debt of many countries to levels close to 100% of GDP, and also by printing money for an approximate value to one sixth of the world production of goods and services the United States, Great Britain, Europe and Japan managed to inject a dose of adrenaline powerful enough to counteract the attacks and tremors of the system. They saved the banks by burying their uncollectible debts; some of them were simply cancelled, others were assumed by the States in the form of sovereign debt, others were buried in financial entities to which the central banks of their respective countries gave a semblance of security by playing their own credibility in it (Mason 2016).

That is what the 'resilience' of the financial system consists of, which is why I favour the imposition of a Financial Transactions Tax (FTT) to mitigate this corrupt system.

rather than 'against'); helping people to access, listen to and have confidence in accurate information (inclusiveness and openness); sharing and working with social networks; building resilience by prioritizing the environment; changing eating habits; switching to renewable energy systems; changing traditional beliefs; and promoting the principle of subsidiarity at the appropriate levels.

In short, because *Towns in Transition* give priority to ecosystems, adopt the most appropriate technologies to meet human needs, and embrace the criterion of sufficiency to replace consumerism, the concept is a good example of the reformist trend of the contemporary economic system in developed countries that seeks reorientation towards sustainable development and takes into account the needs of future generations as well as the necessity to protect and conserve the environment and ecosystems.

5.7.3   The Information Age and Post-capitalism

Paul Mason argues that some post-capitalist experiences are already underway to replace neoliberalism as an economic ideology and capitalism as a dominant system.[25] Such experiences are based on functioning micro mechanisms operating in a self-dependent and spontaneous manner, and not on decrees or macro state policies. For Mason, capitalism is a complex system that functions in a manner "alien to any attempt at control by individuals, governments and even superpowers" (Mason 2016: 16), generating results that are often contrary to those initially intended, because, according to Mason, capitalism is a "kind of organism that learns" and possesses autopoiesis and resilience capacities, as explained by Systems Theory and outlined on previous pages. Mason argues that capitalism constantly adapts to new conditions, its basic survival instinct being to promote technological change. Despite being financed by public funds, the changes introduced by the new information and communication technologies (internet, computers) have been privately appropriated by corporate giants such as Microsoft, Amazon, Google and Apple.

Incidentally, what Mason highlights is comparable to the ideas in the three monumental volumes on *The Information Age* (1996–1998), in which Manuel Castells (formerly a professor at the University of Paris and subsequently teaching at Berkeley) coincidentally pointed out the following:

> Productivity and competitiveness are the commanding processes of the informational/global economy. Productivity essentially stems from innovation, competitiveness from flexibility. Thus firms, regions, countries, economic units of all kind, gear their production relationships

---

[25] The term 'post-capitalism', incidentally, has its antecedents in the ideas of the Austrian economist Drucker (1999), who in 1992 published a book in which the term 'post-capitalism' was used to point out that the transformations suffered by the world economy were of such a nature that – according to him – the basic economic resource was no longer capital, land or labour, but rather *knowledge*. Of course, the traditional factors of production are still there, but they are gradually being replaced by knowledge, and even if the world is still 'in transition' Drucker asserts that in the foreseeable future the world will be living in a 'post-capitalist' society.

to maximize innovation and flexibility. Information technology and the cultural capacity to use it are essential in the performance of the new production functions. In addition, a new kind of organization and management aiming at simultaneous adaptability and coordination becomes the basis for the most effective operating system, exemplified by what I label the network enterprise. Under this new system of production, labour is redefined in its role as producer, and sharply differentiated according to worker's characteristics. A major difference refers to what I call generic labour versus self-programmable labour. The critical quality in differentiating these two kinds of labour is education, and the capacity of accessing higher levels of education; that is embodied knowledge and information. The concept of education must be distinguished from skills. Skills can be quickly made obsolete by technological and organizational change. Education … is the process by which people, that is, labour, acquire the capability constantly to redefine the necessary skills for a given task, and to access the sources for learning these skills. Whoever is educated in the proper organizational environment can reprogram him/herself toward the endlessly changing tasks of the production process. On the other hand, generic labour is assigned a given task with no reprogramming capability, and it does not presuppose the embodiment of information and knowledge beyond the ability to receive and execute signals. These 'human terminals' can, of course, by replaced by machines or by any other body around the city, the country or the world, depending on business decisions. While they are collectively indispensable for the production process, they are individually expendable. …The informational/global economy is capitalist … But capital is transformed as labour in this new economy (Castells 1998: 341–342).

One might wonder, then, what the difference is between Mason and Castells. Regarding the first, the new information technologies have ceased to be compatible with capitalism, hence the prefix 'post' applied to the concept of capitalism. There lies the huge difference between Castells and Mason. In effect, Mason's central thesis is that the capitalist system is no longer able to adapt to the new situation created by Information and Communications Technology (ICT) and Artificial Intelligence (AI) because information technologies are different from any previous technology and there is evidence that they have a spontaneous tendency "to dissolve markets, destroy property rights and disintegrate the relationship between work and wages" (Mason 2016: 17), causing a situation that may constitute a crisis, such as the financial one of 2008 and that of COVID-19 in 2020/21, while Castells wrote his book at a time when ICT looked like an important and very useful invention of capitalism, which was not the case.

In other words, the critics of capitalism argued that in order to maintain the overproduction that characterizes it, the only way the system could survive was to find new markets, which in the past led to colonialist imperialism and the wars of the late nineteenth century. This meant that the principles of the twentieth century were to a large extent determined by the expansion of the colonial territories of the great powers. But now, in the twenty-first century, both because former colonial countries such as India, Malaysia, South Africa and Brazil are emerging as large new economic markets and because of the globalization of the world economy, capitalism has had to invent marketing techniques to promote a frenzied consumerism in order to maintain the desire of masses of consumers to acquire new things. People's wishes and hopes, as well as the rise of internet advertising, mobile telephony and other similar marketing mechanisms, contribute to this quest for acquisition and sales.

Consequently, the downward pressure of prices that this consumerist economy is causing encourages the relentless pursuit of innovation and the constant production of new goods to sell, but, simultaneously and paradoxically, the knowledge put at the disposal of the whole world in an unlimited fashion at almost no cost thanks to ICT goes against the impulse towards individual enrichment, because information is abundant and valuable but very cheap thanks to the internet and AI, and this is problematic for a system based on the accumulation of capital. Thus the tension between unlimited knowledge and an economic model based on private property (not between workers and capitalists as in Marx's time) has become the fundamental contradiction of contemporary capitalism.

So, for Mason, post-capitalism is possible for three main reasons: (1) computer science is blurring the lines that separate work from free time (anyone with a personal computer can now work from home, a phenomenon which has increased considerably thanks to the 2020/21 coronavirus pandemic), which is causing a decrease in salaried work; (2) 'informational goods' are eroding the market's capacity to establish prices, because markets are based on the principle of scarcity but information is abundant, not scarce, as we have just seen; and (3) the spontaneous boom in collaborative production that has led to the provision of goods and organizational services (again favoured by the pandemic) that no longer respond to the dictates of the market, as happened with Wikipedia, to mention one very well-known classic case:

> The largest informational product in the world (Wikipedia) has been produced by 27,000 volunteers who did not charge for their work, destroying the encyclopedia business with a stroke and, according to estimates, depriving the advertising companies of some 3,000 million dollars in income a year. Almost unnoticed, entire swathes of economic life are beginning to move at a different pace between the niches and gaps left by the market system itself. They have proliferated there, many of them as a direct result of the decomposition of old structures by the 2008 crisis, in the form of parallel currencies, time banks, cooperatives and self-managed spaces and professional economists have barely noticed it. New forms of ownership, new legal forms of contracts, a new business and subculture has been emerging over the last ten years. The media has called it 'shared economy' (or collaborative). There is also talk of 'pro-commons' (the Anglo-Saxon 'commons'), and of 'production between equals' but few have bothered to ask what it means for capitalism itself. I believe that it provides us with an escape route, but only if we nurture and strengthen those 'micro' projects through a massive transformation of the practices of our Governments. Such change must be driven, in turn, by a profound change in our conception of technology, property and work itself. When we create the elements of the new system, we must be able to tell ourselves (and others) that this is no longer a survival mechanism, a refuge to shelter from the neoliberal world, as it will be a new way of life in the process of formation (Mason 2016: 18–19).

Therefore, it is absolutely clear that the terrain through which global capitalism transits have already been transformed – a transformation reinforced by the 2020/21 crisis – is increasingly fragmentary and orientated towards small-scale decisions. Furthermore, coinciding with authors such as Richards and Göpel, Mason maintains that temporary work will predominate as well as the multiplicity of small entrepreneurial activities in which the skills and resilience of people are manifested, in addition to collaborative production in which the network technology is used to "generate goods and services that work only if they are free or shared", something that also "defines the route that must be followed to exit the market system" (Mason

2016: 19). Although these new forms will require the support of the State (with a universal basic income, for instance, as proposed by Yanis Varoufakis and others) and demand a reconfiguration of the political parties of the left, the truth is that now there is a new agent of change constituted by all the people who are connected to the internet and operate through social networks, generating electronic commerce and intellectual and academic work and passing across currents of public opinion and political influence networks, including economic endeavours of all kinds, which are those of principal interest to what Mason is calling 'post-capitalism'. This trend opens the doors to cooperative work between equals (P2P: peer to peer) and collaborative non-profit in general, as in the aforementioned example concerning Wikipedia, as well as all kinds of behaviour that, as already asserted, has been reinforced by the changes introduced and propagated by the 2020/21 pandemic. Of course, as it also seems that coronavirus has radically changed the conjuncture and that nobody will again question Keynesian policies or the role that the State has to play, Mason's proposals are inscribed in the framework of classic socio-democratic policies with an eye on the probability that the Labour Party will regain the government as the only political force in a position to implement a return to the welfare state, at least in the United Kingdom, where polls suggest that citizens have realized the enormous mistake made with the infamous *Brexit* and will probably make Boris Johnson pay for the mismanagement of the pandemic, as will hopefully similarly occur in the United States and Brazil for the same reasons.

The final part of Mason's book refers to a series of "maximum level" objectives within which the reduction of carbon emissions posed by the SDGs in the framework of the 2030 Agenda are located. They includes the stabilization of the financial system "so that the ageing of the population, climate change and the overflow of debt do not contribute to trigger a new cycle of abrupt expansion and contraction that would definitively destroy the world economy" (Mason 2016: 153). This is the same problem raised by Varoufakis and referred to on previous pages, for which the remedy proposed by the Greek professor and former Finance Minister is a universal basic income and a surplus recycling mechanism. However, according to Mason this problem could be resolved through the socialization of the financial system in order "to save globalization [while] killing neoliberalism" (Mason 2016: 363), which for him implies the undertaking of radical measures such as the nationalization of the central banks of all countries in order to guarantee sustainability, by which "all decisions would be verified beforehand by elaborating models of their climatic, demographic and social repercussions" (Mason 2016: 161), including restructuring the banking system by converting it into a mixture of public service entities with limited profits with an officially imposed ceiling, and encouraging local and regional non-profit banks, credit cooperatives and P2P providers, while leaving a well-regulated space for complex financial activities, rewarding innovation but penalizing rent-seeking (as Piketty also suggests), ending tax havens, and forcing all large transnational corporations to have a national domicile to guarantee their transparency before the Treasury.

Other quite radical macro-economic reform measures proposed by Mason include the establishment of a *universal basic income* for all people by replacing unemployment insurance and other similar benefits, which, as previously mentioned, is also a key proposal of Varoufakis and his *Democracy in Europe Movement* (DIEM); suppressing or socializing monopolies; nationalizing public services such as water, energy, telecommunications and transport because "the privatization of these sectors over the past thirty years was the way the neoliberals found to make private profit in the 'public' sector" (Mason 2016: 357); and making money from entrepreneurship not from income, which means, among other things, that patents and intellectual property rights expire soon, following a procedure similar to that of medicinal patents, whose maximum duration is twenty years, which, of course, must be a crucial undertaking for global civil society to avoid the private appropriation by transnational pharmaceutical corporations – through patents – of the vaccines against coronavirus, as has been reiterated in several articles by Joseph Stiglitz.

All the measures proposed by Mason have the ultimate goal of achieving zero marginal costs, leading to the reduction of working time so that society and people have more free time and can meet their *needs for creativity* (or leisure) – in other words, activities such as science, sport, painting, sculpture, music, dance, literature, cinema – as the Chilean Manfred Max Neef also argues in his Human Development Matrix of 1986 reproduced in Chap. 2. Thus, according to Mason "the only thing we are trying to do is to move as much human activity as possible towards a phase in which the time of work necessary to sustain a complex and very rich human life on our planet significantly decreases and, instead, the volume of free time available increases. A phase in which, in addition, the distinction between one type of time and the other is further blurred" (Mason 2016: 371).

5.8 How to Evaluate Sustainable Development: Happiness Goes Beyond GDP and Is Based on Quality of Life, Not Quantity Indicators, as Demonstrated by the Paradigmatic Example of Bhutan

Based on the discussions above, it is clear that a re-evaluation of the entire socio-political and economic system is essential because if humankind is now in a condition to realize that work should be regarded as a pleasant and creative activity, that could be one of the benefits of the life-threatening pandemic of 2020/21 and the point of departure for the burial of neoliberalism as an economic doctrine that bears enormous responsibility for some aspects of the crisis. Thus from an optimistic perspective it is possible to say that the times in which we are living offer an incredible opportunity to get rid of neoliberalism and initiate a path to rediscover, for instance, what the Buddhist philosophy said centuries ago: that happiness essentially consists of a feeling of satisfaction with present-time moments of serenity and peace of mind that allow us to enjoy music, food, a dazzling landscape, the smile of a child, the caress of a loved one, the words of a poem, the sound of rain, the beauty of falling snow, the flames in the fireplace, the tweet of a bird in the forest and the marvellous sensations of swimming, skiing, climbing mountains or any sport or artistic activity when you are a good or enthusiastic player or creator. So, if starting from an awareness that the difference between work time and free time is blurred, as Mason says, since the ideal of all creative work (like that of artists, sports people, inventors, scientists, writers, and musicians) is to be pleasant, agreeable and satisfying or even – as happens with masters and champions – capable of guiding you to the supreme bliss, happiness and illumination that the most endowed persons can obtain only through meditation that leads to enlightenment or *nirvana*, it is possible to say that social evolution really can happen and that "living well" means fundamentally enjoying the moments of happiness in our life. And, of course, happiness cannot be measured; it is not a quantitative value but a qualitative one, which is why the classic quantitative parameters of GDP or growth cannot be applied to the concept of Gross National Happiness.

I will now begin this re-evaluation by reviewing the history of the concept of sustainable development. Since the 1990s, the United Nations Development Programme (UNDP) pioneered the establishment of a Human Development Index that included indicators of health and education, to which were added indicators of political freedoms that were widely welcomed to the extent that they improved the way of evaluating social and human development. At the end of the first decade of the twenty-first century, the French government commissioned a group of experts (including Joseph Stiglitz) to produce a report on how to evaluate economic development and social progress. The OECD made remarkable efforts in that direction, and certainly, Bolivia and Ecuador, where the Quechua and Aymara peoples' notion of *good living* (or *buen vivir* in Spanish) was inserted into the new constitutional norms, have also contributed to these changes. In countries like Finland, *quality of life* is considered necessary for sustainability. But perhaps the most extraordinary case is that of the small Kingdom of Bhutan in the Himalayas, which has had the audacity to propose an index of human happiness (GNH: *Gross National Happiness*), which, without a doubt, was the most striking example in a study carried out by the *International Institute for Sustainable Development* (IISD in Canada) and the Bertelsmann Foundation. These institutes investigated thirty-five sustainable development strategies throughout the world, and found that the cases noted above explicitly associated sustainability objectives with the local culture's ideas about what *good living* truly is, or what constitutes living well: *sumak kawsay*, as the indigenous people of the South American highlands call it, a concept I have already explained.

In the case of Bhutan, *Gross National Happiness* was declared a national goal by King Jigme Singye Wangchuck, who declared, "If the government cannot create happiness for its people then there is no sense in the existence of the government" (Göpel 2016: 133). So in the small kingdom of the Himalayas nestled between India and China (population c. 800,000), society as a whole goes hand in hand with good government because Bhutan is now a democratic constitutional monarchy. Those who proposed such a goal included in their consideration the satisfaction of three elements which are extremely important in life: material needs (food, nutrition, housing), social needs (work, health, education), and spiritual needs (meditation, inner peace). Moreover, in order to be happy it is crucial to satisfy the last of these categories, that is, to have enough knowledge of oneself through meditation and to maintain a good relationship with others and with society as a whole.

This means that material development is seen as complementary and of equal importance to spiritual development. Such a mentality or philosophy essentially consists of detachment from material goods, and finds meditation a way to avoid personal suffering and a path to find illumination, so the concept of happiness has a meaning of great spiritual depth, as the New Zealand scholar Ross McDonald of the University of Auckland points out when he says that a philosophy like Buddhism, which is based on the reduction of suffering through meditation to find happiness, is:

> [A] philosophy grounded in an explicit ethical perception – one that aims to facilitate the free expression of compassion, sympathetic joy, loving kindness and equanimity for the sake of all sentient beings. To do this requires the regenerating cultivation of the skilful maturities that underlie these (including … attentional mastery, moral clarity and contributory virtuosity). In Buddhist philosophy, happiness is an inseparable component of these maturities, thus happiness as a goal cannot be practically separated from their preliminary cultivation. In significant part then, the question of how GNH might be maximized involves a more central question relating to how exactly generosity, sympathetic joy, loving-kindness and equanimity can be cultivated at all levels within a national polity and extended to govern economic, ecological and social management. In this sense, then, among the four pillars, Buddhist culture and the specific ideals of ethical inter-relationship provide the most fundamental locus of concern if Bhutan is to move towards its stated aims of achieving a happy society in, and on, its own terms (MacDonald 2015).

The strength of the philosophical approach or world-view that prevails in Bhutan lies in the fact that apart from the fundamental principle of inviting everyone to engage in deep introspection through meditation in order to find enlightenment, nirvana, or *samadhi* (which is just another name for happiness) within ourselves, Buddhism also preaches compassion and constant concern for the welfare of others, since true happiness cannot be found if others (your wife or husband, children, family, friends, the community or the nation as a whole) are suffering or afflicted by any kind of adverse condition, such as poverty, lack of education, poor health, lack of freedom, or the absence of affection – i.e. if their human needs have not been satisfied. Therefore, this approach stems as much from the need to live a life in harmony with nature and society as from the need for our own inner peace and wisdom. This is how the concept of *Gross National Happiness* should be understood.

This philosophy was already explicitly present in the government plan of 1999 called *Bhutan 2020: A Vision for Peace, Well-Being and Happiness*, which contains a long-term strategy because although it recognizes the need for sustainable development, it establishes an adequate balance between the material and the spiritual based on five great objectives. These objectives are human development, cultural development, balanced and equitable social development for the benefit of all, good governance, and the protection of the environment, as stated in the report of the Bertelsman Foundation in 2013 and in the general evaluation made by Göpel (2016), who points out that the plan's greatest strength is that it is based on values and culture because the overall development strategy has always been embedded in environmental programmes and sustainable development strategies. In 2008 the new constitution declared that *Gross National Happiness* was going to be the paramount criterion to evaluate development, introducing a new assessment system in which the GNH index was not intended to quantify happiness as such, but to "orient the people and the nation toward happiness", improving their conditions of life according to a steering committee that established four strategic pillars of policy-planning: (1) sustainable and equitable socio-economic development; (2) environmental conservation; (3) the preservation and promotion of culture; and (4) good governance.

In order to specify and calculate progress, an additional index of nine dimensions was generated, based on living standards, health and education, ecological diversity, and good governance. The category of living standards entails the use of free time, engagement in community activities and respect for cultural diversity, but also requires *psychological well-being*, which refers to the spiritual sphere.

Each dimension has between three and five indicators which are used for taking representative surveys and are analysed according to a total of 124 variables, but – as Göpel (2016) underlines – the most innovative feature of the GNH strategy is that the results of the survey are then embedded in the core of government planning so that the quinquennium policy plans respond to findings that can be differentiated by region, gender, profession, age group and so on in order to decide state interventions and resource allocations. The policies are formulated in multilevel, multi-stakeholder processes based on local consultations and the way that each new law is applied because it must be evaluated in terms of its impacts on the dimensions and indicators of the Gross National Happiness Index.

Regarding energy, the Bhutanese government accentuates the importance of renewables, and as the country is at the foot of the Himalayan range with high altitude mountains, magnificent peaks and glaciers, water is abundant with lots of heavy flows descending from summits. The complete electrification of the country was attained by means of tripling hydro-power generation, which contributed to improved independence in food provision, 100% organic agriculture, a low ecological footprint, economic diversification, a reduction in the use of fossil fuels for transport, and selected tourism. The Bertelsmann Report (with 2013 data) stresses the following:

All of these strategies are assessed from the holistic GNH perspective. Since Bhutan's surveys have only been done twice to date, clear trends within these metrics are hard to map. But on conventional measurement schemes its development is better than other South Asian and comparable states worldwide in terms of poverty reduction, improving HDI rates, constant economic growth and successful environmental protection. The prior GNH framework has also guided a stable and still ongoing transition since the 1970s that involved moving from an absolute monarchy to a constitutional monarchy and finally into a constitutional democracy. This process was initiated by the fourth king, and finished by Jigme Khesar Namgyel Wangchuck, the fifth and current monarch in 2008, when a widely discussed constitution was signed into law and first elections were held. Since the 1980s, Bhutan's life expectancy has gone up by 20 years, its birth rate has gone down from 6.55 to 2.4 children per woman, infant mortality has halved and years spent in school have gone up by seven to the OECD average of 11 years. Its per capita GDP is 13 times higher and unemployment is a mere 4%.

At the same time Bhutan has managed to increase the area under forest cover from 50% in the 1970s to 80% and inserted into its constitution a clause that this must not fall below 60%. The total of its protected nature areas amounts to 50% of the surface of the country and the ecological footprint of each Bhutanese is 0.8 ha, well below the world average of 2 ha. It has pledged to keep its development climate neutral and hydro-power is supposed to be the prime energy source by 2020, even allowing for exports that would reduce Bhutan's still significant dependence on official development aid (Göpel 2016: 134–135).

However, social problems persist and not everything is rosy: in 2002, 12% of the population was still living in poverty (measured according to Western criteria), but it was reduced by 23% in 2007 thanks to economic growth of 8%, which was well distributed and not hoarded by the upper strata. It is interesting to mention a curious government investigation in 2010 referring to its own GNH criteria, which indicates that only 37% of the rural population would have passed the threshold of 'sufficiency in happiness'. Climate change threatens the plans for the construction of additional hydroelectric plants and the risk of flooding remains, since the cities are located at the end of long valleys that descend from the Himalayas; the number of Indian migrants who do the hardest physical labour has also increased (although there are no reports of problems with migrants); and the authorization of television, internet and mobile telephony since 1999 has had social consequences, especially in privileged layers of the population where young people who have received this influence have developed consumer habits and some of them 'complain' about the absence of shopping centres and imported fashion goods, which is not a real problem unless the situation is analysed from the point of view of employment and opportunities for these youths of the new middle classes with their fresh skills and knowledge provided by ICT in the framework of globalization.

Another problem was the strange (for a country like Bhutan) political crisis suffered at the end of the twentieth century, when a small ethnic Nepalese separatist movement settled in the southern provinces and was severely repressed, provoking a wave of refugees. Regarding this issue, it can be said that the Bhutanese government reacted with the same 'realism' of any nineteenth-century European power by using its armed forces, but the fact is that whereas the Westphalian system prevails at world level (and we must remember that Bhutan is a UN member state) territorial issues will continue to exist and the geopolitical condition of Bhutan as a buffer state between the two great regional powers of India and China is a crucial factor in its foreign policy.

However, despite the problems and difficulties that any country may experience, in my view Bhutan is an archetypal success story, and has so far succeeded in keeping the difficult balance between the linearity of sustainable development and the circularity that characterizes natural ecosystem cycles.

Göpel's diagram reproduced in Fig. 5.4 illustrates this new paradigm of well-being and happiness as practised by this Himalayan Kingdom, presenting the cycle between planetary boundaries in the lower part and social happiness in the upper part. It is also possible to see the lines of social dynamics that move through ecological and cultural diversity, psychological well-being, ecosystems conservation, and good governance.

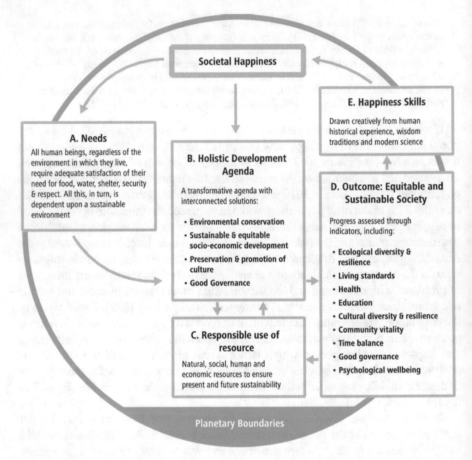

Fig. 5.4 Bhutan: a new development paradigm of well-being and happiness. *Source* Based on Bhutan NDP Steering Committee and Secretariat (2013: 20) and reproduced with permission from Göpel's book (2016: 138)

Undoubtedly, if used properly, new technologies from Western science (the internet, medicine, education, and the use of water resources to generate electricity without damage to the environment or to social needs) will help to meet human needs. Thus, there is no doubt that a holistic and transformative agenda of sustainable development is being applied in Bhutan while preserving and promoting culture and the environment, so that the four spheres of sustainable development (techno-economic, social dynamics, ecosystem, and governance) are adequately integrated in the country's national planning.

There are other pioneering experiences that cannot be addressed at length here, such as the 'commons', a term used to refer to the common goods of humanity: natural elements like the water in oceans, glaciers and polar caps, the oxygen in the atmosphere, water resources, and forests. For Elinor Ostrom and the authors of a book containing seventy-three essays on this subject (Bollier et al. 2012), such goods are part of the common heritage of humanity. This means that even though humankind has the right to these resources, we also have an equal responsibility to protect them. Every generation can make use of them as long as it is careful to ensure that future generations will also be able to do so in accordance with the fundamental principles of sustainable development.

5.9 Boaventura de Sousa Santos

In spite of him not being an expert in the field of economics, the ideas of Santos (2014), one of the founders and leaders of the *World Social Forum*, are so important and influential in the social sciences of contemporary global civil society that the Portuguese scholar and thinker can be regarded as the most important organic intellectual of all social movements struggling for a revolutionary change from world capitalism. Indeed, in this world of appalling social inequalities, thanks to Sousa Santos's theory concerning the *Epistemologies of the South* people are becoming increasingly aware of the multiple dimensions of injustice, whether social, political, cultural, sexual, ethnic, religious, historical, or ecological. Santos argues that *cognitive injustice* also exists, which consists mainly of the way that Western academics and official science ignore and fail to recognize the different approaches to knowledge by which people across the globe run their lives and provide meaning to their existence, and which, according to Santos, constitute a true 'ecology of knowledge' (a concept quoted above).

Therefore, as *cognitive injustice* underlies all the other dimensions of life in the former colonial countries of the *Global South*, social movements and academia must work together to bring about social justice, tackling firstly the huge problem of achieving *global cognitive justice*. Santos's perspective unfolds in inquiries as to how Western modernity fostered the attitudes underlying the long cycle of colonialism and the implantation of global capitalism and how these historical processes profoundly devalued and marginalized the knowledge and wisdom that had been in existence in the Global South for centuries, and that explain why people must fight against '*epistemicide*' (the killing of knowledge). This huge social struggle and undertaking is imperative in order to value the epistemological diversity of the peoples of the world, starting with the recovery and valorization of this different kind of knowledge using four key analytical tools: the *sociology of absences*, the *sociology of emergences*, the *ecology of knowledges*, and *intercultural translation and dialogue*. The transformation of the world's epistemological diversity into an empowering instrument against counter-hegemonic globalization points to a new kind of bottom-up *cosmopolitanism*, in the sense that this concept has been defined in this book, namely referring to collective actions that would promote extensive conversations between humankind, celebrating conviviality, solidarity and life, as opposed to the logic of the market-ridden greed and individualism which implies the destruction of social bonds and ties, a sad situation to which the world's large and small populations are condemned by the dominant forces of the wild and hegemonic capitalism that characterizes neoliberal globalization.

According to Santos, epistemology is now one of the key arenas of the critique of modernity and the Eurocentric bias of knowledge – as discussed also by intellectuals like Dussel (2015) and Mignolo (2012) in 'decolonial theory'. Santos calls the epistemologies of the South that are looking to build a new paradigm in social sciences 'post-abyssal thinking'. These new epistemological theories depart from a critique of Eurocentrism, where it is difficult to find links between truth and politics or to find justice adopted as an overriding concern, and are consequently orientated by the kind of political and economic matters that appeared with Marxism. Shifting the points of view to the countries of the Global South – and globalizing also the framework of analysis – Santos's new Epistemologies of the South are at the centre of his reflection on the north and south colonial division, a reality which permeates both knowledge in modernity and the quest for global justice.

Thus, the point of departure of Santo's political epistemology is a critical characterization of modern reason as 'abyssal thinking', which operates by establishing and radicalizing distinctions between knowledge elaborated in the Global North and the Global South. The epistemologies of the South gravitate, first of all, towards issues of the production of knowledge and, in particular, the ways in which we can think of and transform the social sciences and even the theory and philosophy of law (Santos 2009a). Of course, it is also important to realize that Santos's ideas have fundamental political implications because the problem faced today by social sciences, law and human rights is mainly how to advance the quest for global justice in the midst of the prevailing neoliberalism and neocolonialism. The economic and political injustices that characterize the world situation nowadays result in, and are sustained by, the cognitive injustice that has existed at the core of the production of knowledge since the beginnings of the modern colonization of the world. These injustices are intertwined, and constitute and feed each other. It is within this horizon that Santos states that the struggle for global justice includes the search for epistemic justice, as political resistance needs to be based on "epistemological resistance". Hence, human rights can become, for instance, a new language of emancipation and progressive politics, one able to give impulse and incarnate counterhegemonic social struggles that aim to change the social structures that are accountable for unjust, systemically produced human suffering. However, the capacity of human rights for emancipation is conditioned by the possibilities of being reimagined by an intercultural dialogue between Western and non-Western conceptions of rights. Crucially, when Santos relates human rights to liberation and justice, he is not only considering the issue of social justice in national scenarios, but also the capacity of rights to advance the quest for equality and justice in a global context. Thus, Santos approaches human rights with the aim of strengthening their emancipatory power in the world as a whole by relying on the global appeal and local, culturally grounded legitimacy of human rights, advanced by social movements struggling for justice. Hence, it is clear why he deserves the honour of being listed among the intellectuals and thinkers fighting to transform capitalism by putting an end to neoliberalism.

5.10 Summary

My main conclusion is that the current neoliberal economic model must be changed and that, as we are living a historical epoch similar to that of Polanyi's *Great Transformation*, this change could be facilitated by the social and economic conditions created across the entire world by the pandemic crisis of 2020/21: health appears to be the supreme value that cannot be sacrificed on the altar of neoliberalism. Governments in democratic countries like New Zealand, Iceland, Finland, Norway, Denmark, Germany, France, Costa Rica, Uruguay, South Korea and Taiwan have handled their economies and the health crisis much more efficiently than authoritarian ones like China, Cuba and Vietnam, and incomparably better than countries ruled by populist psychotic leaders, like Brazil and, during the Trump administration, the US; hence the intervention of the State in the economy and society and its role in protecting citizens' health is now less likely to be questioned by the masses. The current conundrum concerns an unsustainable situation insofar as development has been guided by neither the satisfaction of human needs nor respect for terrestrial ecosystems, but by growth and capital accumulation, and that situation is precisely what needs to change. The majority of the world's population lives in poverty, but economic growth, as shown by Thomas Piketty's masterful work about twenty-first century capitalism, does not respond to the interests of national states (as it should) but to the interests of the small world elite of the incredibly wealthy who are neither enterprising nor innovative,[26] but who have become *rentiers* (as Piketty calls them), enjoying a life of leisure and permanent luxurious consumption in imitation of European aristocracies in the times of royal absolutism. Both the Westphalian order (as described in Chap. 1) and globalization (which I will address in Chap. 6) are in crisis, and that situation explains the disarray (Kissinger 2014; Haas 2017) of the supposed 'world order' in current times.

However, the way out of the crisis is not a matter of examining revolutionary alternatives to world capitalism as such, because the time of proletarian Marxist revolutions seems to be over, but rather of finding alternatives to the narrow predominant reductionist conception that sees everything from the point of view of economic growth and capital accumulation. Therefore, what is required is a new peaceful *Great Transformation* (Polanyi 1957) or *Great Planetary Transition* (Raskin 2016), like the one that occurred in Europe when capitalism was consolidated during the eighteenth and nineteenth centuries, first in England after Cromwell and then in France after the Peace of Vienna (1815). This Great Transformation will consist of the adoption of the new holistic paradigm of sustainable development that has been proposed since the 1980s, but which has not been able to become hegemonic due to the ideological influence of so-called 'neoliberalism' (Gramsci; Klein 2014) and the interests of the huge capitalist corporations that take advantage of this situation for their own benefit.

---

[26] Some kind of cosmopolitanism and altruism might be expected from these super-rich. Unfortunately, that global elite of *rentiers* is accustomed to having a 'mainstream economics' type of mindset, and personalities like George Soros or Bill Gates are not frequently found.

Thus, the central issue of the present is that the enormous complexity of the problem of sustainable development (which is simultaneously socio-political, environmental and cultural, as well as economic) has been reduced to a single dimension. That is why I have presented numerous arguments to demonstrate that the efforts that have been made since the 1970s to change the predominant reductionist thinking have not paid off because all governments and decision-makers have, to a large extent, remained under the influence of that biased vision of development that considers development to be exclusively economic and guided by neoliberal ideology. This is a vision that reduces public policies to the promotion of growth, placing the said variable as a central parameter of both evaluation and effective results. If there is no growth there is no development, according to this erroneous, biased and incomplete perspective that is partial, narrow and reductionist.

Indeed, for development to be sustainable, and not exclusively designed to accumulate wealth for the benefit of the few, it is necessary to combine the four spheres of social dynamics, techno-economic factors, good governance, and natural ecosystems (Sachs 2015). For social dynamics to be properly prosecuted, social development (education, health, decent work, the eradication of extreme poverty, the empowerment of women) is required – in short, the satisfaction of human needs (Neef 1986; Brundtland 1987). All this must be conducted within the framework of public policies that are formulated within a democratic system that promotes peaceful and inclusive societies which facilitate access to justice for all "by creating effective, responsible inclusive institutions at all levels", as noted in SDG 16, which clearly alludes to the need for good governance, which in turn implies good political leadership within the rule of law and a democratic political system.[27]

In addition to the appropriate combination of the techno-economic with social dynamics and good governance, it must be borne in mind that articulation with the sphere of natural ecosystems is also fundamental and that this cannot be done properly if those who make decisions in a given state and government, i.e. politicians, do not take them into account. They ought to be aware of the primordial fact that such natural ecosystems, on which all the societies of the world are based, do not follow a linear trajectory, but a circular one, because ecosystems are like the planetary cycles of rotation and translation, such as night and day, autumns, winters, springs and summers, or the rainy seasons in the tropics, as well as vital cycles, wakefulness and sleep or the eternal return of life and death. These ultimately determine the sustainability of any natural ecosystem, be it astrophysical (solar), physical (planetary), micro-physical (atomic or subatomic, quantum), biochemical (cellular), or belonging to a plant, animal or human species (biological).

Therefore, if development is going to be really sustainable, it has to be adequately articulated with the natural ecosystems that are sustainable precisely because they have as a base of support the ecosystems as a whole. In this process, we must take

---

[27] However, democracy depends on the ideology and tendencies of electoral constituencies and, even in the most developed countries, democracy, like the market, can have 'failures', with the disadvantage that they cannot be corrected until the next election. In 1933 Germans elected Hitler, and more recently UK citizens voted for *Brexit*, Americans elected Donald Trump, Brazilians Bolsonaro, Hungarians Orban, Poles Duda and Filipinos Duterte, and so on.

care not to transgress the planetary boundaries, which are not, of course, in extra-terrestrial space, because, as noted earlier, our planet is part of the solar system, so the sun is 'immanent' to the Earth not 'transcendent'. That is why such boundaries are constituted by the limits of sustainability and economic growth, which include but are not limited to the oxygen we breathe; the water we drink; the territory, fauna and flora that feed us; the forests that nourish the rain; oceans; and our marine and terrestrial biodiversity. In other words, we must either respect the planetary boundaries, which have been put at risk by the Great Acceleration of a mode of production on Earth in this era of the Anthropocene, or expose ourselves to the danger of the disappearance of our own species on the planet, as warned by both Lovelock and Ward. After all, *Homo sapiens* has only existed for about 200,000–300,000 years on a planet whose origins go back more than four billion years, and many other 'sister' species of pre-human hominids disappeared before us, which was the fate of even the Neanderthals that co-existed with *Homo sapiens* for a long period of time.

If we continue to transgress the atmospheric frontier – i.e. to emit greenhouse gases from the burning of fossil fuels such as coal, gas and oil, which contribute to the increase in temperatures and the frequency of natural catastrophes (whose origin is climate change) – and to pollute the environment, we run the risk that Lovelock's loving *Gaia* will transform herself into the terrible Medea of Peter Ward's sixth mass extinction and devour our species for not paying her the attention that every loving mother (*Pachamama*) deserves.

However, it is clear that it is not about establishing an autarkic system similar to that of Tikopia in the South Pacific with its three millennia of existence, but about being fully aware of the similarities between our small planet lost in interstellar space and that tiny island lost in the ocean, because the Earth itself is like our island-spaceship whose habitat we must conserve for the benefit of present and future generations, as sustainable development demands, bearing in mind that such an objective will not be achieved without ceasing to produce material goods of unsustainable modalities due to their polluting or predatory effects on the environment. As Maja Göpel says, sustainability means doing things better by discontinuing the use of non-recyclable raw materials, such as plastics, and non-renewable energy sources, such as hydrocarbons, while remembering that they must also be done well, which means calibrating production to the satisfaction of human needs and not to the increase of growth rates or individual enrichment to the detriment of society.

Consequently, another important conclusion is that sustainable development must be evaluated against the parameters of the new holistic and integral paradigm of human development and cyclical sustainability, and should not be based on the parameters of the old dominant economic subsystem, such as growth, per capita income, and linear sustainability. It is important to establish whether the commitments of the government and society are being fulfilled in an appropriate manner. For instance, if a new hydroelectric plant is constructed, it is necessary to negotiate with the local community beforehand so that the prior social right to the use of water for sanitation, in accordance with the corresponding SDG, is not at risk. Members of the community should not be left without benefits in terms of the price they pay for the energy they consume, because water should not be considered an 'externality', as economists call it, since, just like oxygen, it is a common good of humanity. Thus, in order to assess whether the twelfth SDG is adequately met, concerning the guarantee of sustainable consumption and production patterns, it is not only necessary to negotiate with the private sector to produce recyclable materials but to accept the prohibition of those that are not. The European Union and some Latin American countries are pioneering in these issues.[28] Furthermore, the use of advertising to promote consumerism must be discouraged in addition to being drastically regulated, among other reasons because we must reduce the visual pollution in our cities, which is already insufferable.

For the aforementioned to become reality, constitutional and legislative reforms are required. The main purpose of the reference to what has been done in Ecuador and Bolivia in this regard is to highlight the importance of introducing the traditional yet innovative concepts of the indigenous Aymara and Quechua, such as good living (*sumak kawsay*) and *Pachamama* (Mother Earth), into the constitutions of both countries, among other reasons because nature does not possess individual power to defend itself, unless we consider the tectonic fury of earthquakes and volcanoes, hurricanes and tropical storms as an expression of *Pachamama*'s dislike of her own creatures.

But even if we decide to put aside pantheistic world-views such as the 'cosmovision' of the indigenous Quechua, Aymara and Maya, it is no less true that in the face of toxic waste contamination and debris in lakes, rivers, seas and oceans, and the depredation of the tropical forest by logging, livestock and agricultural-industrial companies, a good solution is to grant the exercise of rights on behalf of Mother

---

[28] For instance, concerning plastics, the EU Strategy for Plastics in the Circular Economy (that I have mentioned before) says in its Plan of Action that the EU will restrict intentionally added microplastics, further develop and harmonize methods for measuring unintentionally released microplastics, especially from tyres and textiles, and deliver harmonized data on microplastic concentrations in seawater by applying measures like closing the gaps on scientific knowledge related to the risk and occurrence of microplastics in the environment, drinking water and foods; sourcing, labelling and using bio-based plastics, based on assessing where the use of bio-based feedstock results in genuine environmental benefits beyond reducing the use of fossil resources; and using biodegradable or compostable plastics, based on an assessment of the applications where such use can be beneficial to the environment, and the criteria for such applications. It will help to ensure that labelling a product as 'biodegradable' or 'compostable' does not mislead consumers to dispose of it in a way that causes plastic littering or pollution due to unsuitable environmental conditions or insufficient time for degradation. The EU also says that the Commission will ensure the timely implementation of the new Directive on Single Use Plastic Products and fishing gear to address the problem of marine plastic pollution while safeguarding the single market, in particular with regard to harmonized interpretation of the products covered by the Directive labelling of products such as tobacco, beverage cups and wet wipes, ensuring the introduction of tethered caps for bottles to prevent littering, and new rules on measuring the recycled content in products (European Commission 2020: 12). In Latin America the government of Chile has banned the use of plastic bags throughout the country, according to the *New York Times*: "Chile will become the first country in the Americas to prohibit retailers that use plastic bags, an initiative that aims to protect its more than 8,000 km of coastline. The measure, recently approved unanimously by Congress, gives stores and supermarkets six months to comply with the provisions. Small and medium-sized businesses, such as neighborhood stores, will have up to two years to adhere to the new rules. During that time, they can give a maximum two plastic bags per customer" (*New York Times*, 1 January 2018).

Nature (*Pachamama*) to all those citizens who possess an ecological conscience and culture, even if it is evident that constitutional or legal reforms are not enough to change either mainstream neoliberal economics, as the social upheaval in 2019 against President Lenin Moreno in Ecuador demonstrated, or the contempt of democracy that motivated the *coup d'etat* and resignation of President Evo Morales in Bolivia the same year.

I must say a few more things in conclusion regarding the dominant economic system, i.e. capitalism. Crises like those of 1929, 1973 and 2008 call into question the survival of the system itself if the structural reforms that it needs are not made. Determining whether the systemic crisis derives from the absence of a surplus recycling mechanism, as Varoufakis affirms, or is the result of the process of concentration of wealth among a small world elite that have become rentiers, increasing inequalities at world level, as Piketty argues, is not, in my opinion, the most important question to be decided. The most important issue is for governments, most likely in the framework of a large multilateral negotiation, to implement the *reform of capitalism* measures proposed by both authors, i.e. establish the recycling mechanism proposed by Varoufakis (including his proposal for a universal basic income to be financed by a FTT) as well as the appropriate capital taxes on the super-rich of the world recommended by Piketty.

That means that it is important to resolve the previous dichotomy by making the necessary reforms to adjust the "economic model" to the objectives of sustainable development, otherwise the minimum "sustained growth" required by SDG 8 will not take place. Capitalism has already been transformed by the changes that have taken place in the economy itself. Indeed, since the "end of the millennium", as Castells (1998: 341–343) called the third volume of his work on the Information Age, it is clear that capitalism has entered this new era of information not only because of the 'annihilation' of space and time made by electronic media, but also because of its effects on self-programmable, non-salaried work, and the individual's ability to 'reprogram' when constantly given the changing tasks of the production process. In other words, the economy of knowledge has emerged in this era of post-capitalism.

As noted earlier, Mason (2016) has ideas similar to those of Castells, but while for the former the information era is functional, for the latter it is not. The post-capitalism expressions of current times are especially evident in the micro-economic field, where new information technologies, such as the internet and smartphones, are erasing the borders that previously separated salaried work from free time. In addition, since information is an abundant and no longer scarce commodity, this has had repercussions in the fall in prices, since digital books and all sorts of goods can now be purchased on the internet, but also in the free information provided by the network of free collaborative work which has allowed the rise of Wikipedia. Wikipedia not only bankrupted the former companies that were engaged in producing dictionaries and encyclopaedias (Britannica in English and the UTEHA in Spanish, for instance), but is based on free, collaborative and self-programmable work of more than twenty-seven thousand academic volunteers who do not receive any remuneration, as Mason (2016) reminds us, which is similar to Richards's ideas about social solidarity in the economy or the networks of solidarity amongst workers of the informal sector, or the peasant economy in developing countries.

I have also reviewed ways to acquire a better understanding of the SDGs, and how to face the difficulties that could arise in their implementation by 2030. This involved a brief description of certain paradigmatic changes that in practice could aid the implementation of the SDGS, ranging from those that appear in Maja Göpel's book, such as the strategy of successive small changes that transformed the thinking of the German elites on the question of renewable energy, to the experiments of communities in transition, the common goods of humanity, the remarkable efforts of the small Kingdom of Bhutan in the Himalayas to achieve a *Gross Inner Happiness* capable of replacing the GDP, other trials and social experiments such as those suggested by Howard Richards and John Holloway, and the constitutional law reforms promoted by the indigenous peoples of South America. However, the most important and encouraging of all these changes concerns the circular economy, because the European Union has approved an Action Plan which it has been implementing officially since March 2020.

Finally, I agree with both Santos (2009b, 2014) and Latour (2015) concerning the Epistemologies of the South and the idea that capitalism is much more an ideology than a material reality. As its aim of growth and conquest makes the planet insufficient for capitalism – which is why science fiction films and literature reiterate the oneiric vision of other planets to colonize – we can expect to see pioneering reforms inspired by alternative mindsets leading towards post-capitalism and degrowth. The prevailing wild neoliberal capitalism will be not be subverted but "sucked down", according to Latour. The crucial point is that capitalism must be reformed and transformed in order to reach sustainability or it will disappear. Whether that future looming on the horizon is called 'post-capitalism' or 'renewed socialism' (or any other name) by future generations is a minor issue.

References

Angus, Ian, 2016: *Facing the Anthropocene: Fossil Capitalism and the Crisis of the Earth System* (New York: Monthly Review Press).

Badie, Bertrand; Foucher, Michel, 2017: *Vers un Monde Néo-National? Entretiens avec Gaïdz Minasssian* (Paris: CNRS Editions).

Barié, Cletus Gregor, 2017: "El Buen Vivir y los Derechos de la Naturaleza", in: *Revista Política Internacional* (Guatemala City: Diplomatic Academy).

Braudel, Fernand, 1987: *El Mediterráneo y el Mundo Mediterráneo en la Época de Felipe II* (Mexico City: Fondo de Cultura Económica).

Bollier, David; Helfrich, Silke (Eds.), 2012: *The Wealth of the Commons: A World Beyond Market and State* (Amherst, MA: Levellers Press).

Boulding, Kenneth E., 1966: "The Economics of the Coming Spaceship Earth", in: Jarrett, Henry (Ed.): *Environmental Quality in a Growing Economy: Essays from the Sixth RFF Forum* (New York: RFF Press): 3–14.

Carrera, Fernando; Walter, Juliane, 2009: *La Educación y la Salud en Centroamérica: Una Mirada desde los Derechos Humanos* (Guatemala City: Instituto Centroamericano de Estudios Fiscales [ICEFI]/Serviprensa Centroamericana).

Castells, Manuel, 1998: *The Information Age: Economy, Society and Culture: End of Millennium* (Malden MA/Oxford: Blackwell Publishers Inc.).

Delanty, Gerard; Mota, Aurea, 2018: "Governing the Anthropocene: Agency, Governance, Knowledge", in: *Política Internacional* (Guatemala City: Academia Diplomática) 5 (January–June): 83–110.

Drucker, Paul, 1999: *La Sociedad Post Capitalista* (Buenos Aires: Editorial Sudamericana).

Dussel, Enrique, 2015: *Filosofía del Sur: Descolonización y Transmodernidad* (Madrid: Akal/Inter Pares).

Erkman, Suren, 2001: "Industrial Ecology: A New Perspective on the Future of the Industrial System", *Swiss Medical Weekly*, 131, 37–38: 531–538.

Fox, Jonathan, 2005: "Unpacking 'Transnational Citizenship'", in: *Political Science Review* (Santa Cruz: University of California): 171–201.

Göpel, Maja, 2016: *The Great Mindshift: How a New Economic Paradigm and Sustainability Transformations Go Hand in Hand* (Cham: Springer International Publishing).

Gudynas, Eduardo, 2017: "Ecología Política de la Naturaleza en las Constituciones de Bolivia y Ecuador", in: *Política Internacional* (Guatemala City: Diplomatic Academy).

Habermas, Jürgen, 1988: *Teoría de la Acción Comunicativa: Crítica de la Razón Funcionalista*, Vol. 2 (Madrid: Taurus).

Harari, Youval Noah, 2018: *21 Lessons for the 21st Century* (Miami FL: Penguin Random House).

Klein, Naomi, 2014: *This Changes Everything: Capitalism vs The Climate* (New York: Simon & Schuster).

Kupchan, Charles, 2018: "Trump's Nineteenth-Century Grand Strategy", in: *Foreign Affairs*, 97, 5 (26 September): 116–126.

Latouche, Serge, 1993: *In the Wake of the Affluent Society: An Exploration of Post-Development* (London, Zed Books Ltd.).

Latour, Bruno, 2015: *Face à Gaia* (Paris: La Découverte).

Mason, Paul, 2016: *Postcapitalismo: Hacia un Nuevo Futuro* (Barcelona/Buenos Aires: Paidós).

Mignolo, Walter, 2012: *Decolonial Cosmopolitanism and Dialogues Among Civilizations* (London: Routledge).

Mulet, Edmond, 2016: "La Agenda 2030 y los Objetivos de Desarrollo Sostenible", in: *Revista Política Internacional* (Guatemala City: Diplomatic Academy), 1 (June).

Neef, Manfred-Max; Elizalde, Antonio; Hopenhayn, Martin: *Human Scale Development: An Option for the Future* (Uppsala: CEPAUR, Dag Hammarskjold Foundation).

Nerfin, Marc (Ed.), 1978: *Hacia Otro Desarrollo: Enfoques y Estrategias* (Mexico City: Siglo XXI Editores S.A.).

Nye, Joseph; Keohane, Robert, 1988: *Poder e Interdependencia: La Política Mundial en Transición* (Buenos Aires: Latin American Publisher Group [GEL]).

Oswald Spring, Úrsula, 2016: "The Water, Food and Biodiversity Nexus: New Security Issues in the Case of Mexico", in: Brauch, Hans Günter; Oswald Spring, Úrsula; Bennett, Juliet; Serrano Oswald, Serena Eréndira (Eds.): *Addressing Global Environmental Challenges from a Peace Ecology Perspective* (Cham: Springer International Publishing): 113–144.

Padilla, Diego, 2018: *La Política del Agua. Radiografía Crítica del Estado* (Guatemala City: Universidad Rafael Landívar, Instituto de Estudios sobre el Estado [ISE]).

Piketty, Thomas, 2014: *Capital in the Twenty-First Century* (Cambridge, MA: Harvard University Press).

Polanyi, Karl, 1957: *The Great Transformation: The Political and Economic Origins of Our Time* (Boston, MA: Beacon Press).

Richards, Howard, 2017: *Economía Social Solidaria: Para Cambiar el Rumbo de la Historia. Módulos de un curso en la Universidad Nacional Autónoma de México* (Mexico City: National Autonomous University of Mexico [UNAM]).

Sachs, Jeffrey, 2015: *The Age of Sustainable Development* (New York: Columbia University Press).

Santos, Boaventura de Sousa, 2009a: *Una Espistemología del Sur: La Reinvención del Conocimiento y la Emancipación Social* (Mexico City: Consejo Latinoamericano de Ciencias Sociales [CLACSO]/Siglo XXI Editores).

Santos, Boaventura de Sousa, 2009b: *Sociología Jurídica Crítica: Para un Nuevo Sentido Común en el Derecho* (Madrid: Editorial Trotta; Bogotá: ILSA).

Santos, Boaventura de Sousa, 2010: *Refundación del Estado en América Latina: Perspectivas desde la Espistemología del Sur* (Mexico City: Siglo XXI Editores SA; Bogotá: Universidad de los Andes).

Santos, Boaventura de Sousa, 2014: *Epistemologies of the South: Justice Against Epistemicide* (New York: Paradigm Publishers).

Santos, Boaventura de Sousa, 2016: *La Difícil Democracia: Una Mirada desde la Periferia Europea* (Madrid: Ediciones Akal).

Singer, Paul 1980: *Economía Política del Trabajo: Elementos para un Análisis Histórico-Estructural del Empleo y de la Fuerza de Trabajo en el Desarrollo Capitalista* (Mexico City: Siglo XXI Editores, SA).

Singer, Paul 2002: *Introdução à Economia Solidária* (São Paulo: Editora Fundação Perseu Abramo).

Stavenhagen, Rodolfo 1981: "Siete Tesis Equivocadas sobre América Latina", in: *Sociología y Subdesarrollo* (Mexico City: Nuestro Tiempo): 15–84.

Stavenhagen, Rodolfo (Ed.), 1990: *Entre la Ley y la Costumbre: El Derecho Consuetudinario Indígena en América Latina* (Mexico City: Instituto Indigenista Interamericano).

Stavenhagen, Rodolfo, 2013: *The Emergence of Indigenous Peoples* (Heidelberg: Springer).

Valdez, Fernando, 2015: *El Gobierno de las Elites Globales: Como Se Organiza el Consentimiento* (Guatemala City: Universidad Rafael Landívar/INGEP Cara Parens).

Varoufakis, Yanis, 2015: *El Minotauro Global: EEUU, Europa y el Futuro de la Economía Mundial* (Madrid/Mexico City: Ediciones Culturales Paidós).

Vitón García, Gonzalo, 2017: "Cambio Climático, Desarrollo Sostenible y Capitalismo", in: *Relaciones Internacionales* (Madrid: Universidad Autónoma de Madrid [UAM], GERI), 34 (February–May).

Wallerstein, Immanuel, 1989: *El Moderno Sistema Mundial: I: La Agricultura Capitalista y los Orígenes de la Economía Mundo Europea en el Siglo XVI*, 5th edn. (Mexico City/Madrid/Bogotá: Siglo XXI Editores).

Wallerstein, Immanuel, 2006: *Analysis of World Systems: An Introduction* (Mexico City/Madrid/Buenos Aires: Siglo XXI Editores SA).

Ward, Peter, 2009: *The Medea Hyphothesis: Is Life on Earth Ultimately Self Destructive?* (Princeton: Princeton University Press).

Internet Links

"Carbon Tax Center"; at: https://www.carbontax.org/whats-a-carbon-tax/.

"Chile Ban on Plastic Bags".

"EU Circular Economy Action Plan for a Cleaner and More Competitive Europe"; at: https://ec.eur opa.eu/environment/circular-economy/pdf/new_circular_economy_action_plan.pdf.

Fox, Jonathan, at: https://escholarship.org/uc/item/4703m6bf.

"Global Financial Integrity"; at: http://www.gfintegrity.org/issue/trade-misinvoicing/.

"ICEFI y CIG Difieren por Impacto de Industrias Extractivas", in: *Diario La Hora* Guatemala, 16 July 2017: 31; at: https://lahora.gt/icefi-cig-difieren-impacto-industrias-extractivas.

Ieven, Bram; Overwijk, Jan, 2020: "We Created This Beast: The Political Ecology of COVID-19", in: *Eurozine*; at: https://www.eurozine.com/we-created-this-beast.

Latour, Bruno: "Interview with Diego Milos and Matías Wolff" (4 February 2015).

McDonald, Ross: "Paper Proposal for the Second International Workshop on Operationalizing Gross National Happiness: *What Exactly is the Meaning and Purpose of Gross National Happiness?*" (Auckland: New Zealand Business, Society and Culture Programme, University of Auckland).

"Transition Towns"; at: http://www.transitionnetwork.org.

Wallace-Wells, David [Interview with Jared Diamond]: "The Models Are Too Conservative": Paleontologist Diamond on What Past Mass Extinctions Can Teach Us About Climate Change Today, in: *Intelligencer* (10 July 2017) at: https://nymag.com/intelligencer/2017/07/what-mass-extinctions-teach-us-about-climate-change-today.html.

Chapter 6
Towards a Transnational Cosmopolitan Paradigm and Citizenship

> *While we agree that dialogue between the Earth science and social and human sciences is essential, the argument advanced in this paper seeks to clarify a distinct social theoretical position. The critical interpretative position here is that Anthropocene has become a way in which the human world is re-imagined culturally and politically in terms of its relation with the Earth. This has implications for governance, which point in the direction of Cosmo politics – and thus of a 'Cosmopolocene' – rather than a geologization of the social or, in the post-humanist philosophy, the end of the human condition as one marked by agency.*
> Delanty/Mota (2018: 35)

> *In this era, a new critical theory with a cosmopolitan intent has a crucial task. … Its main claim is that, first of all, the cosmopolitan viewpoint, linked to various realities, detects the* **chasms** *that threaten the beginning of the twenty-first century. Critical theory investigates the contradictions, dilemmas and the unseen and unwanted (unintentional) side-effects of a modernity that is increasingly cosmopolitan. … The main thesis is then that the cosmopolitan perspective opens up negotiation spaces and strategies which the national viewpoint precludes. … In the debate on globalization the main point does not revolve about the meaning of the nation state and how its sovereignty has been subordinated. Rather the new cosmopolitan perspective of the global power field pushes new actors and actors' networks, the power potentials, strategies, and organization forms of de-bounded politics, or, in other words, global civil society into the field of vision. This is why the cosmopolitan critique of nation state-centred and nation state-buttressed politics and political science is empirically and politically crucial.*
> Beck (2003: 54–55)

> *The new challenge for political theory is to go beyond a narrow state-centred approach by considering political communities and systems of rights that emerge at levels of governance above or below those of independent states or that cut across international borders.*
> Bauböck (2017: 81)

© Springer Nature Switzerland AG 2021
L.-A. Padilla, *Sustainable Development in the Anthropocene*,
The Anthropocene: Politik—Economics—Society—Science 29,
https://doi.org/10.1007/978-3-030-80399-5_6

The Kantian ideal of a world citizenship and current thinking
about a 'contextualized universalism' can take us to a 'landed'
cosmopolitism if we consider the real framework of the
international relations development in all its social, cultural,
political and economic dimensions. … Grounds and possibilities
for a transnational citizenship – if not global at least
transnational – surge precisely from the transnationalization of
the social world.
Pries (2017: 173)

6.1   Introduction

On these pages I have presented the idea that the Anthropocene is not just a new geological epoch that replaces the Holocene but must additionally be understood as a normative cultural model based on the transdisciplinary holistic and cosmopolitan paradigm that unifies social and cultural sciences. Regarding the latter, my main goal is to make a contribution to academic understanding about ways to mitigate the *Great Acceleration* of GHG emissions endangering the survival of our species due to climate change threats, which are also closely related to the prevailing neoliberal economic model that misguidedly puts the accumulation of capital and economic growth over human needs and respect for natural ecosystems as the most important purpose of production and even of human individual lives. This is why I argue that the crucial alternative in the face of the global pandemic (aggravated by the dismantling of welfare and health systems in countries under the influence of neoliberalism, such as the US and some European and Latin American countries) and the threats of climate change is to abandon the pernicious worshipping of 'markets' or face the sixth mass extinction that science is foreseeing unless we change the predominant neoliberal mindset and prevailing economic policies.

Evidently, everybody must realize that the way out of the crisis cannot be found in a national manner with isolated national policies. Something we have learned from the COVID-19 pandemic is that international cooperation is vital; developing vaccines and avoiding the increase in contagions and deaths would have been impossible if all countries of the world had not collaborated and shared knowledge, information, medical equipment and all kind of resources, including financial means.[1] This is where the transnational cosmopolitan paradigm that I will be discussing on the following pages enters the scene, because a change in mindset from the conventional nationally or ethnically centred mentality to a world-centric cosmopolitan way of thinking is absolutely necessary. This change in the theoretical approach of

---

[1] This is why Donald Trump's decision to suspend the US payments to the World Health Organization (a political calculation in his search for scapegoats to cover his own incompetence and irresponsibility in dealing with the crisis) was so monstrous and unethical.

social sciences from methodological nationalism to a cosmopolitan focus is fundamental because, as Beck argues, the cosmopolitan critique is not about nation-state
sovereignty or why it has been subordinated to the forces of globalization, but about
the power potentials, strategies and organizational forms of politics *without borders*
('de-bounded') which allow new actors (and networks of actors) from global civil
society to influence the world scenario.

Regarding transnational citizenship, several questions arise: How is it possible
to write about this kind of citizenship when many Latin American states have not
even ended their endless democratic transitions and therefore have permanent *democratic deficits* which are exacerbated by abstention in electoral processes, low citizen
consciousness, and a tendency to favour populism? Is it possible to speak about
transnational citizenship when – as happens in Europe and the US nowadays – significant segments of the population have xenophobic and neo-nationalist ideologies?
Or when authoritarianism seems to be re-emerging?

My response to the queries above is yes, it is possible to open the debate about
transnational citizenship because, in the long term, there is a good chance of implementing the normative cultural model that humanity needs in this time of the Anthropocene. The Westphalian order is already crumbling, and, notwithstanding its avatars,
regional integration processes – like the EU – are still the correct alternative to
neo-nationalism, while holism in social and natural sciences and cosmopolitism are
moving ahead. A glimpse at history over the long term or *longue durée* (Braudel
1987) supports arguments in favour of evolution and progress because there is no
doubt that the absolutism and dictatorial rule of emperors and monarchs of antiquity,
the dark ages and even the beginning of modern times is inferior to contemporary
democratic rule in the majority of nations.

As a result, when the French expert on migratory issues, Wihtol de Wenden
(2017: 109), says that the concept of citizenship is subject to an evolutionary process,
adding that actually we are facing not just economic globalization but also a *human
globalization* that elicits *human mobility*, opening the need to regulate migration, it
is clear that, for practical purposes, the concept of global citizenship is on the right
track. As in the case of the EU, the ever-increasing mobility of the population, which
entails the dissociation between nationality and citizenship and the reconstruction
of identities in a cosmopolitan manner, has been conducive to the establishment of
a European citizenship linked to the values of cultural diversity, non-discrimination,
cosmopolitanism and mobility as a human right.

I will address these themes by beginning with some remarks on Ulrich Beck's
cosmopolitan approach in social sciences, followed by preliminary considerations
to provide context vis-à-vis global civil society as the basis for transnational citizenship. Discussions about globalization, the world-system, the crisis in the Westphalian order, and the emergence of a global civil society that entails the rise of
cosmopolitanism will then follow.

Other issues focused in this chapter are the information era and, of course,
migration and human mobility in the context of globalization. Transnational citizenship, cosmopolitanism and *cosmopolitics* are undoubtedly useful tools for global
governance in these times of the Anthropocene.

6.2   The Cosmopolitan Approach of Ulrich Beck

According to Ulrich Beck, in order to analyse the dynamics of global civil society it is necessary to carry out a methodological shift from "the dominant national perspective to a cosmopolitan perspective" (Beck 2003: 45) because a cosmopolitan theoretical framework questions the belief that both modern society and modern politics are to be understood and organized within the context of a nation state. The cosmopolitan approach essentially means a social science alternative to the "national mystification" of societies and political order based on the traditional "methodological nationalism" that still prevails everywhere, but as national and cosmopolitan perspectives understand sovereignty differently, and as it is evident that national spaces have now become denationalized, this trend has induced not only a change in the meaning of concepts like 'national' and 'international' but also the appearance of a global civil society – and actors – in the contemporary world:

> Global civil society actors can be understood as the agents of a cosmopolitan perspective, even though the phenomenon of global civil society encompasses a diversity of cross-cutting beliefs, prejudices and assumptions. To put it another way, global civil society could be represented as one element of actually existing cosmopolitanism. To grasp the meaning of global civil society, social science must be re-established as transnational science of the reality of denationalization, trans- nationalization and 're-ethnification' in a global age, and this [should be done] at the level of concepts, theories and methodologies as well as organizationally. The fundamental concepts of 'modern society' must be re-examined. Household, family, class, social inequality, democracy, power, state, commerce, public, community, justice, law, history and politics must be released from the fetters of methodological nationalism and reconceptualized and empirically established within the framework of a cosmopolitan social and political science which remains to be developed. This is quite a list of understatements. But nevertheless it has to be handled and managed if social sciences are to avoid becoming a museum of antiquated ideas (Beck 2003: 48).

What are these cosmopolitan realities that intrude and demand a new cosmopolitan methodology capable of preventing social sciences "becoming a museum of antiquated ideas", to quote Beck? First of all, national state borders have become porous, interdependences are growing exponentially, and restrictive immigration policies are trapped in contradictions: on the one hand the rich industrialized countries of the Global North are suffering a "spectacular demographic regression, with ageing populations that threaten to overwhelm pensions and health systems" and reinforce political conservatism – like the growing *neonationalism* and populism of the European extreme right and the former Trump administration – and on the other hand these very countries are busy "building ramparts to ward off both the feared and the real immigration flows from the poorer South". Indeed, as social, economic and political interdependences are growing worldwide, leading to new forms of migration (of both migrant workers and refugees escaping from war and armed conflicts) these circumstances condemn anti-migratory policies as counterproductive and unable to stop human mobility, as happened in the aftermath of terrorist attacks of September 11, because "it is precisely this repressive impulse that undermines the necessary readiness to authorize more immigration, which could counter falling demographic curves and rejuvenate the population" (Beck 2003: 53). Interestingly, that lucid forecast about the ills of "methodological nationalism" was made by Beck in 2003!

Another "cosmopolitan reality" is that certain cosmopolitan issues, such as human rights, have been internalized in nation states. As these standards become part of the universal *jus cogens* widely accepted as enforceable international law – regardless of whether or not its main instruments have been duly subscribed to and ratified by member states[2] – the judges and courts of all UN member states are obligated to accept and apply these norms whether dealing with nationals and foreigners. In other words, to quote Beck (2003: 53), "Human rights are increasingly detached from citizenship status and are no longer bound by national contexts," and "examples of this trend" can similarly be seen in educational curricula; the growing number of bi-national marriages and families; the increasing number of transnational work and private life connections; the growing mobility of communication, information, cash flows and risks (such as the pandemic and climate change catastrophes, products, services, and so on); trade agreements; *diaspora* cultures (the 're-ethnification' or recuperation of ethnic identity, religion, culture and national customs revived in the daily life of people living in guest countries); and what Beck calls the "internationalization of national models of inequality". To put it another way, the permeability of national boundaries entails a distribution of globalization winners and losers (according to the production sectors that are either shielded from the world market or exposed to it) in such a way that in some situations this can produce a contradiction between national and transnational elites "who fight over positions and resources *within* national power spaces" (Beck 2003: 64).

6.3   Transnational Citizenship in the European Union

There is an ongoing academic debate on the topic of transnational citizenship, but my aim in this section is just to highlight the possibilities that this legal category offers for similar processes. As noted in the second chapter of this book, in Latin America and other parts of the world there are *regionalization* processes but not yet processes of regional *integration*, hence it can be argued that at present the EU process is unique and the only true integration process in the world.[3]

---

[2] Thus, for instance, the US has not ratified important international human rights law, such as the 1966 UN International Pact of Economic, Social and Cultural Rights, and a number of UN member states are not parties to the ILO 169 Convention that prescribes the obligation to hold community referendums before authorizing the construction of infrastructure (e.g. for hydropower) or mining in the countryside.

[3] Claims that the EU has a democratic deficit and that it is crucial for the EU's future to strengthen its democratic legitimacy have continued to arise for at least twenty years. For the greater part of its history, citizens have not been at the centre of the European political system. One response to this was to suggest having a category of *Citizenship of the Union* in addition to national citizenship. The Maastricht Treaty (1992) and the *Treaty Establishing the European Community* have integrated the *Citizenship of the Union* into the Treaty of Rome. The rights granted to EU citizens include the

European *transnational citizenship* is a category applied to the collective exercise of rights within the EU, although all persons are simultaneously citizens of their individual EU member countries, which means that EU rights may differ vastly from the individual rights of nationals of each separate state. In other words, even though French, Italian and German citizens can all exercise their rights before, for instance, human rights institutions,[4] the approach through the *European citizens* formula is different. This is also the case when standards originating in the Lisbon Treaty are enforced. Nevertheless, EU transnational citizenship could be seen as the spearhead of a cosmopolitan trend which goes well beyond European regional borders. For Eurosceptics, this is an exception that justifies the rule (that national sovereignties are still fundamental to ensure the preservation of the Union), and for the time being it seems that the US scholar Jonathan Fox is right in his observations about this issue.[5]

Therefore there is a need to differentiate between the notion of global civil society and the concept of *transnational citizenship* because the latter assumes the existence of rights-holders and duty-bearers. In the case of national states these are clearly defined, but that is not the case in the transnational order because it lacks centralized government structures. However, as already mentioned, the European integration system that birthed the European Union has made the most progress in this direction

---

right of free movement, the right to vote in communal elections in all Member States, the right of diplomatic protection and the right to petition the European Parliament. These rights were regarded as the first step towards fully fledged citizens' rights (Efler 2008). More recently, the former Greek Minister of Finance, Yanis Varoufakis, has founded a new political movement (Democracy in Europe Movement, DiEM) with the main objective of promoting the political participation of citizens in European elections in the role of EU citizens (not as nationals of member states of the Union).

[4] In another example, it is interesting to realize that even citizens from a European country which is not a member of the EU but a member of the Council of Europe can present a complaint before the European Human Rights Court if they consider themselves to be victims of a human rights violation in accordance with the terms of the European Convention on Human Rights. The same is possible in the Interamerican system of the Organization of American States (OAS) where in human rights cases citizens can individually sue the State at the Interamerican Human Rights Court headquartered in San José (Costa Rica).

[5] "The term *transnational citizenship* is less expensive than its apparent synonyms, *world citizenship* or *global citizenship*, and is more clearly cross-border than the term *cosmopolitan citizenship*. A longstanding cosmopolitan theoretical tradition calls for 'global' or 'world' citizenship. In contrast, the term transnational citizenship can refer to cross-border relations that are far from global in scope. This is analogous to the distinction between the concepts of global versus transnational civil society. Critics of the concept of *global civil society* argue that it implicitly overstates the degree of cross-border cohesion and joint action in civil society" (Laxer/Halperin 2003). In the context of this debate, the term transnational citizenship would apply most clearly to membership in the EU, a political community that is clearly cross-border yet certainly not global. However, Bauböck (2003), one of the leading proponents of the concept of transnational citizenship, suggests that the European Union is better understood as a *supranational* entity, meaning that individual membership requires citizenship in an EU member state. Indeed, according to Fox (2005) it is still not clear "whether the EU's transnational political experiment is the leading edge of a growing trend or is the exception that proves the rule in terms of the persistent grip of nation states on political sovereignty. So far, the latter seems more likely. Either way, analysts agree that European Union citizenship is still both 'thin' and fundamentally grounded in national citizenship" (Fox 2005: 83).

and is lights years ahead of the regionalization processes of Latin America, Africa and Asia, although that does not mean that the EU formula cannot be a model for other regions of the world.

Returning to transnational citizenship in the EU, a notable example is the fact that European citizens are able to propose bills of law. This is because the Lisbon Treaty gives them the right to petition the European Union Commission, i.e., the entity that is entitled to present bills of law to the European Parliament. The origin of this right is the 2005 failed 'Constitutional Treaty' which was rejected by two different referendums held in France and Holland. Even so, its main guidelines were included in the Lisbon Treaty, which was signed by the Heads of State and ratified by national parliaments in each member country of the Union (without having to be presented to a special referendum), allowing the treaty to be finally approved in December 2007.

On 11 December 2009 the European Commission shared its *European Citizens' Initiative* (ECI) proposal, which was subject to a public consultation process that ended when the Council and the Commission presented a proposal inspired by Swiss direct democracy procedures like the *federal popular initiative* or optional referendum. On 16 February 2011 the European Parliament finally passed the ECI as a mechanism for direct democracy established by the European Union to deepen democracy and provide citizens with the means to participate directly in the creation of European public policies. The procedure requires a minimum of one million EU citizens who are also citizens of 25% of the member states to present to the European Commission certain initiatives that the Commission must present to Parliament. So far, several of these extraordinary and innovative transnational citizenship initiatives have been presented, such as the fraternity initiative to foster student exchange in the framework of the Erasmus Programme, the initiative on the right to water and sanitation, and the initiative focused on halting extremism and xenophobia, as well as initiatives on the free movement of persons within the EU, the reduction of economic inequalities, the strengthening of cultural diversity, and the protection of minorities, that (judging from their titles) are part of a progressive movement designed to improve the legislation enacted by the European Parliament.

The phenomenon of a transnational civil society within the EU where citizens from any member country of the Union can travel, work, reside or vote in any other member state is an expression of transnational citizenship, which is part of the *social integration* process effectively propelled by the policies in the Union.[6]

---

[6] Social integration has been in operation in the labour and economic-entrepreneurial sphere in the European Union since the 1990s. In a well-known jurisprudence case of the epoch, the European Council of Ministers decided that, when operating in EU countries with higher rates of pay (such as France, Germany and Austria), companies from countries with lower rates of pay (such as the United Kingdom, Ireland and Portugal) could not pay workers in those countries according to their national standards, but must pay wages according to the rates and labour legislation in the receiving countries: "The example of the posted workers guideline illustrates well the multiple-level governance nature of social policy: Individual nation-states and the European Union Commission were the main actors. Since 1993, service enterprises can be active all over the EU. One of the consequences has been that companies and even individuals can move to another member country, set up business and deliver

Unfortunately and in contrast, other regions of the world like Latin America have not yet succeeded in gaining the impetus for a social integration process comparable to the EU. Indeed, MERCOSUR, the Andean Pact and the Central American Integration System (SICA) have all failed to establish supranational institutions, so those initiatives consist of regionalization and multilateral cooperation, not integration. However, the absence of social integration does not mean that discussions concerning transnational citizenship are not possible in the context of regionalization, because the interconnectedness brought about by the internet and smartphones in this *information era* (Castells 1997) has facilitated the creation of links and networks between all kind of actors, including individuals and social movements involving religious, cultural and non-governmental organizations. For that reason it is indisputable that not just a regional but a global civil society exists, as pointed out at the beginning of this century by Kaldor et al. (2003). In *Global Civil Society* they refer to the concept as 'normative' but also as a useful way to describe what was then considered an emerging new social reality as a result of the connectivity of global civic action as a counterweight to the narrow notion relating globalization to strictly economic phenomena (Kaldor et al. 2003: 3–4).

Global Civil Society also deals with matters that its authors call "regressive globalization", such as the illegal US war against Iraq in 2003, *Al Qaeda* and jihadist terrorism, but in general terms it is possible to speak of progressive globalization regarding the commercial negotiations that led to the establishment of the WTO; the negotiations for banning biological and chemical weapons that led to the establishment of the OPCW; the surge of transnational peasant movements (such as ASOCODE in Central America and the global social movement called *Via Campesina*); feminist transnational movements; and the creation of the *World Social Forum* (the social alternative to the Davos *World Economic Forum*). Even transnational corporations, so-called 'free trade', and the worldwide financial system can

---

services according to wages and social wages of their country of origin. It is needless to say that, for example, construction companies from Portugal, the United Kingdom and Ireland went to high (social) wage countries such as France, Germany and Austria. Unemployment among construction workers in the high wage countries increased. The unions charged 'social dumping'. This was a problem hard to solve in the European Union because multi-level patterns of policymaking are prone to 'joint-decision traps' (Scharp 1988) in which efficiency and flexibility are subordinated to political accommodation and procedural guarantees. By 1995, sending countries had no incentive to agree on a guideline, because it would have undercut the competitive advantages of firms from their countries. And the main receiving countries, such as France, Austria and Germany had already implemented national laws to regulate the posting of workers from other EU countries within freedom of services (Faist et al. 1998: 7). However, there have been institutional and political solutions to the problem: (1) Because it was an issue of market competition, qualified majority voting and not unanimity applied (Arts. 47 and 55 EC-A). After seven years of bargaining, the Council of Ministers decided in 1996 on the Directive on the Posting of Workers (96/71/EC) – against the votes of Portugal, the United Kingdom and Ireland. According to this directive the companies from abroad have to pay minimum wages and provide working conditions prevalent in the country of activity. (2) One of the prerequisites of this bargain was that the 'losers' such as Portugal, could be partly 'compensated': Logrolling arrangements were possible. For example, Germany aided Portugal in building up youth apprenticeship training" (Faist 2000: 9).

be considered indisputable expressions of this 'global civil society' whose existence in undeniable.

Therefore, global civil society is the root of the cosmopolitan global citizenship which should become the key to addressing all sorts of problems, ranging from human mobility and migration flows to labour rights, working conditions and corporate responsibility in the social and techno-economic spheres. Van Parijs (2018) argues that the European Union has become a community of citizens whose fates are massively affected by the socio-economic institutions that they share, thus cosmopolitanism (not nationalism) is the appropriate mindset. In other words, in the foreseeable future the EU should be governed far less through bargains between member-states and far more through negotiations between cosmopolitan citizens and their representatives. The common *demos* (the people) therefore need to be strengthened through such means as the European Parliament, and it will be necessary to nurture the new transnational structure and dynamics that are needed to improve the Union's operating system. In this respect, the Swiss model could provide inspiration and guidance, as Fischer (2013), Varoufakis (2015) and others have noted.

Since there is no doubt that global civil society could be the point of departure for future transnational citizenship, it is appropriate to review the debate about the nature of globalization for the good reason that a globalized planet constitutes the foundation of world society.

6.4   Globalization, the World System and the Crisis of the Westphalian Order

The *international political system* (IS) has undergone transformations from multipolarity in the nineteenth and early twentieth centuries to bipolarity during the Cold War, but is now again going in the direction of multipolarity. Badie (2012) even argues that the main feature of today's international system is *fragmented apolarity*, because the system essentially consists of a network of interdependencies without a hegemonic centre of power.[7] On an economic level, in spite of the predominance of the financial sector, interdependence favours regionalization, promoting regional integration with supranational institutions, as in the case of the EU, or at least several processes of regionalization with important intergovernmental ties of coordination and cooperation, such as the SICA in Central America, MERCOSUR in South America, the Pacific Alliance, the ASEAN in South East Asia, the African Union, the Asia Pacific

---

[7] Consequently, according to Bertrand Badie, what has existed since the collapse of the USSR (or at least since 1994 after a brief lapse of United States 'unipolarity' at the end of the Cold War) is a "system without a name", precisely because its *fragmented apolarity* can be considered its main characteristic. Badie argues that variables such as a weak degree of inclusiveness and limited deliberative capacity are part of the system's features. Hence, according to the French scholar, multiple centres of real power do not exist at the present time, and even the G7 and the G20 are informal places where a "diplomacy of connivance" is practised, but they are not real power blocs (Badie 2012: 70–76).

Economic Forum (APEC), the Eurasian Economic Union, the Brazil, Russia, India, China and South Africa bloc (BRICS), the Shanghai Cooperation Association, and so on.

Nevertheless, on a social level, the situation has led to the emergence of social movements that question globalization and promote *alternative globalization* through protest rallies against meetings of the G7, the WTO and even the G20. The World Social Forum, created as an alternative to the Davos Economic Forum, is also an expression of "globalization and its discontents" (Stiglitz 2002), and the topic continues to pervade discussions at universities and think tanks (Stiglitz 2002; Mander/Goldsmith 1996; Olea et al. 2004; Bauman 2017; Beck 1998, 2002) around the world. The social crisis provoked by the coronavirus pandemic in 2020 is clear proof that globalization and human mobility are the main culprits of the rapid spread of the virus across the world. This demonstrates the need for local production (of medical equipment, drugs, food etc.) and the importance of internal and local markets, and potentially signals the end of the predominance of "hegemonic globalization".

Furthermore, the transformations of the Westphalian order (which are also at the root of the increased globalization at the end of the twentieth century with the collapse of the communist bloc) show that the globalization process has a nature of its own that cannot be ascribed to the will of political or economic leaders (or their 'agency' capabilities), nor can it be considered the result of changes in the international *political* system, the surge of capitalist modernity as Dussel (2020) argues which is essentially ruled by States. Indeed, globalization has its own *economic* nature that originated in the sixteenth century, due to the discovery of the Americas and Magellan's first journeys around the world, a historical fact which is clearly not the result of policy-planning and was not promoted by particular agents or actors of the international system, whether governments or transnational corporations. Therefore globalization escapes 'agency', as its roots lie in the sixteenth century, and then the Industrial Revolution, and, more recently, the new developments in science and technology (especially in the field of communications), which means that it does not respond to any global 'conspiracy' of the capitalist elites that rule the world economy.

Thus, it does not make much sense to protest or demonstrate against a phenomenon that is not the result of political decisions or government actions, as clearly demonstrated by the fact that the pandemic put a stop to both globalization and economies. Haas (2017), chairman of the US *Council on Foreign Relations*, maintains that although the intangibility of borders or the balance of power continue to be the basic principles of international order, they are questioned by globalization and therefore *sovereign responsibility* must be used to negotiate multilateral agreement on issues like the proper functioning of the internet, the control of epidemics and pandemics, the refugee and migratory crisis, and nuclear disarmament. On the other hand, Joseph Nye wonders if the 'liberal order' will survive the disarray provoked by Donald Trump, and his answer is that the United States could be able to withstand such a misfortune because:

> It has become almost conventional wisdom to argue that the populist surge in the United States, Europe and elsewhere marks the beginning of the contemporary era of globalization and the turbulence may follow in its wake, as happened after the end of an earlier

period of globalization a century ago. But circumstances are so different today that the analogy doesn't hold up. There are so many buffers against turbulence, at both the domestic and the international level, that a descent into economic and geopolitical chaos, as in the 1930s, is not on the cards. Discontent and frustration are likely to continue, and the election of Trump and the British vote to leave the EU demonstrate that populist reactions are common to many Western democracies. Policy elites who want to support globalization and open economy will clearly need to pay more attention to economic inequality, help those disrupted by change, and stimulate broad-based economic growth. … At the same time, military force is a blunt instrument unsuited to dealing with many situations. Trying to control the domestic politics of nationalist foreign populations is a recipe for failure, and force has little to offer in addressing issues such as climate change, financial stability or Internet governance. Maintaining networks, working with other countries and international institutions, and helping establish norms to deal with new transnational issues are crucial. It is a mistake to equate globalization with trade agreements. Even if economic globalization were to slow, technology is creating ecological, political and social globalization that will all require cooperative responses (Nye 2009: 10–16).

As Nye rightly points out, the current situation is comparable to what happened in the 1930s when the crisis caused economic and geopolitical chaos (and the coronavirus crisis of 2020 has aggravated the situation), but that doesn't mean that there is no open field for cooperative responses. On the contrary, the only way out is through multilateral cooperation, as the President of France, Emmanuel Macron, has rightly pointed out in an interview in the *Financial Times* (2020), adding that our "nomadic civilization", with billions of people travelling constantly around the world and the predominance of economics over social needs (such as health and education) and ecosystems – ignoring climate change – has come to an end. Hence discontent and frustration demand substantial political reforms, the reduction of inequalities, and attention being paid to the social sectors harmed by globalization. Far-right nationalistic populism like that practised by Trump is a phenomenon that must be overcome through electoral processes, and military force is obviously useless in the face of the coming dangers of climate change, economic crisis, and pandemics like COVID-19. Multilateral cooperation is absolutely necessary to solve the problems of globalization.

Inspired by Fernand Braudel, Immanuel Wallerstein believes that the world system[8] includes the world economy, and argues that globalization is "something that had been a basic element of the modern world system since it began in the sixteenth century" (Wallerstein 2006: 10) but that excessive emphasis on the study of particular cases prevented the global vision and the examination of history from the perspective of long duration, as claimed by Braudel and the French Annales school.

The foregoing means that if we had studied history with an emphasis on the *long-term perspective,* we would have realized that at least four *turning points* have occurred in the *Modern World System*: 1) the sixteenth century during which the

---

[8] A world system is a temporary space zone that "crosses multiple political and cultural units" and that represents an "integrated area of activities and institutions that obey certain systemic rules". The political units are the nation states which are grouped in the said integrated zone, and their governments are subordinate to the system (Wallerstein 2006).

world economy was formed; 2) the French revolution of 1789 as a world turning point; 3) the social movements of 1968 that constituted the harbinger of what would later become "the long terminal phase of the modern world system in which we find ourselves now" (Wallerstein 2006: 10); and 4) a phase that has been undermining the liberal model unifying the world system. Wallerstein also maintains that the disciplinary separation of the sciences into compartments has prevented a holistic view of the world system and of the long historical duration because:

> The watertight compartments of analysis – which in universities are called disciplines – are an obstacle and not an aid to understanding the world. We have argued that the social reality in which we live and what determines our options has not been that of the multiple nation states of which we are citizens but rather something greater, which we have called the world-system. We have said that this world-system has had many institutions – states and interstate systems, production companies, brands, classes, identification groups of all kinds – and that these institutions form a matrix that allows the system to operate but at the same time stimulates both the conflicts and the contradictions that permeate the system. We have argued that this system is a social creation, with a history, with origins that must be explained, which present mechanisms that must be delineated and for whose inevitable terminal crisis needs to be warned (Wallerstein 2006: 10).

Incidentally, the concept of 'world system' is clearly synonymous with the concept that I have been using throughout this book as the object of study of IR theory, that is to say the *international system* (IS). However, as a subject of study it seems more appropriate to use Wallerstein concept of 'world system' as the object of study of IR since it is a larger concept, tantamount to the planet itself. And, in spite that nations are divided by state boundaries, IR object of study is the multiple relationship the object of study is the multiple relationships between 'national states' separated by imaginary lines (geographical fictional lines as parallels or longitudes) or natural borders (rivers, mountain ranges, peaks, lakes, oceans). In other words, the absence of a holistic approach has prevented academics from going beyond their particular disciplines, otherwise they might have realized long ago that the entire human race is living in this sort of interstellar spaceship called Planet Earth. This is the main reason for arguing that Wallerstein's concept of 'world system' is more appropriate, except for the fact that the main goal of the academic research undertaken in this book is to stress the importance of IR theory. From that perspective, the IS concept may therefore be more useful. But it is just a pragmatic decision, not an epistemological one.

Returning to the French historian Fernand Braudel, in his magnum opus he studies the entire region of the Mediterranean basin at the time of the Spanish monarch Philip II, exhaustively describing and analysing the paths followed by society – farmers and traders, navigators and fishermen, nomads, transhumansts – when forging the 'world economy' of that epoch in a region that encompasses not only southern Christian Europe but also the North African Muslim *Maghreb*, and that can only be understood when the entire geographical area is viewed according to Braudel's vision of the aforementioned *longue durée*:

> The sixteenth century …[was] a dangerous world whose spells and evil spells we had conjured beforehand by fixing those great subterranean and often silent currents whose

meaning is revealed only when we embrace large periods of time with our gaze. The reso-
nant events are often not more than fleeting instants, in which these great destinies are
manifested and which can only be explained thanks to them. We have thus arrived at a
decomposition of history by floors. Or if you like, at the distinction, within the time of the
story, of a geographical time, of a social time and of an individual time (Braudel 1987: 15).

This new approach to history – which largely nourishes and sustains the thinking of
Wallerstein, who, until 2005, was head of the *Fernand Braudel Center for the Study of
Economies, Historical Systems and Civilizations* in New York[9] – is better understood
if we start from the basis that there are three ways of approaching historical studies:
one is the study of man's relationship with the environment, another examines social
history, and a third is the history of events.

For Braudel, the history of mankind's relationship with the natural environment is
an almost immobile history that is slow to flow or transform society, often consisting
of insistent reiterations and incessantly restarted cycles, not linear 'progressions'.
Natural history is in contact with inanimate things, mineral landscapes, agricultural
and with flowers "as if the flowers did not sprout each spring, as their flocks stopped
moving, as if the boats did not have to navigate the waters of a real sea, which
changes with the seasons" (Braudel 1987: 16). Social history is also slow because it
is structural and because, according to Braudel, it concerns:

> Groups and groupings … how this deep sea shakes the whole of Mediterranean life is
> what I have tried to expose … studying successively economies and States, societies and
> civilizations, and trying, ultimately, to clarify better their conception of history, how all these
> deep forces come into action in the complex domains of war. For war is not, as we know, a
> domain reserved exclusively for individual responsibilities (Braudel 1987: 17).

These wise observations of Braudel recall the points made on previous pages
about the way in which military mobilizations were the real trigger of the First
World War, a conflagration that is not attributable to the individual responsibilities of
the politicians or military commanders of the time (even though they were the ones
who ordered the mobilization of troops, as noted by Kissinger) but to the objective
situation in which the armed forces of different countries were geopolitically located.

So the customary history of events alludes to the lives of prominent characters
(Louis XIV, Napoleon, Bismarck, De Gaulle, Churchill, Hitler, Stalin, Roosevelt,
Mao), but it is a kind of

> Agitation on the surface of the waves that raise the tides in its powerful movement. A history
> of brief, quick and nervous oscillations. Ultrasensitive by definition, the smallest weight is
> marked on its measuring instruments …[it] is the most exciting, the richest in humanity and
> also the most dangerous (because) it has the dimension of both their anger as their dreams
> and their (individual) illusions (Braudel 1987: 17).

Hence, it is necessary to study the history of societies in order to better understand the concept of globalization as part of the world system, among other reasons

---

[9] The *Fernand Braudel Center for the Study of Economies, Historical Systems and Civilizations* is
named after the French historian and based at Binghamton University in the State of New York. It
was founded in September 1976 and is one of the most outstanding centres in the USA for advanced
studies in history from the perspective of the dynamics of 'world systems'.

because the very concepts of development and progress are called into question by Wallerstein's theory, since the idea that all societies follow the same path that evolves from the bottom point Z to the top point A in a linear way is simplistic, according to him, and does not adequately explain the complexity of the pathway in which each country or geographical area communicates with the world system and its historical dynamics.

Wallerstein argues that the *world system* is connected by a very complex network of economic exchange relations and all kinds of links (migratory, cultural, commercial) that go beyond national borders and the interstate system based on territories and sovereignty that characterizes the Westphalian order. Therefore, according to his view, taking relative abstractions of a component as separate units of analysis for each 'national society' when in reality they are not isolated from the world system is the wrong way to proceed because it prevents awareness of the multiple connections of the part (the national society) with the whole (the world system). Trying to do this at best produces an incomplete vision of reality and at worst a false, unscientific vision.[10]

In other words, World-systems Theory also seeks to overcome the linear conception of history – and 'development', as its economic expression – according to which every society has an economic dynamic that starts from different stages (or levels) of development. The trouble with the linear concept is that it implies that there are more advanced societies whose 'model' all nations aspire to emulate, hence the description of countries as developing or developed, poor or rich, or third- and first-world. World-systems Theory replaces this type of terminology with explanatory descriptions of the functioning of the centre-periphery economy, in an approach similar to that of Latin American Dependency Theory (Cardoso 1969).

This does not, of course, imply that there are no actors,[11] such as social movements, political parties or responsible citizens, who, aware of the problems of mainstream economics (Göpel 2017), do not propose reforms, especially to matters such as the distribution of wealth or measures that States must take through the fiscal system to redistribute it and reduce social inequalities (Piketty 2014), particularly in developing countries where such inequalities are greater and more serious, but it does mean that national development is not separable from development on a global scale: And, needless to say, it is the source of decolonial theory as proposed by thinkers like

---

[10] Civil society organizations and indigenous movements that in a country like Guatemala, for example, protest against mining and extractive industries do not blame 'globalization' for the presence of this type of economic activity, among other reasons because mining is an activity that has been practised throughout Latin America since the colonial era in close coordination with colonial powers (the world economy of that epoch) and it makes no sense to protest against social-historical phenomena (it would be like protesting against climate change, droughts, earthquakes or floods.) In any case, what is relevant is denouncing the policies that facilitate or induce the occurrence of these natural phenomena and protesting against the said policies, since, in other examples, they increase deforestation, which results in the reduction of rain or the warming of the global climate and in turn determines the increase in the frequency of climatic phenomena such as hurricanes, droughts and forest fires of natural origin, torrential rains and floods, and the melting of polar ice caps and glaciers, etc.

[11] According to IR theory, all the problems of 'agency' are related to the role played by actors.

Quijano (1993) and Dussel (2015) the internal system of a State is connected so closely and complexly with the international system that most of a country's social problems are unsolvable without exchanges of all types or cooperation negotiated in the major bilateral or multilateral arena.[12]

From this perspective, globalization, as we have seen on previous pages, is a phenomenon that has been inherent in the *world system* since the sixteenth century. Nowadays it is better understood from the perspective of its current manifestations, which are the result of the impressive scientific and technological developments exemplified by the new information and communication technologies (ICTs) of the so called *information era* examined above. On a social level there is now a transnational civil society that, as a positive result of globalization, has facilitated the emergence of significant nuclei of a new type of citizenship (people with a *worldcentric mentality* or post-conventional cosmopolitan world consciousness) that rejects ethnocentric nationalism and constitutes the basis of the new transnational social movements that promote *counter-hegemonic globalization* (Santos 2010) and defend causes such as ecology, gender equality, the rights of indigenous peoples, human rights, humanitarian law, sustainable development, the UN 2030 Agenda, the commitments of the Paris COP 21, nuclear disarmament, and other similar causes. It is therefore possible to say that this cosmopolitan world-view and its social movements show that a significant layer of global civil society is emerging with a post-conventional cosmopolitan philosophy or world-view that could also be the bedrock or underpinning for the kind of transnational citizenship pioneered by the EU. Nevertheless, there are some negative expressions of globalization, such as the proliferation of organized crime and other phenomena that can be attributed to the characteristics of the capitalist system, like the growing number of transnational companies whose production is destined for the global market and requires intergovernmental agreements in order to permit merchandise to circulate freely (with reduced or no customs tariffs) or facilitate investments so that corporations may 'relocate' to places where wages are lower, in areas where the use of corrupt procedures is frequent with the purpose of obtaining government contracts which enable the accumulation of capital, the concentration of wealth, and the placing of funds in tax havens: all those negative features can be mitigated by appropriate global regulations negotiated in multilateral UN international fora.

For instance, the main characteristic of free trade agreements is that since the products of *transnational corporations* (TNCs) are destined for the world market, their commercialization has given rise to long negotiations – which were originally carried out under the GATT and now are under the WTO umbrella – to facilitate and regulate trade and reduce (or eliminate) customs tariffs. Given that a large proportion of the goods that are traded constitute exchange within corporate firms or goods such as vehicles or clothes that are assembled and later exported to the huge markets of

---

[12] Transnational organized crime is another piece of evidence: you cannot solve the problem of drug trafficking in isolation in each country. The only way to confront this scourge is through concerted action by all governments that have to deal with criminal structures. Pandemic and epidemic diseases, climate change, food issues, and international migration are other obvious examples that explain why the United Nations proposed the adoption of the 2030 Agenda and the 17 SDGs in 2015.

industrialized countries, all these activities have had an impact on the emergence of the current global *complex interdependence* whereby all actors involved belong to the economic and civil society spheres as well as governments. This is underlined by theoretical contributions, among which those of Robert Keohane and Joseph Nye (senior) stand out.[13]

Other researchers have extensively studied the repercussions of the phenomenon of globalization within the mindset of capitalist elites that seek to strengthen their influence over governments and maintain their *hegemony* in the world economy through the ideological formation of transnational global elites that want to preserve the *status quo*. Using the three countries of the so-called Northern Triangle of Central America – i.e. Guatemala, El Salvador and Honduras – as a case study, Valdez (2015) identifies a phenomenon that he calls the *organization of consent* as the method used by conservative sectors to exercise ideological hegemony. Since the *world system* influences world capitalist elites, the conservative ideology is reproduced at regional level by these powerful elites in order to keep their hegemony, as Gramsci (1931) and Göpel (2017) have explained in their writings.

Instead of developing a favourable attitude (or ideological mindset) to social and political change like the cosmopolitan post-conventional worldcentric conscious-ness described on previous pages, the cultural values and academic formation of these oligarchic groups are still colonial, Eurocentric, conservative and opposed to change. By contrast, and indigineous organizations this ideological change of mentality (or *mindshift* as Göpel calls it) has occurred in the case of leaders and organiza-tions of social movements, largely because most NGO or civil society leaders have received training in human rights, ecology, women's empowerment, humanitarian law, sustainable development and similar issues or have acquired this knowledge via the world interconnectivity and social links that the new *communication technology* (ICT) generates worldwide, that is paradoxically to say thanks to globalization.

However, even though many of the children of these economic elites have studied at American high schools, colleges and universities, their mindset continues to be extremely conservative, retrograde and anchored in the colonial and postcolonial past. As a result, although there are always notable exceptions, the descendants of tradi-tional oligarchies usually reproduce the type of backward and regressive mentality of their parents, whose roots lie in colonial times and in the kind of nineteenth-century liberalism that was never applied, but was just ideology in the worst sense of the concept, somewhat modified at the end of the twentieth century thanks to the neoliberal thinking of the American economic academic centres that predom-inate in the universities or schools where all these youngsters have studied – not necessarily abroad, because the rich people in countries like Guatemala have their own universities where teaching is usually in accordance with the elites' thinking. As a consequence, the influence of the elites on the government and its institutions

---

[13] The Theory of Complex Interdependence states, among other things, that today's international relations are far from being limited to government officials or to the government, and that in fact there are multiple communication channels between civil society actors, with an absence of hierarchy and agenda items such as the diminishing importance of military affairs, etc. (Nye/Keohane 1988).

tends to be negative and contrary to the changes required to promote reforms like those advocated by the *Sustainable Development Goals* (SDGs) of the UN 2030 Agenda. This assertion can easily be corroborated by looking at the statistics of what happened with the former *Millennium Development Goals* (MDGs) in countries like Guatemala, where the deplorable backwardness of the oligarchic elite is the main culprit for their non-implementation, given the fact that they have absolute control over the governmental administration.

As a phenomenon that originated in the sixteenth century, as explained by Braudel's pioneering research on the Mediterranean basin and later confirmed by the rigour and depth of Wallerstein's (1989) historical research, globalization has not only "made the globe smaller",[14] to quote the German philosopher Sloterdijk (2006: 199), who concurs that globalization began with the great discoveries of Spanish and Portuguese navigators during the sixteenth century and English and Dutch explorers during the seventeenth century, but its repercussions for the international political order are giving rise to the crisis of the Westphalian order. The world economy does not require the order based on economies and nation states; on the contrary, national sovereignty is an obstacle to the processes of regional integration.[15]

Consequently, the sovereignty of national governments over their respective territories is currently being diminished by the global economy, which is rooted, according to Braudel, in the Mediterranean basin, and acquired great splendour during the century of geographical discoveries by navigators and explorers like Columbus, Vasco de Gama, Balboa and Magellan. Because the *global economy* is its real base of support, high finance not only does not recognize borders (an elementary issue of which any stockbroker is constantly aware) but operates in such a way that the importance of territories and boundaries is reducing for the benefit and interest of global capitalist elites. This is happening despite *internal* opposition from both the right and the left extremist wings of national political parties and other organizations.[16]

The crisis of the Westphalian order, then, is not only a crisis of the exercise of national political sovereignty when national governments must face the realities of a globalized economy, but also an economic crisis resulting from the very logic of the functioning of the global capitalist economy. This is what both right and left conservatives oppose.

---

[14] The systems of electronic diffusion of the waves make it possible to implement effectively the oblivion of the distances. *Global players* live in a world without distances. From the aeronautical point of view, the ground is reduced to a maximum of fifty hours of flight in a single jet. The orbital flights of the satellites and the *Mir* station, and recently the *International Space Station's* (ISS) rotations around the globe have become routine around the globe in about ninety minutes. For radio and light messages, the Earth has practically reduced itself to an immobile point- it is rotating, as a temporally compact globe in an electronic felt that surrounds it like a second atmosphere. The terrestrial globalization thus progressed so much that it seems odd to demand that it justifies it (Sloterdijk 2006:199).

[15] Consequently, the anti globalization thinking of the nationalist right (Trump, Farage, Le Pen) acquires full meaning.

[16] In France, for instance, both the National Front of Marine Le Pen and *La France Insoumisse* of Jean Luc Melenchon are against globalization.

To reduce labour costs, transnational corporations that operate on a global scale relocate their production to the geographic spaces where it suits them best, precisely because they have no territorial ties, as I will discuss later and as was pointed out lucidly by Bauman (2017). It is also true that they may settle where there is a plentiful supply of the natural resources (hydrocarbons, minerals, productive soils) which are essential for their economic activity. Therefore, every State needs to establish policies that facilitate, hinder or prevent such movements of transnational capital and companies. It is crucial for governments to set out their political positions, engage in dialogues, and create understanding in order to resolve the conflicts that inevitably occur, since otherwise there is no choice but to resort to violent protest and/or repression, and both are undesirable.

The narrative above has social effects that also manifest themselves in a degree of *human mobility* (migration, tourism, refugees, business travel) with no recent historical precedent because this mobility is the result of the supply of global labour, leisure or business at global market level, another normal consequence of globalization.

However, at the same time in the countries that are receiving large migratory flows of workers or refugees, this human mobility is causing social reactions of racism and xenophobia with the consequent disturbances and turbulence in the political sphere because the vote for the extreme right has increased. It is due to this sequence of events that the migratory issue is an additional factor to take into consideration if we want to understand both the crisis of the Westphalian order ('out of order', according to a line on a cover of the journal *Foreign Affairs*) and the crisis of the EU integration process, including phenomena like the mistrust of European citizens regarding the unelected officials of the EU in Brussels, *Brexit,* the election of Donald Trump, andCatalonia's separatist movement in Spain.[17]

In his 2018 speech to the United Nations Donald Trump deliberately signalled a definitive break with the multilateralist consensus that had guided the US grand strategy since World War II, as underlined in a Foreign Affairs article by Charles Kupchan, a distinguished professor of international affairs at Georgetown University and a Senior Fellow of the Council of Foreign Relations. Trump's regressive foreign policy was a return to nineteenth-century isolationism because despite a global range of strategic commitments:

> Trump's instincts – and he does govern by instinct – are unmistakably isolationist … Trump's political ascent clearly rests on his deft ability to appeal to a disaffected electorate by promising to turn back the clock to a more sovereign, whiter, more industrialized and more geopolitically detached United States. Nonetheless, his effort to reorient US strategy using an earlier version of exceptionalism is destined to fail. His isolationist instincts and his attack on multilateralism, globalization, democracy promotion and immigration have provoked passionate opposition at home and abroad. And for good reason. A grand strategy crafted for the nineteenth century is ill-suited to the twenty-first (Kupchan 2018: 63).

---

[17] And most probably also of the European Union because obviously Madrid would not like a re-entry of an independent Catalonia to the EU.

6.5   The Information Era and Globalization

As a phenomenon inherent in the *world system* since the sixteenth century, globalization can more readily be understood if it is studied from the perspective of its historical permanence and its manifestations. To a large degree, these are the result of the amazing development of science and technology, as expressed by the new communication and information technologies which have led to the emergence of the *Information Era*, as Castells (1998) calls it, in which contemporary society is now living.

Clearly, the Information Era has consequences for the economic field, hence the considerable growth of transnational corporations dominating the internet, such as *Google, Microsoft, Facebook, Twitter*, and *Instagram*.[18] Fostered by the interconnectivity facilitated by smartphone technology, this phenomenon has a parallel in global civil society and is leading to the emergence of considerable clusters of citizens with a cosmopolitan world-view. Although they retain their respective cultural identities, which are rooted in their corresponding nations and local communities, many citizens have started to still feel part of a large global community because they have a *worldcentric* or *post-conventional* cosmopolitan mindset or conscience. This is not just because they surpass ethno-centric nationalism but especially because they are the social constituency of transnational social movements that respond to causes such as human rights, climate change, gender equality, nuclear disarmament, indigenous rights, migrant workers' rights, refugees, and similar movements with universal values and a cosmopolitan mentality.

Transnationalization of the economy and the boom in transnational companies whose production is aimed at the world market are other obvious results of globalization. This is because both trade and the requirements to reduce tariffs using free trade agreements have increased considerably. Additionally, as with all goods destined for the world market, the negotiations to regulate world trade – which were originally carried out as part of the GATT framework and are now under the WTO umbrella – have become indispensable for the facilitation and regulation of trade and the reduction (or elimination) of customs tariffs. Among other reasons, this is because not all exchange is an exchange of finished goods destined for sale; rather, as happens in the automotive and clothing industries, a significant number of exchanges are carried out internally in corporations in order to assemble products which are later re-exported to the company's country of origin or elsewhere in the world.

As mentioned before, this phenomenon has had an impact on the emergence of *complex interdependence*, given the number of parties involved and the nature of the transactions. There is also a situation of interdependence not only in the economic

---

[18] But this phenomenon also entails risk, because of the increased manipulation of human behaviour by commercial internet corporations (like Amazon) and the surveillance of individuals by authoritarian regimes (AI), as explained by Hariri (2018).

field but also between civil society entities, national governments and intergovernmental organizations, as highlighted by Interdependence Theory, especially Robert Keohane and Joseph Nye (Senior), which was developed in the 1980s.[19]

In summary, I agree with Braudel and Wallerstein, since the long-lasting historical periods and the formation of the world economy as a phenomenon dating back to the sixteenth century prove that globalization should not be regarded as the result of policies designed by global capitalist elites in specific historical times. Additionally, it was after the fall of the Berlin Wall in 1989 and thanks to the dissolution of the USSR and the inclusion of the former communist bloc in the world capitalist system that the most recent wave of economic globalization took place. It is characterized by the *inclusion* (Badie 2014) of all countries in the *global economy* (Wallerstein 2006), underpinned by *interdependence* (Keohane/Nye 1988) and an increase in the *mobility* of both goods and people in growing migratory influxes (Wihtol de Wenden 2017). All of this has been facilitated by the formidable development of new information and communication technologies (Castells 1998), whose importance is unquestionable.

Therefore, if globalization is an objective phenomenon *irreversible* in nature, what citizens and civil society organizations should do when exercising their right to democratic participation is take action in each specific circumstance in which people's rights are being affected by the expansion of world capitalism in the direction of *counter-hegemonic globalization*[20] (Santos 2009), as explored before. Likewise, reforms to the economic model and the *change of mindset* are required to bring about substantial reforms in both production and the mainstream reformist economic thinking of scholars like Göpel (2017), Mason (2016), Richards (2004), Varoufakis (2012), and Piketty (2014).

On the other hand, it is a fact that the Information Era in which we are now living is decidedly contributing to the dissemination of the neoliberal ideology that informs the dominant paradigm. According to Maja Göpel, the prevailing economic ideology, namely neoliberal *mainstream economics*, determines public policies worldwide. This point of view coincides with that of the Guatemalan author Fernando Valdez,

---

[19] Complex Interdependence Theory states, among other things, that international relations today are far from limited to the official government or government contacts and that there are multiple communication channels between different civil society stakeholders – with an absence of hierarchy in the topics on the agenda, as well as decreased importance in military matters when compared to the past, etc. (Nye/Keohane 1988).

[20] Boaventura de Sousa Santos believes that there is no "isolated entity" called globalization, as it basically consists of a "series of social facts". Whenever these social facts change, so does globalization, which in turn is manifested hegemonically through localized globalism (or vice versa) but which also possesses a dimension of "cosmopolitanism and common inheritance of humankind"; this is positive as it is grounded in social struggles, seen as "counter-hegemonic globalization". This means that in the world "Exclusion and hegemonic processes find different kinds of resistance – grass-roots initiatives, local organizations, popular movements, transnational networks of solidarity, new ways of worker internationalism attempting to counteract social exclusion by opening spaces for democratic participation and community construction, and by offering alternatives to the dominant ways of development and knowledge. In summary, they favour social inclusion. These local/global ties and cross-border activism make up the new transnational democratic movement" (Santos 2009: 229–236).

who has studied the repercussions of the globalization phenomenon for the ideology of the corporate elites in Central American countries, presenting them as social groups who, in the search to strengthen their predominance over governments through the ideological training of transnational global elites, 'organize consent' as a formula to exert their political power. Thus, as seen when referring to the Westphalian system and international relations paradigms taught to the diplomatic and internationalist elites throughout the world, an analogous phenomenon takes place in the field of economics, since, as Valdez states, the modalities in which the national and regional capital elites exercise the hegemony is such that:

> It is related to progressively complex operations to produce consent – with reciprocal influence, this is key – but also with resistance. These are relations that have been forged for many decades, as we shall see, and of which government delegations and civil society – centre, centre-left and left in Europe and Latin America – partake (with strategy or without it, with certain autonomy or none at all, but also more or less 'resignedly') (Valdez 2015).

What is regrettable about the Central American case is that these regional elites,[21] maintain an extremely conservative ideology that opposes modernization and change, even when the offspring of the most powerful families have studied abroad, since, as Valdez remarks:

> The core of the elites is made up of national families … They reproduce transnationality as their own values such as family, culture, rationality, lineage, history and political culture. They are native and more often than not, educated in elite schools where they have met each other, or in universities abroad, obtaining master degrees in business administration that … dictate the corporate philosophy in vogue … and their intellectual surroundings intertwine their influence on State institutions, civil society and political society with greater efficacy proportional to the weakness of national States (Valdez 2015: 64).

So it is neither academic education nor cultural values that have facilitated the change of mindset (Göpel 2017) to a 'world-centred' or cosmopolitan conscience, as might have been expected. Paradoxically, the new modernizing mindset, favourable to sustainable development, for example, has emerged in social leaders or in civil society organizations due to activism on topics such as human rights, the environment, gender equality, the rights of indigenous peoples, and more broadly, globalization. This is not so with corporate elites that maintain the same conservative ideology, whose roots are found in nineteenth-century liberalism, allegedly 'updated' thanks to the previously mentioned 'Washington consensus' but in actuality limited to the commercial liberalization mantra, the privatization of public property, and monetary stability. With a few notable exceptions, always present in these types of processes, corporate elites exert influence on the State and its institutions and tend to diminish and oppose any change required, such as those which form part of the commitments ratified by all world governments in the United Nations as the *Sustainable Development Goals* (SDGs) and the 2030 Agenda.

But, going back to globalization, as previously stated, it is a phenomenon that was initiated in the sixteenth century, according to Braudel's pioneering study (1949)

---

[21] Such as the social responsibility movement within the business and private sector.

and confirmed by the historic investigations of Wallerstein (1989). Not only has the world become smaller,[22] as stated by the German philosopher Sloterdijk (2006), who avers that globalization began with the great discoveries by Spanish and Portuguese sailors in the sixteenth century (followed by the British and Dutch in the seventeenth century), but its political repercussions are contributing to the accelerated transformation of the international system.

Although globalization – a result of the *global economy*, whose roots are found in the Mediterranean basin, as per Braudel's explanation – unquestionably gained immense glory during the century of great geographic discoveries by sailors and explorers such as Columbus, Vasco da Gama, Balboa and Magellan, in the present its ability to exert dominion on the sovereignty of national governments in their respective territories has gradually declined. One of the may reasons is that since its real foundation is the global economy, high finance not only disregards borders (a basic principle of which any stockbroker is permanently aware) but also operates in such a way that national borders have seen their importance reduced because world markets determine the economic interdependence and growing mobility of goods – free trade – and people ('human capital'). Just like corporate executives, global elites and tourists, workers move to places where there are job openings. Human mobility has increased, and this requires due legal regulation of the immigration phenomenon and the implementation of policies aimed at informing citizens of the importance of foreign workers for the corresponding national economies, as well as to reduce the influence of mistaken perceptions that seek to assign to migrants and refugees the responsibility of economic problems like unemployment when extensive research data shows that immigrants in fact make a substantial contribution to the receiving countries.[23]

---

[22] "The systems of electronic diffusion of the waves make it possible to effectively implant the forgetfulness of distances. Global players live in a world without distances. From an aeronautical point of view, the Earth is reduced to a maximum of fifty flight hours in a single jet. The orbital flights of the satellites and the Mir Station, then recently the round-the-globe rotations of the International Space Station (ISS) have become accustomed to circling the globe in about ninety minutes. For radio and light messages, the Earth has practically reduced to a still point – it is rotating, like a temporally compact globe in an electronic field that surrounds it like a second atmosphere. Terrestrial globalization has progressed to such an extent that it seems odd to demand that it be justified" (Sloterdijk 2006: 199).

[23] According to Roberto Savio, statistics of 2018 indicate that the total immigrant flow was 50,000 people, compared with 186,768 in 2017, 1,259,955 in 2016, and 1,327,825 in 2015. Such an astonishing difference could mean that there is a manipulation of information and data. Savio says that a survey of 23,000 citizens of France, Germany, Italy, the United Kingdom, the United States and Sweden showed an enormous amount of disinformation among European citizens. For instance, in five European countries, people believed that immigrants were three times higher than in reality. Savio asserts that Italians have the erroneous idea that immigrants form 30% of the population when they form only 10% and that the Swedes are closer to the reality because they think immigrants account for 30%, when in fact they account for 20%: "And Italians think that 50% of the immigrants are Muslim, when in fact it is 30%, conversely, 60% of the immigrants are Christian, and Italians think they are 30%. And in all six countries citizens think that immigrants are poorer and without education or knowledge, and therefore a heavy financial burden. Italians think that 40% of the immigrants are jobless, when it is close to 10% – no different from the general rate. The seventh

The crisis of the Westphalian system is not only a crisis in the exercise of political sovereignty due to the democratic deficits of each country's own specificities, but, above all, the consequence of the opposition between the political sphere and the economic one – geopolitics against geo-economics, geopolitics against globalization, geopolitics against sustainable development. Humanity and the cosmopolitan segment of global civil society are tired of this nonsensical confrontation. As a result, humankind is encountering the challenge of fixing the former incompatibility or facing the enormous breakdown (Raskin 2016) that will lead the world to barbarization and the abyss. The incompatibility between geopolitics and globalization is a consequence of the operating logic of the world capitalist economy embedded in the *world system* (Wallerstein 1989), and this phenomenon should be subjected to inter- and trans-disciplinary research.

As a consequence, if the world tourism industry can move thousands of persons in cruise ships or *charter jets;* if transnational companies can outsource their production to geographic areas that better serve their interests including the mobility – place of residence – of TNC's CEOS; if thousands of Chinese workers can work on the OBOR project; if Egyptians, Indians and Filipino labourers can work in the oil-producing states of the Persian Gulf – all of which is properly regulated by the immigration authorities – it is absurd (except for racism, xenophobia or *'aporophobia'*) that no proper immigration regulations are in place in the United States and the rest of the world to ensure the free mobility of the labour force. Of course, it still not known how the pandemic of 2020/21 will affect human mobility for the foreseeable future, but at least for the people already installed in a guest country proper regularization of their migrant status is needed and must be resolved promptly.

6.6   Human Rights, Mobility and the Migration Crisis

It is clear that the job market and the demographic needs of the economies in the industrial countries of the Global North are the factors driving the growing mobility of the workforce worldwide. Unless global capitalism decides to invest in the Global South to create employment, which would be problematic and is therefore improbable, this phenomenon will continue in the long term, among other reasons because most citizens of the host countries prefer not to do the kind of work that immigrant labour undertakes. Legal regulations should therefore be in place to address mobility and not criminalize migrant workers who lack the proper documentation, partly because the

---

report on the economic impact of immigration, from Foundation Leonessa that based its research on the statistics of the Italian Institute of Statistics, presented some totally ignored facts. The 2.4 million immigrants in Italy have produced 130 billion Euros, or 8.9% of the Gross Internal Product: an amount larger than the GDP of Hungary, Slovakia and Croatia. In the last 5 years, companies started by immigrants have become 25.8 of the total, with 570,000 companies, or 9.4% of the total. The director of the Pension Italian system, Boeri, said in Parliament that immigrants give to the system 11.5 billion, more than what they cost. He also stressed that Italy is having a demographic crisis, with seven births for eleven deaths" (Savio 2018).

said state policies violate, in certain cases, international treaties (i.e. the UN Convention on Refugees and Asylum [1951] and the Convention on the Rights of Migrant Workers and their Families [1990]), but mainly because anti-immigration policies are clearly and absolutely contrary to the UN Universal Declaration of Human Rights which states everyone has the right to freedom of movement and residence within the borders of each State. Equally, any person (states Article 13 of the Declaration) has the right to leave any country, including his or her own, and to return to his or her own country.[24] Obviously, if you have the right to leave your country, other countries are obliged to let you enter or travel through their territories. While proper regulations are required to facilitate freedom of movement and residence, the detention and deportation of migrant workers who cannot present proper migration documents is a violation of human rights that must be denounced as such by the judiciary both within and outside the country responsible for that kind of action.

The demographic phenomenon is causing the economies of industrial and rich countries to require a growing number of international labourers to replace people of retirement age because there are not enough native nationals to take their place. It is a fact that the contribution by foreign workers to the receiving countries is positive and convenient. Consequently, it is essential for skilled, unskilled and semi-skilled labourers to be granted work permits and multiple entry visas in their passports so that they can enjoy free mobility (enforcing their human right to leave and re-enter their countries in accordance with the 1948 Universal Declaration of Human Rights) and travel to their countries of origin whenever they choose, in the same way that they can freely send monetary remittances to their families and relatives. Therefore, the reform of migration policies and the establishment of new rules assuring some kind of 'transnational citizenship' – or at least some kind of permanent multiple-entry visa of the type granted to technical personnel, executives and CEOs of TNCs – for all long-lasting or constantly moving workers is absolutely vital. Regarding this crucial matter of our times, Savio says:

> The statistics are clear. Each year there are 300,000 fewer working people. Of the 80.6 million Germans, only 61% are of working age. In 2050, this will shrink to 51%, and the number of those older than 65 will increase from 21% to 33%. The birth rate in Germany is 1.5%; to have a constant population you need a birth rate of 2.1%. The huge influx of immigrants has increased the birth rate to a modest 1.59%. Immigrants tend to imitate local trends and do not have many children. Therefore, it is clear to all that in two decades productivity will decline dramatically (some say by 30%), because of fewer people working, and there will not be enough tax-payers to keep the pension and social security system going. It will be the end of the German locomotive. The same consideration applies to all of Europe, which has a statistical birth rate of 1.6, which means it will lose close to one million people per year. The UN division of Population Statistics considers that Europe should have an influx of 20 million immigrants to retain its course. … We have plenty of data about the positive

[24] Article 13 of the United Nations Universal Declaration of Human Rights (1948) states that everyone has the right to freedom of movement and "residence within the borders of each State" and also the right "to leave any country, including his own, and to return to his country". This right, incidentally, needs to be differentiated from the right to asylum stated in Article 14, which refers to how the right to asylum "may not be invoked in the case of prosecutions genuinely arising from non-political crimes or from acts contrary to the purposes and principles of the United Nations".

impact of immigration. The last is a very complex study of over 30 years of immigration, done by the very respected CNRS, the National Centre of Scientific Investigation in France, published by *Science Advances*. ... The dream of people who come to Europe to escape hunger or wars is to get a job as soon as possible, pay taxes and financial contributions to ensure their stability and future, and work hard (Savio 2018).

Savio's analysis can easily be extrapolated to the US to solve the migration crisis unchained in the western hemisphere by the outrageous policies of Donald Trump. The fact is that if regulating the mobility of persons in specific social categories (i.e. the upper and middle classes) poses no problem, then – as emphasized by Cortina (2017) when she refers to *aporophobia* or the rejection of the poor as a challenge for democratic societies – why is it so difficult to recognize the right to mobility of poor people? Why they are criminalized? Why do they constantly suffer such terrible discrimination everywhere? Are migrants the equivalent of contemporary slaves? How can the rejection of the poor be explained if there is employment for them in rich countries? Why do migration authorities in the US refuse to regularize the migrant status of people who already have a job? How can this constant violation of their human rights be explained? Is it because they are poor or because of racism? Or both? In the US even asylum seekers and refugees are suffering from this brutality and hatred, which increased with the signing of the special agreements that Trump imposed on corrupt and weak governments.[25]

Is it possible to start a multilateral negotiation process in order to solve this artificial migration crisis provoked by ideology and unscrupulous politicians? The *Global Compact for Safe, Regular and Orderly Migration* signed at the Inter-governmental Conference held in Morocco in December 2018 is undoubtedly an advance in that direction. It strived to recognize the "right to mobility" of workers and was initially implemented after a long history of UN-sponsored meetings, starting with one held in Geneva in 2004.[26] Organized by fourteen international organizations and NGOs,

---

[25] It is important to bear in mind that there are two different aspects to this issue: 1) the criminalization of people who enter the US without proper documents or overstay their visa authorization; and 2) the problem of asylum-seekers and refugees. Concerning the former, it is evident that the current US migration policy of criminalizing both workers and refugees opposes the UN Declaration of Human Rights. With regard to the latter, when Trump decided to use migration issues as a banner among his WASP voters in his re-election campaign in 2020, he applied the UN Refugee Convention (1951) category of (supposedly) "third secure country" to countries like Guatemala ruled by a weak and corrupt president, so Guatemala will have to receive non-Guatemalan asylum-seekers who must wait for months and years while the US authorities make a decision about their applications. To this humiliation, Guatemalans arriving at the US border to ask for asylum must add the inhuman decision taken by Trump in another executive order to separate children from their parents as a way of deterring the wave ('caravans') of refugees fleeing criminal violence in other Central American countries (like Honduras). Even though an honourable judge stopped that inhuman and merciless policy, children have waited for weeks and months before being reunited with their parents and families, some of them already deported to Guatemala. Some children have also died in the custody of American migratory officials because of inadequate health care and lack of medicines. Since 2020 the COVID-19 pandemic has increasingly been used as a pretext for these 'express deportation' policies against asylum-seekers.

[26] The antecedents include the *Convention on the Protection of the Rights of Migrant Workers and Members of their Families* (1990) – unfortunately signed and ratified by only a few States.

the main result of that conference was the establishment of the *Global Forum on Migration and Development* (GFMD), and in the following years meetings were held in Brussels (2007), Manila (2008), Athens (2009), Puerto Vallarta (2010), Geneva (2011), Mauritius (2012), and Istanbul (2015). On September 2016 the UN General Assembly unanimously adopted a *Declaration for Refugees and Migrants* which recognized the need for more cooperation in order to manage migration effectively. However, the most important outcome of the Declaration was that it set off a process leading to the negotiation of the Global Compact for Migration, which was approved in Marrakesh in December 2018 following another resolution adopted by the General Assembly (6 April 2017) which decided on the modalities and timeline for the nego-tiations of the so-called 'compact' in three consultation phases (April–November 2017) after six sessions in Geneva and Vienna, another phase (December 2017–January 2018) that led to a first draft (called *zero draft*), and a final one (February-July 2018) in New York. The Canadian lawyer and jurist Louise Arbour was appointed by the UN Secretary-General Antonio Guterres as his Special Representative for International Migration and asked to work with nations and stakeholders to develop the draft text that on 10 December 2018 was finally approved by 164 nations and endorsed afterwards by the General Assembly (19 December 2018). 152 countries voted in favour of endorsing the resolution, but unfortunately, and as expected, Trump voted against, as did Hungary, Israel, the Czech Republic and Poland, while another twelve countries abstained.

The Compact has twenty-three objectives and commitments. These include collecting and using data to develop an evidence-based migration policy, ensuring that migrants have identity cards, enhancing availability and flexibility for regular migration, encouraging cooperation for tracking missing migrants and saving lives, ensuring migrants can access basic services, and making provisions for both the full inclusion of migrants and social cohesion. The agreement also says that within their sovereign jurisdiction, governments may distinguish between regular and irregular migration status while they determine the legal and policy measures for the imple-mentation of the Compact, taking into account the different national realities, policies, priorities and requirements for entry, residence and work, in accordance with inter-national law. In order not to criminalize migrants, the agreement correctly makes no distinction between 'illegal' and legal migrants, but it does distinguish between regular and irregular ones, and affirms the right of states to distinguish between those categories. Signatories are obliged to prevent irregular migration but the Compact does not make any distinction between economic migrants and refugees.

The 2018 Marrakesh conference and the *Global Compact* are undoubtedly remarkable achievements, but the US is not a signatory to the agreement and the Compact is not legally binding, which compromises its enforcement. What is the best way in the foreseeable future to address these issues in the multilateral UN scenario? Would it be possible to organize a world summit on migration issues with the aim of signing a legally binding treaty? In my view, Governments should take a stand on these important matters to enable the UN to summon an international summit on immigration worldwide, like the ones that have already taken place on issues like the environment (the Rio summit), human rights (the Vienna summit), population (the Cairo summit), women (the Beijing summit), and social development (Copenhagen). Such a conference could be the starting point for multilateral negotiations aimed at reaching agreements, such as the ones achieved in Paris and New York in 2015 on climate change and sustainable development. Holding a multilateral conference of this nature is perfectly feasible, as demonstrated by the *new global governance of migration diplomacy* (Wihtol de Wenden 2017).

6.7 Human Mobility in the Context of Hegemonic Capitalist Globalization

Globalization clearly has social effects that are manifesting themselves in an unprecedented *human mobility* phenomenon which responds to the demands of job markets globally. Unfortunately, the migratory influx is causing regrettable xenophobic reactions in the receiving countries, which are predictably accompanied by unrest and disorder in the political sphere. The social aspect of migration is an additional factor in the crisis of the Westphalian order, and has ramifications for regional integration processes, as occurred in the European Union with *Brexit*. The fear of terrorism, associated with the fact that Islam is the religion of most refugees and migrants from the Middle East, is undoubtedly also a component of these collective concerns.

The social aspect is also related to demographic factors because although globalization is the more important objective factor driving human mobility, as Wihtol de Wenden notes, the demographic influence is also significant, given that, on average, Africans and Latin Americans are younger than Europeans and Americans, that overpopulated China looks with envy at underpopulated Russia, that Southern and Eastern Europe are now receiving migrants because Polish workers – for instance – going to Germany or the UK are substituted by Ukrainians or Byelorussians, and so on:

> Mobilities, whether temporary, pendulum or permanent, affect more particularly the main lines of the world, where the differences in wealth, standard of living, demographic profiles, political regimes are most glaring. The Mediterranean is one of these major fault lines: 50% of the population on the south shore is under 25 years old and a third are unemployed; on its European shore, it is subject to an ageing demography. The median age (age which divides the population equally into two groups) has increased in sixty years from 28 years to 41 years in Italy and 39 years in France, while in sub-Saharan Africa it is today 19 years. Another major divide is formed by the border between Mexico and the United States, the largest in the world according to the number of illegal crossings and undocumented migrants living in the United States (some 11 million), and then by that separating Russia from China, where the tension between populations, territories and underground resources is particularly acute. Certain regions of the world have recently experienced great migratory changes, going from the status of the country of departure to that of host country: this is the case in Southern Europe … and Eastern Europe, where a migration chain from east to west is happening (Poles go Germany, the United Kingdom or Ireland to work, while the Ukrainians and the Belarussians go to Poland); and in Romania, hit by the health care drain, that is to say the exodus of the health professions (the doctors, nurses and carers of old people who have left to work in Western Europe are replaced by Moldavians). The transition is also brutal in

Morocco, Mexico and Turkey, because these large emigration countries have become host and transit countries for migrations coming from sub-Saharan Africa, the Near and Middle East, or the Central American countries (Wihtol de Wenden 2017: 26–28).

The immigration issue brought about by human mobility – which in turn is one of the results of globalization, as already noted – is eroding the ability of the State to exercise its sovereignty, because new 'global' practices, such as the operation of transnational networks for migration flows with constant coming and going to the country of origin, have led people to make mobility a way of life.[27]

Additionally, the proliferation of new types of transnational actors – ranging from financial entities to multiple organizations championing human rights, feminists, farmers, indigenous and social movements with different missions – tend to establish transnational links in order to have greater political presence as part of their counter-hegemonic globalization practices (Santos 2009). Viewed from a complex holistic approach, the *nation state* consequently no longer seems the most appropriate analysis unit in an international system, even though the classic geopolitical realist paradigm is contrary to this kind of approach. However, there are still some exceptions.[28]

Concerning the right to vote, transnational constituency in electoral processes needs to adapt to the new situations arising from mobility. Social transnational movements are keen to defend the rights of refugees and undocumented persons, but the issue is complex. Should migrants be entitled to vote in the receiving country

---

[27] During a recent trek through the mountains of Cuchumatanes (Guatemala) in 2018, I had the opportunity to interview an indigenous peasant who owned beasts of burden which were carrying food purchased with his savings from trips made as a temporary worker to the USA, where his two oldest sons live and work, without having regular migratory status. Worker mobility – which no wall will stop – is quite expensive for migrants because traffickers charge amounts ranging from US $2,000 to US $6,000 for freight from countries such as Guatemala. Those who emigrate often have to get into debt with loan sharks or banks, or mortgage their properties to afford the trip, and when the trip succeeds, the debt is paid off thanks to the higher wages paid in the US (for instance, in 2020 the average rate of pay for a bricklayer in the US was $26/hour). According to an investigation into migration and remittances in Guatemala undertaken by the International Organization for Migration (IOM) in 2016, 61.7% of migrant workers explained that their reason for migrating was the "search for better revenues", compared to 37% "just looking for a job". In summary and according to the IOM (2017: 18–28), the US $7.27 billion received in family remittances in 2016 (which increased to around $10 billion in 2019) explains the increasing migratory flows from a country like Guatemala to the US. Clearly, cross-border mobility is a strategy adopted by individuals who are in search of upward mobility (de Wenden 2017: 19–20). Another interesting figure from the IOM's investigation is that out of more than 2.3 million Guatemalan immigrants to the US, about 700,000 – roughly 30% – are in an irregular situation, which is another important reason for demanding the regularization of that 'minority' of Guatemalans who lack proper documents (IOM 2016).

[28] In my opinion, within the military international subsystem and among the nuclear powers the classic *balance of powers* approach continues to be absolutely valid. For instance, Trump's decision to quit the Intermediate-Range Nuclear Forces (INF) Treaty was immediately followed by Putin's decision to restart the construction of intermediate range missiles. This dynamic may be regrettable, but it is perfectly understandable when viewed through geopolitical and global *balance of power* lenses. A similar reaction occurred in Moscow in 2014 when Ukraine's interest in joining NATO was immediately followed by Russia's retort over the Crimean peninsula and the Kremlin's support for the separatist rebellion against Kiev in the Donbass region.

during election processes? Should they be allowed to vote in their country of origin if they rarely live there any more? What should happen if they have dual nationality? What about different situations due to ethnicity and religion? Concerning advanced regional integration processes like the EU, it is clear that transnational citizens now play a greater role in reconfiguring the international scenario, thanks to the European Citizens' Initiative (ECI) and the Supranational Constitution imposed on nation states on the initiative of citizens exercising their transnational rights.

Consequently, there is no question that governments will be faced with 'retaliation' by global civil society organizations and movements against the inter-state Westphalian order if nation states misguidedly continue trying to secure territories through repression, deportation and the violation of migrants' human rights, especially since migration is partially prompted by the need for labour forces as a consequence of capitalism. This is an instance of what I referred to earlier as the clash between geopolitics and geo-economy as the core characteristic of the present time. Added to that, the previously mentioned "tribulations of capitalism" (expressed as constant economic crises that require reforms like those proposed by Varoufakis and Piketty to climb out of the quagmire caused by the growing inequality and concentration of wealth and the lack of mechanisms to recycle surplus produce) indicate that it is in the interest of the world elites ruling the economic system (the G7 and G20, among others) to change from the mainstream economic paradigm to the UN sustainable development one. Overcoming the existing contradictions between social dynamics and the political sphere means addressing the challenge of sustainable development, which entails the proper management of the social, techno-economic and natural ecosystem spheres (Sachs 2015). In order to promote the worldwide constitution of a multilateral agreement on human mobility, governments will have to accept a global public policy negotiated and accepted at international assemblies like the Global Forum on Migration and Development that led to the signing of the *Global Compact for Migration* in 2018.

Each State should then begin by promoting reforms to their national legislation on migration and, to be worthy of the name and to make them truly *genuine*, they should forsake hypocritical policies that keep the front door of the house open to the wealthy, while the poor are only allowed access through the service door. All of this happens while undocumented migrants remain under the Damocles sword threat of flash detentions and deportations separating them from their family and children, as happens in the US at the times of Trump. This is a threat that also serves as a cruel formula for exploitation and *de facto* restriction of human rights and labour benefits, as it leaves employees absolutely defenseless before their employers. However, if we examine specific situations and compare migratory policies of the EU with that of the US, it is evident that Europeans have tried to stay within the margins of their own legality and political correctness. Despite the European economic crisis and the waves of refugees fleeing from the armed conflicts in Syria and sub-Saharan Africa, it is undeniable that European policies are by far less intolerant and have better standards of compliance with human rights.

Wihtol de Wenden's findings regarding asylum (2017: 35–36) show that in 2015 330,000 out of 1.2 requests for protection were granted. The number increased in 2016 to 717,400 (70% of successful applications being received by Germany) so even with all the existing problems due to the refusal of Hungary, Slovakia and Poland to accept refugees, and the increase in racism and xenophobia in neonationalist political parties, in general terms the migration policies implemented by the European Union are much more advanced and correct than those of the United States. The US government's migratory policy needs to be reformed as president Biden promised considering how greatly America benefited from its late-nineteenth and early-twentieth-century policy of encouraging the migration of European workers who were the force behind America's economic development.[29] But the benefits of migration are not just limited to the receiving country. By way of explanation, even when migration is instrumental in preserving the economic stability of countries of origin since it reduces the pressure on the domestic market caused by unemployment, it is because of remittances (which in 2017 alone totalled $430 billion worldwide – more than triple the funds from international donors in development organizations) that their contribution is so crucial for keeping the economy of the receiving countries afloat. As Wihtol de Wenden so poignantly states:

> Development often drives migration. The rapid modernization of agrarian structures, some-times induced by global development programmes, favours a rural exodus towards the urban peripheries of the large metropolises of the South. Many countries in the South are confronted with the situation that Europe experienced in the 19th century, when the economic take-off caused a rural exodus and massive urbanization which often resulted in emigration abroad (Italy, Germany, United Kingdom). Development can also mean distancing people from their home states when they are undemocratic, corrupt, poor and offer no prospects. Migrants increasingly educated, urban and informed then decide to make a success of their life elsewhere by migrating (Wihtol de Wenden 2017: 46).

This explains the reasons that allows to understand why the immigration of labourers is desirable the immigration of labourers is desirable for the proper functioning of the economy, as happens in Germany and other EU countries that are in need of a young population, and therefore human mobility is the response to a labour market demand that cannot, and should not, be stopped for ideological motivations or electoral demagogy, as Trump did in the United States. This phenomenon is also tied to transnational social networks, as revealed by the ties created directly between citizens in the country where migratory influx originates and those that receive them.

---

[29] For instance, at the end of nineteenth century more than a million Swedes emigrated to the US, as described in the 1990s Swedish movie *The Immigrants* starring Liv Ullmann and Max von Sydow. Yves Lacoste, a French expert on geopolitics, says that migratory flows from countries such as Germany, Ireland, Italy and France demonstrate that workers were attracted by a better paid *job market* or by the promise of land. According to Lacoste, a survey conducted by the Federal Census Office included a question about the American people's ancestors by nationalities of origin, which revealed that in 23 of the 50 states involved – the entire Midwest and the West, plus Pennsylvania and Florida – approximately 42.8 million Americans said they were of German descent; 30 million, Irish; 24 million, English; 15.7 million, Italian, 8.3 million, French; 25 million, African; and so on. Lacoste also alluded to the last twenty years, when immigration to the US of people from Latin America and Asia has been of some 40 million and 10 million, respectively. After the collapse of the USSR, there was also an increase in the immigration of Russians and other former communist European countries, including a significant number of top-level engineers and scientists who left their countries after the dismantling of State-sponsored think tanks (Lacoste 2009: 37).

Hence, migrants are making governments proceed to the *de facto* opening of their borders (even if it is through the back door), since globalization causes a structural demand for labour for demographic reasons. The drop in fertility rates means that the economically active population in developed countries is growing smaller. Each year there are fewer young people entering the job market to replace those who retire, which means that the native population in industrialized countries is not reproducing itself at the required rate to meet the demands of the job market. For that reason it is essential to turn to foreign labour, especially for those jobs that locals reject.[30]

6.8 Transnational Citizenship Could Solve the World's Migration Crisis

The concept of citizenship needs to be differentiated from that of nationality, even if their meanings could be considered equivalent in general terms. Depending on the country and its legislation, nationality is a legal status obtained through *jus sanguini* (if either or both parents is a native of the country where the birth takes place) or *jus soli* (the person is a citizen of the country where the birth takes place).[31] However, this section explores the concept of citizenship as a category distinct from the concept of nationality, which is closely related to the exercise of political rights when a person comes of age, but must be differentiated from citizenship, as will be explained next.

Following the Spanish scholar Suarez and the French philosopher Baliba/Ruiz (2018: 121–139) recently claimed that in order to understand the concept of citizenship, one must begin with the concept of the subject and the way in which its meaning changed as a result of the French Revolution and the French *Declaration of the Rights of Man and of the Citizen*. The concept of *man*, whose meaning in this context is intrinsically related to that of *subject*, was the legal status of individuals during the monarchical *Ancien Régime*. Thanks to the Revolution, it was transformed into 'legal subject' (*sujet de droit*; literally, 'subject of law'). When asked "What comes before the subject?", the expected response was: the citizen. Like the inhabitants of the Ancient Greek or Roman *polis*, citizens are free (not slaves, servants or subjects) and *equal before the law*, and entitled to enjoy the parallel *right to liberty* as well the

---

[30] According to the quoted IOM report (2016), 21.5% of Guatemalans working in the USA are occupied in construction; 19.3% in services such as trade, accommodation (hotels), and restaurants; 8.8% in mining and industry; 4.7% in financial and insurance entities; 1.9% in transportation; and 15.7% in "other services" (i.e. house cleaning, gardening and similar).

[31] It is interesting to mention again the case of Guatemala because its Constitution establishes that those born in the former Central American Federation (which included the Captaincy General of the Kingdom of Guatemala during the colony: Guatemala, El Salvador, Honduras, Nicaragua and Costa Rica) have the right to a Guatemalan nationality *of origin* (as if *jus sanguini*) if they reside in Guatemala and state this before the competent authorities. It could therefore be asserted that, for historical reasons, the Guatemalan Constitution has an extremely evolved and progressive stance, since citizens from Central American countries have political rights usually reserved for nationals, such as the right to vote in general elections.

right to equality. During the French Revolution, the latter was understood, among other things, as the right to abolish aristocratic privileges.

Hence, one of the main achievements of the French Revolution was to put an end to absolutism and the privileges of the aristocracy (and the clergy), which is why the Revolution can be regarded as a true social and political cataclysm for the nobility and the privileged classes. Even though social inequality persisted *de facto* (due to the differences between the wealthy and the poor), giving a political meaning to the values of *liberty* and *equality* – albeit mainly in the realm of legislation – was innovative and revolutionary in the best sense of those words.

More than half a century passed before *fraternity* – the revolutionary aspiration to equality from a social point of view – began to gain formal recognition in Europe, thanks to the social movements of 1848. The positive and material realization of this aspiration is still under construction everywhere, especially in the so-called developing countries. It has accordingly taken longer for economic and social rights to gain *formal recognition* in other continents, countries and regions, such as Latin America, where the popular aspiration to social equality did not begin to be accepted by the constitutional norms until the Mexican Revolution and the Mexican Constitution of 1917. In Guatemala these rights were introduced as constitutional norms for the first time in the 1945 Constitution, and the same thing happened during the twentieth century in all Latin American constitutions. More time was required for the international community to approve the Universal Declaration of Human Rights (United Nations 1948), the International Covenant on Civil and Political Rights (which was adopted by the UN in 1966 and entered into force in 1976), and the *International Covenant on Economic, Social and Cultural Rights* (1966). The signatories accepted that the second generation of human rights had to complement the first generation of civil and political rights whose international covenant was also approved by the United Nations in 1966 as part of the universal aspiration to provide specific content to the broad principles of *equality* and *fraternity* propounded during the French Revolution.

Liberty is a right (and a value) closely tied to the right to equality, since it is liberty which is diminished and hampered if a person does not have *equal rights* – witness the plight of undocumented migrants in the United States. Their lack of documents (a passport with a visa allowing multiple entries, like the one granted to Turkish workers in Germany[32] or Egyptians in the Gulf countries,[33] to mention two examples from other areas) prevents them travelling back to their countries of origin when they want; even travelling internally in the US is seriously limited because they are unprotected and exposed to detention as if they are criminals. Obviously, this has nothing to do with demanding equality related to second-generation rights, i.e. economic, social and cultural rights,[34] or political rights; rather, it concerns, as a minimum, the human right to leave your own country and reside in the country of your choice, which must be accompanied by an identification document or work permit equal to that of any citizen in any democratic political system. It is the elementary claim of all human beings to be treated as equal before the law, i.e. the equality afforded by the *"right to have rights"*, as stated by Suarez Ruiz:

> Tensions manifested as subjectivism result from the noncompliance of equality. … The lack of what Balibar calls *equaliberty* [*l'égaliberté*] – summed up by the fact that liberty cannot exist without equality or equality without liberty – causes a sort of insurgent subjectivity that can be defined as a constituent part of citizenship (Balibar 2011: 62). Thus, an individual differentiates himself from others when, due to his social condition (like a migrant worker) he

---

[32] For example, information randomly searched on the internet about residence permits for relatives of Turkish workers in Germany states: "This residence permit is for Turkish workers and members of their families (spouse or child) already living and working in Germany residence title in the conditions of the EEC/Turkey Association Agreement. Specifically, this residence in Germany is for Turkish people of these circumstances: Family members of a Turkish worker; a spouse or a child of a Turkish worker who has worked in Germany constantly for minimum 3 years from entering Germany, or otherwise the child was born in Germany and lived within a family union; a child of a Turkish worker who has worked in Germany not more than 3 years, who has finished professional training in Germany; Turkish workers having worked for minimum 3 years for the same employer; Turkish workers having worked for 4 years in the same profession. The fee to apply for this residence permit is 28.8 Euro (for persons aged 24-over), and is 22.8 Euro (for those aged up to 24 years). In the meantime the application is free for persons benefiting from SGB II, XII, or asylum seekers." See at: https://visaguide.world/europe/germany-visa/residence-permit/turkish/.

[33] Another random example from the internet shows the requirements for a residence/work permit in the UAE for Egyptian workers: "Egyptian nationals applying for a new employment residence permit in Dubai sponsored by a company in the Dubai Airport Free Zone or the Dubai International Financial Centre Free Zone must now personally appear at a designated UAE visa service centre in Cairo to obtain an entry visa. … While there may be slight differences in the application process depending on the jurisdiction of the Egyptian national's employer in the United Arab Emirates, generally, the following is a description of the new process for affected Egyptian nationals to enter the United Arab Emirates: 1. The UAE sponsoring company must obtain a pre-approved entry permit on behalf of the Egyptian worker; 2. The foreign national must register for and attend a medical examination at an approved medical centre in Egypt; 3. The foreign national must pick up their medical results in person; 4. The UAE sponsoring company or the foreign national must schedule an appointment at the UAE visa service centre in Cairo through the Ministry of Foreign Affairs and International Cooperation (MOFAIC)'s electronic portal; and 5. The foreign national must personally pick up their UAE entry visa at the UAE visa service centre in Cairo.

[34] As expected, the USA is the only signatory which has not ratified the International Covenant on Economic, Social and Cultural Rights, which, along with the International Covenant on Civil and Political Rights, is one of the two main international UN instruments on the matter of human rights. Some countries, such as Saudi Arabia and some Arab oil monarchies of the Gulf, have never signed it, and other countries, such as South Sudan, which joined the UN after its entry into force in 1976, and have not yet acceded.

cannot exercise the condition of equality and, therefore, of declared liberty. … Regarding this issue, Balibar is especially interested in the plight of the undocumented and he understands their struggle as a citizen's struggle for the right to have rights (Suarez Ruiz 2018: 126).

In countries like the US, where unauthorised migration is criminalized, irregular (not 'illegal') migrants – or non-documented workers, as they must be called according to the United Nations[35] – have no rights; they are non-existent from the legal standpoint. In other words, their human rights have been utterly violated. This is why they seek "the right to have rights". It is interesting to see that the concept of citizenship, as understood by Balibar, diverges from the traditional notion that reduces citizenship to a legal situation or *status* granted by the legislation in each nation state. His concept of citizenship is dynamic, not structural, and linked to Jeffrey Sachs's definition of sustainable development as the art of managing social dynamics alongside the techno-economic sphere and the natural ecosystems from the political sphere:

> Balibar lashes out against the vision of citizenship as a mere status, for to him, as for Jacques Rancière, citizenship is above all an 'action' in the sense used by Arendt: it is a recreation of life within the plurality that allows a man to recognize himself as different from others and to belong to the human community, but also in the sense of making himself heard and being part of collective action to pursue the attainment of rights. The citizen, for them, will not have a fixed status that is obtained passively. Nor will he be tied to a pre-existing identity group in society. It is his 'actions' that will give substance to the citizenship.

> On the contrary, thinking about citizenship as a status, as Balibar says that politicians and jurists do, implies thinking of it as mere nationality, as if citizens are merely patient subjects. It would mean conceiving citizenship as something passive, and therefore, a subject subjected to a stable identity. This is contrary to the constant constituent processes from which he understands that trans-individual identities are built. This means forgetting that modern citizenship is something subjects have built through their 'actions', either through insurrection, as in the case of the French Revolution, or through resistance, as in the case of the American War of Independence. To these political events we add the Neo-Zapatist rebellion. Political action is required for people to cease being subjects; rights have to be won through different collective practices brought about by modern citizenship (Suarez Ruiz 2018: 127).

---

[35] The Global Migration Group is an inter-agency entity comprised of sixteen United Nations agencies and other international organizations dealing with migration. It defines a "migrant in an irregular situation" as "any person who, by reason of entry without documents or the expiry of the visa, lacks legal status in a country of transit or destination. The term applies to migrants infringing the rules for entering the country and any other person unauthorized to remain in the country of destination". Further, the 1975 General Assembly of the United Nations approved Resolution 3449 (XXX), which requests that all member states, United Nations organs and specialized agencies utilize in all official documents the term "non-documented or irregular migrant workers" to define workers who "illegally and/or surreptitiously enter another country to obtain work". Furthermore, in its General Observation No. 2 (2013) on the rights of migrant workers in irregular situations and their families, the Committee for the Protection of the Rights of All Migrant Workers and their Families expressed the opinion that expressions like "in irregular situation" or 'undocumented' were the appropriate terminology to use when referring to the situation of the said workers. The insistence of the Trump administration on using the term 'illegal' to describe migrant workers in an irregular situation is inadequate and needs to be avoided, as it tries to stigmatize and criminalize those persons by relating their irregular migration situation to crimes typified and sanctioned by criminal law (United Nations 2014).

Viewing transnational citizenship as part of a process to "make yourself heard and from the collective action pursue the attainment of rights" underlines the humanitarian aspect of designing a foreign policy for all interested governments which not only complies with the obligation to grant diplomatic and consular protection to all migrant workers, including those currently classified as irregular, but also supports the constitution or consolidation of transnational networks of citizens 'without papers' who should insist on being granted the requisite documents and having their human right to work in the country of their choice recognized, in accordance with the Universal Declaration of Human Rights.[36]

Another consideration is that the globalized world market needs workers to migrate for economic reasons linked to globalization and the predominance of capitalist markets in the world economy. Consequently, regularizing migrant workers – or granting refuge/asylum whenever applicable – should be recognized as a way to overcome the disjuncture between geopolitics and geo-economics. Because principles are at stake, the protection of human rights could be the *leitmotiv* of foreign policy in countries where migratory flows originate. Multilateral negotiations must be prioritized, fostering intense and proactive work in large UN forums like the General Assembly, the Human Rights Council, the ECOSOC, and the *International Labour Organization* (ILO), including regional organizations like the UE, the OAS and the African Union in order to fulfil the commitments agreed in the *Global Compact* of 2018, laying the groundwork for a possible global public policy covenant on migration by 2030.[37]

Naturally, claiming rights like the 'right to have rights', which is an essential principle of the value of *equality* originating in the French Revolution, is not the same as obtaining citizen *status*, even though Arendt and Balibar may be correct in the sense that citizenship should not be considered a legal status but a constructive process that takes place thanks to the action – and especially the collective action – of individuals wherever they reside or work.

---

[36] The Universal Declaration states that (1) Everyone has the right to freedom of movement and residence within the borders of each State. (2) Everyone has the right to leave any country, including his own, and to return to his country. See at: http://www.un.org/en/udhrbook/pdf/udhr_booklet_en_web.pdf.

[37] The *Global Compact for Safe, Orderly and Regular Migration* negotiated at the UN was proposed by two facilitators, the Permanent Representatives of Mexico (Ambassador Juan José Gómez Camacho) and Switzerland (Ambassador Jürg Lauber), who stated that "A Global Compact expresses our collective commitment to improving cooperation on international migration. Migration has been part of the human experience throughout history, and we recognize that it is a source of prosperity, innovation and sustainable development in our globalized world, and that these positive impacts can be optimized by improving migration governance. The majority of migrants around the world today travel, live and work in a safe, orderly and regular manner. Nonetheless, migration undeniably affects our countries, communities, migrants and their families in very different and sometimes unpredictable ways. It is crucial that the challenges and opportunities of international migration unite us, rather than divide us. This Global Compact sets out our common understanding, shared responsibilities and unity of purpose regarding migration, making it work for all" (Global Compact 2018: 2). Therefore could Mexico and Switzerland lead the way towards the signing of a world covenant on migration?

However, it is no less true that the said collective action is aimed at obtaining recognition from the state authorities of migrant workers' need for a *legal status*. Developing countries (i.e. those on the periphery of the capitalist *world system*, as Wallerstein calls it) must be interested in their own people – if they are facing an irregular immigration situation – being recognized as having "the right to have rights" and obtaining at least the legal status of residents with the right to work (work permits) complemented by the *right to mobility*, allowing them to travel to their countries of origin and return to their receiving country or be granted a multiple entry visa in their passports, as previously recommended in this chapter. A significant element of the debate on transnational citizenship focuses on this very important matter. For instance, the American scholar Jonathan Fox says:

> One of the problems that arises when we look for citizenship in the transnational public sphere is that claiming rights is not the same as gaining citizenship. Most claims are not enforceable, which underscores the big difference between the widely resonant notion of the 'right to have rights' and the actual winning of those rights. … The claiming of rights is necessary but not sufficient to build citizenship. Along the lines of the state-versus-society based dimensions of citizenship described above, one could pose a distinction between a rights-based approach and an empowerment-based approach. Empowerment, in the sense of actors' capacity to make claims, is distinct from rights, defined as institutionally recognized guarantees and opportunities. They do not necessarily go together. Institutions may nominally recognize rights that actors, because of a lack of capacity to make claims, are not able to exercise in practice. Conversely, actors may be empowered in the sense of having the experience and capacity to demand and exercise rights, while lacking institutionally recognized opportunities to do so. Rights and empowerment can each encourage the other, and indeed they overlap in practice, but they are analytically distinct. In other words, some must act like citizens (claim rights) so that others can actually be citizens (have rights), but acting like a citizen is not the same as being a citizen. If this distinction makes sense, then most of transnational civil society falls far short of transnational citizenship (Fox 2005: 176).

As previously noted, acting as a citizen and being able to claim rights is not the same as just being a citizen who has rights passively. According to Bauböck, the empowerment of individuals is different from the sole recognition of constitutional rights. People are empowered and evolve into citizens through collective action aimed at obtaining from state institutions recognition of their rights.[38] Fox is clearly correct when he states that one of the problems arising from the concept of citizenship in the transnational public sphere is that claiming rights is not the same as actually being a

---

[38] Migrant workers in an irregular situation in the US are clearly in the position of claiming *the right to have rights* (as they are *non-existent* people in legal terms), but through collective action they will be empowered. Empowerment – in the sense of actors' capacity to make claims – is distinct from rights or "institutionally recognized guarantees and opportunities". For instance, if, despite being endowed with the human right to education at primary and high schools, young people in an irregular migratory situation in the US cannot go to universities or get jobs when they leave school due to their lack of regular migratory status, it is the empowerment-based approach (in the sense of having the capacity to demand and exercise rights) that works. Indeed, the movement of 'dreamers' succeeded in obtaining from the Obama Administration the executive order called Deferred Action for Children's Arrivals (DACA). It is therefore clear that, as Fox (2005) argues, rights and empowerment can be mutually beneficial despite overlapping in practice and being analytically distinct.

citizen, for in order to attain citizenship, rights need to be enforced – implemented – which presupposes the support of state authorities. Consequently, the "right to have rights" (the ideal situation) is not the same as effectively exercising them in practice, since the first statement is normative (or in the realm of the idealist paradigm) while the second is empirically tangible (and in the realm of reality). Fox adds that the problem lies in the will of the stakeholders and is agency-driven, which explains why "most of transnational civil society falls far short of transnational citizenship" (Fox 2005: 177).

Fox's last statement is absolutely true. It is a fact that the majority of migrant labourers in global civil society (not middle-class tourists, TNC personnel, CEOs or rich people belonging to the international *jet set*, of course) do not have transnational citizenship status, which is driving these 'potential citizens' to take action in order to have their rights recognized and ratified by the authorities in the receiving country. In other words, obtaining legal residence, work permits or documents such as passports with a multiple entry visa could be the empowerment-based approach to citizenship that is needed by irregular migrants.

Consequently, if people looking for asylum, refuge or employment abroad have their human rights violated by racism, xenophobia or criminalization (including the former Trump Administration's horrendous and atrocious orders that separate children from their parents), they must empower themselves by acting together in their struggle for transnational citizenship.

6.9 Rainer Bauböck's Perspective

The Austrian academic Rainer Bauböck asserts that citizenship can be defined as the status of full and egalitarian membership with rights and obligations in a self-governed political community based on certain values. Narrowly stated, it is regarded as equivalent to nationality, but since international migration has caused an imbalance between territory and citizenship, the result is that emigrants living abroad and immigrants arriving in the receiving country generate problems for their own legal status from two standpoints: the state from which they emigrate (how far can the intervention of a foreign country go to protect the rights of its own nationals?) and the receiving country (the rights of immigrants to legal residence and employment). This is all related to the convenience of defining the three key concepts of this debate – international, multinational, and transnational – in greater detail, even if these are not 'canonical' definitions but purely heuristic, as stated by Bauböck (2002a, b).

From Bauböck's perspective, the concept 'international' refers to the relationship between states and intergovernmental organizations such as the UN and the OAS; 'multinational' is to do with the different national communities within nation states – e.g. Canada, Belgium, Spain and the United Kingdom, as they are composed of different nationalities, like the Walloons and the Flemish in Belgium; Castilians, Galicians, Catalans and Basques in Spain; Scottish, Welsh and English in the

United Kingdom; French and English communities in Canada, and so on;[39] and the concept of 'transnational' refers to people (individuals or legal persons) with simultaneous affiliations or ties to several states in separate geographic locations and also to different social and corporate practices (NGOs, social movements, corporations) which transcend the borders of sovereign states. Hence citizenship clearly concerns *transnational* relations (Bauböck 2002a, b: 7).

The phenomenon of transnationality is the result of globalization and in that sense the world economy is confronting geopolitics and the principle of territoriality. In spite of the sacrosanct banner of individual freedom and free trade typical of neoliberalism, restrictions on the mobility of workers and poor people (not tourists, CEOs or rich people) are still the norm. Consequently, the needs of the world's job market are not being met because they lack the kind of arrangement granted to merchandise via free trade agreements. Herein lies the clash between geo-economy and geopolitics.

Given that globalization is an irreversible historic phenomenon, it is clear that national governments must adapt to it. This explains the importance of updating or reforming immigration policies, as countries like Germany and the rich petroleum monarchies of the Gulf have done. The obvious purpose is to adapt their economies to the new realities of global labour markets. Countries which fail to adapt risk having a lagging immigration policy, potentially due to spurious ideological motivations, as in the case of the Trump Administration in the United States.

Returning to the political legal sphere, since the transnational perspective of immigration connects societies by sending people to receiving societies, and since this happens not only through flows of economic and cultural exchanges but also because of "overlapping membership frontiers", to use Bauböck's phrase, this condition of simultaneously belonging to two societies is liable to be reflected in the burdens and obligations of migrants – weights that will become heavier if the receiving state does not acknowledge the legality of their presence in the guest country. Bauböck asserts:

> Drawing on the discussion so far, we can sum up … three points: foreign residents must be protected by universal human rights; after a certain time of legal residence they must be given opportunities to naturalize; and the public conception of national unity should be civic rather than ethnic so that immigrants of whatever origin can be recognized as full members. Each of these requirements can be argued from a liberal interpretation of the international and multinational perspectives on migration; none refers to transnational affiliations. The added value of a transnational conception of citizenship can be shown by considering how it would extend each of the three minimal liberal norms. First, instead of merely offering migrants protection of their human rights it would establish a presumptive equality of rights for citizens and permanent resident foreign nationals. … Second, it is not enough to allow immigrants to

---

[39] However, multinationality is not necessarily expressed in that way in the political system, as countries like the United States, Brazil and Germany claim to be 'mononational' while being federal states, whereas Russia, Switzerland and India are federal and admit to being multinational, and other countries (e.g. Bolivia) are 'plurinational' without being federal. In another example, in spite of being *de facto* plurinational, Guatemala has a unitarian mononational type of political regime, as does Mexico, although it is a federation. Evidently, the fears of certain politicians in mononational states such as the UAE is that greater immigration openness will lead to multinationality, something that has been forewarned by scholars such as Huntington (2004).

change their nationality. Naturalization should be seen as an individual entitlement that does not require renunciation of a previously held nationality. Multiple nationality is a formal legal expression of transnational citizenship. ... Third, a transnational conception goes beyond civic nationalism in providing arguments for a more robust version of multiculturalism. The main difference with mainstream liberalism is not so much about recognizing certain cultural minority rights for immigrant communities, but about the need to transform the public culture of the society in response to immigration. Instead of pretending that liberal democracies can be neutral towards linguistic difference and historic collective identities, they should provide for multilingual public services and rewrite their public histories so that they include the collective memories and myths of immigrant communities (Bauböck 1998b, 2002a, b: 17–18).

Seen from this perspective, transnational citizenship goes beyond the obligation to respect the human rights of foreigners (e.g. granting them the option of naturalization and facilitating their integration – beneficial measures which must be maintained) because it also offers the possibility of obtaining some political rights, the right to obtain the new nationality without losing the former, and recognition of the cultural rights of immigrants, thereby enriching the national culture of the guest state with the collective memory and myths of migrant communities.

Naturally, Bauböck's analysis expresses a European vision. Enforcement of the Universal Declaration of Human Rights on the American continent alone would mean progress of great importance,[40] as it is of fundamental importance to respect the rights to not be detained arbitrarily, to move freely and choose your residence in the territory of a foreign country,[41] to seek asylum from persecution, to change nationality, and

---

[40] Several articles in the UN Universal Declaration of Human Rights are applicable to migrants: Article 3: "Everyone has the right to life, liberty and security of person"; Article 6: "Everyone has the right to recognition everywhere as a person before the law"; Article 7: "All are equal before the law and are entitled without any discrimination to equal protection of the law. All are entitled to equal protection against any discrimination in violation of this Declaration and against any incitement to such discrimination"; Article 9: "No one shall be subjected to arbitrary arrest, detention or exile"; Article 11(1): Everyone charged with a penal offence has the right to be presumed innocent until proved guilty according to law in a public trial at which he has had all the guarantees necessary for his defense; Article 11(2): No one shall be held guilty of any penal offence on account of any act or omission which did not constitute a penal offence, under national or international law, at the time when it was committed. Nor shall a heavier penalty be imposed than the one that was applicable at the time the penal offence was committed; Article 13: "Everyone has the right to freedom of movement and residence within the borders of each State. Everyone has the right to leave any country, including his own, and to return to his country"; Article 14: "Everyone has the right to seek and to enjoy in other countries asylum from persecution"; Article 15: "Everyone has the right to a nationality. No one shall be arbitrarily deprived of his nationality nor denied the right to change his nationality"; Article 23: "Everyone has the right to work, to free choice of employment, to just and favorable conditions of work and to protection against unemployment"; Article 28: "Everyone is entitled to a social and international order in which the rights and freedoms set forth in this Declaration can be fully realized".

[41] Some analysts have pointed out an omission in the wording of Article 13 of the Declaration, as the right to exit and return to your own country is not complemented by an obligation from the other states to grant a visa or entry permit to emigrants. However, since the Declaration states that everyone has the right to freedom of movement and residence within the borders of each State and that *"Everyone has the right to leave any country, including his own, and to return to his country"*, it would be illogical if the right of movement (or 'mobility') does not includes the right of not

to work, including "the free choice of employment, just and favourable conditions of work and protection against unemployment" (prescribed also by the UN International Covenant on Economic, Social and Cultural Rights of 1966). Not doing so is not only contrary to human rights but to the realities of globalization. There should be no interference in the work situation of an economic migrant just because they entered with no documents.[42] Obviously, if an individual entered a country without the corresponding legal documents, the proper administrative penalty (most likely pecuniary) should be applied, and then there should be a standard procedure for obtaining any missing documentation. Lack of documentation should not be a reason for arbitrary detention, deportation or loss of employment, and even less for family separation, forcing fathers and mothers to abandon their children as happened in the US during Trump's administration.

On the other hand, related to the matter of naturalization, which is also included in Bauböck's transnational perspective, although I agree in principle with the idea that the right to obtain the nationality of a receiving country, and dual nationality whenever applicable, is inherent in transnationality – which means the implicit right

---

being denied residence abroad if you haven't perpetrated any criminal offences, and migratory irregularity is not a crime. Hence it is clear that the Declaration implicitly assumes that the right to enter and exit different state territories and to move within them cannot be denied. Obviously, national governments also have the rights to regulate immigration flows and to demand a passport with a visa whenever applicable, but when no documents are available this should not be a crime but an administrative fault. Consequently, both the applicable sanctions and the solution to the problem have to be made in the realm of civil – not criminal – legislation. Indeed, there is difference if the lack of documents is evidenced at a border crossing or port of entry (maritime or airport), since in this case the authorities of any sovereign state do have the right to refuse entry to an undocumented person and deport him/her to his/her country. However, if the said situation does not take place at a border crossing, a technical violation of the human right to freedom of movement (such as detaining a person in the street because of his/her appearance and asking for ID documents) might occur. If there is no further infringement of the penal code of that country, the person cannot be detained even if they lack ID documents. In this hypothetical scenario, if the receiving country authorities imposed an arbitrary detention, this would be a violation of human rights. As an example of the seriousness of the problem for one Central American country alone (Guatemala), a report by the International Organization for Migrations (IOM) states: "The Migration Policy Institute estimated that, in 2015, there were 704,000 Guatemalans living irregularly in the United States of America. With this survey, the IOM investigated the immigration status of the persons residing in the United States, through their relatives and confirmed that 73.0% are in a regular situation. This information differs from estimates previously reported, as these estimates were based on the 2010 census and the bias of the irregular population that does not report this information. But 91.0% of Guatemalans residing in Spain are regularized, as well as most of those in Italy (67.4%) and Mexico (57.0%)" (IOM 2016: 39).

[42] Economic migrants go to the USA because it is a job market, as proven by the fact that, according to the same IOM report quoted above: "Of the total persons sending remittances, 68.0% were working at the time of their parting and 30.0% were unemployed or not economically active, and 2.0% did not provide information. Of those unemployed, 28.1% worked in agriculture; 13.2% in construction, handicrafts and operators; 4.4% as machine operators, fork lift operators, drivers; and 54.3% had other occupations." The same report confirms that the majority of emigrants to the United States have found a job there that will allow them to regularize their immigration status: "The Guatemalan population living abroad and sending remittances are operators, artisans, mechanics and construction workers (29.7%); they provide services or perform direct sales (18.1%); work as

of human mobility in this era of globalization – for pragmatic reasons I do not think it is advisable to grant nationality as one of the prerogatives of transnational citizenship prerogatives, at least in the US. My opinion is based on the fact that both European policies and European migration legislation are too advanced and it is not advisable to extrapolate them to other parts of the world. For instance, I imagine that if Egyptian workers in the Gulf countries became nationals of the receiving country they would quickly surpass the local inhabitants.[43] In the case of Latin American workers in the US, although scholars like Huntington rationalize racism on supposed sociological and cultural grounds, the fact is that there are racist considerations at work in a sizable proportion of the white Anglo-Saxon Protestant US citizenry that is opposed to giving US nationality to 'Hispanics' and afraid of having in the future more cultural and ethnic plurality in the local, state or federal iterations of US government.

In any case, transnational migrant status translates to 'pendular' mobility between migrants' country of origin and their country of employment. This kind of pendularity is convenient for both countries, because if people can move freely most voluntarily choose to return to their country of origin (and family remittances attest to this), a fact that also means that simple regularization of their immigration status would facilitate the pendular mobility. Even so, introducing the topic of naturalization is not advisable in the US, where racism and the political demagogy of individuals have gone hand in glove with the rejection of multiculturality. This has been bolstered by personalities of the academic world, such as the aforementioned Professor Huntington (2004),[44]

---

unqualified labour (18.7%); work in agriculture (6.5%); as machine, fork lift operators or drivers (3.7%) or in other occupations (6.7%). 17.3% do not know their occupation (IOM 2016: 41).

[43] Bauböck himself acknowledges this problem when providing the following explanation to what happens in those countries: "As the situation of immigrant workers in the Arab Gulf States illustrates, a permanent segregation of minorities of immigrant origin goes mostly hand in hand with their social and political disempowerment, which prevents any challenge to the hegemonic conception of national identity" (Bauböck 2002a, b: 11).

[44] Incidentally, the "zero tolerance" policy of family separations in countries like the United States not only causes lifelong trauma for child victims of it but is also a serious human rights violation. From May to June 2018, more than 2,500 children were separated from their parents due to the "zero tolerance" ordered by Trump. Forced to back-pedal and return the children to their parents and relatives, by August 2018 more than 500 minors remained under federal custody, among other reasons because more than 400 parents had already been deported to their countries of origin, according to the Spanish newspaper El País in its weekly digest of 18 August 2018. A carefully planned system of transnational nationality would aid international mobility by regularizing immigration status and issuing work permits, so, as Bauböck points out, it should not be an excessively complex bureaucratic procedure to request a letter from an employer to verify and legalize the residence of any migrant worker and grant them the requisite work permit. When such policies are not implemented in countries like the United States it is because most irregular migrants are poor and indigenous (not white), and because of the fear of an increase in the Hispanic population and the political repercussions which an increased non-'WASP' (White Anglo-Saxon Protestant) population might have on elections in the future. From this perspective, Donald Trump's stance can be interpreted as a racist reaction by the most conservative sectors of the American constituency in 'response' to the progressive constituency that placed Barack Obama in the White House. Given that problem, policies need to be implemented to start changing the mindset of the still numerous segment of the United States population which opposes multiculturality. A sociological mindshift – analogous to the one Maja Göpel advocates for the economy – is most likely to be achieved through education.

and what Bauböck himself describes as a risk which migrants are willing to take vis-à-vis the political reality of the receiving countries has in many countries gone beyond a risk, as large segments of their population are already inclined towards xenophobia, racism and nationalism.[45]

Nevertheless, it is worth remarking that obtaining the nationality of the receiving country is not of such crucial interest to economic migrants as the *mobility* to allow them to come and go to their countries of origin to visit family and verify that the remittances sent to them have been used wisely. Clearly, granting legal status to economic migrants is in the interests of both the country of origin and the receiving country. In several Central American countries remittances are now the main source of foreign currency. In Guatemala, for instance, the IOM reported that they amounted to US $7,273,365 in 2016, more than $9 billion in 2018, and more than $10,000 in 2019, compared to US $1 billion from textile exports and $1.8 billion from coffee, rubber, sugar and cardamom combined. In fact, remittances – which exceed foreign investment and foreign aid by far, and equal the State budget, making up 11.2% of Guatemala's GDP – are so important that preserving migrant mobility and ensuring the legality of migrants' residence in the US is fundamental to the national interest, even if in the long term the reliance on remittances is not sensible. Thus, I agree with Bauböck when he says:

> Emigration can be an asset in two ways: first, by contributing to economic development in the sending country through the flow of remittances from emigrants while they live abroad and through their upgraded skills and investment capital when they return; second, if an emigrant community promotes trade, cultural and political interests of the sending state in the host society. The first strategy relies on temporary emigration. This is obvious if the goal is to induce successful emigrants to return in order to invest their savings and their human capital. But the same is true for remittances. In order to keep them flowing, the

---

Bauböck stresses that "the public conception of national unity should be civic rather than ethnic so that immigrants of whatever origin can be recognized as full members" and "transform the public culture of the society in response to immigration." A transnational citizenship formula that does not involve naturalization would help to ease the fears and reservations of the conservatives highlighted by Huntington (2004).

[45] "Nevertheless, a multinational interpretation of the immigrant experience has a strong presence in political and media discourses. In receiving societies this perspective may result in three different nationalist approaches to integration: First, a racist discourse on the danger of *Überfremdung* [The term is untranslatable. Literally it means 'over-alienation'. It was part of the standard repertoire of National Socialist discourses on ethnic minorities and has recently re-emerged in the anti-foreigner campaigns of the Austrian Freedom Party] – the swamping of host nations by immigrant cultures. Immigrants are perceived as representing alien nations whose intermingling with the host nation should be prevented by stopping new entries, enforcing return and by keeping them segregated while they stay. Second, a culturalist discourse on the need for geographic dispersal and cultural assimilation as the precondition for political integration. In this view, distinct national identities of immigrants are not immutably given by their origins and descent, but they will be consolidated if immigrants form their own communities and maintain their cultural traditions. Third, a liberal discourse that promotes a civic national identity that can be shared by populations of native and immigrant origin. Immigrants are free to maintain distinct cultural and ethnic identities as long as they accept that these will not be promoted through state-sponsored programs of multiculturalism, but they are expected to abandon all political loyalties that tie them to their nations of origin" (Bauböck 2002a, b: 12).

sending country must be interested in delaying family reunification and naturalization in the receiving state. The second goal assumes, however, permanent settlement and integration abroad. When emigrants turn from low-skilled workers into self-employed businessmen, they can open markets in the receiving country for consumer goods produced in the sending state. And they can influence the foreign policy of their host state towards their country of origin once they have become voting citizens with their own representatives in mainstream parties, parliaments and high public office. These considerations should make it evident that, just as there is there is no realistic scenario for transforming immigrant into national minorities, so there is also little danger that sending states will mobilize their emigrants as a nationalist irredenta that threatens the integrity of the receiving polity. In the context of labour and refugee migration from poorer states into wealthy democracies the former will be either interested in keeping migration temporary and encouraging return, or they will promote the economic and political integration of their nationals in order to maximize these emigrants' clout as mediators and lobbies for the sending state's concerns (Bauböck 2002a, b: 13).

It is obvious that Bauböck is chiefly writing about the experience of migrant workers in the European Union, especially in Austria and Germany. However, his considerations perfectly explain the behaviour of Central American migrants in the United States. I have mentioned the role family remittances play, but another example pertaining to Guatemala is US Congresswoman Norma Torres, whose country of origin is Guatemala, and who has played an outstanding part in supporting the fight against corruption and in providing protection and aid to migrant workers, including efforts by Democratic representatives in the US congress for an immigration reform. The local evening newspaper *La Hora*, which has a weekly section dedicated to the millions[46] of Guatemalan migrants in the US, has published items about those who have been able to open small businesses in their receiving country and the success stories of others who, upon their return to Guatemala, have invested in the country's development. Facilitating work mobility by granting legal residence and work permits to those who could then become 'transnational citizens' (or whatever the name) is in the interests of both societies, as it favours the social dynamics that foster sustainable development in the sending country while facilitating the temporary nature of their stay in the receiving country.

Naturally, solidarity, mutual aid networks and transnational communities (Portes 1996)[47] are also crucial in explaining the globalization process of the job market. These are exactly the type of networks that attract new workers. Similarly, economic transfers to the country of origin ('family remittances') help to explain why migrants have become important entrepreneurs by promoting actions such as the 'nostalgia marketing' of products ranging from local cuisine to handicrafts related to cultural customs and traditions. Family reunions – for those who wish to remain in the receiving society – and inter-ethnic marriages are factors contributing to the evolution of the concepts of nationality and citizenship, as stated at the beginning of the chapter.

---

[46] According to estimates by the *National Institute of Statistics* and the IOM (2016: 36), 2,301,175 Guatemalans out of approximately 16,545,589 Guatemalan inhabitants) are currently living and working in the United States.

[47] For example, de Wenden (2017: 92–93) mentions the case of the Mourides community and the role it played in the Senegal diaspora in France.

However, the concept of transnational citizenship is much more complex and needs to be differentiated from other concepts and specific individual situations.

Something similar happens with the concept of cosmopolitanism. It is perfectly feasible to drive forward a new vision of global democracy, such as that of Richard Falk, who says that global citizenship should correlate to a global democracy in which representatives from global civil society liaise with the UN or are involved in the creation of a global parliament, as has been suggested by Federico Mayor, former UNESCO Director. The ideas of academics like David Held, Daniele Archibugi and Rainer Bauböck about cross-border governance with a cosmopolitan vision could also be implemented. There are already some international law instruments that can be used at multilateral level to fulfil that end. For instance, governments could give full force to the *Convention on the Protection of the Rights of All Migrant Workers and Members of Their Families*[48] – on which Mexico worked intensely in long, multilateral negotiations at the end of the twentieth century – by upholding it through the implementation of legal procedures aimed at obtaining a new transnational citizenship that does not necessarily imply the granting of nationality in the receiving country.

Furthermore, it is clear that the 'diasporas'[49] of German, Dutch and Norwegian retirees who relocate to the sunny Spanish coast and the upper social strata of multinational corporate executives would also benefit from transnational citizenship and the benefits it entails. This type of fortunate social sector, composed of people who comply from the outset with the receiving country's immigration procedures, does not experience any problems with the legality of their residence or source of income. It is a matter of ethics (and justice) to enlarge that kind of positive discrimination to the poor segments of global civil society. Transnational citizenship could be a good way to solve the problem, because if the richer segments of global civil society are already enjoying it, there is no reason why this category could not be applied to migrants with frequent mobility (because they visit and regularly send money to their families in their countries of origin) whose only fault is to enter the host country without documentation (or have expired tourist visas). This category of the migrant population does not aspire to citizenship (regarded as equivalent to nationality in the United States) even though that would regularize their status. To obtain citizenship of the US or any other country in the world you need a period of legal residence, but if pendular mobility characterizes some types of migrant, rather than being awarded the benefit of citizenship, they could be granted 'transnationality', enabling them to travel to their country of origin (with a multiple-entry visa in their passport) while retaining their foreign status.

---

[48] The International Convention on the Protection of Migrant Workers and Members of Their Families (ICPMW) was adopted by the United Nations General Assembly in Resolution 45/158, dated 18 December 1990, which entered into force early in the twenty-first century, although even now it has not been ratified by the majority of countries receiving a significant influx of migrants.

[49] The concept of *diaspora* refers to people who leave their original homeland and settle elsewhere, but preserve the traditions and culture of their country of origin. In some cases, after residing legally in the host country for generations, the descendants of the original settlers return to their forebears' country of origin, having recovered their nationality on the basis of *jus sanguini*.

This kind of transnational citizenship (that does not confer the right to naturalization or to obtain the nationality of the guest country) could eventually include some political rights for foreigners, as happens in Europe, where some countries with a broader, progressive and cosmopolitan vision have granted foreign residents the right to vote. This is completely appropriate and fair when foreigners pay the same taxes as their neighbours, since policies enforced by the elected authorities directly affect them due to their place of residence. In Central American region, as already noted, for political reasons related to regional history, the Guatemalan Constitution grants nationality of origin (with no need to start a naturalization administrative procedure) to anyone born in any of the former Central American federation states who resides in Guatemala and is interested in obtaining it.[50]

6.10   Ludger Pries's Approach

Ludger Pries is a distinguished German scholar who has worked in the US, Brazil and Mexico and is the current director of the postgraduate programme of sociology, migration and transnational studies at Ruhr-Universität in Bochum, Germany. He is interested in the concept of *social space* and his main focus of research is the way in which national societies are internationalized in the host country. For him there are at least seven 'ideal types' of this internationalization of a social relationship: (1) internationalization; (2) re-nationalization; (3) supra-nationalization; (4) globalization; (5) glocalization; (6) diaspora; and (7) transnationalization. The first type essentially refers to the intergovernmental relations between sovereign states; the second deals with the conservative 'reaction' of national groups to international institutions and globalization; the third with regional integration processes like the EU; the fourth with the strengthening of global frameworks and economic activity – financial flows, climate change, information and communication technologies, universal human rights; the fifth with the links between local and global phenomena, such as the global warming caused by local *greenhouse gas emissions* (GHE); the sixth with refugees and persecuted people who live in a foreign country but retain their cultural relationship with their original country or religion (like the Jews or political groups in exile) as well as political or religious elites (e.g diplomats in the case of the former and the Catholic Church in the case of the latter) living and working abroad for professional reasons; and finally there is *transnationalization,* a concept defined as "strong social and lasting frameworks rooted in diverse geographical spaces with no clear centre-periphery relationship" (Pries 2017: 129–130).

---

[50] The dissociation between nationality and citizenship is one of the most significant changes to political rights that has occurred in Europe, as the right to vote in local elections has been granted to foreign citizens residing legally in fifteen European countries, including Ireland, Sweden, Denmark, the Netherlands, Finland, Luxembourg, Belgium, Switzerland, the United Kingdom and some countries in Central and Eastern Europe. Additionally, by virtue of the 1992 Maastricht Treaty, citizens from any member country of the European Union are allowed to vote in local elections from their place of residence, even if they are not nationals (de Wenden 2017: 114).

According to Pries, the seven types of internationalization briefly outlined describe different constellations of geographic and social spaces but they should not be viewed as mutually exclusive, as substitutes for one another, or as consecutive, because these social and geographical spatialities co-exist and are reciprocally influent. Even though it is true that globalization can lead to the formation of neo-national groups, it would be a mistake to believe that national states can perish because of globalization or be transformed into a new kind of de-territorialized cyberspace in perpetual flow like the internet.

Pries also says that new and complex configurations in social and geographical spaces can be developed because the national social sphere does not disappear in an enormous globalized space, but differentiates between distinct patterns of geographic and social spatial relations. Social scientists have the challenge of investigating these tendencies theoretically and empirically. For Pries, transnationalization leads to a framework of durable relationships between individuals, social actors and corporations that share social practices, symbolic systems, and both plurilocal and plurinational trans-boundary artefacts. For him, the nub of this issue consists of giving a meaning to the entanglement of daily transborder interactions between individual and collective actors from global civil society, such as international NGOs and the ensemble of diverse emerging transnational expressions of cultural, ethnic and social phenomena, like migrants and migration; political lobby groups; cultural, religious and ethnic mobilization groups; economic, commercial and entertainment flows (films, music, pop stars); and transnational corporations. According to Pries, while the globalization discourse was initially dominated by the economy and diminished the importance of nation states in favour of processes and international orders (world economy, world citizens, global civil society, cosmopolitanism), in the new contemporary perspective of transnational studies the relationship between national and transnational processes is not a zero-sum game but complementary and of reciprocal differentiation, which highlights the big difference between the globalization discourse at the end of the twentieth century and the research on transnationalization during first decades of the twenty-first century (Pries 2017: 159–161).

Pries also asserts that whereas during modernity science was orientated by rationalism and positivism, in postmodernity it is characterized by scepticism, relativism, a general suspicion of reason, and acute sensitivity to the role of ideology in asserting and maintaining political and economic power. This essentially means that positivist science itself is questioned, because science has its own limitations and has lost its monopoly on what, from an epistemological point of view, should be regarded as the truth, as happens in Boaventura de Sousa Santos' *ecology of knowledge.*

The latter is also related to the erroneous idea assumed by positivism and rationalism that religion was an expression of backwardness, which is probably due to the increase in Pentecostal mega churches and small fundamentalist and fanaticist sects that exist in all religions, but also to the belief that religion is an expression of human *spiritual needs* (as explained in the matrix of needs and satisfiers of the Chilean scholars Neef, Elizalde and Hopenhayn [1989: 33] described in Chap. 4) which cannot be satisfied by consumerism. Similarly, Buddhist philosophy responds to a spiritual need that material economic satisfiers cannot fulfil (Wilber 2018).

Moreover, these spiritual needs are apparent from the fact that the ancient (pre-Columbian) religious 'cosmovision' of indigenous people persists to this day, as noted in the analysis of the constitutional reforms of Ecuador and Bolivia. It also continues in Guatemala (Matul/Cabrera 2007), a country where the importance of the Mayan pre-Columbian cosmovision is undeniable. Citing a study by Popkin (2005: 675–676), Pries (2017: 341–342) notes that some of these peoples have not adopted the laicist vision of the guest society in, for instance, Los Angeles, New York and Chicago, but have re-interpreted their own sociocultural roots, incorporating the new experiences of life in the US in a new identity and wider autobiographic concept. Migrant workers develop a "new mental map of identity" and in the midst of their new circumstances find cultural elements and cultural values that were supposedly forgotten and buried. Thus, a rediscovery of their culture and traditions was found by Popkin in these indigenous groups far from their country of origin, which is equivalent to a 're-territorialization' of Guatemalans in the United States.

Another interesting instance of the transnationalization of social space and the cultural world is Doña Rosa's Mexican family, which has lived in the city of Yonkers (New York State) since the beginning of the 1960s but has succeeded in keeping its links with the Mexican state of Puebla. Pries argues that social-labour international migration does not normally occur in an isolated manner but in the framework of a chain of social networks. Doña Rosa's relatives decided to establish a Mexican food shop and a restaurant in Yonkers, creating, thanks to their entrepreneurship, new structures not only in their host country but also in Mexico, where they established a hotel in La Mixteca, the region they came from. For Pries, the interest of this case lies in the fact that in spite – or because – of having US citizenship, a significant number of migrant workers in this Mexican family are attracted to the land of their ancestors, as currently there are at least four generations of the same family, most of them born on US soil,[51] who reproduce their cultural background in terms of food, customs, religion, way of living and language, and some of whom decided to return more or less permanently to Mexico to invest in a new kind of family business. Of course, this phenomenon occurs thanks to their transnational citizenship or dual nationality, which allows free movement between the two countries. Additionally, the family is bilingual and its members are perfectly integrated and good citizens of both Mexico and the United States. Therefore, according to Pries:

> [The] case of Doña Rosa's family is a comparatively complex example of sustainable development and transnational relations strengthened in a transnational social space of relative stability. This situation is, for the two younger generations [of Mexican immigrants] a natural component of their way of life. For them and for the majority of the people interviewed, the fact of belonging to the two countries looked like an extension of the options for negotiations and personal strategies as well as a differentiation of the possibilities of choosing a place to live. In this sense, this international migration example is completely transferable to other migratory relations. Actually there are a considerable number of case studies that clarify

---

[51] Pries's example of the successful integration of Mexican migrant workers in US society demonstrates that the Trump Administration's intention to deny *jus soli* to Hispanic immigrants is not only racist but absolutely nonsensical.

different aspects of daily transnationalization through international migration (Pries 2017: 52–60).

There is insufficient room here to do more than refer briefly to the many other cases that Pries presents in his book, relating to, for instance, Cabo Verde, Malaysia, Bangladesh, Turkey, Poland, and the already mentioned example of the Guatemalan indigenous Mayans in the US, but those cases demonstrate not only the importance but also the practicality (for political, social, cultural, economic and even security reasons) of promoting and strengthening this path of development that, in order to be sustainable, includes transnational citizenship for foreign nationals everywhere.

6.11   Cosmopolitanism and Global Civil Society

In this section I will briefly address the predicament of cosmopolitanism as a political philosophy and its relation to the topics presented in this chapter, to be precise the migratory flows worldwide and the theme of transnational citizenship as part of the globalization phenomenon, including the cultural model of the Anthropocene in the context of sustainable development. Beck (2003) says that in order to analyse the dynamics of global civil society, a methodological shift from the dominant national perspective to a cosmopolitan perspective is necessary. He questions the notion that sociological and political modernization needs to be understood as a nation-state-organized society and points out that the supposition that humanity is naturally divided into a limited number of nations which organize themselves internally as nation states and set external boundaries to distinguish themselves from other nation states, presupposing that cultural boundaries and state boundaries coincide and that social action occurs primarily within this division and only secondarily across it, is just a hypothetical assumption.

If the concept of *global civil society* is to be considered the sociological basis for cosmopolitanism[52] in the political sphere, the very concept of 'cosmopolitanism' can be contextualized, and from my standpoint is tantamount to the concepts of 'cosmopolitics' (Delanty/Mota 2018), 'cosmopolitan democracy' (Archibugi 2011) and 'cosmopolitan order' (Held 2005). The thinking of authors such as Göpel

---

[52] Smith defines cosmopolitism as follows: "Cosmopolitanism is a perspective in the study of international ethics that takes as its starting point the idea that all persons belong to a universal community of humanity and enjoy equal moral status as citizens of the world (Lu 2000). The idea that we belong to a community of humanity has been interpreted variously as a claim about the primary object of our sense of identity or allegiance (Nussbaum 1997), the contours of the moral landscape and the duties it imposes on us (Beitz 1999), and the appropriate structure of our political or legal arrangements (Held 1995). The contention that citizens of the world enjoy an equal moral status has also been interpreted in different ways: as a claim that all persons deserve to be shown some kind of equal respect or a claim that their interests deserve to be treated equally in a more substantive sense" (Smith 2007: 27–31); 'Cosmopolitanism', Jan 2013; at: http://internationalstudies.oxfordre.com/view/10.1093/acrefore/9780190846626.001.0001/acrefore-9780190846626-e-133.

(2016), Klein (2014) and Sachs (2015) is also perfectly compatible with cosmopolitanism. Thus a cosmopolitan perspective like *cosmopolitics* could be the best way to understand the extremely complex problems facing decision-makers when trying to manage the four spheres of social dynamics, techno-economic factors, natural ecosystems and governance.

In other words, there is no doubt that governments must cooperate and work together in order to manage the techno-economic sphere effectively alongside the spheres of natural ecosystems and social dynamics. Hence, if we want to deal with the management of those four spheres in an appropriate manner, undoubtedly this is going to be less difficult if the ruling classes are democratic and have a cosmopolitan vision. It goes without saying that democratic rulers orientated by cosmopolitanism will be in a much better condition to deal with complex global issues that authoritarian ones. A cosmopolitan philosophy is also the proper way to address the problems of governance and the consolidation of institutions able to promote democracy and the rule of law, as required by SDG 16. As the Earth System itself is not constrained by national borders, this will require a fundamental rethinking of governance in a cosmopolitan direction, which essentially means re-defining boundaries in terms of "planetary boundaries". Thus public policies designed with a cosmopolitan orientation could contribute effectively to the accomplishment of the 17 SDGs and the management of the four sustainable development spheres, as illustrated by Fig. 6.1 (Sachs 2015).

The geopolitical sphere could be transformed into a planetary cosmopolitan sphere where, as a priority, all national governments could establish not the classic territorial security (geopolitical) but human security (planetary security) – as the prefix '*geo*' suggests – in the quest to protect and provide safety to all citizens of the world. In an article about the principles of cosmopolitism the English scholar Held (2005) argues that the principles must be shared universally and become the basis for the protection of all persons in the *moral kingdom of humanity*. The principles must be divided into three groups: 1) personal responsibility and accountability; 2) the basis for the way in which individual actions are transmitted in the collective sphere; and 3) the procedures to prioritize the pressing needs of the population as part of the preservation of natural ecosystems. The latter is fundamental, as it relates to the principle of sustainability and coincides with Sachs's views about the need for all economic and social development to be consistent with the stewardship of natural resources, for they are irreplaceable and non-substitutable.[53]

---

[53] "The first principle is that the ultimate units of moral concern are individual human beings, not states or other particular forms of human association. …The second principle (human agency) is the ability not just to accept but to shape human community in the context of the choices of others and of Principle 3, of personal responsibility and accountability, (Principle 4) of consent (a commitment to equal worth and equal moral value, along with active agency and personal responsibility requires a non-coercive political process. … Principles 4 and 5 ('a legitimate public decision is one that results from consent; this needs to be linked with voting at the decisive stage of collective decision-making and with the procedures and mechanism of the majority rule.) The sixth principle recognizes that if the decisions at issue are trans local, transnational or trans regional then political associations need not only to be locally based but also to have a wider scope and framework of operation. … The seventh

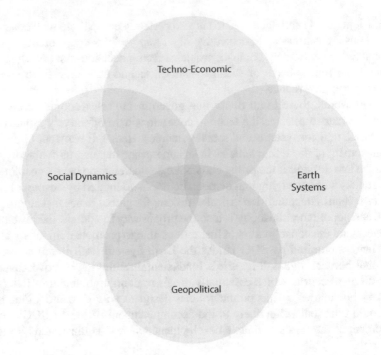

Fig. 6.1 Sustainable development: the challenge of managing four global-scale, complex, inter-connected systems. *Source* IIASA/OeAW Public Lecture Series, Lecture 4: The Age of Sustainable Development, in: Sachs (2014)

Other academics who have presented their views about cosmopolitanism include the Italian academic Daniele Archibugi, who refers to the importance of building a cosmopolitan democracy precisely because independent and sovereign states, even when sovereign by virtue of their constitutions and internal legislation in force or COVID-19, are *de facto* not independent when they have to confront the natural threats that all humankind face today – climate change, pandemics such as AIDS and other contagious diseases, trade wars, terrorism, migratory flows etc. – given that these natural and social phenomena originate outside national borders and happen globally. Thus, national authorities lose control of the political measures required to tackle them and need to agree with other states on how to address them. Archibugi points out that it is therefore not possible to maintain the minimum standards of global governance without international cooperation from a cosmopolitan democracy perspective.

---

principle is a leading principle of social justice: the principle of the avoidance of human harm and the amelioration of urgent need … and the eighth and final principle is the principle of sustainability. which specifies that all economical and social development must be consistent with the stewardship of the world core resources – by which I mean resources that are irreplaceable and non-substitutable". See Held (2005) "Principles of Cosmopolitanism Order", *in: Anales de la Cátedra, Francisco Suárez,* 39: 153–169; at: file:///C:/Users/IRIPAZ/Downloads/1030-1747-1-PB%20(1).pdf.

This also means that governments are finding it increasingly difficult to make democratic decisions since global issues do not foster citizen participation. There is a need to review the basic principles of the democratic exercise which has been based on the existence of geographic communities. This means that local has to connect to global properly for it is a fact that many of the local social networks are connected and active transnationally, as shown in the studies of Ludger Prius in the previous section. These kinds of exchanges are constantly increasing, as demonstrated by the so-called "nostalgia market" (food, handcrafts, textiles and all sort of cultural objects taken by and sold to immigrants in the US) and in all kinds of relationships between communities located in different countries that nowadays are facilitated by the use of smartphones, the internet, air transport and so on in a way that reinforces the need for cosmopolitan democracy. Even though the confluence of biotechnology with info-technology is worrying and might lead humankind into dangerous situations in the foreseeable future, as Harari (2018) has stressed, in general terms it is possible to affirm that the positive outcomes of these technologies prevail over the negative and dangerous ones.

Although Argentinian and from the decolonial theoretical perspective, the American scholar Walter Mignolo, who is very critical of Western colonialism, shares cosmopolitan views in general terms, he also underlines their limitations but especially of the roots of the racist colonial ideology, that considers both indigenous and black people as "sub-humans" (even if the Spanish crown decided only black people were going to be slaves). He describes himself as a "local cosmopolitan", saying that one should not attempt to universalize principles – as Kant did – because they are local in origin and application, not universal, contrary to the thinking of ancient Greek and Roman philosophers. From Mignolo's perspective, thought has to be relocated in its own historical context (e.g. eighteenth-century Europe). Thus, although Kant can be considered one of the greatest predecessors of cosmopolitanism, it is important[54] to remember that there have been others before and after him.[54] This is why Mignolo correctly asserts:

> Cosmopolitanism was a term re-invigorated by progressive humanists of liberal, postmodern and Marxist bent. However, it ran parallel to the discourse of neoliberal globalization promoting the end of nationalism and the beginning of a marvellous borderless world. Scholars promoting cosmopolitanism were explicitly against neoliberal globalization. But above all, cosmopolitanism was mainly a concern of Western intellectuals and scholars. I did not encounter any interest in Bolivia and Ecuador for example, and I wonder what cosmopolitanism may mean in the Middle East or in Central Asia. Cosmopolitanism, like globalism, was also unidirectional and centrifugal; both were global designs imagined and discussed within Western Europe and the US for the rest of the world (Mignolo 2011: 46).

---

[54] Among the first to remember the importance of Kant on the topic was Habermas (2003: 11–16), who argues that the challenges posed by the great catastrophes of the twentieth century (the two World Wars) and by globalization have invigorated the Kantian ideals of cosmopolitanism, though he acknowledges that the Kantian vision of the eighteenth century cannot be repeated today. Habermas's basic observation is that the Kantian ideal of world citizens' rights is tremendously relevant today, just as it was in Kant's time. Habermas also asserts that in order to reconstruct Kant's idea of cosmopolitanism, it is essential to think like Kant but also to question his ideas, as Mignolo, and before him the German philosopher Apel (1922–2017), did.

During colonial times theologians discussed whether or not Indians had a soul or not – assuming they were at all human – and this paved the way for the famous controversy between the Dominican friar Bartolomé de las Casas and the theologian Sepúlveda, whose main point was the possibility – or not – of reducing the natives to slavery. The Spanish Crown finally gave its blessing to the stance adopted by the former by issuing the 'New Laws of the Indies', forbidding the enslavement of Indians, which was a positive result of that debate (Padilla 1974). Obviously, Kant did not consider the Las Casas-Sepúlveda historical debate, given the ignorance then prevalent in Eastern Prussia about the realities of the Spanish colonial world, and, as Mignolo emphasizes, this colonialist vision was unaware of the enormous difference between indigenous cultures and Western culture:

> While in Europe the Council of Trent was setting the stage for a bloody scenario that would consume Western Christian Europe until the Peace of Westphalia (1648), ending the Thirty Years' War that piggy-backed on the eighty years' war between Spain and the Netherlands, legal theologians at the University of Salamanca were starting their long journey to solve two interrelated problems: to what extent Indians in the New World were human and to what extent, as a consequence, they had property rights. Far from the mind of Castile was to just think for a minute that property rights are not universal, and that in the Inca and Aztec civilizations, as well as in other existing communities in the Caribbean, natives do not regard land as property but as Mother Earth (*Pachamama*, was the name in Aymara and Quechua, and Gaia the name in Ancient Greece). It is possible – not a myth or fantasy – to understand that Pachamama or Gaia is the energy thanks to which life on Earth is possible. It is just common sense. To see instead 'land' and 'natural resources' as only a commodity is not common sense but the alienation of a civilization built on the idea of private property, including what Pachamama and Gaia generate and the epistemology of the South. Francisco de Vitoria and his followers were confronting, during the second half of the sixteenth century, issues of a history parallel to and intertwined with the internal history of Europe that Kant framed in terms of nation state and national characteristics (Mignolo 2011: 23).

The above quotation is reminiscent of Jean-Jacques Rousseau's well-known discourse of 1755 on private property as the origin of inequality between people. It is interesting to note that his ideas were in circulation many years before Marx and Engels tackled the subject. According to Mignolo, the reason why cosmopolitanism has been manifesting itself as a trend in critical thinking towards neoliberalism and globalization is as much to do with criticism of nationalism needs – tantamount to the "Buen Vivir" of the indigenous people- and the search for an intercultural dialogue as with the so-called hegemonic globalization defined by de Sousa Santos (2009). Having already discussed that, let us now explore Mignolo's explanation for the appearance of cosmopolitanism:

> 'Cosmopolitanism' was a buzzword in the late nineties and continues to be in the first decade of the twenty-first century. Why such widespread interest in 'cosmopolitanism'? I see four main motivations: 1) One was the previous widespread concerns and limits of 'national thinking'. Nationalism was what cosmopolitanism was trying to overcome. Cross-cultural and planetary dialogues were argued as ways towards the future, instead of leaping to defend and enclose the borders of nations. Immigration contributed to the surge of cosmopolitanism. Nationalists saw immigration as a problem; cosmopolitans as an opening toward global futures. 2) The second motivation was the need to build arguments that, moving away from nationalism, did not fall into the hands of neo-liberal and economic globalization. That kind of

global world was not what cosmopolitans liked to support at the end of the twentieth century. Thus, one of the strands of cosmopolitan thinking, confronting globalization, was caught between honest liberalism opposed to neo-liberal globalization and a renovated Marxism that saw new global players invited to think about cosmopolitanism beyond the international proletarian revolution. 3) A third motivation, related to the first two, was to move away from closed and monocultural conceptions of identity supporting State designs to control the population by celebrating multiculturalism. At this level, cosmopolitanism focused on the individual: the person was invited to see herself as an open citizen of the world, embodying several 'identities'. In a word, it was a liberal conception of cosmopolitanism born out of dissent simultaneous with the formation in Europe of the modern nation-states. That legacy has been translated into an ideal of flexible and open cultural citizenship simultaneous with the process of neo-liberal globalization. 4) The fourth motivation, compatible but also distinct from the second, was the legal proposal putting on the agenda 'cosmopolitanism from below', that was eventually connected with the agenda of the World Social Forum (Mignolo 2011: 15–16).

On the other hand, according to Archibugi's ideas about democratic cosmopolitanism, the foreign policy of democratic states is no more virtuous than that of non-democratic states, which means that, in practice, it can also be aggressive and exclusively aligned to relatively small but very influential economic groups, as actually happens in America, where, to cite just a few obvious examples, the influence of the Koch brothers in Trump's decision to leave the Paris commitments on climate change and fossil fuels is evident, as well as the NRA's influence in stopping reformist legislation concerning gun control, and the armaments industries' influence in the Pentagon regarding the decision to quit the INF treaty. This is why Archibugi defends the idea of reforming the *International System* (IS) in order to establish a cosmopolitan democracy in which individuals can take on a global-scale role while being "citizens of the world" to foster causes such as peace, the environment and human rights concurrently with Mignolo's assertions about the role of individuals in international relations being one of the legacies of cosmopolitanism because it fosters the ideal of a culturally open transnational citizenship. Even if it is parallel to the predominant neoliberalism, it brings citizens to the forefront as agents of political change in a similar way to the 'mindshift' proposed by Maja Göpel, since both are connected to the World Social Forum agenda (as Mignolo points out), which could align with Boaventura de Sousa Santos's ideas about counter-hegemonic globalization.

From a different angle, it is perfectly possible to agree with Mignolo's comments about the non-universality of the capitalist conception of property rights because indigenous cultures have a quite different perspective regarding this issue, as previously noted, but there is no doubt that in general terms the Universal Declaration of Human Rights is applicable to all peoples of the world and thus to all citizens from a cosmopolitan democracy perspective. For instance, nobody can question the universality of the right to life. The matrix of human needs developed by Manfred Max Neef (1986) is similarly an appropriate framework for clarifying the means to satisfy both human and sustainable development.

Individuals who have embraced Göpel's mindshift play a crucial role in the path to a cosmopolitan global democracy, as do bottom-up social movements that are convinced of the importance of the rule of law at world level (including new instruments like the SDGs and the 2030 Agenda, the COP 21 agreements, the International Court of Justice, the International Criminal Court, transnational citizenship, and "the role of stake-holders," as Archibugi calls them). Having a common language or culture allows them to work together to defend their rights (such as the human rights that must be respected regardless of the territory where you live or transit, as currently happens with migrant workers and refugees) and solve problems that go beyond borders and are not territorial, such as pandemics and epidemics, natural disasters, climate change, and the migration of minors. This trend has resulted in a considerable increase in transnational stakeholders such as the *Vía Campesina* organization, which promotes food security and opposes 'corporate advantages' and agro-industries, and it could also lead to new types of organizations for global civil society, such as a Global Parliament or a United Nations Parliamentary Assembly.[55] Badie's (2004) critique on the 'powerlessness of power', the role that humiliation plays in international relations, and the diplomacy of connivance in 'oligarchic' groups such as the G7 and, to a certain extent, the G20, and his criticism of the multilateral system (Badie 2012) are useful for its reform.

---

[55] On the notion of a 'World Parliament', Archibugi states: "The dream of a world parliament is very old, coming back to the fore in particular in the last years (see Falk/Strauss 2003). Such an institution will be the natural and most effective way to bring together the peoples of the Earth, allowing them to deliberate on common issues. It is unlikely that such an organ will have effective powers (at least in the short term), but even as a forum of global public opinion it could have an important role in identifying what the real and the imagined differences are among various civilizations. Such a new institution should complement the UN General Assembly. In the last decade, such a proposal has been supported by a variety of authorities and institutions (for a list of the endorsers, see the Campaign for the Establishment of a United Nations Parliamentary Assembly, at: http://en.unpaca mpaign.org/news/374.php). The basic function of a World Parliament is to allow individuals to have voice and representation in global affairs that is not associated with the voice and representation of the government of the state they belong to. This, in turn, is based on the assumption that the agendas of governments, even when democratically elected, do not necessarily correspond to the interests and the will of their population. A common forum of the citizens of the world is more likely to find workable solutions in cases of controversies. Some of these plans have envisaged a Parliament made of about 600 deputies with a criterion of representation that will favour delegations elected in small nations. According to the Charter, the UN General Assembly can establish such an organ. Such a legislative Assembly should not necessarily be involved in every aspect of global political life, but rather it would concentrate on the most relevant issues either for their impact on global life (such as the environment) or for their political significance (such as major violations of human rights). On other occasions, the World Parliamentary Assembly might limit itself to providing suggestions on what would be the most appropriate constituency to address issues that cut across borders. There are many transitional devices that can lead to the establishment of a directly elected World Parliament. The three principal ones are: i) the formation of an Assembly of the few thousand of International Non-Governmental Organizations recognized by the UN; ii) a Parliament made of representatives nominated by National Parliaments. The European Parliament, prior to the first direct election in 1979, followed this route; iii) a treaty between a selected number of like-minded states, in the hope that other states will follow. The institution of the ICC has followed this route" (Archibugi 2018: 10–11).

6.12 Cosmopolitanism and Governance in the Anthropocene

Chapter 3 explained why the field of geological sciences finally accepted the notion of the Anthropocene – proposed at the beginning of the century by the Dutch Nobel Prize Winner in Chemistry, Paul Crutzen – as a new geological era attributable to the intervention of human beings on the planet. Although it is widely accepted that it began during the early industrialization era in eighteenth-century England, it was reinforced and expanded exponentially after the so-called 'Great Acceleration'. Aside from the footprint left behind by the atomic explosions in 1945, the subsequent testing of nuclear devices in the 1950s, the increased ecological footprint left by socio-economic trends in fishing, agriculture, the construction of dams, population growth, urbanization, deforestation, and the use of fertilizers, etc. (the techno-economic and social dynamic spheres in Jeffrey Sachs's graph), the Great Acceleration had repercussions for natural ecosystems, triggering indicators on the presence of carbon dioxide, methane, and nitrogen oxide in the atmosphere, an increase in the temperature of the Earth's surface, and the melting of the glaciers, etc., which led scientists to accept the early Anthropocene as a new geological era subsequent to the Holocene, despite the fact that in geological terms 12,000 years is a fairly short period of time.

What are the implications of this new geological era? According to Delanty/Mota (2017), the notion of the Anthropocene opened new roads for investigation, because the resulting rapprochement of the social and the natural sciences also assumes a new interpretative framework – transdisciplinary – with greater implications, from the ontological and epistemological point of view, for all sciences, since:

> The social and human sciences have much to contribute to the advancement of a theoretical approach to the Anthropocene, but they also have much to learn from the natural sciences. As part of that endeavour, greater theoretical clarification is needed. A possible outcome may be an integrated history of the Earth system and the human world, since one of the obvious directions of the emerging paradigm – at least within Earth science – is that the human societies and the Earth have now forged a tenuous unity as well as a consciousness of that unity. The presuppositions of modernity are now once again called into question with the emergence of an entangled conception of nature and society, Earth and world (Delanty/Mota 2017: 91).

For this reason, Delanty and Mota assert that the Anthropocene is a new 'cultural model' in at least four areas relevant to social theory: the issue of temporality; the nature of subjectivity and the role played by 'agents' (or stakeholders) in social processes; the problem of knowledge; and lastly, a new way of understanding governance. The first issue, i.e. temporality, is related to the geological force on the planet. The authors state they agree with the idea that the bulk of interpretations allow for an informal definition in which the term is not tied to any specific origin, including that of geological time, because, from an encompassing perspective of both scientific fields (the natural and the social), the concept has experienced the type of opposition that concepts in social sciences normally experience. The matter of temporality is important because the challenge of governing the self-destructive forces unleashed by modernity depends, to a large extent, on better understanding of the said period.

There seems to be a growing consensus that the historical point of departure of the Anthropocene is modernity (Polanyi's *Great Transformation*) combined with the establishment of capitalism as the dominant mode of production (Marx) and the 'Great Acceleration' as the origin of the largest footprint that human activity has left on the planet:

> A major question for the present day, then, is whether or not the cultural and political currents of modernity can be harnessed to challenge the self-destructive forces that the modern age has unleashed in creating the age of the Anthropocene. In this sense, the challenge of governing the Anthropocene – or transforming it into a positive political project – is also about overcoming the limits of modernity whose presuppositions, it has been much noted, have been based on the separation of human history from natural history (Latour 1993; Rossi 1984). For this reason, the problem of temporality lies very much at the heart of the problem. Modernity begins with the presumption of a rupture of human history from nature and is based on a logic of human autonomy and a capacity for the radical transformation of the present in the image of an imaginary future, to take an influential formulation of the nature of modernity associated with the writings of Castoriadis (1987). But if this condition is also the dystopic condition of the Anthropocene, the transformative powers of human agency will need to be considerably rethought. An alternative account, then, would be to see the advent of the Anthropocene moment not only as a product of modernity or a condition coeval with modernity, but as a condition that can be challenged by the affirmation of modernity and a cultural model that can be located within the modern as opposed to some post or non-modern condition. Whether or not modernity has the capacity to bring about a further transformation is clearly one of the major questions for the present day. From a theoretical perspective, despite the entwinement of modernity and the Anthropocene, and the resulting transformation in the temporal horizons of human and planetary life, the radicalization of the modern condition is probably the best way to envisage a livable future. The Anthropocene can be seen as the outcome of the instrumental rationality of modernity, to invoke the Weberian concept of rationality, but modernity always entails more than this condition that has often been equated with capitalism. There can be little doubt that capitalism has been one of the main drivers of the Anthropocene, as argued by Chakrabarty (2009, 2014) and Hornborg and others, though positions will differ on whether it is the predominant force. The current situation is best seen as shaped by contradictory tendencies. This is also what Polanyi noted about the Great Transformation: the instrumentalizing forces of the market are checked in a 'double movement' by the assertion of social protection. Something like this is not quite in evidence in respect of the Great Acceleration, but the theoretical terms of analysis can be fruitfully applied. Such a double movement, when applied to the Great Acceleration, would have to require a very different kind of political subject than the one Polanyi had in mind (Delanty/Mota 2017: 91).

Therefore, if the Anthropocene is a product of modernity and of the rupture of human history from nature, as modernity -in my view—can not bring about a further transformation of society, transmodernity (Dussel: 2020, 2015) is the best way to overcome capitalism. The other fundamental issue concerns elucidation of the role played by human beings, or the issue of 'agents' and subjectivity in the processes of social and economic change, including the ecological footprint. The postulates of critical theory (Adorno, Horkheimer, Habermas) indicate that with modernity and rationalism the intent of dominion over nature was parallel to the dominion over society by capitalist elites and was spurred on by capitalism as the predominant economic system. Notwithstanding the ethical purposes of critical theory, it is important to be aware that social emancipation means affirming the autonomy of the individual before the State but not before nature, because the latter is the starting point of modernity, both the School of Frankfurt and the decolonial theory criticizes modernity. According to the erroneous perspective of philosophers like Rousseau and Hobbes, humanity had to step out of the *state of nature* (which was good in Rousseau's opinion – a "paradise lost" like the biblical myth – whereas for Hobbes the state of nature was bad, *homo homini lupus*) in order to sign a *social contract* establishing *Leviathan* (the State) according to Hobbes, or Society – the social pact – according to Rosseau in such a way that the very foundations of modern liberal thinking are based on this political mythology.

The scientific paradigmatic vision of the Anthropocene and contemporary natural sciences are potent reminders that we have to go back to basics by recognizing that we belong to the decolonial theory of academicians like Dussel, Grosfoguel, Santos, Mignolo, scientist Lovelock or the animal kingdom and have never stepped out of Nature, as the Israeli scholar Harari (2014, 2016) keeps reminding us in their books. Furthermore, if we view Nature not as an inert entity but as a vital force like Lovelock's *Gaia* or the *Pachamama* of the Andean indigenous peoples, and accept that humanity belongs to Earth and not the other way around, and that humankind is an inseparable part of Gaia – i.e. that we are immanent and not transcendent to Nature – we will realize that the fundamental error of modernity lies in believing we are different and superior or that *Homo sapiens* exerts any kind of authorization from God to dominate and exploit Nature, as the dualist philosophy of Descartes postulate.

As stated by the French philosopher Latour (2015), the idea that human will is entirely social and cultural must be abandoned. According to the new scientific paradigm, society and culture do not exist 'outside' the natural realm but emerged when human beings began to act upon nature as a biomorphic force and later as a *geophysical force* (this is why we are leaving indelible marks in the atmosphere and the oceans and on the surface of the planet). Humans will transform themselves, but it will happen inside the natural kingdom, not outside it. Human subjectivity (conscience, thought, will) takes shape and is moulded permanently in a co-evolution process with nature. The Anthropocene is not only a natural condition of the Earth, but also a 'cultural model', meaning that notwithstanding its natural roots, the Anthropocene is a cultural interpretation category used to give a new significance to the physical and biological reality, requiring a normative approach that implies values such as responsibility, truth and justice. That this is why this model has repercussions and consequences for the field of political action and governance and the establishment of a global democratic society (which translates into *cosmopolitics* and the *cosmopolocene*). To survive, it requires adequate management of the four spheres of sustainable development presented in Jeffrey Sachs's diagram (Fig. 6.1). Hence, Delanty and Mota maintain that:

> As most clearly outlined by Strydom … the Anthropocene is not merely a natural condition of the earth, but is also a cultural model in so far as it is a category of cultural interpretation or sense making (see also Strydom 2015). In terms of the previously mentioned dynamic between human experience and interpretation, the significance of the Anthropocene in one important respect is that it is an interpretation of a new dimension of human experience,

namely the perception that human beings are part of nature and that the quintessence of human life does not reside in the promethean domination of nature. In Strydom's analysis, there is a double logic to this. The Anthropocene in becoming more than a concept is also a cultural model in which contemporary society today seeks to interpret itself by recourse to cognitively structured referents, such as responsibility, truth, justice. In other words, the notion of the Anthropocene now captures a wider domain of experience and interpretation about the present and future of the world. It incorporates within it evolutionary thresholds of learning and thus has acquired a strongly normative and critical character. In the present day, this is beginning to have an impact in the ways in which human subjectivity is constituted. In terms of agency, it points in the direction of a conception of agency that is no longer predicated on the destructive separation of society and nature. We discuss this further below in relation to cosmopolitics and the challenge of governing the Anthropocene through a 'Cosmopolocene', which is one of the direct outcomes of what might be more generally termed the Anthropocene complex (Delanty/Mota 2017: 92).

Nevertheless the authors realize that by emphasizing the importance of the role played by humans, they could have fallen into a certain 'anthropocentrism', because if the planet is transformed by human agency through science and technology, man's destiny is to be the lord and master of nature, as proposed by rationalism and positivism during the times of Polanyi's Great Transformation and at the height of modernity, which would be an erroneous assumption. Still, they remind us that the Holocene (the end of the final interglacial period) began some 12,000 years ago without human intervention and that a new ice age could happen again because geological evolution is subordinated to Earth ecosystems transformations that occur in periods of thousands and even millions of years. Humans do not have the ability to intervene in such lengthy processes. In the short term humans can accelerate the pace of climate change, as is happening right now with the phenomenon that we have decided to call the Anthropocene, but humanity has no clue at all about what kind of changes our planet will experience at the end of this millennium that in geological periods is just medium term. It could be that climate change (the global warming initiated by the Holocene) is currently only delaying the onset of a still-very-distant future ice age. Other phenomena, such as the extinctions that happened in the geological past of the Earth, and the risk of being impacted once again by meteorites – such as the one leading to the disappearance of dinosaurs 65 million years ago – could be added to these considerations that seek to raise awareness of the impacts of our behaviour and increase our humility, because even if it is true that we are agents of change in nature, we are also subjects of nature and depend on her. By being part of the natural world we will never be able to wholly dominate it. This perspective is something that the post-humanist school of thought will explore with pleasure as Latour (2013, 2014) and Haraway (2015) can confirm.

Although it is wise to be aware of the limits of our free will and of natural determinants, we must also be mindful – as are other authors like Hornborg (2014), Chakrabarty (2014) and Moore (2015) – that industrialization and the capitalist mode of production are largely responsible for *greenhouse gas* (GHG) emissions; this means that corporate elites – not humankind as a whole – are the ones making the decisions about global policy (hegemonic globalization, to use Santos's phrase), which is pernicious to the environment and must be fought against by social movements as part of the counter-hegemonic globalization that I have referred to repeatedly. Also, of course, there needs to be a search for alternative means of production and social dynamics (a solidarity social economy, transition movements, a common goods alternative, the *Via Campesina* movement proposals, transnational citizenship for migrant workers etc.) and reforms to the capitalism system (Varoufakis, Piketty, Mason), as presented in previous chapters.

It is also worth distinguishing between capitalism – the economic system characterized by the accumulation of capital and the private appropriation of earnings – and the Anthropocene as a result of the human footprint on natural ecosystems. Although not all human actions on the environment are attributable to capitalism (the use of firewood by poor farmers to prepare meals, for example, or the absence of knowledge and techniques to properly dispose of the garbage and waste of poor people) – even if it has become the dominant system worldwide (Wallerstein 1989) and been a decisive factor in the Great Acceleration of the 1950s – capitalism has its own history. Consequently, it is also key to view the Anthropocene as part of the socio-economic and cultural processes and not as the result of a de-politicized 'human condition'. Every effort to achieve the *mindshift* advocated by Göpel depends on this. Delanty and Mota similarly identify this as essential to governance in the Anthropocene:

> The signs of such political formation in *Homo sapiens* are evident in shifts in consciousness since the 1980s. Before the mid-1980s discussion of ecological issues was almost entirely confined to the world of science and there was little if any awareness of major planetary crisis, as opposed to ecological devastation. Since the Chernobyl explosion and the discovery of the Ozone hole in the Antarctic in 1986 that changed with the rise of climate politics and global environmental movements, such that by the 1990s environmental issues entered the agenda of every government and have entered into new orders of governance around sustainability as the leitmotif and the search for renewable sources of energy. There can be little doubt that these developments in governance around green politics have been accompanied by shifts in self-understanding even if they do not as yet translate into a new kind of democracy, which is still confined to the increasingly incapacitated politics of the Right and Left. Relevant too are new social movements addressed at planetary politics or political movements challenging neo-liberal projects in particular places throughout the world (Delanty/Mota 2017: 99).

These shifts in self-awareness have not yet translated into a new type of democracy (the global democracy to which cosmopolitism aspires, for instance). This is because it is not only something that has to be encouraged by a democratic regime which relies on citizen participation in electoral processes and on other ways of expressing the new thinking about the environmental predicament, but is also related to governance. It is clear that the current system of national states is not designed to carry out actions that require, as a minimum, regional approaches, as in the case of climate change and sustainable worldwide development. Given the planetary nature of the problems which need to be solved within the sphere of governance, democracy extends not only beyond the discourse of the Right and the Left but is positioned on a global stage. Even if governance in the Anthropocene does not mean the disappearance of political ideologies (liberalism, socialism, republicanism), such ideologies are insufficient to address challenging global problems.

These are the reasons which have led Delanty and Mota to suggest that *cosmopolitics*, a political construction derived from a cosmopolitism which differs from post-colonial Eurocentrism, is gaining recognition and influence in political philosophy and social sciences. *Cosmopolitics* must continue to promote actions to increase biodiversity and encourage North-South dialogues to reduce GHG emissions while respecting the interests of a different kind of industrialization in the Global South and being mindful that science and technology, though key to the future, will require alternative technologies to provide renewable energy while reorganizing governance in societies. In summary:

> The notion of cosmopolitanism invokes not only a global response to climate change, but one that goes to the core of the problem in linking the human polis with the cosmic order of planet. It challenges the reduction of solidarity and loyalty to narrow conceptions of human community kept apart in different spaces and times as well instrumental forms of rationality that are divorced from the more substantive forms of rationality. Moreover cosmopolitanism is also a normative and critical idea that counter-opposes an alternative to the present while at the same time seeking in the present the sources to make possible a better future. The deliberative conception of democracy fits very well into the cosmopolitan tradition in that both are underpinned by communication as the medium by which political issues are handled. The basis of cosmopolitanism is that in the encounter with the other, the self undergoes change. This can only come about when self and other engage in communication, which can be said to be constitutive of subject formation. This cosmopolitan sensibility accords with the deliberative understanding of democracy. Moreover, it affirms the centrality of agency and an ethic of care and responsibility. For all these reasons, the political challenge of the Anthropocene is very much one that can be cast in the terms of cosmopolitics (Delanty/Mota 2017: 100).

Finally, Delanty and Mota underline the importance of being aware that the problems which sustainable development seeks to solve cannot be addressed within the narrow space of national borders. For this reason, the *mindshift* that Göpel advocates, the *Great Transition* of the Global Scenario Group (GSG), and a change of paradigm are absolutely indispensable. The SDGs and climate change necessitate a global approach which takes into account planetary boundaries. This means reorientating governance in a cosmopolitan direction and attempting to redefine the Westphalian order, despite the risk of a clash with the type of geopolitics practised by sovereign states around the world. Planetary boundaries should not be regarded as a substitute for the territorial borders of sovereign states, but as a way to share sovereignty in order to guarantee global security in a new approach to geopolitics which recognizes that the planet itself goes beyond the artificial boundaries of nations:

> The current discourse of sustainability will undoubtedly prove to be inadequate when it comes to addressing the challenges since it is locked in a hopeless compromise between the instrumental appropriation of nature and inadequate conservation measures without offering a viable vision for a future that will inexorably increase production to meet the new consumption demands in the developing world. In this scenario the reduction of poverty comes at a cost that cannot be met within the prevailing system, which was predicated on the assumption that only recently has it become evident that western civilization was only able to enjoy its privileged ascent because of poverty elsewhere, much of which was contributed to, if not caused, by the West. This has now changed and has opened up new political scenarios that require fundamentally new thinking that recognizes that climate change as well as other

manifestations of the Anthropocene cannot be contained within national orders of governance. The Earth system is not itself constrained by the human-made boundaries of nations. This implies a possible geopolitics (Clark 2014). Such politics would require a fundamental rethinking of governance in a cosmopolitan direction, for instance in re-defining boundaries in terms of what Rockstrom et al (2009) have referred to as 'planetary boundaries' (Delanty/Mota 2017: 101).

If the future's new political scenarios entail this new thinking which acknowledges that climate change and other Anthropocene manifestations cannot be contained within national orders of governance, among other reasons because the Earth system is not constrained by the imaginary boundaries of nations – as is evident from photographs of the planet taken from space – this means that humankind needs a completely new approach to global governance or institutional cosmopolitanism.

6.13 Institutional Cosmopolitanism

Institutional Cosmopolitanism (Cabrera 2018) is the title of an interesting collection of recent contributions to this academic debate which deal with the need to institutionalize the cosmopolitanism attitudes that Delanty and Mota have dubbed 'cosmopolitics'. In his chapter about institutional cosmopolitanism and national governments' duties towards suprastate institutions, Cabrera tackles the subject of assessing the legitimacy of global governance institutions to determine whether they fully comply with human rights obligations. The involvement of ordinary citizens in such assessments would be beneficial, particularly in the case of, for instance, the migration issue, since the right to migrate is enshrined in the Universal Declaration of Human Rights (1948), offering the perfect opportunity to evaluate the legitimacy (and good governance) of migratory institutions at world level and to assess whether they are really working towards respect for human rights. Another example is the campaign for the foundation of supra-state institutions, such as the proposed UN Parliamentary Assembly. The philosophy of cosmopolitanism as a way of thinking is thus fundamental to the implementation of and respect for human rights, but may also entail political responsibilities for world citizens, encouraging people to embrace responsibilities and uphold a 'global ethic' favourable to the fulfilment of human rights.

Leif Wenar, another contributor to the volume, notes that sovereignty as an expression of popular will is ineffective in the majority of national states because the rich ruling classes are rarely representative of ordinary citizens or ethnic groups – like indigenous peoples – especially in former colonial countries. Projects for more global institutional integration seem premature as long as popular sovereignty is not more strictly realized at nation-state level. Therefore, as another author (Richard Beardsworth) upholds, the internalization of cosmopolitan principles in states is essential in a world faced with urgent existential threats where national interests must be aligned – in every nation's own self-interest – with the interests of humanity if cosmopolitan results are wanted, an opinion that is shared by another contributor (Richard Shapcott) in a chapter in which he maintains that the constitutional transformation of nation states must be promoted first.

In this regard, Simon Caney's chapter about global governance procedures and outcomes introduces criteria for the assessment of institutions, making a distinction between a procedural perspective (of input legitimacy) and an instrumental perspective which evaluates institutions based on their outcomes. Instead of viewing the two models as irreconcilable, he shows the flexibility and indeterminacy of both principles, indicating that those involved in assessing institutions need not be faced with an opposing choice but can rather aim for both fair and workable approximations.

Mathias Koenig-Archibugi's contribution on the relationship between international organizations and democracy maintains that when assessing institutions from a cosmopolitan perspective, it is essential to define 'the people' and to check whether they actually rule according to criteria proposed by Koenig-Archibugi in a matrix which includes three categories – the *demos-category* (composition, performance), the *input category* (elections, participation, deliberation) and the *outcome category* (types, range) – and gives cosmopolitan democrats tools for evaluating the democratic (or not) functioning of the UN, the IMF, the WTO, and any other international organization.

The book also includes a very interesting case study of the city of Medellin in Colombia, as urban agglomerations are considered both drivers and miniatures of the globalized world. Fonna Forman and Teddy Cruz describe the transformation of Medellín, a valley city with two million inhabitants which was once known to be very dangerous, and how cities can internalize cosmopolitan principles to fight inequality. Multiple policies transformed it, especially the previously marginalized neighbourhoods on the surrounding mountainsides. Of core importance were not only the infrastructure investments to increase connectivity, but also the active and egalitarian participation of communities, the reclaiming of public spaces, and the reform of administrative institutions to enable open feedback between the periphery and the centre.

In his chapter "Demos-cracy for the European Union: Why and How", Philippe Van Parijs favours a more demos-centred approach to governing the EU in the European Parliament, as opposed to the more fragmented 'demoi'-centred way designed to increase the role of national parliaments. Van Parijs presents the Swiss confederation as a practical example which in the future could enable the EU to take more action without risking representativeness (a policy also suggested by Joschka Fisher [2013], as expounded earlier). Catherine Lu's chapter on cosmopolitan justice, democracy and global governance in the foreseeable future outlines the classic cosmopolitan attitude towards an institutional cosmopolitanism that must ultimately entail a form of world state with a world government, while Daniel H. Deudney's contribution recalls the historical cosmopolitan waves of the Stoics in ancient times, the Kantian Enlightenment, and the modern wave that he calls "hyphenated cosmopolitanism". He recognizes the centrality of a defined locality but also sees "Planetary Earth" as a new place and identifier of human political locality.

6.14   Paul Raskin and the Global Scenario Group

As mentioned on previous pages, in his essay "Journey to Earthland" Paul Raskin argues that evolution reveals two sweeping macro-transformations. One occurred 12,000 years ago when gatherers and hunters settled and agriculture began, initiating the rise of early civilization. The second macro-transformation was the evolution from agriculture to the modern age, including the Industrial Revolution in England during the eighteenth century. In present times, the modern age is confronting a deep structural crisis produced by contradictions such as the search for constant economic growth on a planet with limited resources, the upsurge of political nationalism in an interdependent globalized world, the widening gap between rich and poor, the exclusion of a considerable social majority, and an alienated and oppressive way of life dedicated to consumerism. All these contradictions are leading to the end of modernity, according to post-moderns. As a result, human conditions are heading towards a third macro-shift with far-reaching implications, which means that history has entered a *Planetary Phase of Civilization*, with the social system moving from the highly local to global in a extremely rapid process characterized by:

> The complexification and enlargement of society [that] also quickens the pace of social evolution. Just as historical change moves more rapidly than biological change (and far more rapidly than geological change), so, too, is history itself accelerating. ... The Stone Age endured about 100,000 years; Early Civilization, roughly 10,000 years; and the Modern Era, now drawing to a close, began to stir nearly 1,000 years ago. If the Planetary Phase were to play out over 100 years, this sequence of exponentially decreasing timespans would persist. Whether this long pattern of acceleration is mere coincidence or manifestation of an underlying historical principle, the fact remains that the vortex of change now swirls around us with unprecedented urgency. ... The Planetary Phase is entangling people and places in one global system with one shared destiny. Observers highlight different aspects – economics, corporations, climate change, health, technology, terrorism, civil society, governance, culture – all introduced by the modifier 'global.' Looking through specialized windows, economists see 'globalization,' technologists spotlight digital connectivity, environmentalists foreground the transformation of nature by human action, and geologists proclaim the arrival of the Anthropocene, a new geological age. Heterodox social scientists suggest other sobriquets: the Econocene dominated by the false ideology of neoclassical economics or the Capitalocene defined by capitalist relations of production and power. Meanwhile, visionary philosophers and theologians point to signs of an emerging global ethos, while realpolitik types see only clashes of civilizations and great powers (Raskin 2016: 11–14).

According to Raskin, the signals of a global shift were already present in the 1980s even though capitalism had become predominant in the global economy since the fall of communism, inducing Fukuyama (1992) to proclaim that capitalism is in crisis again. Hence the "Planetary Phase" in which we are living raises awareness of interdependence, of the importance of understanding the world-as-a-whole as a primary arena for the contending forms of consciousness that will determine whether the *Planetary Phase* will be an era of social evolution or devolution, environmental restoration or degradation.

Bearing in mind that globalization is not the end of local communities and nations, which will continue to be the main source of people's identity, the 'Earthland' of the future will consist of a sort of outer circle site of unfolding cultural and political struggles. Although some countries are still pre-modern, if history really is moving rapidly beyond modernity – as postmodern philosophy argues – the planetary phase is a historical phenomenon that requires a systemic response. The environmental stress that exacerbates poverty and incites conflict and economic instability could result in a movement of desperate underclasses that not only degrade the environment but also seek access to affluent countries – as can be seen everywhere with the migration flows – precipitating a reaction that undercuts cooperation and multilateral negotiations. Thus, the mounting pressure on the structure of the whole social and ecological system puts its resilience (capacity to recover) at risk because this cycle of violence and disorder is coupled with abrupt climate change that entails food shortages and provokes mass migration and shortages of resources like water, oil, arable land, and so on. Therefore, the worst scenario entails humanity suffering a collapse, a fall into a chaotic situation called *barbarization* or *breakdown* that will provoke the establishment of a sort of 'fortress world' where elites retreat to protected enclaves (as already happens in multiple cities of poor countries under siege by organized crime), with an impoverished majority of people excluded and institutional collapse within nation states. Hence, a new 'Dark Age' could be descending over the entire planet.

How can we avert this terrible scenario? For Raskin, the current trend and driving force of systemic change could go in the directions of sustainable development or a conventional policy reform of capitalism in accordance with market forces, or – the best alternative – initiate the *Great Transition* towards post-capitalism and a new holistic and cosmopolitan paradigm (*eco-communalism*). But an important question concerns who the leading actor – the 'driving force' – of this transformation could be. For Raskin, none of the principal actors now on the global stage are strong enough candidates to pioneer this Great Transition because of their narrow concerns and myopic vision. The United Nations, for instance, relies on the cooperation of member states reluctant to give up their sovereign prerogatives and national interests, so they are not able to respond adequately to the crises and challenges of this planetary phase. Neither transnational corporations worried only about increasing the gains for shareholders nor institutionalized civil society organizations competing for donor funds are good candidates, being "ill-prepared for the larger project of conceptualizing and advancing a coherent system shift". Only a "new global citizens movement" capable of orchestrating "a vast cultural and political rising able to redirect policy, tame corporations and unify civil society" (Raskin 2016: 32) might succeed.

> The contemporary world stage is missing this critical actor, but it is stirring on a planet bubbling with intensifying crises and shifting consciousness. A harbinger is the army of engaged people working on a thousand fronts for justice, peace, and sustainability. … Exhausted and frustrated, many activists burn out, while many more concerned citizens never find a meaningful way to engage a crisis so amorphous and overwhelming. A global movement, were it to develop, would speak especially to this growing band of the disempowered: to their minds, with a unifying perspective; to their hearts, with a vision of a better world; and to their feet, with an organizational context for action (Raskin 2016: 32–33).

This is where Raskin's ideas connect with Maja Göpel's *Mindshift*, Gerard Delanty's and Aurea Mota's *cosmopolitics,* and Boaventura de Sousa Santos's *counter-hegemonic globalization,* not forgetting Polanyi's *Great Transformation,* discussed above. Any substantial change in human history has been the result of a change of paradigm, which means a change to the prevailing mindsets in the majority of people because ideas – mental models – are the leverage for the transformation of society, not material 'productive forces, as Marx believed, and still less the market's so-called 'invisible hand' of neoliberalism. However, mindsets are the result of philosophies, ideologies, religions, social practices and scientific paradigms; they are social constructions (Wendt) and don't fall from the sky. Consequently, one of the main tasks of this *Great Transition* planetary phase is to identify agents of change (the *dramatis personae* who could drive forward this new world-view), such as a *global citizens movement.* However, it is crucial to be aware that in order to operate efficiently, this *agent of change* will have to be inspired by both the holistic paradigm and the world-view provided by cosmopolitanism. Therefore, an appropriate theoretical formulation and a strengthening of the unifying perspective of cosmopolitanism are key requirements for converting this worldwide social movement into a structured network of activists and leaders. Raskin consequently asserts that peoples and nations will have to awake as citizens of a planetary community, like:

cosmopolitan taproots sprouting in the crumbling foundations of the Modern Era ... [because] shared risks and a common fate urge collective responses that transcend fractions, political arrangements and truncated social visions. Augmented interdependence kindles modes of association and currents of thought attuned to the superordinate configuration of Earthland. ... Philosophers have long dreamed of a time when the ring of community would encircle the entire human family. This universal vision has captivated the social imagination since the fifth century BCE, when Socrates proclaimed, 'I am a citizen, not of Athens, or Greece, but of the world.' Two centuries later, the Stoics built an ethical framework that centred on the notion of cosmopolis – a world polity in harmony with reason and the universe. From this ancient font, the cosmopolitan idea mutated and evolved through the millennia, as visionaries pondered its meaning and world- changers pursued its promise. The lofty dream refused to die even as the sorry saga of the disputatious human species made One World seem a mere pipe dream. But criticism from philosophic sceptics and ideological opponents did not stop the quest for world civilization, which reached new prominence in the humanism and universalism of the Enlightenment. After a lull in the nineteenth century, the prophetic search resumed in the mid-twentieth, a response both courageous and desperate to the ambient sense of cultural exhaustion in an era rocked by world war and threatened by nuclear holocaust.

Heretofore, the cosmopolitan sensibility evolved in a sphere of ideas detached from the material sphere of actual history Cosmopolis floated in rarefied ether above the divided turf where it must be built. Now, the Planetary Phase brings the once quixotic dream into the practical realm, anchoring the ethos of human solidarity in the logic of the contemporary condition – if the alarms of danger and bells of promise it sounds rouse the global citizen from slumber. As connectivity globalizes in the external world, so might empathy globalize in the human heart (Raskin 2016: 60–61).

Being a global citizen essentially means being a member of a political community that grants rights (like fundamental liberties) to a person while simultaneously requiring him or her to fulfil responsibilities and obligations (such as paying taxes), although it is possible to differentiate historical categories following the extension of rights in different epochs. Hence, *civil citizenship* granted freedom and property rights to individuals in the eighteenth century, while *political citizenship* spread democracy in the nineteenth century, and the election of authorities amongst nations and *social citizenship* granted welfare (social and economic rights) to the working classes in the twentieth century after waves of mobilization against ruling classes and privileges:

> Thus, civil citizenship codified the triumph of entrepreneurial classes over feudal interests, political citizenship nullified the divine rights of monarchs and the despotism of powerful elites, and social citizenship was won by associated workers in their long battle with laissez-faire capitalism. Of course, it has taken many decades for women and excluded subgroups to gain these entitlements, an unfinished struggle in many countries (Raskin 2016: 61–62).

In present times, as the *Planetary Phase* unfolds, it is initiating the construction of a fourth wave of global citizenship. As in the nineteenth and twentieth centuries, this new broader concept has emotional and institutional dimensions. A person becomes a 'citizen of the world' when his or her concerns extend to the whole of humankind and the Earth's biosphere. This will permit an outpouring of world consciousness that will gradually penetrate people's mindsets, allowing the construction of the institutional platform and the functional apparatus that will be needed to democratize the international system. At the same time, global governance will no longer be focused on the Westphalian system's balance of power, and neither will surrender to sectoral or national interests – including those of transnational corporations – permitting global governance to be bounded by the planetary interests and rights of everyone: "The vision of an organic planetary civilization lies before us as possibility and exigency. We may never reach that distant shore, but what matters most is imagining its contours and travelling in its direction. The quest for a civilized Earthland beckons us, the journey its own reward and privilege" (Raskin 2016: 113).

6.15   Summary

This last chapter has shown that globalization, the global economy, the crisis of the Westphalian order, and the new geological era – the Anthropocene – are opening the door to a new cultural model with a new type of governance that is more democratic and cosmopolitan – a new vision of the world – for now we understand that humankind as a whole not only depends on Mother Earth (*Pachamama* or *Gaia*), but that the human species itself, in being part of her, must care for and respect the planet as a whole, for it is our 'common house', as Pope Francis called it in his Encyclical *Laudato Si'*. The Anthropocene is not only a new geological epoch but a new paradigm, according to which a close, inseparable interrelationship exists between natural sciences and social sciences. Hence separating them into individual scientific disciplines requires a comprehensive and holistic methodology. This is also the rationale for creating a new global democratic model of governance designed to establish a transnational citizenship based on cosmopolitanism and *cosmopolitics* that must transcend the Westphalian order in order to overcome both the global and the national state political crisis and guarantee, as a minimum, the accomplishment of the SDGs by the year 2030, in accordance with the commitments made by all UN member states in New York in the renovated pledge of 2015.

Similarly, the crisis of the Westphalian order assumes, in the socio-political sphere, that we are faced with a considerable reduction in the capacity of national states to keep their territories under control because the expansion of the world market has made national borders less and less safe and 'porous', resulting in the defensive/security-prone reaction of the conservative sectors of independent States. These conservative groups are opposed to immigration influxes, free exchange, and alleged political interference by the 'international community' in violation of national sovereignty (and they reject as interference in internal affairs cosmopolitan movements for the defence of human, labour, indigenous, environment and gender rights, and even the fight against corruption and impunity). As a result, they wield their poorly understood concept of national sovereignty as their authority and justification for making decisions on behalf of governments and ignoring popular will.

This can be interpreted as a call to address the dilemma of globalization by adopting stances analogous to those adopted in response to climate change and by fostering multilaterally negotiated policies to mitigate the negative effects, because that is precisely what is necessary for the Anthropocene to be governed with an adequate *cosmopolitics* approach. For instance, one of the most visible effects of globalization is that of human mobility and the resulting increase in migration flows, so transnational citizenship and awareness of cosmopolitanism aimed at establishing a global democracy are feasible solutions to this global predicament.

Additionally, if both globalization and interdependence presuppose the economic mobility of goods and people, the free movement of workers travelling to job markets would be solved by regulating the human dynamic (for instance, by instituting transnational citizenship). Thus, one of the main objectives of national states' foreign policy could be to establish a global regulatory framework or at least adhere to and respect the terms of the 2018 Global Compact for Migration. Freedom to emigrate, transmigrate or immigrate anywhere on the planet, or *human mobility*, should be regarded as a key factor in the satisfaction of human needs and consequently as a crucial aspect of sustainable development and human rights. Since the UN Universal Declaration of Human Rights specifically mentions the right to migrate, it is clear that implementing such objectives will require new and multilaterally negotiated policies in the United Nations to regularize migration flows and put an end to policies that criminalize migrant workers and infringe human rights. Full implementation of the *Global Compact for Safe, Orderly and Regular Migration* will be an important step in that regard. However, it would be prudent to keep migration on the agenda of global civil society for the foreseeable future and to hold a world summit on migration to promote the adoption of a binding international treaty on migratory affairs.

In brief, even if the classic IR paradigms (realism/idealism) of the Westphalian order were able to explain how the political dimension of the international system operated in the past, they are unable to explain the social and economic dimensions of the present. This is why we need the holistic approach of global governance in the

Anthropocene; climate change, loss of biodiversity, and economic unsustainability are threats that do not recognize borders. The United Nations system requires an in-depth restructuring in order to allow representatives from every continent and subcontinent access to the Security Council as permanent members. Exclusive clubs must disappear and the multilateral system must be democratized via the logic of solidarity. A change of mindset is undoubtedly necessary to address world problems, as are counter-hegemonic globalization policies inspired by a cosmopolitan world-view whose main concern – in addition to the protection of human rights, ecology, and other analogous demands – should be the defence and protection of global public goods: oceans, including freshwater reservoirs of the polar ice caps and mountain glaciers, rivers, lakes, water reservoirs, the oxygen we breathe, forests, biodiversity in general, or, in sum, protection of the rights of the planet we inhabit and share collectively. I am aware that this is just a declaration of principles that could guide cosmopolitan citizens and like-minded governments, but when individuals mentally define a situation as real, their vision can become reality and be manifested in social circumstances. Therefore, it is vital to democratically convince the majority of world citizens to adopt a cosmopolitan mindset, to enlarge the audience worldwide, and to demonstrate the value of cosmopolitan philosophy if we want to turn our ideals into reality.

References

Badie, Bertrand, 2004: *L'Impuissance de la Puissance: Essai sur les Nouvelles Relations Internationales* (Paris: Fayard).

Badie, Bertrand, 2012: *Diplomacy of Connivance* (New York: Palgrave Macmillan).

Badie, Bertrand, 2014: *Le Temps des Humiliés: Pathologie des Relations Internationales* (Paris: Odile Jacob).

Bauböck, Rainer, 2002a: *How Migration Transforms Citizenship: International, Multinational and Transnational Perspectives*, IWE-ICE No. 24 (Wien: Österreischische Akademie der Wissenschaften und Europaische Integration).

Bauböck, Rainer, 2002b: *Transnational Citizenship: Memberships and Rights in International Migration* (Aldershot: Edward Elgar).

Bauböck, Rainer, 2003: "Towards a Political Theory of Political Transnationalism", in: *The International Migration Review* (Staten Island, New York), 37,3: 700–723.

Bauman, Zygmunt, 2017: *La Globalización: Consecuencias Humanas* (Mexico City: Fondo de Cultura Económica [FCE]).

Beck, Ulrich, 1998: *¿Qué es la Globalización? Falacias del Globalismo; Respuestas a la Globalización* (Buenos Aires/Barcelona: Paidos).

Beck, Ulrich, 2002: *Alternatives to Economic Globalization: A better World is Possible: A Report of the International Forum on Globalization* (San Francisco: Berrett-Koehler Publishers).

Beck, Ulrich, 2003: "The Analysis of Global Inequality: From National to Cosmopolitan Perspective", in: Kaldor, Mary; Anheier, Helmut; Glasius, Marlies (Eds.): *Global Civil Society* (Oxford/New York: Oxford University Press).

Braudel, Fernand, 1949: *La Méditerranée et la Monde Mediterranéen à l'Époque de Philippe II* (Paris: Librarie Armand Colin).

Braudel, Fernand, 1987: *El Mediterráneo y el Mundo Mediterráneo en la Época de Felipe II* (Mexico City: Fondo de Cultura Económica).

Cabrera, Luis (Ed.), 2018: *Institutional Cosmopolitanism* (Oxford: Oxford University Press).

Delanty, Gerard (Ed.), 2012: *Routledge Handbook of Cosmopolitanism Studies* (New York/ Oxford: Routledge).

Delanty, Gerard; Mota, Aurea, 2017: "Governing the Anthropocene: Agency, Governance, Knowledge", in: *Revista Política Internacional* (Guatemala City: Academia Diplomática), 5.

Dussel, Enrique (2015): *Filosofías del Sur. Descolonización y Transmodernidad*, México, Editorial Akal.

Dussel, Enrique (2020): *Siete Ensayos de Filosofía de la Liberación. Hacia una fundamentación del Giro Decolonial*, Madrid, Editorial Trotta.

Faist, Thomas, 2009: *Social Citizenship in the European Union: Residual, Post National and Nested Membership?* Working Paper 17/2000 (Bremen: Institut for Interkulturelle und Internationale Studien [InIIS]).

Falk, Richard, 1994: "The Making of Global Citizenship", in: van Steebbergen, Bart (Ed.): *The Condition of Citizenship* (London: Sage).

Flores Olea, Victor; Flores Abelardo, Mariña, 2004: *Crítica de la Globalidad: Dominación y Liberación de Nuestro Tiempo* (México: Fondo de Cultura Económica).

Fukuyama, Francis, 1992, *The End of History and the Last Man* (London: Penguin).

Kaldor, Mary; Anheier, Helmut; Glasius, Marlies (Eds.), 2003: *Global Civil Society* (Oxford/New York: Oxford University Press).

Lacoste, Yves (2009): *Geopolítica: La Larga Historia del Presente* (Madrid: Editiorial Sintesis).

Laxer, Gordon; Halperin, Sandra, 2003: *Global Civil Society and its Limits* (New York: Palgrave Macmillan).

Mander, Jerry; Goldsmith, Edward, 1996: *The Case Against the Global Economy and For a Turn Toward the Local* (San Francisco, CA: Sierra Club Books).

Mignolo, Walter, 2011: "Cosmopolitan Localism: A Decolonial Shifting of the Kantian's Legacies", in: *Localities*, Vol. 1: Cosmopolitan Localism.

Nye, Joseph Jr., 2017: "Will the Liberal Order Survive? The History of an Idea", in: *Foreign Affairs*, 96,1 (January–February): 79–84.

Padilla, Luis Alberto, 1974: 'Essai d'Interprétations Historique des Sources Philosophiques du Droit Positif de la République du Guatemala' (Unpublished PhD thesis, University of Paris II, Pantheon Sorbonne).

Pries, Ludger (2017): *La Transnacionalización del Mundo Social: Espacios Sociales más allá de las Sociedades Nacionales* (Mexico City: DAAD & Colegio de México).

Rousseau, Jean-Jacques (1994 [1755]), *Discourse on the Origin and Foundations of Inequality Among Men* (Oxford: Oxford University Press).

Sachs, Jeffrey (2016): *La Era del Desarrollo Sostenible* (Barcelona: Editorial Deusto).

Stiglitz, Joseph, 2002: *Globalization and its Discontents* (New York/London: Norton).

Suarez Ruiz, Hero, 2018: "El Concepto de Ciudadanía en Etienne Balibar y la Nueva Estrategia Zapatista: Respuestas a la Gobernanza", in: *Oximora, Revista Internacional de Ética y Política*, 12 (January–June): 121–139.

Van Parijs, Philippe, 2018: "Demos-cracy for the European Union: Why and How", in: Cabrera, Luis (Ed.): *Institutional Cosmopolitanism* (Oxford: Oxford University Press).

Wihtol de Wenden, Catherine, 2017: *La Question Migratoire au XXIe Siècle: Migrants, Refugiés et Relations Internationales* (Paris: Presses de Sciences Po Les Presses).

Wihtol de Wendem, Catherine: *Faut-il Ouvrir les Frontières* (Paris: Presses de Sciences Po).

Internet Links

Archibugi, Daniele; Urbinati, Nadia; Zürn, Michael; Marchetti, Raffaele; Macdonald, Terry; Jacobs, Didier, 2010: "Global Democracy: A Symposium on a New Political Hope", in: *New Political*

Science, 32,1 (March): 83–121; at: http://www.danielearchibugi.org/downloads/papers/2017/11/new-political-science.pdf; doi: https://doi.org/10.1080/07393140903492159.

Bummel, Andreas: *The Composition of a Parliamentary Assembly at the United Nations: A Background Paper of the Committee for a Democratic UN*; at: https://en.unpacampaign.org/files/2010seats_en.pdf

Campaign for a United Nations Parliamentary Assembly World Parliament.

Deakin University, 2012 at: http://dro.deakin.edu.au/eserv/DU:30049755/brown-empiricalworld-evid-2012.pdf.

Fox, Jonathan, 2005: "Unpacking Transnational Citizenship", in: *Annual Review Political Science*, 8: 171–201; at: https://escholarship.org/content/qt4703m6bf/qt4703m6bf.pdf.

Global Compact for Safe, Orderly and Regular Migration; at: https://www.un.org/pga/72/wp-content/uploads/sites/51/2018/07/migration.pdf.

Global Migration Group (2017): *International Migration and Human Rights: Challenges and Opportunities on the Threshold of the 60th Anniversary of the Universal Declaration of Human Rights*; at: https://globalmigrationgroup.org/migration-and-human-rights.

Macron, Emmanuel: *"We Are at a Moment of Truth"*, in: *Financial Times* (17 April 2020).

International Organization for Migration (2017): *Encuesta sobre Migración Internacional de Personas Guatemaltecas y Remesas 2016* (Guatemala City: OIM); at: https://onu.org.gt/wp-content/uploads/2017/02/Encuesta-sobre-MigraciOn-y-Remesas-Guatemala-2016.pdf.

Migratory Policies; at: https://visaguide.world/europe/germany-visa/residence-permit/turkish/.

Savio, Roberto, 2018: "Lots of Myths and Little Reality".

Smith, William: "Cosmopolitanism"; in: *Oxford Research Encyclopaedia*; at: https://doi.org/10.1093/acrefore/9780190846626.013.133.

United Nations, 1948: "Universal Declaration of Human Rights".

Epilogue

This book was originally conceived as a way to address the problem of the Anthropocene and its relationship with sustainable development, international relations theory and the call for a new cosmopolitan philosophical vision capable of addressing the present-day major challenges for humankind, including the risk of collapse endured by our species while dealing with the complex relationship that nature and terrestrial ecosystems have with the economy, society, and global governance. The COVID-19 pandemic of 2020/21 has revealed the imminence and reality of that threat. Nobody really believed in it before. Nobody would ever have thought that a minuscule microbe that cannot be seen without a powerful electronic microscope would put the entire world in such disorder or such a critical situation. Fortunately, science and states have come to our aid, not religion or the politicians who reject state intervention in the economy and worship 'markets' as if they were gods. Certainly, the pandemic is the result of the kind of relationship that societies around the world have established with nature and biodiversity. The foolish neoliberal ideology that sees 'growth' and personal gain as the foremost objectives of human life is absolutely wrong, and one of the lessons of the traumatic experience lived through the ominous years of 2020 and 2021 is that health and the protection of nature are much more important than production and the mobility of goods and people, as proved by the dramatic halt in economic activities ordered by governments that, when facing an existential danger for the population, realized that the protection of life is the foremost objective of every state. The second great lesson of the pandemic is that without international cooperation it will be impossible to defeat the virus. The insanity of the irresponsible behaviour of Trump, who decided not to pay the US financial contribution to the World Health Organization, repealed the Obama care measures, and deported migrants infected with the virus, is obvious and needs no commentary, among other reasons because it would be beneficial for the vaccine to be put at humanity's disposal through the WHO, circumventing pharmaceutical corporations' greed and voracity. Consequently, although the concept of the Anthropocene relates to the new geological epoch in which humanity has been living since the

© Springer Nature Switzerland AG 2021
L.-A. Padilla, *Sustainable Development in the Anthropocene*,
The Anthropocene: Politik—Economics—Society—Science 29,
https://doi.org/10.1007/978-3-030-80399-5

Great Acceleration of the second half of the twentieth century, the cultural model of cosmopolitanism related to this epoch essentially means that humankind must put to work the knowledge derived from the lessons of the traumatic experience suffered in 2020 and start behaving more respectfully towards nature if our species wants to survive and avoid Lovelock's "revenge of Gaia". Therefore, the Anthropocene is also a cultural model that goes beyond natural and social sciences, setting forward a normative framework for global governance, as well as a holistic methodology overarching particular disciplines and looking for the kind of transdisciplinary knowledge needed to face both present and future menaces to the survival of humankind. In other words, the great challenge of this century is to avoid looking backwards and promote instead a new normality that was not begun in the year 2000 but is really being inaugurated two decades later, thanks to the traumatic experience of the health crisis.

In addition to holism and transdisciplinarity, a cosmopolitan philosophical worldview is crucial for tackling the present-day problems. Global governance cannot function adequately within the old and anachronistic Westphalian order, and even though the United Nations is an indispensable tool, its institutions and organisms must be reformed to improve world governance. But right now it is absolutely essential to redesign the mechanisms for dealing with world affairs. The cosmopolitan vision offers a philosophy opposed to nationalism, an outdated ideology responsible for the twentieth century's two world wars and the Balkan confrontations in the 1990s. It is impossible to deal with world problems in a national manner; states need to cooperate because we live on a single planet that has no material territorial borders. The imaginary lines separating national states were invented by politicians and warriors to stop wars in the past; they provided a practical solution to interstate conflicts in the nineteenth century, but that strategy is not fruitful anymore. The regional integration process which is taking down frontiers inside the European Union is a good example of the route that must be followed by the rest of the world, because it demonstrates that integration is the most suitable pathway to peace and overcoming territorial disputes, nationalism, underdevelopment, illiberal and undemocratic governments, and so on. As cosmopolitanism is essentially based on human rights, no UN member state will be in conflict with such a universal philosophy in spite of its European inception in the French revolution, Kantian rationalism and postmodernism. Even Walter Mignolo, an American scholar born in Argentina radical critic of colonialism, says that cosmopolitanism originally focused on the individual, because the person was invited to see him- or herself as a free citizen of the world embodying several identities (Mignolo 2011; 2012). From my perspective, that is why cosmopolitanism is perfectly compatible with Santos's *Ecology of Knowledge*, the protection of local cultures, and the defence of indigenous rights and national identities while simultaneously promoting world citizenship. The pandemic of 2020/21 is also a good demonstration of this point of view because even though each country dealt with the disease in its own way – some with better results that others, e.g. Iceland, New Zealand, Norway, Denmark, Germany, Finland, Taiwan, South Korea, Costa Rica, Uruguay and even China and Vietnam compared with Italy, Spain, Sweden, the United States, Russia, Brazil, Peru and Chile – the fact remains that without international cooperation the crisis could have been worse and that nationalist policies like those followed by Trump complaining against China and the World Health Organization only aggravated the crisis.

Another important theme addressed in these pages concerns globalization. Even though it is a man-made phenomenon, it is crucial to understand the impossibility of reversing the process, as the world economy has entered a post-capitalist stage based on the new technologies of communication and information, which have been reinforced thanks to the health crisis. Furthermore, as world markets are the destiny of all transnational corporations' output (aeronautics, computers, entertainment, vehicles, pharmaceuticals, tourism, and so on), the territorial principle – which is the fundamental precept of the Westphalian order and the ideology of nationalism – cannot function as it did in the nineteenth and twentieth centuries. Geopolitics will not be the main interest of the foreign policy of national states – not because it will not continue to be important, but because twenty-first century issues are distinct and are placing the survival of our species at risk because of anthropogenically-induced climate change. Certainly, nuclear weapons continue to be a threat, and their capacity for destruction is also a great risk to humanity; however, both the invisible virus and global warming are much more dangerous and their consequences are much more catastrophic. A nuclear war would ultimately leave some regions untouched by material destruction, if not by radioactivity. However, the rise in the sea level will affect every country (and city) with maritime coasts, which means the majority. The pandemic is again a good example because coronavirus did not require a passport to infiltrate borders, a fact that essentially means that the world will continue to deal with health issues that will require global cooperation, not national isolation. And even though geography facilitated the success of national policies in countries like New Zealand, Taiwan, Japan, Cuba and Iceland, that was a matter of topography, not the result of any 'nationalist' approach. It is simply not possible to solve Richard Hass's problem of a *world in disarray* by re-establishing a Westphalian kind of *world order* – as Kissinger suggested – and even less possible to solve it via a change of regime in countries that do not accept American hegemony, because globalization is a formidable obstacle. This means that geo-economics is confronting geopolitics former US president. In that battle geopolitics will probably be the loser, simply because military means are completely ineffective against non-territorial world issues like pandemics and climate change. However, this does not mean that irresponsible personalities like Trump could not eventually unchain a war against, say, China for geopolitical reasons.

Therefore, in the foreseeable future, the reinterpretation of state sovereignty must include the awareness that in this century it is not the protection of territory that is at stake, but the protection of planetary boundaries. This means that the crucial function of contemporary international relations is global and human security, not national security. For that reason, sovereignty must include 'Earth's sovereignty', based on the concept of *Pachamama* embedded in the constitutions of Ecuador and Bolivia. The rights of Mother Earth as the threshold of contemporary security are not territorial – Westphalian – but entail a worldwide planetary security that includes territories, peoples and the planet itself. This is tantamount to saying that water, glaciers, forests, wildernesses, oceans, polar caps, biodiversity, and all natural life – not just human life – are of foremost importance and as valuable as people's lives and the territorial integrity of nation states. In conclusion along these lines, national governments must learn to share sovereignty – as in regional integration processes like that of the EU – in order to protect the rights of Mother Nature, *Gaia* or *la Pachamama* – the name is unimportant – otherwise Planet Earth could 'take revenge' (as Lovelock has warned) and get rid of us, which means that *Homo sapiens* could disappear from Earth. Hence, respecting the rights of Gaia is also the best way to prevent the sixth mass extinction – which is already under way – encompassing our species as well, as predicted by palaeontologists like Peter Ward.

Having said that, it must be recognized that security problems are related to geopolitics and that multilateralism is the only way to go beyond the profound crisis of the Westphalian order. As president Biden did. This means that global public policies ought to be negotiated by the United Nations and regional organizations in terms of a diplomacy of solidarity which goes beyond cooperative diplomacy in order to deal with traditional world security problems, such as terrorism, transnational organized crime, cyber security, internal armed conflicts, refugee and migratory flows, and so on. In my view, the doctrine of responsibility to protect must be enlarged to include human and planetary security, and states must be held to account if they do not meet their obligations with regard to matters of public health and climate change.

Needless to say, other security problems related to geo-economics must be negotiated with a view to the protection of human security. For instance, vaccines against malignant viruses must be put at the disposal of all nations freely without the restrictions of patents and intellectual property. Since sustainable development essentially consists of micro- or macro-managed social dynamics and techno-economic factors that are connected with ecosystems, making linear development compatible with the cycles of nature, it is clear that humanity must not trespass planetary boundaries. Ocean pollution and greenhouse gas emissions must be terminated and controlled, and multilateral agreements must prioritize the reduction of world inequalities, food security, the improvement and strengthening of public health and educational systems, the empowerment of women, respect for indigenous peoples' rights, the mitigation of vulnerabilities, adaptation to climate change, the eradication of corruption, the rule of law, respect for human rights – fundamental liberties as well as political, economic, social and cultural rights – the deepening of participatory democracy, and the regularization of migratory movements, including the right of human mobility. The right to leave your own country and work abroad, as well as the right to return to your homeland and human mobility, must be respected, and unless you commit a crime, arbitrary detentions for migratory reasons must be banned from state policies. Transnational citizenship could be a solution to the current world migratory crisis.

Capitalism must be reformed, and neoliberalism already shows signs of coming to an end. The UN Sustainable Development Agenda 2030 is the most appropriate option for solving major problems in the social and environmental spheres. The macro-economic field requires an entire overhaul along the lines recommended by well-known experts like Varoufakis and Piketty, though this has yet to be implemented. In contrast, the micro-economic field, with fully functioning multiple reforms and innovations, has long since been part of the civil society domain. Serge Latouche (1993) mentions various examples in his analysis of the so-called 'informal sector' of the economy in developing countries, but can reforms such as transition towns, the social solidarity economy, the economy of the commons, and the circular economy of the EU 'suck down' capitalism, as Bruno Latour (2015) imagines? No one knows. But small reforms and the rule of law are the keys to improving the way the economy functions. As the business sector is embedded in capitalism, such reforms would enable it to maintain its preoccupation with growth and the accumulation of capital while respecting both national and international law, in particular the rules designed to regulate markets, avoid tax evasion and corruption, and guarantee better income distribution through fiscal reforms, making production compatible with ecosystems (waste and plastics must be put under control), decarbonizing energy by forsaking fossil fuels, and reinforcing renewables like solar, wind, geothermic, tide, hydro and even nuclear power in its fusion form in all countries, not just the exemplary ones like Germany, Costa Rica and Bhutan. Henceforth opposing hegemonic globalization has to be a matter of social prioritization, respect for ecosystems, and the enforcement of international laws which regulate markets and transnational corporations in the framework of a post-capitalist economy looking to establish a renewed form of democratic socialism.

Not surprisingly, the democratization and legitimization of the international system require a new kind of multilateralism, a true cosmopolitan vision that could prevent and reduce the danger of internal ethnic cleansing and interstate warfare and violence, especially in the 'flash-points' of Eurasia, the South and East China Seas, the Middle East, and even Latin America, where geopolitics still confronts the trend towards regional integration. As previously stated, geopolitics and the territorial principle must be transformed. The contemporary concept of global security based on the preservation of planetary boundaries must be adopted, and the absence of sustainable development must be regarded as a threat to peace ecology because climate change is caused by the accumulation of greenhouse gases in the atmosphere. Last but not least, positive peace based on the eradication of structural violence is also fundamental.

In synthesis, in today's world, sustainable development must be the answer to the multiple socio-economic, environmental and political crises that are threatening the survival of our own species. Thus, if the anthropogenic deterioration of climate is aggravated by the ecological threats linked to the *Great Acceleration*, urgent measures must be taken – the COP 21 commitments as a minimum – to stop an existential danger that could include humanity in the sixth mass extinction. *Cosmopolitics* is not a utopia that would never become reality. When a person acknowledges the reality of a situation, the *social consequences* of this perception can be transformed in *reality*. Thus, despite the fact that our contemporary world is still mainly Westphalian and neoliberal and that, unfortunately, neo-nationalism is reappearing, at the same time globalization and trans-nationalization have already introduced important changes in global civil society. If we want to avoid the collapse of civilization, sustainable development complemented with a substantial reform of capitalism, a return to the welfare state, the regulation of markets, imposing taxes on the super-rich, establishing a universal basic income, allowing everybody to enjoy a decent standard of living based on Buen Vivir (*good living*) – not personal gain – and the non-material dimensions of fulfilment such as meditation and spirituality, the quality of life, social solidarity, the preservation of ecosystems, and the protection of the environment are all fundamental issues.

Then it will just be a matter of time before we attain the collective change of mentality and philosophy that will lead the majority of world citizens to redefine reality according to the new holistic and cosmopolitan paradigm with all its anticipated factual and social consequences. Nevertheless, engaging in these topics at world level requires a reassessment of how ethics, international relations, economics, sociology, natural sciences and ecology can work together towards sustainability, using cosmopolitanism, holism and transdisciplinarity as methodological tools that will help to overcome the current world crisis. Escaping extinction by embracing sustainability – that is the great challenge of our century.

Bibliography

Books

Aguilera Peralta, Gabriel; Torres, Edelberto, 1998: *Del Autoritarismo a la Paz* (Guatemala City: FLACSO).

Alker, Hayward; Gurr, Ted Robert; Rupesinghe, Kumar, 2001: *Journeys Through Conflict: A Study of the Conflict Early Warning Systems Research Project of the International Social Science Council* (Boulder/New York/Oxford: Rowman & Littlefield Inc.).

Angus, Ian, 2016: *Facing the Anthropocene: Fossil Capitalism and the Crisis of the Earth System* (New York: Monthly Review Press).

Archibugi, Daniele; Held, David (Eds.), 1995: *Cosmopolitan Democracy: An Agenda for a New World Order* (Cambridge: Polity Press).

Archibugi, Daniele; Held, David (Eds.), 1998: *Re-Imagining Political Community: Studies in Cosmopolitan Democracy* (Cambridge: Polity Press).

Aron, Raymond, 1984: *Paix et Guerre entre les Nations* (Paris: Calmann-Lévy).

Aron, Raymond, 1996: *Une Histoire du XXe Siècle: Anthologie* (Paris: Plon).

Attali, Jacques, 2006: *Une Brève Histoire de l'Avenir* (Paris: Fayard).

Badie, Bertrand, 1999: *Un Monde sans Souveraineté* (Paris: Fayard).

Badie, Bertrand, 2004: *L'Impuissance de la Puissance: Essai sur les Incertitudes et les Espoirs des Nouvelles Relations Internationales* (Paris: Librairie Arthème Fayars).

Badie, Bertrand, 2011: *La Diplomatie de Connivence* (Paris: La Découverte).

Badie, Bertrand, 2012: *Diplomacy of Connivance* (New York: Palgrave Macmillan).

Badie Bertrand, 2013: *La Fin des Territoires: Essai sur le Désordre International et sur l'Utilité Sociale du Respect* (Paris: CNRS Éditions).

Badie, Bertrand, 2014: *Le Temps des Humiliés: Pathologie des Relations Internationales* (Paris: Odile Jacob).

Badie, Bertrand; Foucher, Michel, 2017: *Vers un Monde Néo-National? Entretiens avec Gaïdz Minasssian* (Paris: CNRS editions).

Bauböck, Rainer, 2017: *Transnational Citizenship and Migration* (London: Routledge).

Bauman, Zygmunt, 2017: *La Globalización: Consecuencias Humanas* (Mexico: Fondo de Cultura Económica).

Beck, Ulrich, 2006: *Reinventar Europa: Una Visión Cosmopolita* (Barcelona: Centro de Cultura Contemporánea de Barcelona).

Beck, Ulrich; Grande, Edgar, 2006: *La Europa Cosmopolita: Sociedad y Política en la Segunda Modernidad* (Madrid: Ediciones Paidós Ibérica).

© Springer Nature Switzerland AG 2021
L.-A. Padilla, *Sustainable Development in the Anthropocene*,
The Anthropocene: Politik—Economics—Society—Science 29,
https://doi.org/10.1007/978-3-030-80399-5

Beck, Ulrich, 1998: ¿Que es la Globalización? Falacias del Globalismo, Respuestas a la Globalización (Barcelona: Paidós).

Beck, Ulrich, 2012: "Global Inequality and Human Rights: A Cosmopolitan Perspective", in: Routledge Handbook of Cosmopolitan Studies (London/New York, Routledge).

Bollier, David; Helfrich, Silke (Eds.), 2012: The Wealth of the Commons: A World Beyond Market and State (Amherst, MA: Levellers Press).

Brauch, Hans Günter; Oswald Spring, Úrsula; Bennett, Juliet; Serrano Oswald, Serena Eréndira (Eds.), 2016: Addressing Global Environmental Challenges from a Peace Ecology Perspective (Cham: Springer International Publishing).

Braudel, Fernand, 1949: La Méditerranée et la Monde Mediterranéen à l'Époque de Philippe II (Paris: Librarie Armand Colin).

Braudel, Fernand, 1987: El Mediterráneo y el Mundo Mediterráneo en la Época de Felipe II (Mexico: Fondo de Cultura Económica).

Brundtland, Gro Harlem et al., 1989: Our Common Future (New York: United Nations).

Brzezinski, Zbigniew, 1998: El Gran Tablero Mundial: La Supremacía Estadounidense y sus Imperativos Geoestratégicos (Barcelona: Paidós).

Capra, Fritjof, 1975: The Tao of Physics (Berkeley: Shambala).

Capra, Fritjof, 1982: The Turning Point (New York: Simon & Schuster).

Capra, Fritjof, 1996: The Web of Life (New York: Anchor Books).

Carlsnaes, Walter; Smith, Steve, 1995: European Foreign Policy, the EC and Changing Perspectives in Europe, Sage Modern Politics Series, vol. 34 (London: Sage).

Castells, Manuel, 1998: The Information Age: Economy, Society and Culture: End of Millennium (Malden, MA: Blackwell Publishers Inc.).

Castoriadis, Cornelius, 1975: L'Institution Imaginaire de la Société (Paris: Seuil).

Cortina, Adela, 2010: Ética Aplicada y Democracia Radical (Madrid: Tecnos).

Dabène, Olivier, 2012: "Explaining Latin America's Fourth Waves of Regionalism: Regional Integration of a Third Kind", Paper for the "Waves of Change in Latin America History and Politics" panel, 2012 Congress of the Latin American Studies Association (LASA), San Francisco, CA.

Dalai Lama, 2005: The Universe in a Single Atom: The Convergence of Science and Spirituality (New York: Broadway Books).

Delanty, Gerard (Ed.), 2012: Routledge Handbook of Cosmopolitan Studies (London/New York: Routledge).

Delanty, Gerard; Mota, Aurea, 2017: "Governing the Anthropocene: Agency, Governance, Knowledge", in: European Journal of Social Theory, 20,1: 9–38.

Delanty, Gerard; Mota, Aurea, 2018: Política Internacional No.5 (Guatemala City: Academia Diplomática).

Diamond, Jared, 2007: Colapso: Por Qué Unas Sociedades Perduran y Otras Desaparecen; (Mexico DF: Random House Mondadori).

Dussel, Enrique (2015): Filosofías del Sur. Descolonización y Transmodernidad, México, Editorial Akal.

Dussel, Enrique (2020): Siete Ensayos de Filosofía de la Liberación. Hacia una fundamentación del Giro Decolonial, Madrid, Editorial Trotta.

Drucker, Paul, 1999: La Sociedad Post Capitalista (Buenos Aires: Editorial Sudamericana).

Falk, Richard, 1994: "The Making of Global Citizenship", in: van Steebbergen, Bart (Ed.): The Condition of Citizenship (London: Sage).

Flores Olea, Victor; Mariña Flores, Abelardo, 2004: Crítica de la Globalidad: Dominación y Liberacion en Nuestro Tiempo (Ciudad de México: Fondo de Cultura Económica [FCE]).

Friedman, George, 2015: Flashpoints: The Emerging Crisis in Europe (New York/London: Doubleday).

Foucault, Michel, 2014: Seguridad, Territorio, Población (Mexico: Fondo de Cultura Económica).

Fukuyama, Francis, 1992, The End of History and the Last Man (London: Penguin).

Fukuyama, Francis, 2015: *Political Order and Political Decay: From the Industrial Revolution to the Globalization of Economy* (London: Farrar, Straus and Giroux & Profile Books Ltd.).

Galtung, Johan, 1980: *Peace and World Structure: Essays in Peace Research, Volume IV* (Oslo: International Peace Research Institute Oslo [PRIO]; Copenhagen: Christian Ejlers).

Galtung, Johan, 1981: "Contribución Específica de la Irenología al Estudio de la Violencia", in: Galtung, Johan: *La Violencia y Sus Causas* (París: UNESCO).

Gebser, Jean, 1985, 1991: *The Ever-Present Origin* (Athens: Ohio University Press).

Giacalone, Rita, 2016: *Geopolítica y Geoeconomía en el Proceso Globalizador* (Bogotá: Acontecer Mundial, Ediciones Universidad Cooperativa de Colombia).

Göpel, Maja, 2016: *The Great Mindshift: How a New Economic Paradigm and Sustainability Transformation go Hand in Hand* (Cham: Springer International Publishing).

Habermas, Jürgen, 1988: *Teoría de la Acción Comunicativa: Crítica de la Razón Funcionalista*, vol. II (Madrid: Taurus).

Harari, Yuval Noah, 2014: *De Animales a Dioses: Breve Historia de la Humanidad* (Mexico: Penguin Random House/Grupo Editorial SA de CV).

Harari, Youval Noah, 2016: *HomoDeus: Breve historia del Mañana* (Barcelona: Penguin Random House).

Haas, Richard, 2017: *A World in Disarray. American Foreign Policy and the crisis of the old order*, (Penguin Press, New York).

Harari, Youval Noah, 2018: *21 Lecciones para el Siglo XXI* (Barcelona: Penguin Random House).

Haushofer, Karl, 2012: "Los Fundamentos Geográficos de la Política Exterior", in: *Geopolitica Revista de Estudios sobre Espacio y Poder*, 3,2: 329–336.

Held, David, 1995: *Democracy and the Global Order* (Cambridge: Polity Press).

Held, David, 2006: *Models of Democracy* (Stanford: Stanford University Press).

Holloway, John, 2002: *Cambiar el Mundo sin Tomar el Poder: El Significado de la Revolución Hoy* (Mexico: Universidad de Puebla).

Holsti, Kalevi, 2016: *Kalevi Holsti: A Pioneer in International Relations Theory, Foreign Policy Analysis, History of International Order, and Security Studies* (Cham: Springer International Publishing).

Huntington, Samuel, 1996: *The Clash of Civilizations and the Remaking of World Order* (New York: Simon and Schuster).

Huntington, Samuel, 2004: *¿Quiénes Somos? Los Desafíos de la Identidad Nacional Estadounidense* (Barcelona: Paidós).

Kaldor, Mary; Anheier, Helmut; Glasius, Marlies (Eds.), 2003: *Global Civil Society* (Oxford/New York: Oxford University Press).

Kaplan, Morton, 1957: *System and Process in International Politics* (New York: John Wiley and Sons).

Kaplan, Robert D., 2012: *The Revenge of Geography: What the Map tells us about Coming Conflicts and the Battle against Fate* (New York: Random House).

King, Alexander; Schneider, Bertrand, 1991: *Questions de Survie: La Révolution Mondiale A Commencé* (Paris: Calman-Lévy).

Kissinger, Henry, 1996: *La Diplomacia* (Mexico: Fondo de Cultura Económica [FCE]).

Kissinger, Henry, 2014: *World Order* (New York: Penguin Press).

Klein, Naomi, 2014: *This Changes Everything* (New York: Simon and Schuster).

Klein, Naomi, 2015: *Esto lo Cambia Todo: El Capitalismo contra el Clima* (Barcelona: Ed. Paidós).

Kohlberg, Lawrence, 1984: *The Psychology of Moral Development: The Nature and Validity of Moral Stages: Essays on Moral Development*, vol. 2 (New York: Harper & Row).

Kohlberg, Lawrence; Gibbs, John; Lieberman, Marcus, 1983: *A Longitudinal Study of Moral Judgment*, Monographs of the Society for Research in Child Development, 48,1–2, Serial No. 200 (Chicago: University of Chicago Press).

Kuhn, Thomas, 1962: *The Structure of Scientific Revolutions* (Chicago: University of Chicago Press).

Kuhn, Thomas, 1989: *Que son las Revoluciones Científicas y Otros Ensayos* (Barcelona: Paidós, Ibérica).

Lacoste, Yves, 2009: *Geopolítica: La Larga Historia del Presente* (Madrid: Editorial Síntesis).

Laszlo, Ervin, 2006: *Science and the Reenchantment of the Cosmos: The Rise of the Integral Vision of Reality* (Rochester, VT: Inner Traditions).

Laszlo, Ervin, 2010: *Chaos Point 2012 and Beyond: Our Choices between Global Disaster and a Sustainable Planet* (Charlottesville, VA: Hampton Roads Publishing Co.).

Latour, Bruno, 2004: *Politiques de la Nature: Comment faire entrer les Sciences en Démocratie* (Paris: La Découverte).

Latour, Bruno, 2015: *Face a Gaia: Huit Conférences sur le Nouveau Régime Climatique* (Paris: La Découverte).m

Lovelock, James, 2007: *La Venganza de la Tierra. Por qué la Tierra está rebelándose y cómo Podemos todavía salvar a la humanidad* (Barcelona, Planeta)

Mander, Jerry; Goldsmith, Edward, 1996: *The Case against the Global Economy and for a Turn toward the Local* (San Francisco, CA: Sierra Club Books).

Martin, Hans Peter; Schumann, Harald, 1998: *La Trampa de la Globalización: El Ataque contra la Democracia y el Bienestar* (Madrid: Taurus, Grupo Santillana de Ediciones).

Marx Karl, 1974: *El Capital: Critica de la Economía Política*, 3 vols. (Mexico: FCE).

Maslow, Abraham, 1982: *La Amplitud Potencial de la Naturaleza Humana* (Mexico: Trillas).

Maslow, Abraham, 1994: *La Personalidad Creadora* (Barcelona: Kairós).

Maslow, Abraham, 1998: *El Hombre Autorrealizado: Hacia una Psicología del Ser*, (Barcelona, Kairós).

Mason, Paul, 2016: *Postcapitalismo: Hacia un Nuevo Futuro* (Barcelona/Buenos Aires: Paidós).

Matul, Daniel; Cabrera, Edgar, 2007: *La Cosmovisión Maya* (2 vols.) (Guatemala City: Liga Maya de Guatemala/Real Embajada Noruega en Guatemala/Amanuense Editorial).

Mearsheimer, John J., 2014: *The Tragedy of Great Power Politics* (New York/London: Norton & Co.).

Merle, Marcel, 1991: *La Crise du Golfe et le Nouvel Ordre International* (Paris: Economique).

Mignolo, Walter, 2011: "Cosmopolitan Localism: A De-colonial Shifting of the Kantian's Legacies", in: *Localities*, 1: *Cosmopolitan Localism*: 11–46.

Mignolo, Walter, 2012: *Decolonial Cosmopolitanism and Dialogues among Civilizations* (London: Routledge).

Morin, Edgar, 1999: *Relier les Connaissances: Le Défi du XXIe siècle* (Paris: Seuil).

Morin, Edgar, 2011: *La Vía para el Futuro de la Humanidad* (Barcelona: Espasa Libros).

Nerfin, Marc; et al., 1977: *Hacia otro Desarrollo: Enfoques y Perspectivas* (México: Siglo XXI Editores S.A.).

Nerfin, Marc; et al., 1978: *Another Development, Approaches and Strategies* (Uppsala: Dag Hammarskjold Foundation).

Nye, Joseph; Keohane, Robert, 1988: *Poder e Interdependencia: la Política Mundial en Transición* (Buenos Aires: Grupo Editor Latinoamericano [GEL]).

Nye, Joseph S. Jr., 2011: *The Future of Power* (New York: Public Affairs & Perseus Group).

Oswald Spring, Úrsula, 2016: "The Water, Food and Biodiversity Nexus: New Security Issues in the case of Mexico", in: Brauch, Hans Günter; Oswald Spring, Úrsula; Bennett, Juliet; Serrano Oswald, Serena Eréndira (Eds.): *Addressing Global Environmental Challenges from a Peace Ecology Perspective* (Cham: Springer International Publishing): 113–144.

Padilla, Luis Alberto, 2009: *Paz y Conflicto en el Siglo XXI: Teoría de las Relaciones Internacionales* (Guatemala City: IRIPAZ).

Padilla, Luis Alberto, 2012: *Guatemala: Relaciones Internacionales y Contexto Geopolítico Mundial*, in: *Guatemala: Historia Reciente*, 5 vols. *(1954–1996)* (Guatemala City: FLACSO).

Padilla, Luis Alberto, 2018: *Human Rights and Radical Democracy*, in: Oswald Spring, Úrsula; Serrano Oswald, Serena Eréndira (Eds.): *Risks, Violence, Security and Peace in Latin America* (Cham: Springer International Publishing).

Padilla Vassaux, Diego, 2019: *Política del Agua en Guatemala: Una Radiografía Crítica del Estado* (Guatemala City: Universidad Rafael Landivar, Instituto de Investigación y Proyección sobre el Estado [ISE]/Editorial Cara Parens).

Papa Francisco, 2015: *Laudato Si' Sobre el Cuidado de la Casa Común Carta Encíclica* (Guatemala City: Editorial Kyrios).

Piaget, Jean, 1982: *The Moral Judgment of the Child* (London: Trench, Trubner & Co Colby).

Piketty, Thomas, 2013: *Le Capital au XXI Siècle* (Paris: Seuil).

Piketty, Thomas, 2014: *Capital in the Twenty-First Century* (Cambridge: Harvard University Press).

Polanyi, Karl, 1957: *The Great Transformation: The Political and Economic Origins of Our Time* (Boston, MA: Beacon Press).

Polanyi, Karl, 1983: *La Grande Transformation: Aux Origines Politiques et Economiques de Notre Temps* (Paris: Éditions Gallimard).

Pries, Ludger, 2017: *La Transnacionalización del Mundo Social: Espacios Sociales más allá de las Sociedades Nacionales* (Mexico City: Deutscher Akademischer Austauschdienst [DAAD]/El Colegio de México, Centro de Estudios Internacionales).

Ricard, Matthieu; Xuan Thuan, Trinh, 2000: *L'Infini dans la Paume de la Main: Le Moine et l'Astrophysicien* (Paris: Fayard).

Robbins, Paul, 2012: *Political Ecology: A Critical Introduction*, 2nd edn. (Oxford: Wiley Blackwell).

Rostow, W.W., 1962: *Las Etapas del Crecimiento Económico* (México: Fondo de Cultura Económica [FCE]).

Rousset, David, 1987: *Sur la Guerre: Sommes Nous en Danger de Guerre Nucleaire?* (Paris: Éditions Ramsay).

Sachs, Jeffrey, 2005: *The End of Poverty: Economic Possibilities for Our Time* (New York: Penguin Books).

Sachs, Jeffrey, 2011: *The Price of Civilization: Reawakening American Virtue and Prosperity* (New York: Random House).

Sachs, Jeffrey, 2015: *The Age of Sustainable Development* (New York: Columbia University Press).

Sachs, Jeffrey, 2016: *La Era del Desarrollo Sostenible* (Barcelona: Ediciones Deusto).

Sanahuja, José Antonio; Sotillo, José Angel, 2007: *Integración y Desarrollo en Centroamérica: Más allá del Libre Comercio* (Madrid: La Catarata).

Santos, Boaventura de Sousa, 1992: "A Discourse on the Sciences", in: *Review*, 15,1 (Winter): 9–47.

Santos, Boaventura de Sousa, 2009: *Una Epistemología del Sur: La Reinvención del Conocimiento y la Emancipación Social* (Mexico City: Consejo Latinoamericano de Ciencias Sociales [CLACSO]/Siglo XXI Editores): 229–236.

Santos, Boaventura de Sousa, 2009: "La Ciencia Moderna como parte de una Ecología de Saberes", in Santos, Boaventura de Sousa (Ed.): *Una Epistemología del Sur: La Reinvención del Conocimiento y la Emancipación Social* (Mexico City: Consejo Latinoamericano de Ciencias Sociales [CLACSO]/Siglo XXI Editores): 186–191.

Santos, Boaventura de Sousa, 2010: *Refundación del Estado en América Latina: Perspectivas desde una Epistemología del Sur* (Bogota: Universidad de los Andes/Mexico City: Siglo XXI Editores).

Santos, Boaventura de Sousa, 2014: *Democracia al Borde del Caos* (Mexico City: Siglo XXI Editores).

Searle, John R., 1997: *La Construcción de la Realidad Socia* (Madrid: Ediciones Paidós Ibérica).

Serbin, Andres (Ed.), 2018: *América Latina y el Caribe frente a un Nuevo Orden Mundial: Poder, Globalización y Respuestas Regionales* (Barcelona: Coodinadora Regional de Investigaciones Económicas y Sociales [CRIES]/Icaria Editorial).

Sloterdijk, Peter, 2006: *Le Palais de Cristal: À l'Intérieur du Capitalisme Planétaire* (Paris: Maren Sell Éditeurs).

Stiglitz Joseph, 2002: *Globalization and its Discontents* (New York/London: Norton).

Stiglitz, Joseph, 2006: *Making Globalization Work* (New York/London: Norton).

Teilhard de Chardin, Pierre, 1955: *El Fenómeno Humano* (Madrid: Taurus Ediciones, S.A.).

Tocqueville, Alexis de, 1992: *De la Démocratie en Amérique*, 2 vols. (Paris: Gallimard).

Tocqueville, Alexis de, 2011: *The Ancient Régime and the French Revolution* (Cambridge: Cambridge University Press).

Todd, Emmanuel, 2002: *Après l'Empire: Essai sur la Décomposition du Système Américain* (Paris: Gallimard).

Torres, Edelberto, 1969: *Interpretación del Desarrollo Social Centroamericano* (Santiago de Chile: Editorial Pla).

Torres, Edelberto, 1998: *Historia General de Centroamérica* (San José, FLACSO/ Comunidades Europeas).

Torres, Edelberto, 2007: *La Piel de Centroamérica: Una Visión Epidérmica de Setenta y Cinco Años de su Historia* (Costa Rica: FLACSO).

Torres: Edelberto, 2011: *Revoluciones sin Cambios Revolucionarios: Ensayos sobre la Crisis en Centroamérica* (Guatemala City: F&G Editores).

Valdez, J. Fernando., 2015: *El Gobierno de las Elites Globales: Como se Organiza el Consentimiento: La Experiencia del Triángulo Norte* (Guatemala City: Editorial Cara Parens).

Varoufakis, Yanis, 2015: *El Minotauro Global: Estados Unidos, Europa y el Futuro de la Economía Mundial* (Madrid/ Mexico City: Ediciones Culturales Paidós).

Vernadsky, Vladimir Ivanovich, 1997: *La Biosfera* (Madrid: Fundación Argentaria).

Vitón García, Gonzalo, 2017: "Cambio Climático, Desarrollo Sostenible y Capitalismo", in: *Revista Relaciones Internacionales*, 34 (February–May): 100.

Wallace, Allan, 2003: *Buddhism and Science* (New York: Columbia University Press).

Wallerstein Immanuel, 1989: *El Moderno Sistema Mundial I: La Agricultura Capitalista y los Orígenes de la Economía Mundo Europea en el Siglo XVI*, 5th edn. (Mexico City/Madrid/Bogotá: Siglo XXI Editores).

Wallerstein, Immanuel, 2006: *Análisis de Sistemas Mundo: Una Introducción* (Mexico City/Madrid/Buenos Aires: Siglo XXI Editores SA).

Ward, Peter, 2009: *The Medea Hyphothesis: Is Life on Earth Ultimately Self-Destructive?* (Princeton: Princeton University Press).

Wilber, Ken, 2005: *Sexo, Ecología y Espiritualidad: El Alma de la Evolución* (Madrid: Gaia Ed).

Wilber, Ken, 2007: *Integral Spirituality* (Boston/London: Integral Books [Shambala]).

Wilber, Ken, 2018: *The Future of Religion* (Boulder, CO: Shambala).

Withol de Wenden, Catherine, 2017: *La Question Migratoire au XXI Siècle: Migrants, Réfugiés et Relations Internationales* (Paris: Presses de Sciences Po).

Withol de Wenden, Catherine, 2017: *Faut-il Ouvrir les Frontières?* (Paris: Sciences Po).

Academic Reviews, Papers and Reports

Aguilera, Gabriel, 2016: "El Regionalismo Centroamericano entre la Unión y la Integración", in: *Revista Oasis* (Universidad del Externado de Colombia), 24 (July–December): 89–105.

Archibugi, Daniele, 2010: *The Hope for a Global Democracy* (Rome: Italian National Research Council).

Archibugi, Daniele, 2011: *Cosmopolitan Democracy: A Restatement* (Rome: Italian National Research Council).

Barié, Cletus Gregor, 2017: "Nuevas Narrativas Constitucionales en Bolivia y Ecuador: El Buen Vivir y los Derechos de la Naturaleza", in: *Revista Política Internacional*, 3 (January–June): 48–67.

Bauböck, Rainer, 2002: *How Migration Transforms Citizenship: International, Multinational and Transnational Perspectives*, IWE-ICE No. 24 (Wien: Österreischische Akademie der Wissenschaften und Europaische Integration).

Bauböck, Rainer, 2003: "Transnational Citizenship: Memberships and Rights in International Migration; Towards a Political Theory of Migrant Transnationalism", in: *The International Migration Review*, 37,3: 700–723.

Bauböck, Rainer, 2008: *Stakeholder Citizenship: An Idea Whose Time Has Come?* (Washington, DC: Migration Policy Institute).

Carrera, Fernando; Walter, Juliane, 2008: *La Educación y la Salud en Centroamérica: Una Mirada desde los Derechos Humanos* (Guatemala City: Instituto Centroamericano de Estudios Fiscales [ICEFI]/Serviprensa Centroamericana): 16–45.

Cavanagh, John; Mander, Jerry, 2002: *Alternatives to Economic Globalization: A Better World is Possible: A Report of the International Forum on Globalization* (San Francisco, CA: BK Berretr-Koehler Publishers Inc).

Chakrabarty, Dipesh, 2008: "The Climate of History: Four Theses", in: *Critical Inquiry* (University of Chicago), (Winter 2009).

Chakrabarty, Dipesh, 2016: "Whose Anthropocene? A Response", in: Emmet, Robert; Lekan, Thomas (Eds.): *Revisiting Dipesh Chakrabarty: Four Theses of Climate History: RCC Perspectives: Transformations in Environment and Society*, 2: 102–113.

Crutzen, Paul; Stoermer, Eugene, 2000: "The Anthropocene", in: *Global Change Newsletter* (International Geosphere-Biosphere Programme (IGBP), International Council of Science), 41: 17–18.

Dabène, Olivier, 2012: *Explaining Latin America's Fourth Waves of Regionalism: Regional Integration of a Third Kind*, Paper for the "Waves of Change in Latin America History and Politics" panel, 2012 Congress of the Latin American Studies Association (LASA), San Francisco, CA, 25 May.

Delanty, Gerard; Mota, Aurea, 2017: "Governing the Anthropocene: Agency, Governance, Knowledge", in: *European Journal of Social Theory*, 20,1.

Delanty, Gerard; Mota, Aurea, 2018: "Governing the Anthropocene: Agency, Governance, Knowledge", in: *Revista Política Internacional* (Guatemala City: MINEX, Academia Diplomática), 5.

De Lombaerde, Phillippe, 2016: "Teorizando el Regionalismo Latinoamericano", in: *Revista Política Internacional* (Guatemala City: Academia Diplomática), 2: pp. 89–99.

Faist, Thomas, 2000: *Social Citizenship in the European Union: Residual, Post National and Nested Membership?* (Bremen: University of Bremen, Institut für Interkulturelle und Internationale Studien [InIIS]).

Gilligan, Carol, 1977: "In a Different Voice: Women's Conceptions of Self and of Morality", in: *Harvard Educational Review*, 47,4: 481–517.

Gómez Camacho, Juan José, 2017: "Paz Sostenible: Un Nuevo Paradigma para el Trabajo de Naciones Unidas", in: *Revista Política Internacional* (Guatemala City: Academia Diplomática), 4 (July–December),).

Göpel, Maja, 2017: "Shedding Some Light on the Invisible: The Transformative Power of Paradigm Shifts", in: Henfrey, Thomas; Maschkowski, Gesa; Penha Lopez, Gil (Eds.): *Resilience, Community Action and Societal Transformation* (21 June): 120.

Göpel, Maja, 2017: "Shedding Some Light on the Invisible: The Transformative Power of Paradigm Shifts", in: *Revista Política Internacional* (Guatemala City: MINEX, Academia Diplomática), 5: 38–57.

Gudynas, Eduardo, 2017: "Ecología Política de la Naturaleza en las Constituciones de Bolivia y Ecuado", in: *Revista Política Internacional* (Guatemala City: MINEX, Academia Diplomática), 2,4 (July–December): 140–141.

Gutiérrez, Edgar, 2016: "La Cicig: Un Diseño Nacional y una Aplicación Internacional", in: *Revista Política Internacional* (Guatemala City: MINEX, Academia Diplomática), 1 (June): 26–36.

Held, David, 2010: "Global Democracy: A Symposium on a New Political Hope", in: *New Political Science*, 32,1: 83–121.

Held, David, 2005: Principles of Cosmopolitanism Order, in: *Anales de la Cátedra Francisco Suarez* (Granada: University of Granada), 39: 153–159.

Hettne, Bjorn, 2002: "El Nuevo Regionalismo y el Retorno a lo Político", in: *Revista de Comercio Exterior*, 52,1 (November): 58–75.

Holloway, John 2002: "Cambiar el Mundo sin Tomar el Poder: El Significado de la Revolución Hoy", in: *Revista de la Universidad de Puebla* (Mexico).

Johnson, Alan; Pleyers, Geoffrey, 2008: "Globalización, Democracia y Mercados: Una Alternativa Socialdemócrata: Entrevistas con David Held", in: *Revista Sociológica* (Madrid), 23,66 (January–April): 187–224.

Ladaria Luis; Cardenal Tukson, Peter et. al., 2018: "Oeconomicae et Pecuniariae Quaestiones: Consideraciones para un Discernimiento Ético sobre Algunos Aspectos del Actual Sistema Económico y Financiero", in: *Bolletino Sala Stampa de la Santa Sede* (Vatican City: Press Offices of the Holy See and of the Vatican).

Latour, Bruno, 2017: "Waiting for Gaia", Paper delivered at the French Institute in London, November 2011, on the occasion of the launch of the Political Science in Arts and Politics Programme (SPEAP) [translated by Silvina Cucchi], "Esperando a Gaia: Componiendo el Mundo mediante las Artes y la Política", in: *Revista Política Internacional* (Guatemala City: MINEX, Academia Diplomática), 3.

Mackinder, Halford, 2010: "El Pivote Geográfico de la Historia", in: *Revista Geopolítica: Revista de Estudios sobre Espacio y Poder* (Madrid: Universidad Complutense de Madrid), 1,2: 301–319.

Malamud, Andres, 2011: "Conceptos, Teorías y Debates sobre la Integración Regional", in: *Revista Norte América*, 6,2 (July–December): 220–241

McDonald, Ross: "What Exactly is the Meaning and Purpose of Gross National Happiness?", Paper Proposal for the Second International Workshop on Operationalizing Gross National Happiness (Auckland, New Zealand: University of Auckland, Business, Society and Culture Programme, MER Department).

Meadows, Donella; Meadows, Dennis; Randers, Jorgen; Behrens, William; et al., 1972: *Los Límites del Crecimiento: Informe al Club de Roma sobre el Predicamento de la Humanidad* (Mexico City: Fondo de Cultura Económica).

Mearsheimer, John, 2014: "Why the Ukraine Crisis is the West's Fault: The Liberal Delusions that Provoked Putin", in: *Foreign Affairs*, 93,5 (September/October 2014): 77–89.

Mearsheimer, John; Walt, Stephen, 2016: "The Case for Offshore Balancing: A Superior US Grand Strategy", in: *Foreign Affairs*, 95,4 (July/August).

Mulet, Edmond, 2016: "La Agenda 2030 y los Objetivos de Desarrollo Sostenible", in: *Revista Política Internacional* (Guatemala City: MINEX, Academia Diplomática), 1 (January–June): 151–158.

Nye, Joseph Jr., 2017: "Will the Liberal Order Survive? The History of an Idea", in: *Foreign Affairs*, 96,1 (January–February): 10–16.

Organization of International Migration [OIM], 2016: *Survey on International Migration of Guatemalans and Remittances* (Guatemala City: OIM).

Organization of International Migration [OIM], 2017: *Encuesta sobre Migración Internacional de Personas Guatemaltecas 2017* (Guatemala City: OIM, February).

Portes, Alejandro, 1996: "Transnational Communities: Their Emergency and Significance in Contemporary World Systems", in: Korseniewics, Patricio; Smith, William: *Latin America and the World Economy* (Westport, Conn: Greenwood Press).

Padilla, Luis Alberto, 2015: "El Conflicto de Ucrania a la luz de los Paradigmas Clásicos de la Teoría Internacional", in: *Espacios Políticos: Revista de la Facultad de Ciencias Políticas* (Guatemala City: Universidad Rafael Landívar).

Padilla, Luis Alberto, 2015: "Neutralidad y Equilibrio de Poder en el Conflicto de Ucrania", in: *Espacios Políticos: Revista de la Facultad de de Ciencias Políticas y Sociales* (Guatemala City: Universidad Rafael Landívar).

Padilla, Luis Alberto, 2016: "Asia Pacífico, Eurasia y la Nueva Rivalidad Geopolítica de China con Estados Unidos", in: *Revista Polìtica Internacional* (Guatemala City: MINEX, Academia Diplomática), No. 1 (January–June): 91–107.

Padilla, Luis Alberto, 2001: "Peace Making and Conflict Transformation in Guatemala", in: Alker, Hayward; Gurr, Ted Robert; Rupesinghe, Kumar, 2001: *Journeys Through Conflict: A Study of the Conflict Early Warning Systems Research Project of the International Social Science Council* (Boulder/ New York/Oxford: Rowman & Littlefield Publishers Inc.): 56–78.

Padilla, Luis Alberto, 1974: *Essai d'Interprétations Historique des Sources Philosophiques du Droit Positif de la République du Guatemala* (PhD thesis, Université de Paris II, Pantheon Sorbonne [unpublished]).

Piñera, Sebastián, 2018: *Presidential Statement to Present the Immigration Reform*, (Santiago de Chile: Palacio de la Moneda, Government of Chile, 9 April).

Popkin, Eric, 2005: "The Emergence of Pan Mayan Ethnicity in the Guatemalan Transnational Community Linking Santa Eulalia and Los Angeles", in: *Current Sociology*, 53,4 (Monography 2): 675–706.

Richards Howard, 2017: *Economía Social Solidaria: Para Cambiar el Rumbo de la Historia. Módulos de un curso en la Universidad Nacional Autónoma de México* (Mexico City: National Autonomous University of Mexico [UNAM]).

Rojas Aravena, Francisco, 2017: "América Latina: Altas Incertidumbres Globales y Difícil Recuperación Económica", in: *Revista Política Internacional* (Guatemala City: MINEX, Academia Diplomática).

Rosenthal, Gert, 2016: "Participación de Guatemala en el Consejo de Seguridad de Naciones Unidas 2012–2013", in: *Revista Política Internacional* (Guatemala City: Academia Diplomática), 1 (January–June): 9–25.

Ruddiman, William, 2005: "How Did Humans First Alter Global Climate?", in: *Scientific American*, 292,3: 46–53.

Sanahuja, José Antonio, 2012: "Regionalismo Postliberal y Multilateralismo en Sudamérica: El Caso de UNASUR", in: *Anuario Regional*, 9 (Buenos Aires: Coordinadora Regional de Investigaciones Económicas y Sociales [CRIES]).

Sanahuja, José Antonio, 2018: "Crisis de Globalización, Crisis de Hegemonía: En Escenario de Cambio Estructural para América Latina y el Caribe", in: Serbin, Andres (Ed.): *América Latina y el Caribe frente a un Nuevo Orden Mundial: Poder, Globalización y Respuestas Regionales* (Barcelona: Coordinadora Regional de Investigaciones Económicas y Sociales (CRIES)/ Icaria Editorial).

Solís, Luis Guillermo, 2016: "Reflexiones en torno a Guatemala, Costa Rica y la Integración Centroamericana: Lecture of the President of Costa Rica at the State University of San Carlos, Guatemala City, February 2016", in: *Revista Política Internacional* (Guatemala City: MINEX, Academia Diplomática), 1 (January–June): 181–190.

Suarez Ruiz, Hero, 2018: "El Concepto de Ciudadanía en Etienne Balibar y la Nueva Estrategia Zapatista: Respuestas a la Gobernanza", in: *Oximora, Revista Internacional de Ética y Política*, 12 (January–June): 121–139.

United Nations, 2014: *The Economic, Social and Cultural Rights of Migrants in an Irregular Situation* (New York/Geneva: Office of the High Commissioner of the United Nations for Human Rights).

Vergara, Walter; Rios, Ana R.; Galindo, Luis M.; Gutman, Pablo; Isbell, Paul; Suding, Paul H.; Samaniego, Joseluis, 2014: *El Desafío Climático y de Desarrollo en América Latina y el Caribe: Opciones para un Desarrollo Resiliente al Clima y Bajo en Carbono* (CEPAL, BID, WWF), in: *América Latina Cambio Climático y Desarrollo*, pdf).

Villagrán, Carlos Arturo, 2016: "Soberanía y Legitimidad de los Actores Internacionales en la Reforma Constitucional de Guatemala: El Rol de CICIG", in: *Revista Política Internacional* (Guatemala City: Academia Diplomática, June), 1: 37–57.

Walt, Stephen; Mearsheimer, John, 2016: "The Case for Offshore Balancing: A Superior US Grand Strategy", in: *Foreign Affairs*, 95,4 (July/August): 70–83.

WBGU [German Advisory Council on Global Change], 2011: *World in Transition. A Social Contract for Sustainability* (Berlin: WBGU).

Wendt, Alexander, 2005: "La Anarquía es lo que los Estados hacen de Ella: La Construcción Social de la Política de Poder", in: *Revista Académica de Relaciones Internacionales* (Madrid: Grupo de Estudio de Relaciones Internacionales [GERI], Universidad Autónoma de Madrid, Facultad de Derecho [GERI-UAM]), 1 (March): 1–47.

Wihtol de Wenden, Catherine, 2018: "Las Nuevas Migraciones", in: *Política Internacional* (Guatemala City: Academia Diplomática), 5.

World Commision on Environment and Development, (Brundtland Report), 1987: *Our Common Future* (New York: Oxford University Press).

Zalasiewicz Jan; Williams, Marc; et al., 2008: "Are We Now Living in the Anthropocene?", in: *GSA Today,* 18,2 (February): 4–8.

Newspapers, Internet Sources and Interviews

Archibugi, Daniele; et al. 2010: "A Symposium on a New Political Hope", in: *New Political Science,* 32,1 (March); at: http://www.danielearchibugi.org/downloads/papers/2017/11/new-political-science.pdf.

Badie, Bertrand: "Conferencia".

Beck, Ulrich, 2009: Critical Theory of World Risk Society: A Cosmopolitan Vision, in: *Constellations,* 16,1: 3–22; at: https://onlinelibrary.wiley.com/doi/abs/10.1111/j.1467-8675.2009.00534.x; https://doi.org/10.1111/j.1467-8675.2009.00534.x

Brubacker, Rogers: "Between Nationalism and Civilisationism: The European Populist Movement in Comparative Perspective", at: http://www.tandfonline.com/doi/full/10.1080/01419870.2017.1294700.

CEPAL: "La Economía del Cambio Climático en Centroamérica, Síntesis", 2010.

Delanty, Gerard, 2012: *Routledge Handbook of Cosmopolitanism Studies* (New York/ Oxford: Routledge); at: http://dro.deakin.edu.au/eserv/DU:30049755/brown-empiricalworld-evid-2012.pdf.

Efler Michael, 2010: "European Citizens' Initiative: Legal Options for Implementation below Constitutional Level"; at: European%20citizens%20initiative_Michael%20Efler.pdf.

Encíclica Papal, 1967: *Populorum Progressio;*.

Fischer, Joschka, 2013: "Peligra la Unidad Europea", in: *El País,* 3 May.

Fischer, Joschka, 2016: "Interview"; at: https://www.euractiv.com/section/euro-finance/interview/joschka-fischer-stabilise-the-eurozone-to-defuse-hurricane-brexit/.

Fischer, Joschka, 2016: "Declaraciones", at: https://www.euractiv.com/section/euro-finance/interview/joschka-fischer-stabilise-the-eurozone-to-defuse-hurricane-brexit/.

Fox, Jonathan, 2005: "Unpacking "Transnational Citizenship", in: *Political Science Review,* 8: 171–201 (Santa Cruz, University of California), at: https://escholarship.org/uc/item/4703m6bf.

Holloway, John; Matamoros Ponce, Fernando; Tischler, Sergio, 2015: Cambiar el Mundo sin Tomar el Poder: El Significado de la Revolución Hoy, in: *Revista Digital Herramienta,* 57; at: www.herramienta.com.ar.

Jiang Su, Liu; Dietz, Thomas; Carpenter, Stephen, et al., 2007: "Complexity of Coupled Human and Natural Systems", in: *Science,* 317, 5844: (14 September): 1,513–1,516.

Kupchan, Charles, 2018: "Trump's Nineteenth Century Grand Strategy", in: *Foreign Affairs* (26 September); at: www.foreignaffairs.com.

Lafontaine, Oskar, 2015: "Let's Develop a Plan B for Europe", in: *International Journal of Socialist Renewal* http://links.org.au/node/4573 and http://science.sciencemag.org/content/317/5844/1513.full.

Latour Bruno, 2015: "Interview by Diego Milos and Matías Wolff".

Naciones Unidas, 2014: "Decenio Internacional para la Acción: El Agua Fuente de Vida"; at: http://www.un.org/spanish/waterforlifedecade/iwrm.shtml.

Navarro, Vinçenç, 2013 "Por qué la socialdemocracia no se recupera en Europa?"; at: www.vnavarro.org.

Petro, Nicolai N., 2018: "Russia's Mission", in: *Journal of International Affairs* (November), American Committee for East-West Accord [ACEWA]); at: https://eastwestaccord.com/nicolai-n-petro-russias-mission/.

Poch, Rafael, 2017: "Adiós Unión Europea", in: *La Vanguardia* (February).

Sachs, Jeffrey: "The Age of Sustainable Development", Lecture at the *International Institute for Applied Systems Anaylisis* (IIASA); at: https://www.youtube.com/watch?v=ksZaAqRA5qg.

Sachs, Jeffrey, 2017: "La Democracia Resquebrajada de Estados Unidos", in: *El País* (Spain) and *El Periódico* (Guatemala), (30 July): 12–13.

Savio, Roberto, 2018: "Lots of Myths and Little Reality", in: *Facebook* @robertosavioutopia.

Scott, Margaret; Alcenat, Westenley, 2008: "Revisiting the Pivot: The Influence of Heartland Theory in Great Power Politics" (Macalester College, 9 May); at: file:///E:/ GEOPO-LITICA/Mackinder'sPivot%20in%20US%20Foreign%20Policy_Alcenat_and_Scott.pdf.

Scott Smith, Fresno, 2017: "Difieren por Impacto de Industrias Extractivas", in: *Diario La Hora* (27 June): 25.

Sloterdijk, Peter, 2006: *Le Palais de Cristal: À l'Intérieur du Capitalisme Planétaire* (Paris: Maren Sell).

Smith, William, 2017: "Cosmopolitanism", in: *International Research Encyclopaedia* (International Studies Association and Oxford University Press); at: http://internationalstudies.oxfordre.com/view/10.1093/acrefore/9780190846626.001.0001/acrefore-9780190846626-e-133.

Ward, Peter, 2017: "The Models Are Too Conservative: Paleontologist Diamond on What Past Mass Extinctions Can Teach Us About Climate Change Today", in: *NY Magazine* (17 July); at: http://nymag.com/daily/intelligencer/2017/07/what-mass-extinctions-teach-us-about-climate-change-today.html.

Weisbrot, Mark; Lefebvre, Stephan; Sammut, Joseph, 2014: *Did NAFTA Help Mexico? An Assessment After 20 Years* (Washington, DC: Center for Economic and Policy Research).

Will, Steffen; Broadgate, Wendy; Deutsch, Lisa; Gaffney, Owen; Ludwig, Cornelia, 2015: "The Trajectory of the Anthropocene: The Great Acceleration", in: *The Anthropocene Review* (16 January): 1–18; at: https://doi.org/10.1177/2053019614564785.

On the International Peace Research Association (IPRA)

Founded in 1964, the *International Peace Research Association* (IPRA) developed from a conference organized by the Quaker International Conferences and Seminars in Clarens, Switzerland, 16–20 August 1963. The participants decided to hold international *Conferences on Research on International Peace and Security* (COROIPAS), which would be organised by a Continuing Committee similar to the Pugwash Conferences. Under the leadership of John Burton, the Continuing Committee met in London, 1–3 December 1964. At that time, they took steps to broaden the original concept of holding research conferences. The decision was made to form a professional association with the principal aim of increasing the quantity of research focused on world peace and ensuring its scientific quality.

An Executive Committee including Bert V.A. Röling, Secretary-General (The Netherlands), John Burton (United Kingdom), Ljubivoje Acimovic (Yugoslavia), Jerzy Sawicki (Poland), and Johan Galtung (Norway) was appointed. This group was also designated as Nominating Committee for a 15-person Advisory Council to be elected at the first general conference of IPRA, to represent various regions, disciplines, and research interests in developing the work of the Association.

Since then, IPRA has held 27 biennial general conferences, the venues of which were chosen with a view to reflecting the association's global scope. IPRA, the global network of peace researchers, held its 25th General Conference on the occasion of its 50th anniversary in Istanbul, Turkey in August 2014, where peace researchers from all parts of the world had the opportunity to exchange actionable knowledge on the conference's broad theme of *Uniting for Sustainable Peace and Universal Values*.

© Springer Nature Switzerland AG 2021
L.-A. Padilla, *Sustainable Development in the Anthropocene*,
The Anthropocene: Politik—Economics—Society—Science 29,
https://doi.org/10.1007/978-3-030-80399-5

The 26th IPRA General Conference took place from 27 November to 1 December 2016 in Freetown, Sierra Leone on the theme: *Agenda for Peace and Development: Conflict Prevention, Post-Conflict Transformation, and the Conflict, Disaster and Development Debate.* The 27th IPRA General Conference took place in Ahmedabad, India, 24–27 November, 2018 on the theme *Innovation for Sustainable Global Peace.* The 28th IPRA General Conference will be in Nairobi, Kenya.

On IPRA http://www.iprapeace.org/

On the IPRA Foundation: http://iprafoundation.org/

IPRA Conferences, Secretary Generals and Presidents (1964–2021)

IPRA general conferences	IPRA secretary-generals/presidents
1. Groningen, The Netherlands (1965)	1964–1971 Bert V.A. Röling (The Netherlands)
2. Tallberg, Sweden (1967)	1971–1975 Asbjorn Eide (Norway)
3. Karlovy Vary, Czechoslovakia (1969)	1975–1979 Raimo Väyrynen (Finland)
4. Bled, Yugoslavia (1971)	1979–1983 Yoshikazu Sakamoto (Japan)
5. Varanasi, India (1974)	1983–1987 Chadwick Alger (USA)
6. Turku, Finland (1975)	1987–1989 Clovis Brigagão (Brazil)
7. Oaxtepec, Mexico (1977)	1989–1991 Elise Boulding (USA)
8. Königstein, FRG (1979)	1991–1994 Paul Smoker (USA)
9. Orillia, Canada (1981)	1995–1997 Karlheinz Koppe (Germany)
10. Győr, Hungary (1983)	1997–2000 Bjørn Møller (Denmark)
11. Sussex, England (1986)	2000–2005 Katsuya Kodama (Japan)
12. Rio de Janeiro, Brazil (1988)	2005–2009Luc Reychler (Belgium)
13. Groningen, the Netherlands (1990)	2009–2012 Jake Lynch (UK/Australia)
14. Kyoto, Japan (1992)	Katsuya Kodama (Japan)
15. Valletta, Malta (1994)	2012–2016 Nesrin Kenar (Turkey)
16. Brisbane, Australia (1996)	Ibrahim Shaw (Sierra Leone/UK)
17. Durban, South Africa (1998)	2016–2018 Úrsula Oswald Spring (Mexico)
18. Tampere, Finland (2000)	Katsuya Kodama (Japan)
19. Suwon, Korea (2002)	2019–2023 Christina Atieno (Kenya)
20. Sopron, Hungary (2004)	Matt Meyer (USA)
21. Calgary, Canada (2006)	**Presidents**
22. Leuven, Belgium (2008)	The first IPRA President was Kevin Clements (New
23. Sydney, Australia (2010)	Zealand/USA, 1994–98)
24. Mie, Japan (2012)	His successor was Úrsula Oswald Spring (Mexico,
25. Istanbul, Turkey (2014)	1998–2000)
26. Freetown, Sierra Leone (2016)	
27. Ahmedebad, India (2018)	
28. Nairobi, Kenya (2021)	
29. Trinidad & Tobago (2023)	

© Springer Nature Switzerland AG 2021
L.-A. Padilla, *Sustainable Development in the Anthropocene*,
The Anthropocene: Politik—Economics—Society—Science 29,
https://doi.org/10.1007/978-3-030-80399-5

On CLAIP

The *Latin American Council of Peace Research* (usually known by its Spanish acronym CLAIP) is a regional organization of the *International Peace Research Association* (IPRA) and was established in Mexico in 1977 under the umbrella of the Seventh IPRA General Conference. In addition to scientific research in the field of peace, among its purposes CLAIP promotes a culture of peace and peace education in the region, as well as the establishment of networks of institutions, organizations and individuals involved in peace research, including peace education, human rights, sustainable development and positive peace (Galtung). CLAIP facilitates the exchange of knowledge and experience in meetings and at conferences that are normally held under the auspices of different academic institutions and involve the task of publishing the results of these investigations and experiences in academic journals and books. At its first conference, held in Oaxtepec (Mexico) 1977, more than 150 researchers, peace education activists, students and university professors participated. At CLAIP's conferences the papers, presentations and discussions usually focus on issues such as international relations, political science, and social sciences in general, including the role of the military and armed forces; internal armed conflicts; respect for human rights and humanitarian law; peace processes; the situation of social groups such as indigenous peoples, working classes, and Afro-American peoples; women and gender subjects; children and youth topics; demography; urbanization; migration and human mobility issues; climate change and environmental problems; culture and the arts; and nuclear non-proliferation.

For the last forty years CLAIP has been under the coordination of Secretary-Generals from different Latin American countries, among them Herbert de Souza (Brazil), Jorge Serrano (Mexico), Antonio Cavalla (Chile), Luis Alberto Padilla (Guatemala), Nielsen de Paula Pires (Brazil), Úrsula Oswald Spring (Mexico), Sara Rozemblum de Horowitz (Argentina), Laura Balbuena (Perú), Diana de la Rúa (Argentina), and Eréndira Serrano Oswald (Mexico). At the last CLAIP Conference, held at the University of São Paulo (Brazil, September 2019), the distinguished jurist, peace education activist and Argentinian Judge of Peace Marité Muñoz was elected

© Springer Nature Switzerland AG 2021
L.-A. Padilla, *Sustainable Development in the Anthropocene*,
The Anthropocene: Politik—Economics—Society—Science 29,
https://doi.org/10.1007/978-3-030-80399-5

as Secretary-General, and the Honduran peace education professor of the Faculty of Social Sciences at the *National Autonomous University of Honduras* (UNAH), Dr Esteban Ramos Muslera, was elected as Co-Secretary-General, taking into account the commitment made by the Honduran University *Institute for Democracy, Peace and Security* (IUDPAZ) to organize the 12th CLAIP Conference in Tegucigalpa in 2021 and to promote the publication of CLAIP's first academic scientific review.

It is also important to note that CLAIP's mission to organize the region's peace researchers and activists in order to hold conferences and exchange knowledge has been enriched with the vision of sustainable and transformative positive peace. This basically consists of the permanent planetary preoccupation as well as the conservation of Earth ecosystems, and includes supporting the struggles of citizens and international NGOs against corporations and economic pressure groups that do not support policies of decarbonization, and calling for a change in both public and private sector policies regarding renewable energy as a replacement for fossil fuels. Sustainable development, the SDGs and the UN 2030 Agenda are an important part of CLAIP's vision in the decade to come, as well as the promotion of a new paradigm based on the understanding that our species is an integral part of Mother Nature (called Pachamama by indigenous peoples) and that *buen vivir* (a good and harmonius life) is the foremost aim of sustainable development.

On IRIPAZ

The *Peace Research and International Relations Institute* (commonly known by its Spanish acronym IRIPAZ) is a Guatemalan academic institution organized by a group of Guatemalan scholars with the main purpose of conducting research in the field of *positive peace* (Galtung) and sustainable development. It also studies issues relating to peace processes and conflict resolution as well as international relations and both regionalization and integration processes in Central and Latin America. It has the legal status of a non-profit foundation. Established in February 1990, the Institute has a significant track record of conducting research and providing training in educational projects on peace education and human rights, and is member of the *Latin American Peace Research Council* (CLAIP) and the *International Peace Research Association* (IPRA). As an independent non-profit foundation, for a decade (the 1990s) IRIPAZ received financial support from the *Foreningen Orjangarden AGNI* in Sweden as well as grants for specific projects from the OAS, USAID, the Norwegian Agency for Development Cooperation, the Friedrich Ebert and Friedrich Naumann foundations and other benefactors. For more than a decade the Institute has published a pioneering Guatemalan review of peace research, *Estudios Internacionales*, and, among other academic activities, should be mentioned the first Latin American Conference on International Relations and Peace Research that took place in Guatemala City in August 1995 under the auspices of Landívar and San Carlos Universities, with the support of UNESCO, UNDP, CLAIP, and IPRA, and the cooperation of the embassies of Spain, France, Sweden, and Mexico. A book containing contributions from participants such as Johan Galtung, Rodolfo Stavenhagen, Peter Wallensteen, Andrés Serbin, Alain Rouquie, Augusto Ramírez Ocampo, Vincent Fiças, Rodrigo Borja, Nielsen de Paula Pires, Úrsula Oswald Spring, Peter Stania, Kjell Ake Nordquist, Gabriel Aguilera, René Poitevin, Edelberto Torres, Francisco Villagrán Kramer, Gert Rosenthal, Manuel Angel Castillo, and Rigoberta Menchú – among others – was published under the title *Peace Building, Culture of Peace and Democracy*. The Institute has published several other books, among them *Peace and Conflict in the Twenty-First Century: Theory of International Relations* (2009), and

© Springer Nature Switzerland AG 2021
L.-A. Padilla, *Sustainable Development in the Anthropocene*,
The Anthropocene: Politik—Economics—Society—Science 29,
https://doi.org/10.1007/978-3-030-80399-5

contributed papers like "Peace Making and Conflict Transformation in Guatemala" (in the 2001 volume, *Journeys Through Conflict* by Hayward Alker and Ted Robert Gurr [Boston: Rowman & Littlefield]) on peace education, conflict resolution, indigenous culture, human rights, and environmental issues. IRIPAZ is currently engaged – among other endeavours – in a national campaign to promote the United Nations' SDGs and the 2030 Agenda.

On the Author

Luis Alberto Padilla (Guatemala) has a PhD in Philosophy of Law from Paris University (Pantheon-Sorbonne) and a BA in Legal and Social Sciences from the University of San Carlos in Guatemala. He is a professor of International Relations Theory and Global Geopolitics at the universities of San Carlos and Rafael Landívar. He is founder and President of the Directive Board of the *Institute of International Relation and Peace Research* (IRIPAZ). He was Secretary General of CLAIP and editor of the *Journal of International Studies* (IRIPAZ). He is a former Ambassador of Guatemala at the United Nations in Geneva and Vienna and also chief of mission in The Netherlands, Russia, Austria and Chile.

Dr Padilla was also Director of the Diplomatic Academy of the Foreign Ministry from 2014 to May 2019 and during that period has additionally been director of the Journal of the Academy *International Politics*. He has published multiple articles and books, such as *Peace and Conflict in the Twenty-First Century: Theory of International Relations*.

© Springer Nature Switzerland AG 2021
L.-A. Padilla, *Sustainable Development in the Anthropocene*,
The Anthropocene: Politik—Economics—Society—Science 29,
https://doi.org/10.1007/978-3-030-80399-5

Address: 30 avenue 8–81 El Prado, Guatemala City

E-mail: luis.alberto.padilla@iripaz.org

Website: www.iripaz.org.

Printed in the United States
by Baker & Taylor Publisher Services